BIOPHOTONS

Biophotons

edited by

Jiin-Ju Chang
Institute of Biophysics, Chinese Academy of Sciences, Beijing, China
International Institute of Biophysics, Hombroich, Germany

Joachim Fisch
Institute of Optical Devices, Technical University Ilmenau, Germany

and

Fritz-Albert Popp
International Institute of Biophysics, Hombroich, Germany

SPRINGER-SCIENCE+BUSINESS MEDIA, B.V.

A C.I.P. Catalogue record for this book is available from the Library of Congress.

ISBN 978-90-481-5033-5 ISBN 978-94-017-0928-6 (eBook)
DOI 10.1007/978-94-017-0928-6

Printed on acid-free paper

TABLE OF CONTENTS

PREFACE

From June 19 to 25, 1997, twenty-two scientists of eleven nations met at the Technical Universität Ilmenau, and the International Institute of Biophysics (Neuss, Germany), in order to present and to discuss the most recent results in the interdisciplinary field of 'biophoton research' and 'biophotonics'. Up to now this meeting was the most frequented conference on this topic, impressively showing the increasing importance of this research, which concerns an extremely weak but permanent photon emission from biological systems. This photon flux of up to several hundreds of photons per second and per square centimeter surface area of the living system, continously covering the spectral range from at least 260 to 800 nm, is neither heat radiation (which, in the infrared region, roughly follows the Boltzmann distribution) nor ordinary „bioluminescence" which is assigned to photochemical reactions in a few species like deep-sea fish and fireflies. Rather, the term „biophotons" refers to the phenomenon of single photons indicating their classification in terms of quantum optics as well as to the biological origin of the light source. It seems that this term is now becoming common after periods of rather confusing terms, ranging from 'mitogenetic radiation' to 'dark luminescence', 'ultraweak photon emission', 'low level luminescence', 'chemiluminescence', 'bioluminescence'. Reflecting the nature of this phenomenon -the discovery of which can be traced back to the Russian biologist Alexander Gurwitsch in 1923 - the research requires interdisciplinary cooperation, covering fields like quantum optics, biochemistry, molecular biology and engineering. Thus, international participation and widely distributed interdisciplinary scientific contributions characterize the topics and the discussion of this conference at Ilmenau and Hombroich. Most of the outstanding scientists working in this field presented new and exciting experimental results on photon emission as well as on non-linear absorption of weak light intensities, attempting to find a common frame of understanding biophoton emission in its basic nature. Profound discussions are devoted to the probable light source(s) within the cells, the possible coherence of the radiation field within the body, the connection to non-classical (squeezed) light, the interaction of biophotons and biomolecules including metabolic events, species-specific effects as well as the interference with basic biological functions like defense mechanisms against parasites, immunological and repair activities. At this conference evidence of non-substantial biocommunication was presented for the first time as well as whole-body photon counting which reveals biological rhythms indicating their real origin in the biophoton field. Further topics treated are the common nature of „delayed luminescence" (d.l.) and biophoton emission (bpe), where d.l. refers to a long-time light rescattering after exposure of the biological system to external light, then flowing smoothly down to the quasi-stationary bpe. Evidence is shown that both effects are linked. Besides problems of basic research on this field, various applications like germination capacity, correlations of light reflection to malignant cell growth, new optical tracer techniques in molecular biology have been presented and discussed at this conference. In view of the rather complicated but rapidly developing technique of measuring single light quanta in the optical range, some experimental as well as theoretical papers have been devoted to the questions of the sensitivity of optical detectors and signal/noise-ratio at the limit of single-photon counting. The spectrum of topics ranges from technical questions about basic scientific documentations to descriptions of the history up to the development of consciousness, connecting coherence and the nature of life. The last topic was presented by Fanchon Fröhlich, the wife of the late Herbert Fröhlich who was the first to point to the basic connection between long-range coherence and biological regulation. We are happy to have Fanchon Fröhlich as the Honorary Member of the International Institute of Biophysics.

These Proceedings can be looked upon as a consequent continuation of reports on biophoton conferences, as, for instance, B. Jezowska-Trzebiatowska et.al. (eds.): Photon emission from biological systems. World Scientific, Singapore-London 1987; F.A.Popp et al.: Biophoton Emission. Multi-author Review. Experientia 44 (1988), 543-630; F.A.Popp et. al. (eds.): Recent Advances in Biophoton Research and its Applications. World Scientific, Singapore and London 1992; L.Beloussov and F.A.Popp (eds.): Biophotonics, Proceedings of the International Confererence, September, 28 - October, 2, 1994 at the Moscow State University, Bioinform Services, Moscow 1995.

Very special thanks are due to the colleagues of the Deutsche Akademie für Photobiologie and Phototechnologie and the Fachgebiet Lichttechnik of the Technical University Ilmenau for the perfect preparation and organization of the meeting in Ilmenau. We like to express our gratitude in particular to Mr. Menz for his personal engagement in providing all the success of this conference.

We are grateful to Sophie Cohen and Yan Yu for their willingness and patience in bringing the manuscript to its final stage. We appreciate the constant support and advice of Dr. Mariette de Jong and her assistant. We would like to express our thanks to Heinrich Müller for his hospitality at the Conference and his friendship. Last but not least we thank Dr. Gisela Draczynski for her constant engagement which allowed to establish this efficient work

We are facing the modern and fast-expanding field of biophotons, where this exciting topic of modern life sciences will become a focal point of interdisciplinary scientific research in revealing, probably a basic, up to now widely unknown channel of communication within and between cells, stimulating thus a new scientific approach to understanding the nature of life.

Institute of Biophysics, Chinese Academy of Sciences, Beijing
Fachgebiet Lichttechnik, Technische Universität Ilmenau
International Institute of Biophysics, Raketenstation, Hombroich

February 1998

J.J.Chang, J.Fisch and F.A.Popp

INTRODUCTION

Since the discovery of „mitogenetic radiation" by Alexander Gurwitsch about 75 years ago almost two generations of scientists have passed. More and more it turns out that the original ideas of the discoverer, who postulated the existence and the cell growthstimulating function of single ultraviolet photons, are confirmed by the present science which

- has been sucessful in showing evidence of „ultraweak cell radiation" in the visible and UV-range,
- is becoming aware now of the basic role of „photonic communication" in Biology and modern Physics (Quantum Optics, „non-classical light").

If one takes into account that true theories cannot be proven at all, but only wrong ones can be rejected, 75 years of unsuccessful objections, rejections, and suppressions of Gurwitsch's visions and a suddenly increasing agreement to his ideas is a rather strong argument in favour of this important pioneer.

This is one of the conclusions which can be drawn from the most recent papers on this field.

The variety in the presentation impressively reflects the interdisciplinary character.

1. We start with a paper of G. Renger who bridges the gap between conventional Photo- Biophysics and Biophoton Research.
2. The contribution of R.Vogel and R. Süßmuth is a foray into the investigation of „low-level-luminescence", where a puzzling nonlinear absorption of weak light has been confirmed experimentally now also in the case of bacteria.
3. M.Hiramatsu discovered a surprising correlation between biophoton emission and defense response in plants which will be reported in the third paper.
4. X. Shen and coworkers confirm the exciting non-substantial biophotonic communication of blood cells during phagocytosis. This throws a completely new light onto the immunological reactions in blood.
5. R. van Wijk and his colleague introduce a new powerful tool of optical tracers in molecular biology.
6. H.Niggli shows evidence of the link between photorepair and biophoton emission, in particular of the interaction of biophotons with the DNA.
7. X. Shen and F.A. Popp report about a coincidence method which is used for calculating the photo counts statistics (PCS) of a weak light source under non-stationary conditions. There is new indication of the coherence of biophotons.
8. V.L.Voeikov and his colleague demonstrate nonlinear chemical effects at low concentrations of the reactants, pointing to a completely new field of chemistry and biochemistry.
9. F. Musumeci and coworkers investigate the problem of the coherence of biophoton emission in the case of an unicellular algae.
10. L.Beloussov and his collegue demonstrate nonlinear biophoton effects on developing eggs and embryos, confirming further the theory of Beloussov's grandfather (Alexander Gurwitsch).

11. The eleventh chapter the book focuses briefly onto technical and theoretical questions of the instrumentation. W.Heering is the expert for introducing the reader to the rather difficult problems of demonstrating significant effects under the condition of lowest intensities.
12. Using the example of a „whole-body biophoton counter", the group around D.Gall and J.Fisch show evidence of biophoton emission of the human skin and investigate the problem of the sensivity of the multiplier techniques for measuring single quanta.
13. Chapter 13 is another excursion into a new field of application: the first report on systematic biophoton measurements of the human body and the correlation of biophoton emission to biological rhythms, revealed by S.Cohen and F.A.Popp..
14. The extraordinary optical properties of human tissue invite to applications in cancer diagnosis. Experimental results on human melanoma, using a fluorescence imaging technique, are reported by B.W. and S.Chwirot and coworkers.
15. The high sensitivity of single-photon counting techniques allows the measurement of electric currents in fluids with the highest reliability. The conductivity measurement („electroluminescence") provides a powerful new tool of detecting microbial contamination and, in general, small changes in the quality of fluids. J.J.Chang and F.A.Popp present a basic report.
16. The question of biological organization, based on non-substantial communication, is discussed from Chapter 16 on. J.J.Chang and F.A.Popp start with the experimental evidence of partially synchronous flickering of dinoflagellates.
17. B.W.Chwirot investigates the crucial problem of local and non-local information originating from biophotonic and autofluorescence analysis.
18. F.A.Popp and J.J.Chang discuss the informational character of biophotons, the role for biochemical reactivity as well as for growth regulation and spatio-temporal organization in living sytems.
19. B.Zeiger presents a new and fundamental biophysical theory of the germination capacity of seeds which helps to judge the importance of investigating entropic parameters for a deeper understanding of life.
20. Q.Gu points to the remarkable possibility that biophotons have to be traced back to „non-classical light", where the quantum character is necessary for displaying highest signal/noise-ratio.
21. R.P.Bajpai follows Gu's arguments in investigating carefully the dynamics of „delayed luminescence".
22. G.J.Hyland, coworker and follower of Herbert Fröhlich in England, presents an overview of the theoretical implications and functions of coherent fields in biology.
23. M.Lipkind, one of the last students of Alexander Gurwitsch, and now a well established virologist, discusses the modern view of field theories and cohence in terms of Gurwitsch's morphogenetic field theory.
24. M.Bischof has been working on the history of biophysics for a long time. He reports about the development of field theories and holistic approach to biology over the last centuries.
25. Lastly, Fanchon Fröhlich, wife of the late Herbert Fröhlich, gives an impression of the deep and wide implications of coherence for a new understanding of life.

PHOTOPHYSICAL REACTIONS IN CELLS

G. RENGER

Max-Volmer-Institute
Technical University, Berlin
Str. des 17. Juni 135
10623 Berlin, Germany

1. Abstract

The present report briefly summarises general principles of the interaction between solar radiation and living matter. In general, a photon flux can be used either as a carrier of different levels of information (enzyme activation, regulation of gene expression, photomorphogenesis, phototaxis, phototropism, process of vision) or of Gibbs energy as the driving force for biological processes. The main part of this contribution deals with the process of solar energy exploitation by water cleavage in oxygen evolving photosynthesising organisms. Special emphasis is paid to the adaptation to different illumination conditions by suitable *antenna systems* and the transformation of electronically excited states into electrochemical Gibbs energy by charge separation in the *reaction centres*. It is also shown that electron-hole recombination in the oxygen evolving Photosystem II leads to formation of the lowest excited singlet state of chlorophyll and radiative emission either as delayed luminescence or as thermoluminescence.

2. Introduction

Photophysical processes comprise the generation and reactions of electronically excited states of atoms and molecules. In cells there exist numerous compounds that absorb radiation in the wavelength region of UV, visible and near infrared (UV/VIS/NIR) light. Fig. 1 schematically summarises the possible pathways of generation and subsequent reactions of an excited singlet state at a pigment molecule P. State ${}^1P^*$ can be formed via two different pathways: a) by light absorption and b) by a chemical reaction. The former process is the fastest reaction of photophysics with a time constant of about 1 fs while the latter is slower by many orders of magnitudes. There exist several routes for the subsequent decay of ${}^1P^*$. The radiative relaxation gives rise to light emission. In the case of ${}^1P^*$ generation by light absorption the fluorescence emission is normally slightly shifted to longer wavelength due to the Stokes shift; the emission from a chemically formed excited electronic state is referred to either as chemiluminescence or as

J. J. Chang et al. (eds.), Biophotons, 1-17.

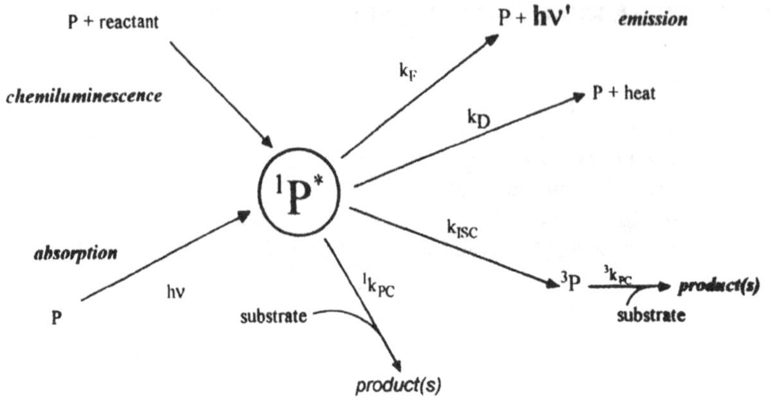

Figure 1. Simplified scheme of generation and reactions of an electronically excited state (for explanation, see text).

bioluminescence, depending on the nature of the underlying process. Another decay route of the $^1P^*$ state is the radiationless dissipation via a cascade of vibrational levels leading to heat production. Spin-orbit coupling gives rise, via intersystem crossing, to formation of triplet states with a comparatively long lifetime. A great variety of chemical reactions can originate from electronically excited states including intramolecular nuclear rearrangements (e.g. cis-trans isomerisation), bond fission (photodissociation) or intermolecular interaction with reactants (e.g. photo-redox reactions). The scheme of Fig. 1 is rather simplified because it does not explicitly comprise higher electronically excited states, excited state absorption etc.

2.1. TYPES OF PHOTOPHYSICAL EVENTS IN CELLS

In cells the properties of photophysical processes can be tuned in a delicate manner by incorporation of pigments into proteins. This unique material offers the possibility to optimise distances between reactants at a subnanometer scale, to modulate the energetics of optical and/or redox transitions and to separate spatially definite reactions by incorporation of pigment protein complexes into membranes. In this way practically any degree of specificity, efficiency and regulatory control could be achieved for a particular process under the selective pressure of evolutionary development.

Before illustrating this point for the most important process of solar energy exploitation by photosynthetic water cleavage, a more general scheme will be briefly discussed for the interaction between living matter and electromagnetic radiation in the UV/VIS/NIR region. As schematically shown in Fig.2, a photon flux characterised by its density and spectral distribution can induce the formation of electronically excited states in pigments (symbolised by P_i^*). Two basically different reaction sequences can be induced by P_i^*:

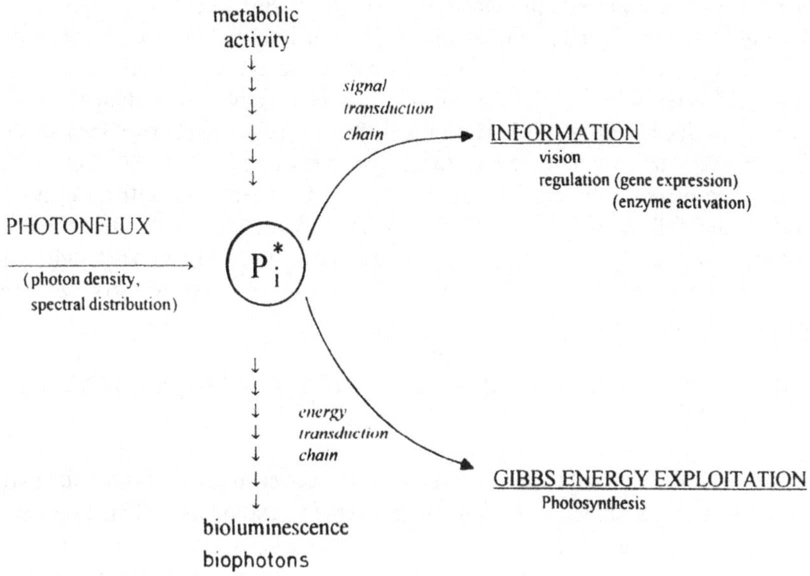

Figure 2. Simplified scheme of the transformation of photon fluxes in biological systems and generation of electrochemically excited states by metabolic activity (for details, see text).

i) a signal transduction chain using the *information* content of the photon flux or ii) an energy transduction chain using the *Gibbs energy* content of solar radiation for its storage in form of energy-rich products.

The information transfer occurs at different levels of functional and/or structural complexity. Comparatively simple regulatory processes by light triggering are enzyme activation and gene expression although in many cases the mechanism of signal transduction is not yet clarified (for reviews on light regulation processes in the photosynthetic apparatus, see Gray, 1996, Haag and Renger, 1997 and references therein). More complex reaction sequences are involved in phototaxis, photomorphogenesis and phototropism (see Häder and Tevini, 1987 and references therein). The most sophisticated information transfer by light is the process of vision in higher organisms (for a review, see Stieve, 1983). It has to be emphasised that the photochemical reaction itself is comparatively simple. In the case of vision it is a cis-trans isomerism of the protein bound chromophor retinal but the subsequent signal chain induced by the primary event is rather complex.

The second type of processes induced by light absorption of biological organisms is the exploitation of solar radiation as a unique source of Gibbs energy in order to satisfy the indispensible energetic demand for the development and sustenance of almost all living

matter. The direct transformation of solar energy occurs only in photoautotrophic organisms while all other systems energetically depend on the chemical use of substances that are ultimately produced with light as driving force.
Apart from the use of photon fluxes as a source of information and/or Gibbs energy, biological organisms are also able to produce electromagnetic radiation either via the conventional pathway of enzyme catalysed reactions leading to bioluminescence (for short reviews, see McElroy, 1983, Hastings, 1986) or via metabolic reactions which give rise to ultraweak biophoton emission (for review, see Popp et al., 1992). The origin and the unique properties of biophotons will be discussed in most of the articles of this book. Therefore the following considerations will not address this latter topic but rather concentrate on the first part of the interaction between electromagnetic radiation and biological organisms, in particular the use of solar radiation as driving force for metabolic activity.

2.2. BIOLOGICAL SOLAR ENERGY EXPLOITATION VIA PHOTOSYNTHETIC WATER CLEAVAGE

The invention of a molecular device that enables water cleavage by visible light was the cornerstone in the evolution of biological solar energy exploitation. This event occured 2 - 3 billion years ago and had two consequences of paramount importance for the development of highly organised living matter: (1) it allowed the huge water pool on the earth's surface to become available as a hydrogen source for the biosphere, and (2) the formation of molecular oxygen as photosynthetic "waste product" led to the present day aerobic atmosphere. The latter result opened the road for much more efficient exploitation of the Gibbs energy content of food via aerobic respiration (for a review, see Renger, 1983) and simultaneously led to the formation of the essential protective ozone layer in the stratosphere.
Fig. 3 shows a scheme of the structural and functional hierarchy of photosynthetic activity in higher plants. Under illumination all leaves emit a characteristic red fluorescence that originates from the radiative decay of the lowest excited singlet state of chlorophyll. The quantum yield of this fluorescence is of the order of a few percent depending on the functional competence of the photosynthetic apparatus. Therefore this emission can be used as a powerful diagnostic tool for noninvasive screening of algae and plants (see Renger and Schreiber, 1986, Govindjee, 1995 and references therein). After cessation of illumination the photosynthesising organisms continue to emit light with virtually the same spectral properties as prompt fluorescence but at much weaker intensity. This delayed light emission is due to the formation of excited chlorophyll singlet states via back reactions of oxidising and reducing equivalents (*vide infra*).
The overall process leads to water cleavage into molecular oxygen and hydrogen bound to nicotinamide adenine dinucleotide phosphate (NADP). The lower part of Fig. 3 shows that the transformation of solar radiation into Gibbs energy takes place in integral pigment protein complexes referred to as Photosystem I (PS I) and II (PS II). Both photosystems provide the driving force for the electron transfer from water to $NADP^+$ and the coupled ATP formation. The PS I and PS II complexes are anisotropically incorporated into the thylakoid membrane. Accordingly, the light induced charge

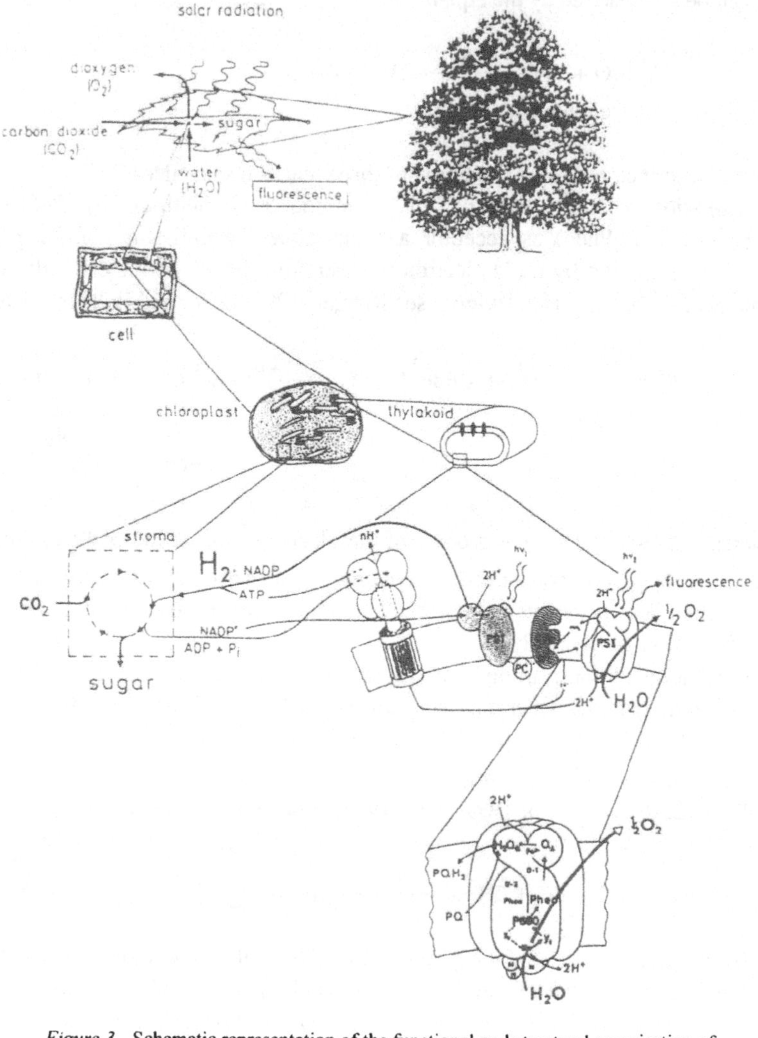

Figure 3. Schematic representation of the functional and structural organisation of photosynthesis in higher plants.

separation is a vectorial process and the protolytic reactions coupled with electron transfer steps at the donor and acceptor side give rise to formation of a transmembrane electrochemical potential difference of protons. This difference, referred to as proton motive force, energetically drives the ATP synthesis (for a review, see Gräber, 1997). A closer inspection of Fig. 3 reveals that the essential steps for the use of water as biological hydrogen source take place in PS II. Therefore the following considerations will concentrate on the reactions in this complex. A scheme of the structural and functional organisation of this complex is presented at the bottom of Fig. 3. The overall reaction can be summarised by the equation:

$$2H_2O + 2PQ \xrightarrow{4h\nu} 2PQH_2 + O_2 \qquad (1)$$

where PQ is the molecule plastoquinone-9.

The process described by Eqn. (1) comprises three reaction sequences:

a) photooxidation of a special Chl-a compound (symbolised by P680) with pheophytin-a (Pheo) as acceptor and subsequent stabilisation of the primary charge separation by rapid electron transfer from $Pheo^-$ to a specifically bound plastoquinone (Q_A) (for reviews, see Renger, 1992, Diner and Babcock, 1996):

$$P680PheoQ_A \xrightarrow{\langle\varepsilon\rangle} {}^1P680^*PheoQ_A \xrightarrow{k_{PC}} P680^{+\bullet}Pheo^{-\bullet}Q_A \xrightarrow{k_{stab}} P680^{+\bullet}PheoQ_A^{-\bullet} \qquad (2)$$

where $\langle\varepsilon\rangle$ symbolises the process of the electronic excitation of P680 either by direct light absorption or excited state energy transfer (*vide infra*), k_{PC} and k_{stab} are the rate constants of primary charge separation and its subsequent stabilisation, respectively.

b) cooperation of four strongly oxidising redox equivalents (⊕) in a manganese containing unit, the water oxidising complex (WOC), where two water molecules are oxidised to oxygen and four protons via a sequence of four univalent redox steps with $P680^{+\bullet}$ as driving force (for reviews, see Debus, 1992, Rutherford et al., 1992, Renger, 1993, 1997, Britt, 1996, Haumann and Junge, 1996).

$$4\oplus + 2H_2O \xrightarrow{WOC} 4H^+ + O_2 \qquad (3)$$

c) cooperation of two reducing equivalents (Θ) at a PQ-molecule transiently associated with a special site (Q_B-site) thus giving rise to plastoquinol formation under proton uptake (for reviews, see Crofts and Wraight, 1983, Lavergne and Briantais, 1996):

$$2\Theta + PQ + 2H^+ \xrightarrow{Q_B-\text{site}} PQH_2 \qquad (4)$$

This reaction pattern readily shows that only reaction sequence (a) comprises photophysical and photochemical processes. Therefore the following description concentrates on this sequence.

The light induced generation of the primary cation-anion radical pair $P680^{+\bullet}Pheo^{-\bullet}$ requires the electronic excitation into state ${}^1P680^*$ and subsequent electron ejection

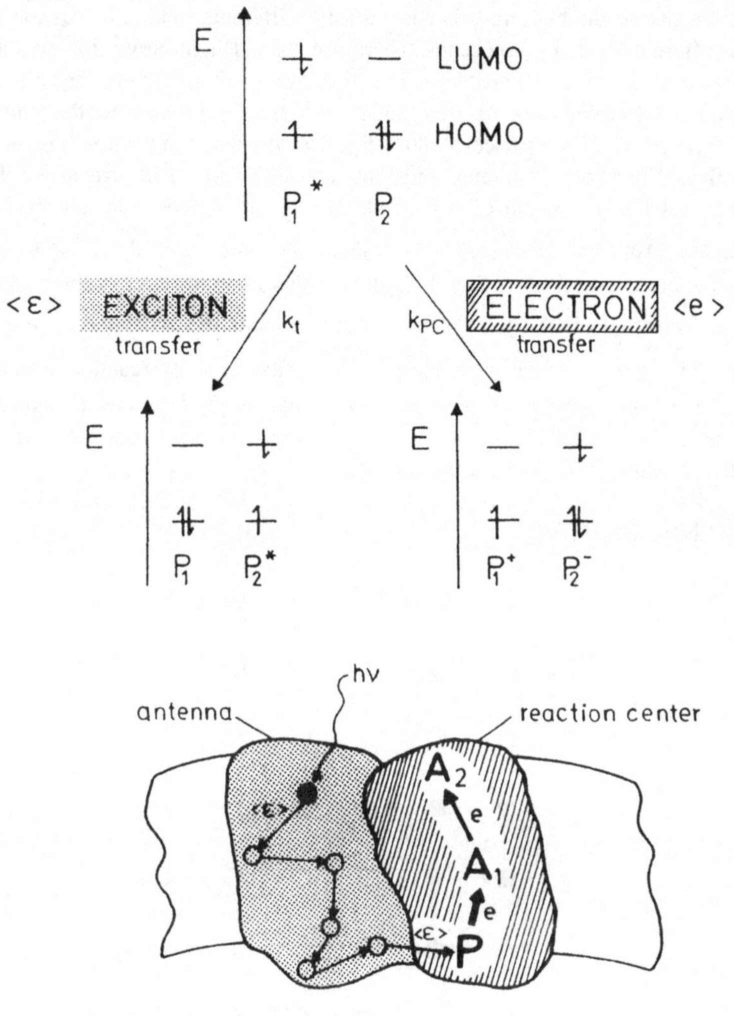

Figure 4. Electronic coupling between two pigment molecules P_1 and P_2 of the same molecular orbital structure. The interaction of the electronically excited molecule ${}^1P_1^*$ with the ground state molecule P_2 can either lead to exciton or electron transfer (for further details, see text). The bottom part shows a schematic representation of the principles of an antenna system and reaction centre in photosynthesising organisms.

with Pheo as acceptor. Light absorption by P680 alone would not provide a suitable adaptation of photosynthetic organisms to different illumination conditions. The necessity of such a mechanism is easily understandable. In a tropical rainforest the intensity and spectral distribution of solar radiation at the top of a tree canopy markedly differs from that at the bottom. Likewise, drastic differences exist for aquatic systems living at different levels in the oceans. Therefore special light harvesting systems have been developed to solve this problem. The pigments of these systems absorb the light and funnel the electronically excited states with high efficiency to the photoactive pigment P_{RC} (in PS II component P680 is P_{RC}) of the reaction centres. The principles of the different reactions of antenna pigments and P_{RC} can be illustrated for the most simple case of the interaction of two identical pigments. Depending on the mode of coupling, the excited state of pigment ${}^1P_1^*$ undergoes two basically different types of reaction that are summarised in Fig. 4 (top): i) radiationless transfer of the excited state leading to formation of ${}^1P_2^*$ or ii) electron transfer from ${}^1P_1^*$ to P_2 thus giving rise to formation of the ion radical pair $P_1^{+\bullet}P_2^{-\bullet}$. The first type of reaction provides the functional basis of antenna systems, the second one is the key step of excited state transformation into electrochemical Gibbs energy in the reaction centres, as schematically shown in the bottom part of Fig. 4.

2.3. ANTENNA SYSTEMS

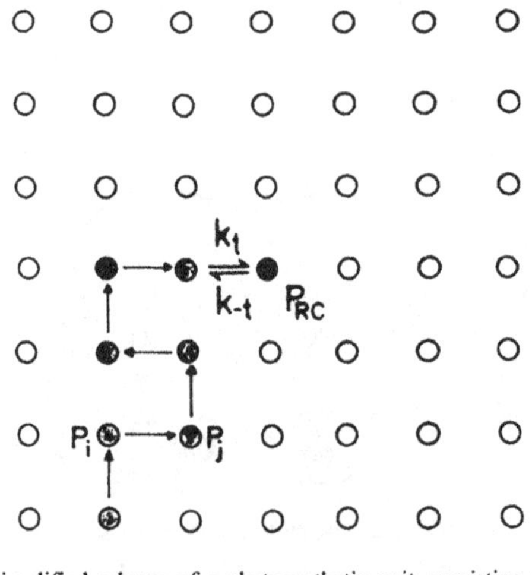

Figure 5. Simplified scheme of a photosynthetic unit consisting of N pigment molecules P_1 ... P_N. The central molecule is assumed to act as a photochemical trap, P_{RC}, in the forward and reverse direction, respectively (marked in black, the pigments participating in a particular excitation energy migration hopping path are marked in grey).

The general features of the function of pigment arrays as antenna systems can be illustrated by the idealised model of a regular lattice of pigment molecules as shown in Fig. 5. In most systems the pigment-pigment interaction is weak enough so that an excited state can be considered to migrate through the ensemble via an incoherent stochastic hopping process (although not precise in terms of solid state physics, this type of excited state is often designated as "localised" exciton). In this case the excited state (exciton) is always localised on a particular chromophore during its lifetime and transferred in a random walk pathway from pigment to pigment within the array until it becomes trapped by the photoactive pigment P_{RC} of the reaction centre. The rate constant of each elementary step of exciton transfer from excited pigment $^{*}P_l$ to a neighbouring pigment P_j is given by the Förster type mechanism (Förster, 1948, 1965). If, on the other hand, the interaction between the pigments is sufficiently strong an exciton coupling arises and the excited state is delocalised all the time of its existence over the whole ensemble.

In reality the antenna systems are composed of pigment-protein complexes with a well structured array of the chromophores, and therefore the actual pathway of exciton migration depends on the peculiarities of pigment-pigment interaction in each subunit and its mutual interaction with other subunits and the reaction centre complex. A variety of antenna systems have been evolved in photosynthetic organism to achieve optimal adaption to different illumination conditions. Two examples of quite different types of antenna systems in oxygen evolving photosynthetic organisms are schematically shown in Fig. 6.

Recent progress in protein crystallisation and X-ray structure analyses has led to detailed information on the geometry of the pigments within the protein matrix (for an antenna system of anoxygenic purple bacteria, see Freer et al., 1996; for an example of higher plants, see Kühlbrandt, 1994). This provides the basis for mechanistic considerations on the mode of exciton transfer in particular pigment protein complexes and their mutual interaction. The most abundant antenna complex is LHC II which contains about 50% of the total Chl content of the thylakoid membrane (for reviews, see Jansson, 1994, Paulsen, 1995). A structural analysis reveals that the 12 (13) chlorophyll molecules in each monomeric subunit are arranged in two layers with a comparatively short van der Waals distance (order of 5Å) between the chromophores within each layer (see Fig. 6, bottom left and Kühlbrandt, 1994). These short distances are indicative of strong pigment coupling. It is therefore not surprising that time resolved spectroscopy revealed very fast exciton energy transfer steps (time constants of the order of 150 fs are found for the transfer from Chl-b to Chl-a, Bittner et al., 1994). Furthermore, recent analysis led to the conclusion that exciton coupling can give rise to delocalisation of the excited states among the chlorophyll molecules of a monomeric unit (Schödel, et al., 1996). The pigment array in the antenna complexes permits efficient light harvesting and transfer to the photoactive pigment P680 of the PS II reaction center (*vide infra*). Based on the above mentioned findings the overall process of light harvesting comprises both exciton delocalisation among few tightly coupled chromophors within a pigment protein subunit and incoherent hopping steps between the subunits and in subunits with weakly coupled chromophors.

Apart from the efficient funneling of excitons to the photoactive pigment, the antenna complexes have also to regulate superfluous excitation at high light intensities in order to protect the organisms from photoinduced destruction under light stress. One way to achieve this goal is the formation of dissipative channels for radiationless decay. It was

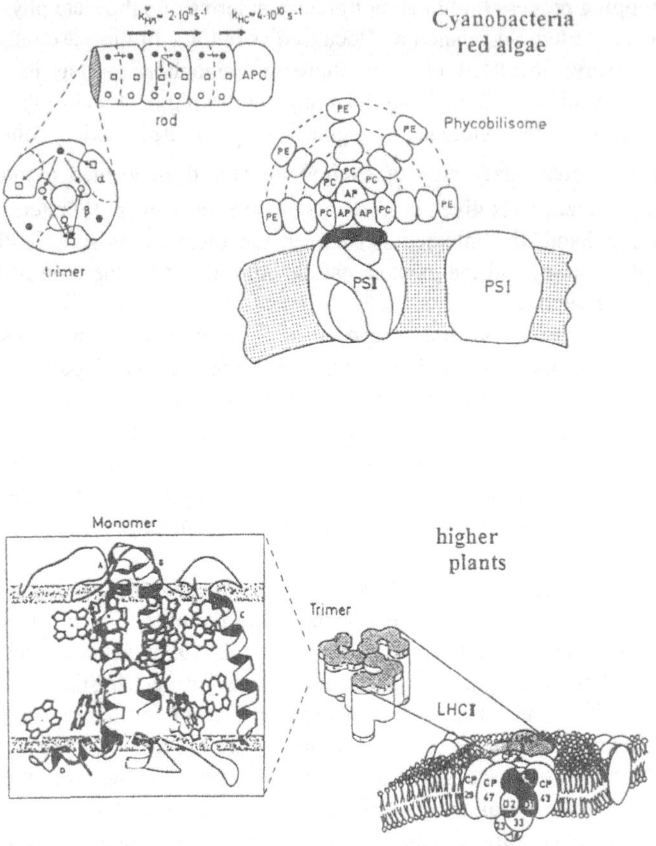

Figure 6. Antenna systems in oxygen evolving cyanobacteria, red algae (top) and higher plants (bottom). The top part shows a hemispherical phycobilisome associated with the PS II complex via the linker protein symbolised by a black area (modified after Gantt, 1986). The left hand side shows a top and side view of a particular rod typical for cyanobacterial phycobilisomes (modified after Holzwarth, 1989). AP = allophycocyanin, PC = phycocyanin, PE = phycoerythrin, circles and squares symbolise phycocyanobilin chromophers bound to different aminoacid residues and subunits (α,β).
In the bottom part the organisation membrane bound antenna complexes is schematically shown (Renger, 1992). The most abundant LHC II is organised in trimers. The structure and the pigment array is depicted in the left side of the bottom part (modified after Kühlbrandt, 1994).

found that aggregation of LHC II gives rise to accelerated fluorescence decay which is indicative of the formation of quenchers for excited states (Mullineaux et al., 1993,

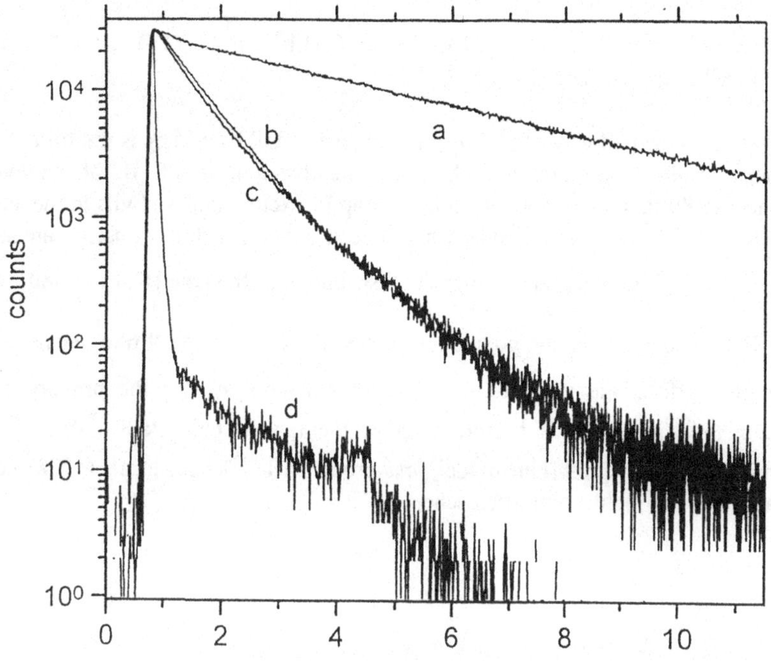

Figure 7. Single photon counting fluorescence decay of solubilised (a) and two different forms of aggregated (b,c) LHC II complexes, (d) represents the system response function (for experimental details and explanation, see Vasil'ev et al., 1997).

Vasil'ev et al., 1997). A typical feature of the fluorescence curves in solubilised and aggregated LHC II is depicted in Fig. 7. Based on different lines of evidence it was concluded that aggregation of LHC II complexes causes enhanced dissipation by formation of quenching states (see Horton et al., 1996 and references therein).
Another most important protective function has to be established in order to prevent oxidative photodestruction of the photosynthetic apparatus. The indispensible intersystem-crossing reaction causes the population of the Chl-triplet state that gives rise to sensitised formation of reactive singlet oxygen $(^1\Delta_g O_2)$. Therefore, all antenna systems contain carotenoids (Car) which exert a dual protective function: they enable a rapid triplet energy transfer from ^{3}Chl to Car and subsequent radiationless decay of ^{3}Car and simultaneously Car react rapidly with singlet oxygen under its decay to the triplet ground state $(^3\Sigma_g O_2)$ (for a review, see Siefermann-Harms, 1987).

These few selected examples reveal that the antenna system of photosynthesising organisms permits appropriate adaption to a great variety of different illumination conditions.

2.4. TRANSFORMATION OF ELECTRONICALLY EXCITED STATES INTO ELECTROCHEMICAL GIBBS ENERGY

The key step in the pathway of photons into quanta of Gibbs energy is the formation of the primary radical pair and its subsequent stabilisation. In PS II the photoactive component P680 functions as a very shallow trap of excited chlorophylls in the antenna (these chlorophylls are symbolised by Chl_A) because their spectral properties are similar to those of P680. Therefore, an equilibrium distribution of the type $\left[Chl_A^* P680\right] \rightleftharpoons \left[Chl_A\, {}^1P680^*\right]$ determines the population probability of the state ${}^1P680^*$. This exciton equilibration affects the value of the overall rate constant for the primary charge separation (k_{PC}^{trap}) that leads to formation of the radical pair $P680^{+\bullet}Pheo^{-\bullet}$. The exciton-radical pair equilibrium model first proposed by Schatz et al. (1988) can be summarised by the highly simplified scheme of Fig. 8.

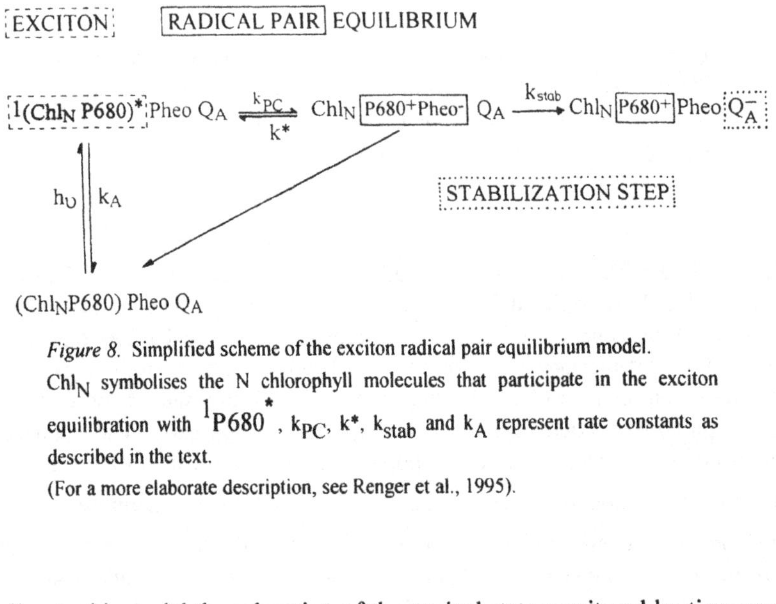

Figure 8. Simplified scheme of the exciton radical pair equilibrium model. Chl_N symbolises the N chlorophyll molecules that participate in the exciton equilibration with ${}^1P680^*$, k_{PC}, k^*, k_{stab} and k_A represent rate constants as described in the text.
(For a more elaborate description, see Renger et al., 1995).

According to this model the relaxation of the excited state monitored by time resolved fluorescence decay curves should exhibit a biphasic kinetics. The parameters of these kinetics permit an extraction of the rate constants k_{PC}^{trap}, k^* and k_2. A typical trace of

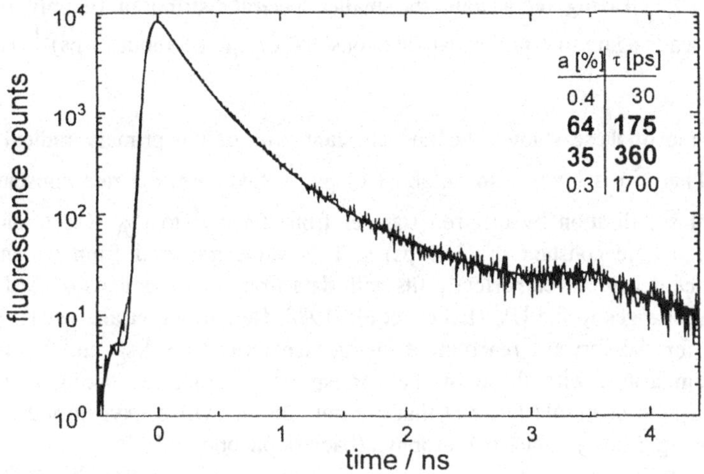

Figure 9. Single photon counting fluorescence decay trace in PS II membrane fragments with functionally competent reaction centres (for experimental details, see Renger et al., 1995).

fluorescence decay curves measured in PS II membrane fragments with a functionally competent reaction centre is shown in Fig. 9. As expected the data are well described by a biexponential decay (the small contributions of 30 ps and 1.7 ns can be ignored; for further analyses, see Renger et al., 1995). A numerical fit of these data lead to the following rate constants k_{PC}^{trap} = $(230\ ps)^{-1}$, k* = $(2.3\ ns)^{-1}$ and k_{stab} = $(300\ ps)^{-1}$ with k_A kept constant at $(3\ ns)^{-1}$ (a variation of this value within a realistic range does not significantly affect the calculated rate constants k_{PC}^{trap}, k_{stab} and k*, (see Vasil'ev et al., 1996). The exciton-radical pair equilibrium model predicts that k_{PC}^{trap} is depending on the spectral properties and number of antenna pigments that participate in the exciton equilibrium. Based on the assumption that the latter process is sufficiently fast, the molecular rate constant of the primary charge separation, k_{PC}, can be calculated by using the simple relation $k_{PC} = k_{PC}^{trap}/p_{^1680^*}$ where $p_{^1680^*}$ is the population probability of state $^1P680^*$ when one singlet exciton exists in the PS II complex with its associated antenna. The PS II membrane fragments used for the measurements of Fig. 9 contain about 240 antenna chlorophylls. Accounting for the spectral differences between P680 and antenna chlorophylls, a rough approximation (Renger et al., 1995) leads to a rate constant k_{PC} of about $(3\ ps)^{-1}$. According to the exciton-radical pair equilibration model the rate constant k_{PC}^{trap} is expected to increase when the number of antenna chlorophylls decreases. Therefore, corresponding measurements were performed in PS II core complexes with only 50 - 60 chlorophylls per P680. A value of $(130\ ps)^{-1}$ was found for k_{PC}^{trap}. Taking into account the smaller spectral distribution (mainly due to lack of Chl b) leads again to a molecular rate constand of k_{PC} of about $(3\ ps)^{-1}$. (Renger et al., 1995).

As a resumé of these studies the rate constant k_{PC} of the primary radical ion pair $(P680^{+\bullet} Pheo^{-\bullet})$ is inferred to be about $(3\ ps)^{-1}$. Likewise, the rate constant for the subsequent stabilisation by electron transfer from $Pheo^{-\bullet}$ to Q_A was found to take place with a rate constant of $(300\ ps)^{-1}$. This value gathered from an analysis of fluorescence decay curves perfectly fits with data from measurements of flash induced absorption changes in the UV (Eckert et al., 1988, Bernarding et al., 1994). Both rate constants for the forward reaction of charge separation, i.e. k_{PC} and k_{stab}, exhibit striking similarities with those of the corresponding processes taking place in the reaction centres of purple bacteria that are not able to evolve oxygen and where Chl (Pheo) are replaced by bacteriochlorophyll (bacteriopheophytin).

This phenomenon illustrates that the same principles are used by all photosynthesising organisms for the transformation of solar radiation into Gibbs energy, i.e. the reaction centres were optimised already at early stages of the evolutionary development. The only new "discovery" was the selection of the photoactive pigment P680 with its strongly oxidising cation radical $P680^{+\bullet}$ as the indispensible prerequisite for the exploitation of water as electron source in all oxygen evolving organisms. The underlying principles of this unique property of $P680^{+\bullet}$ still remains to be clarified.

2.5. DELAYED LUMINESCENCE AND THERMOLUMINESCENCE OF PS II

The reaction sequence (b) leading to water oxidation takes place via a sequence of four univalent redox steps and therefore comprises the intermediary storage of oxidising redox equivalents (holes) (for reviews, see Debus, 1992, Rutherford et al., 1992, Renger, 1993, 1997). Likewise, the reductive pathway leading to PQH_2 formation with Q_A^- as electron donor (see Crofts and Wraight, 1983, Lavergne and Briantais, 1996) requires the intermediary storage of one reducing equivalent (electron). These electrons and holes can recombine in comparatively slow reactions that lead with very low quantum yield to the reformation of $^1P680^*$ in the dark. This gives rise to a very low intensity delayed light emission (for a review, see Lavorel, 1975). The reaction requires thermal activation because the electrochemical energy of the electron-hole pair is below that of the excited singlet state $^1P680^*$. Accordingly, this recombination can be triggered by heating the samples thus leading to a radiative emission referred to as thermoluminescence. The shape of these "glow curves" provides invaluable information on the functional integrity of PS II (for a review, see Inoue, 1996 and references therein). It has to be emphasised that these emissions are characterised by the chlorophyll spectrum and therefore markedly differ from the spectral properties of biophotons emitted from photosynthesising organisms (see Popp et al., 1992 and references therein).

3. Concluding remarks

The brief description of photophysical processes in cells has shown that biological systems are able to use photon fluxes as carrier of information in a great variety of signal transducing chains and to exploit the energy content of solar radiation as unique source to satisfy the Gibbs energy demand of almost all living matter. Photosynthetic organisms use sunlight directly as driving force for bioenergetic processes. Different antenna systems have been evolved to permit an optimal adaption to varying illumination conditions. On the other hand, the reaction centres evolved for transformation of electronically excited states into electrochemical Gibbs energy exhibit striking similarities in their structural and functional organisation. This mode of biological optimisation is not only an exciting topic of basic reseach but could also inspire the construction of biomimetic systems for solar energy exploitation in technical devices.

4. Acknowledgements

The financial support by Deutsche Forschungsgemeinschaft and Fonds der Chemischen Industrie of the author's own research cited in this manuscript is gratefully acknowledged.

5. References

Bernarding, J., Eckert, H.-J., Eichler, H.J., Napiwotzki, A. and Renger, G. (1994) Kinetic studies on the stabilisation of the primary radical pair $P680^+Pheo^-$ in different photosystem II preparations from higher plants. *Photochem. Photobiol.* **59**, 566-573.

Bittner, T., Irrgang, K.-D., Renger, G. and Wasielewski, M.R. (1994) Ultrafast excitation energy transfer and exciton-exciton annihilation processes in isolated light harvesting complexes of photosystem II in green plants. *J. Phys. Chem.* **98**, 11821-11826.

Britt, R.D. (1996) Oxygen Evolution, in D.R. Ort and C.F. Yocum, C.F. (eds), *Oxygenic Photosynthesis: The Light Reactions*, Kluwer Academic Publishers, Dordrecht, pp. 137-164.

Crofts, A.R. and Wraight, C.A. (1983) The electrochemical domain of photosynthesis. *Biochim. Biophys. Acta* **726**, 149-185.

Debus, R.J.(1992) The manganese and calcium ions in photosynthetic O_2 evolution. *Biochim. Biophys. Acta* **1102**, 269-352.

Diner, B.A. and Babcock, G.T. (1996) Structure, dynamics and energy conversion efficiency in photosystm II, in D.R. Ort and C.F. Yocum (eds), *Oxygenic Photosynthesis: The Light Reactions*, Kluwer Academic Publishers, Dordrecht, pp. 213-247.

Eckert, H.-J., Wiese, N., Bernarding, J., Eichler, H.J. and Renger, G. (1988) Analysis of the electron transfer from Pheo⁻ to Q_A in PS II membrane fragments from spinach by time resolved 325 nm absorption changes in the picosecond domain. *FEBS Lett.* **240**, 153-158.

Förster, Th. (1948) Intermolecular energy transference and fluorescence. *Annalen Physik* **2**, 55-75. [in German]

Förster, Th. (1965) Delocalised excitation and excitation transfer, in O. Sinanoglu (ed) *Part II.B.1 of Modern Quantum Chemistry: Istanbul Lectures. Part III, Action of Light and Organic Crystals*, Academic Press, New York, pp 93-137.

Freer, A., Prince, S., Sauer, K., Papiz, M., Hawthornwaite-Lawless, A., McDermott, G. and Cogdell, R. (1996) Pigment-pigment interactions and energy transfer in the antenna complex of the photosynthetic bacterium *Rhodopseudomonas acidophila*. Structure **4**, 449-462.

Gantt, E. (1986) Phycobilisomes, in L.A. Strachlin and C.J. Arntzen (eds) *Photosynthetic membranes and light harvesting systems, Photosynthesis (II)*, Springer, Berlin, pp. 260-268.

Govindjee (1995) Sixty-three years since Kantsky: Chlorophyll *a* Fluorescence. *Aust. J. Plant Physiol* **22**, 131-160.

Gräber, P. (1997) Bioenergetics, in P. Gräber and G. Milazzo (eds) *Biochemistry: Principles and Practice*, Birkhäuser Verlag, Basel, Vol 4, pp 486-535.

Gray, J.C. (1996) Regulation of expression of nuclear genes encoding polypeptides required for the light reactions of photosynthesis, in D.R. Ort and C.F. Yocum (eds), *Oxygenic Photosynthesis: The light reactions*, Kluwer Academic Publishers, pp. 621-641.

Haag, E and Renger, G (1997) Chloroplasts, in P. Gräber and G. Milazzo (eds), *Treatise on Bioelectrochemistry, Vol. 2: Bioenergetics*, Birkhäuser Verlag, Basel, pp. 212-273.

Häder, D.-P. and Tevini, M. (1987) *General Photobiology*, Pergamon Press, Oxford.

Hastings, J.W. (1986) Bioluminescence in bacteria and dinoflagellates, in Govindjee, J. Amesz and D.C. Fork (eds), *Light emission by plants and bacteria*, Academic Press, New York, pp. 363-398.

Haumann, M. and Junge,W. (1996) Protons & charge indicators in oxygen evolution, in D.R. Ort, and C.F. Yocum (eds) *Oxygenic Photosynthesis: The Light Reactions*, Kluwer Academic Publishers, Dordrecht, pp. 165-192.

Holzwarth, A.R. (1989) Applications of ultrafast laser spectroscopy for the study of biological systems. *Quarterly Reviews of Biophysics* **22**, 239-326.

Horton, P., Ruban, A.V. and Walters, R.G. (1996) Regulation of light harvesting in green plants. Indication by nonphotochemical quenching of chlorophyll fluorescence. *Annu. Res. Plant Physiol. Plant Mol. Biol.* **47**, 655-684.

Inoue, Y. (1996) Photosynthetic thermoluminescence as a simple probe of photosystem II electron transport, in J. Amesz and A.J. Hoff (eds), *Biophysical techniques in photosynthesis, Kluwer*, Dordrecht, pp. 93-107.

Jansson, S. (1994) The light harvesting chlorophyll a/b-binding proteins. *Biochim. Biophys. Acta* **1184**, 1-19.

Kühlbrandt, W. (1994) Structure and function of the plant light-harvesting complex, LHC II. *Curr. Biol.* **4**, 519-528.

Lavergne, J. and Briantais, J.-M. (1996) Oxigenic Photosynthesis: The Light Reactions in C.A. Yocum and D.R. Ort (eds) *Advances in Photosynthesis*, Vol. 4, Kluwer Academic Publishers, Dordrecht, pp. 265-287.

Lavorel, J. (1975) Luminescence, in Govindjee (ed) *Bioenergetics of Photosynthesis*, Academic Press, New York, pp. 223-317.

McElroy, W.D. (1983) Bioluminescence, in W. Hoppe, W. Lohmann, H. Markl and H. Ziegler (eds), *Biophysics*, Springer, Berlin, pp. 559-569.

Mullineaux, C.W., Pascal, A.A., Horton, P. and Holzwarth, A.R. (1993) Exciton-energy quenching in aggregates of the LHC II chlorophyll-protein complex: a time-resolved fluorescence study. *Biochim. Biophys. Acta*, **1141**, 23-28.

Paulsen, H. (1995) Chlorophyll a/b binding proteins. *Photochem. Photobiol.* **62**, 367-382.

Popp, F.A., Li, K.H. and Gu, Q. (eds) (1992) *Recent Advances in Biophoton Research and its Applications*, World Scientific, Singapore.

Renger, G (1983) Biological Energy Conservation, in W. Hoppe, W. Lohmann, H. Markl and H. Ziegler (eds.), *Biophysics*, Springer, Berlin, pp. 347- 371.

Renger, G. (1992) Energy transfer and trapping in Photosystem II, in J. Barber (ed), *Topics in Photosynthesis, The Photosystems: Structure, Function and Molecular Biology*, Elsevier, Amsterdam, pp. 45 - 99.

Renger, G. (1993) Water cleavage by solar radiation - an inspiring challenge of photosynthesis research. *Photosynth. Res.* **38**, 229-247.

Renger, G. (1997) Mechanistic and structural aspects of photosynthetic water oxidation. *Physiol. Plant.* **100**, 828-841.

Renger, G. and Schreiber, U. (1986) Practical applications of fluorometric methods to algae and higher plants, in D.C. Fork, Govindjee and J. Amesz. (eds), *Light Emission by Plants and Bacteria*, Academic Press, New York, pp. 587-619.

Renger, G., Eckert, H.-J., Bergmann, A., Bernarding, J., Liu, B., Napiwotzki, A., Reifarth, F. and Eichler, H.J. (1995) Fluorescence and spectroscopic studies on exciton trapping and electron transfer in photosystem II of higher plants. *Austr. J. Plant Physiol.*, **22**, 167-181.

Rutherford,.A.W., Zimmermann, J.-L. and Boussac, A. (1992) Oxygen evolution, in J. Barber (ed), *The Photosystems: Structure, Function and Molecular Biology*, Elsevier, Amsterdam, pp 179-229.

Schatz, G., Brock, H. and Holzwarth, A.R. (1988) Kinetic and energetic model for the primary processes in photosystem II. *Biophysical Journal* **54**, 397-405.

Schödel, R., Hillman, F., Schrötter, T., Voigt, J., Irrgang, K.-D. and Renger, G. (1996) Kinetics of excited states of pigment clusters in solubilized LHCII: Photon density dependent fluorescence yield and transmission. *Biophysical Journal* **71**, 3370-3380.

Siefermann-Harms, D. (1987) The light harvesting and protective functions of carotenoids in photosynthetic membranes. *Physiol. Plant* **69**, 561-568.

Stieve, H. (1983) Photoreception and its Molecular Basis, in W. Hoppe, W. Lohmann, H. Markt and H. Ziegler (eds) Biophysics, Springer, Berlin, pp. 709-729.

Vasil'ev, S., Bergmann, A., Redlin, H., Eichler, H.J. and Renger, G. (1996) On the role of exchangeable hydrogen bonds for the kinetics of $P680^{+\bullet}\ Q_A^{-\bullet}$ formation and $P680^{+\bullet}\ Pheo^{-\bullet}$ recombination in Photosystem II. *Biochim. Biophys. Acta* **1276**, 35-44.

Vasil'ev, S., Irrgang, K.-D., Schrötter, T., Bergmann, A., Eichler, H.J. and Renger, G (1997) Quenching of chlorophyll a fluorescence in the aggregates of LHCII: steady state fluorescence and picosecond relaxation kinetics. *Biochemistry* **36**, 7503-7512.

WEAK LIGHT EMISSION FROM BACTERIA AND THEIR INTERACTION WITH CULTURE MEDIA

R. VOGEL* AND R. SÜßMUTH

Institute of Microbiology, University Hohenheim,
70593 Stuttgart-Hohenheim, Germany
* *also at International Institute of Biophysics, Station Hombroich,*
41472 Neuss, Germany

Abstract

In this paper we present a survey on our investigations on ultraweak photon emission from bacterial cultures. We report on some aspects of the chemiluminescence of culture media, the interaction between culture media and cells and its consequences for the light emission. We observed a sudden decrease of the light emission of a wide variety of bacterial cultures, that occurs at, in general, low cell densities of about 10^7 cells/ml. Probable explanations point to active centers located in the cell membrane, where the inactivation of light generating species or its precursors takes place. Results on cell lysis after phage infection and on DNA-free minicells support this hypothesis. Finally we examined more in detail the emission pattern of strains of the lactic acid group and report on results concerning the underlying mechanisms of light generation in these strains.

1. Introduction

Light emission from bacterial cultures is usually connected with bioluminescence, a relatively bright photon emission of more than 10^6 photons / s cm^2, that results from certain enzymatic activities as luciferin-luciferase reactions. However, there is also a second, weak photon emission of in general less than 1000 photons / s cm^2. In contrast to the first one, that is restricted to cultures of only a few strains as e.g. *Vibrio fisheri*, the latter is a general phenomenon. As such a weak photon emission has also been observed from almost all higher organisms examined up to now[25, 30], this work extends the field to bacteria as the most simple organisms in order to study some basic aspects of this emission and its interrelations to the growth of bacterial cultures.

Most commonly weak photon emission is ascribed to oxidation reactions resulting in a weak chemiluminescence, at least partly due to the generation of excited carbonyl groups as intermediate products or the formation of singlet molecular oxygen $^1O_2$ [5, 11, 27]. A different approach is coming from Popp and co-workers. They developed a model basing on the chromosome of the cells and ascribing it a functional role as a photon storage and as a source of a coherent photon field with regulatory functions[21, 22].

J. J. Chang et al. (eds.), Biophotons, 19-44.

Corresponding to this model an emission pattern dependent on the cell division cycle has been found in the light emission of generative cells of larch[7] and synchronized yeast cultures[16].

Light emission from bacteria is a topic that has been attempted several times[13, 24, 28, 29], but it had been always restricted to single or only a few strains, such that comparisons between different strains could not be made without reservation. Moreover the setup of most experiments aimed at the response of the cells on externally exposed stress and varied considerably between the different experimenters. In this paper, we will not put that much emphasis on experimental details, but present results on a larger selection of strains under different culture conditions and try to give a more general view of the topic.

All experiments presented here were carried out with the single photon counting equipment[18, 25] at IIB, Neuss, or, when stated, with a Wallac scintillation counter (LSC) at the University Hohenheim.

2. Emission of culture media and dependence on composition and preparation

Considering light emission of liquid cultures of micro organisms or cell cultures it has to be taken into account that there is always a background radiation coming from the culture medium. This radiation is dependent on the composition of the medium, but also on the temperature and the duration of the heating while autoclaving the media[6, 17]. The age of the medium and the length of the period between opening the bottle (after autoclaving, the medium is in general under anaerobic conditions, which are removed by opening the bottle and exposing the medium to air) and the start of the measurement may also be important (see Fig 2). The emission of a medium can be reduced substantially by autoclaving certain components separately in order to avoid Maillard's reaction. In general, the emission of the medium raises with increasing complexity of the medium.

The emission is ascribed to auto-oxidation of the medium[12, 17] by compounds formed during sterilization of the medium. The strong dependence of the emission intensity on the sterilization procedure points to the generation of reducing compounds that lead to chemiluminescence under exposure to oxygen by the continuous formation of free radicals[12]. A direct dependence of the intensity on certain medium components could not yet be determined, but it seems to be connected with the occurrence of e.g. Maillard reaction and its reaction products.

But as shown in Fig. 1, even non-autoclaved, filtered media show chemiluminescence. The generation of superoxide radical and its following product hydroxyl radical are partially responsible, peroxide is assumed to be involved [6, 12]. However, addition of superoxide dismutase and of catalase to the medium leads to no substantial decrease of the emission intensity (data not shown).

For growth enhancement anti-oxidants, as ascorbic acid or sodium-pyruvate, have to be added to some media or are already added to components of the media by the manufacturer[10]. In M17 ascorbic acid has to be added explicitly. M17 autoclaved without ascorbic acid shows a 3-fold higher emission compared to M17 autoclaved with ascorbic acid. Moreover the emission could be successfully diminished by addition of

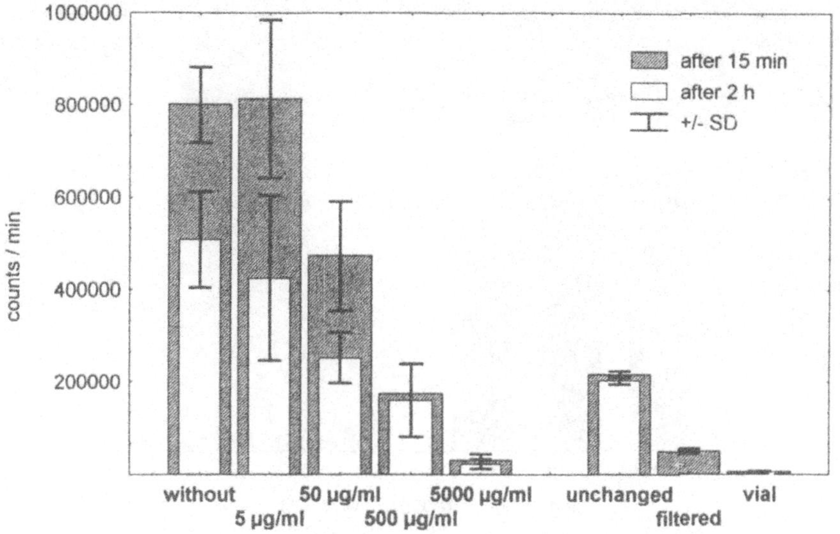

Figure 1. M17 was autoclaved without ascorbic acid (which is usually added at a concentration of 500 µg/ml before autoclaving) for 15 min. at 121°C. Ascorbic acid was sterile filtered and added to the medium in the stated concentration and the light emission was measured. In comparison the data for unchanged M17 (with ascorbic acid added at a concentration of 500 µg/ml before sterilization), that was either sterilized by autoclaving or sterile filtration (measured in LSC)

ascorbic acid after autoclaving the medium (Fig. 1).

As shown in Fig. 2, no emission can be detected from anaerobic culture medium[13]. This indicates that oxygen is necessary either for the generation of the oxidants or for the generation of excited states following oxidation reactions and leading to the light emission, or both.

The spectral distribution of the emission of e.g. M17, as shown in Fig. 14, extends at least from 200 to 800 nm (which is the sensitivity range of the equipment) with 75 % of the overall measured emission between 500 and 700 nm. Fluorescence spectroscopy revealed that there is a maximum in the excitation spectrum between 400 and 500 nm, probably due to the excitation of singlet carbonyl groups, causing a broad maximum in the emission spectrum around 520 nm, that extends into the red range of the spectrum due to multiple re-absorption and re-emission processes (data not shown).

3. Interaction of bacterial cells with the medium emission

For these and the following experiments, small inocula (0.5-5µl) of aerobically grown, early stationary phase cultures were brought into 15 ml cuvettes with fresh medium immediately before the measurement started. The cultures resumed growth almost immediately with no significant lag period. Cell density was monitored by turbidimetric

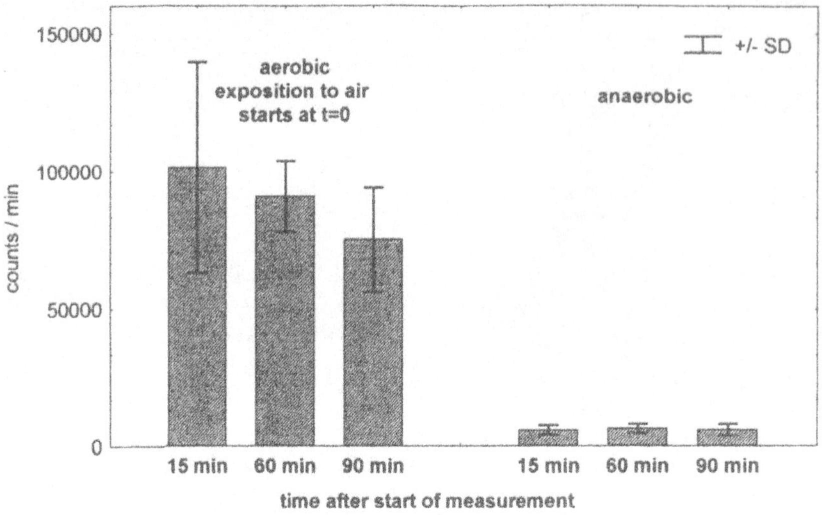

Figure 2. Emission of M17 under aerobic and anaerobic conditions (measured in LSC)

measurements at 605 nm in a parallel sample under same conditions as in the measuring chamber. The turbidity of the culture is given as the optical density (OD).

3.1 MEASUREMENTS ON *E.COLI* B IN DIFFERENT CULTURE MEDIA

Fig. 3 shows the emission of cultures of the gram-negative strain *E.coli* B in M17, HNB and casamino acids medium. It can be observed that at low cell densities the cells do not alter the chemiluminescence of the medium significantly. At a turbidity of about 0.005 ($5 \cdot 10^6$ cells/ml), however, the emission starts to decrease and at OD 0.020 (roughly $2 \cdot 10^7$ cells/ml) it has dropped to a minimal level. This level can be looked upon as almost zero emission, as, nevertheless that the dark count rate of the detectors has been already subtracted, there is still a weak thermic emission from the measuring chamber itself.

In M17, this effect is not as perfect as in HNB, there is still an increased emission at OD 0.020, which is dropping now at a slower rate, possibly due to the increasing turbidity of the sample. In minimal medium, the drop is, due to the low emission of the medium, not as marked. Therefore, to ensure statistical significance, we calculated the mean of the data of three independent measurements. At the crossing-over to the stationary growth phase, there seems to be an increase of emission, which might be luminescence due to the cells or their metabolism itself, as stated previously under other conditions[28].

The decrease of the emission can not be ascribed to only light scattering in the growing culture. This can be readily if one considers, that the intensity of light passing through a scattering medium follows [4]

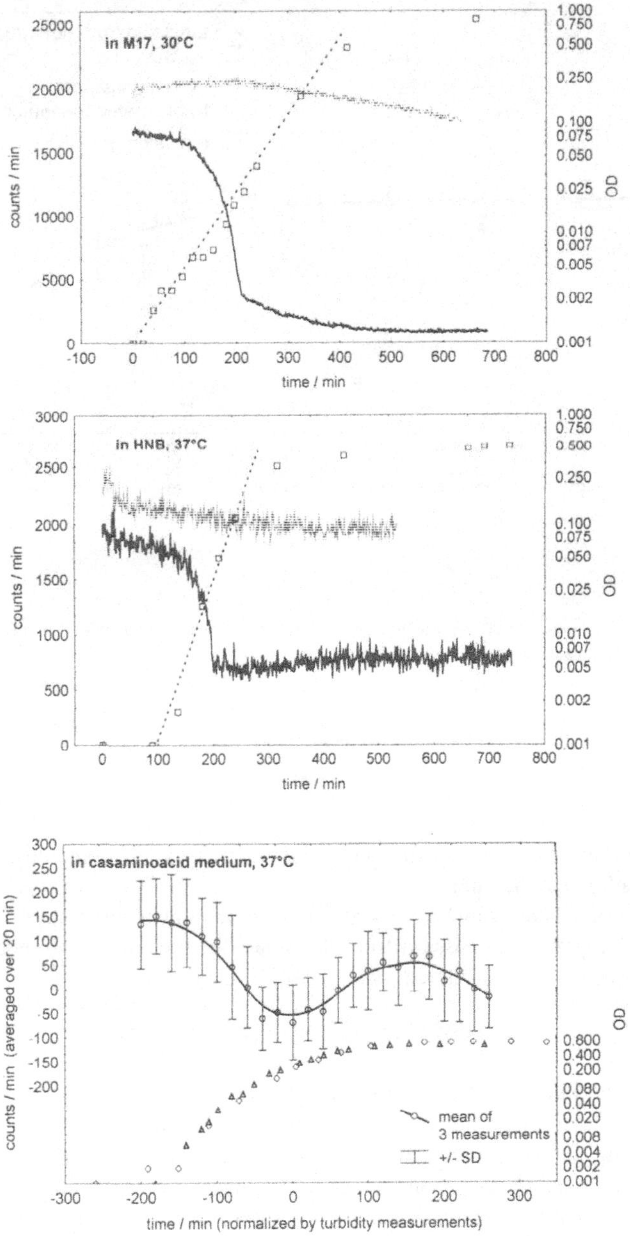

Figure 3. Emission of cultures of E.coli B in different media (solid line) and emission of the sterile culture medium alone (dotted line), the points represent turbidity measurements on a parallel sample (in the lower graph, the emission of the medium of about 650 cpm is for simplicity already subtracted)

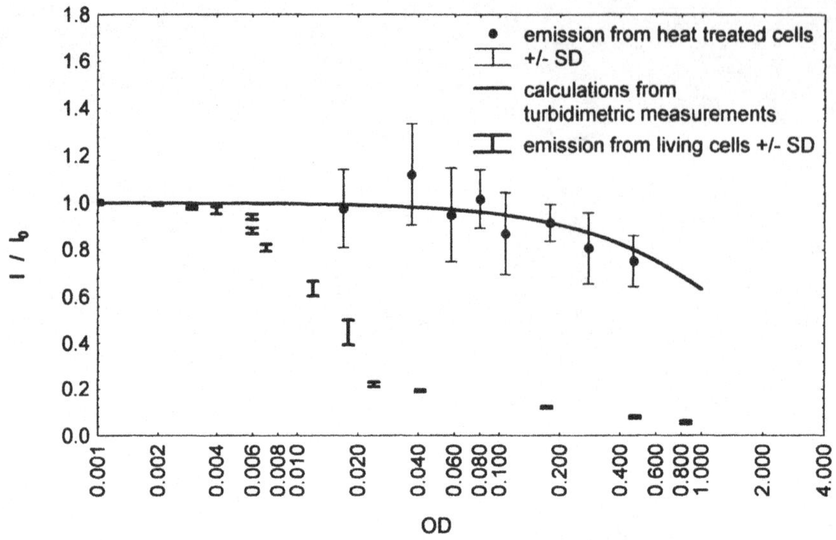

Figure 4. Comparison between the emission of viable and heat treated cells of E.coli B in M17 and calculations as described in the text

$$I = I_0 \exp(-\tau d) \ ,$$

with I_0 as the initial intensity, τ as the optical density of the medium and d as the length of the light path in the medium.

If the medium itself can be considered as a homogeneous source of light, it is necessary to integrate the emission from the area A of the source perpendicular to the direction of propagation considered over the width d of the cuvette:

$$I = \int_0^d dx \rho A \exp(-\tau x) = \frac{\rho A}{\tau}\left(1 - \exp(-\tau d)\right),$$

where ρ is the intensity density of the source. τ is directly accessible by the turbidimetric measurements and ρA can be determined by measuring the intensity before inoculation. Fig. 4 shows a comparison of the measured emission of a growing culture of viable cells in dependence on the optical density, compared to the emission of cultures with cells that were inactivated by heat-treatment (viability less than 0.001%) and the calculations from the model described above. Obviously, the calculations seem to coincide with the control measurements on the heat-treated cells. This is an observation, that is not trivial, as the turbidity measurements are done on a macroscopic sample, while the calculations consider a microscopic scale. The emission of the viable cells, however, differs considerably from the emission of the killed cells and can not be explained by pure light scattering.

3.2 MEASUREMENTS ON OTHER STRAINS

Figure 5 shows results of measurements on other Gram-negative enteric bacteria, as *Serratia marcescens* and *Proteus vulgaris*, and on the unusual Gram-positive coccus *Deinococcus radiodurans*, that is highly resistant to radiation, all growing in HNB. Similar behavior has been observed with the Gram-positive, endospore-forming bacterium *Bacillus subtilis* and the enteric bacterium *E.coli* DK6, as well as with *E.coli* C600 with a plasmid carrying the *lux* gene and *Vibrio fisheri*, both bioluminescent at higher cell densities (all Gram-negative, data not shown),. It is striking that the cross-over to the stage of low emission occurs for the strains described here at low cell densities of in the order of 10^7 cells/ml, corresponding to turbidity values of 0.020 ± 0.010.

In order to simplify the notation in the following, we will use the term *"quenching"* for this active elimination of the light emission by the cells. It should not be confused with the term, as it is used in spectroscopy for the inactivation of electronically excited states.

3.3 HOW CAN THE QUENCHING BE EXPLAINED?

Defining an ability of the cells to cancel the light emission of the medium, one can introduce the amount of quenched light, $\Delta I = I_{medium} - I_{culture}$ and plot it on a logarithmic scale as in Fig. 6. As ΔI forms a straight line in the log-plot, the quenching ability increases exponentially with growth time. The decrease of the emission of the culture can therefore be described by the equation

$$I_{culture} = I_{medium} - c\exp(t/\tau),$$

with a time constant τ that is roughly in the range of the generation time for all strains examined. Thus, the quenching ability is a linear function of the turbidity of the culture and therefore of the cell density or the total cell surface.

There are different mechanisms that may serve as a possible explanation of the light quenching on different levels:

- the requirements for the auto-oxidation of the medium are impaired by the growth of the cells, e.g. by the consumption of the oxygen dissolved in the medium or the consumption of the substrate
- the generation of electronically excited states in the medium, that would lead to chemiluminescence, is impaired, e.g. by the presence of radical scavengers or anti-oxidants produced by the cells
- the cells are interacting with the electronically excited states (quenching in a narrower sense)
- the cells are directly interacting with the electromagnetic light field (e.g. by destructive interference or a kind of active absorption of the photons)

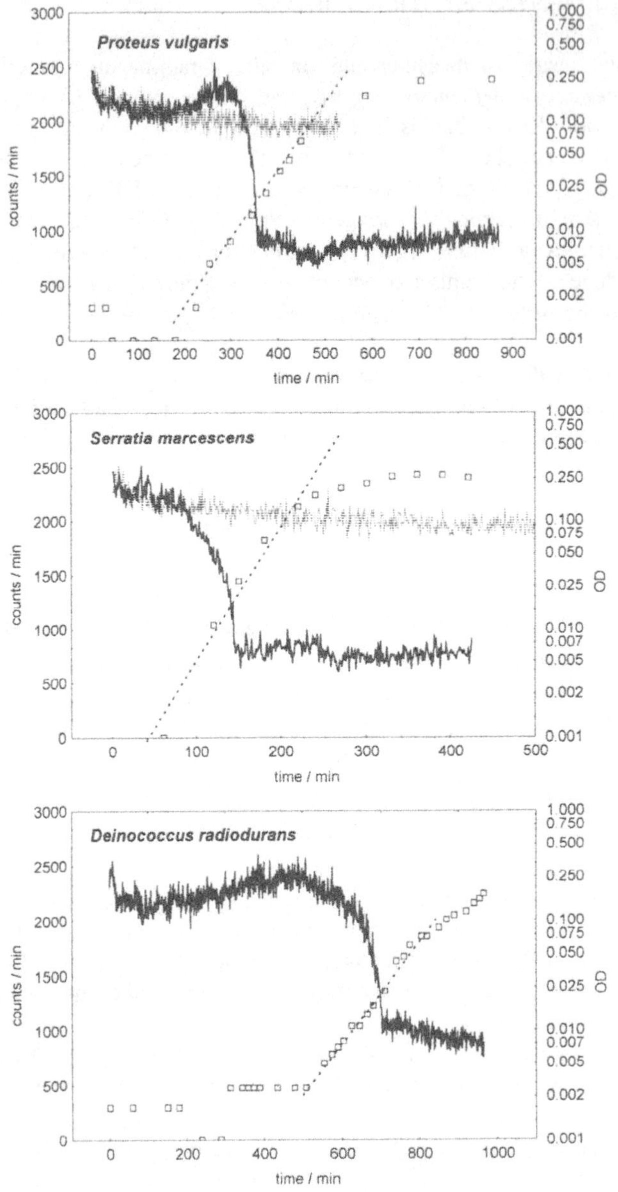

Figure 5. Light emission during growth of cultures of Proteus vulgaris, Serratiamarcescens and Deinococcus radiodurans (solid line) in HNB at 37°C andemission of the sterile culture medium (dotted line), the points represent turbiditymeasurements on a parallel sample

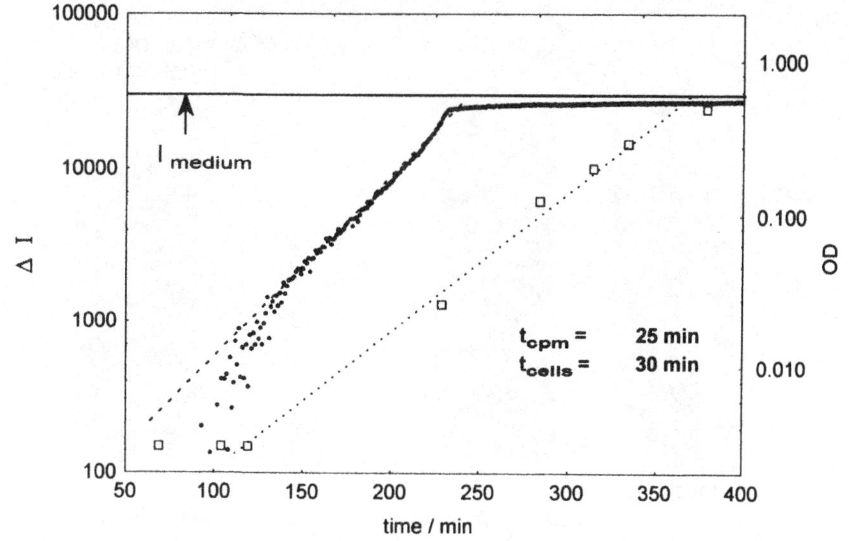

Figure 6. Quenching ability $\Delta I = I_{medium} - I_{culture}$ and turbidity of E.coli B in M17 in a log-plot

3.3.1 O_2-concentration in the medium.

To check, whether the decrease of the light emission is only due to a decrease of the level of oxygen in the medium, we measured the O_2-concentration in the medium during growth of the cells. As expected, the concentration decreases due to consumption of O_2 with an increasing rate during growth and reaches a zero level at OD 0.080 for *E.coli* B and at OD 0.180 for *Enterococcus faecalis* (that lacks cyanide sensitive respiration), both growing in HNB (data not shown).

3.3.2 Removal of the cells.

In another experiment, we let a culture grow to a turbidity of 0.100 and removed the cells from the culture medium by passing it rapidly through a cascade of sterile filters. Measuring the intensity of the culture before inoculation, after inoculation (OD 0.002) and after having grown to a turbidity of 0.100, we observe the expected quenching of the emission (see Fig. 7). The emission drops from 80000 counts / min. (cpm) to 9000 cpm (dark count rate of approximately 5000-7000 cpm not subtracted).

After removal of the cells (which takes only about 1 min.) the intensity increases immediately to almost the level before inoculation. This suggests, that a consumption of certain components of the medium, that might serve as a substrate for auto-oxidation, is not the reason for the decrease of emission. Moreover, it excludes the possibility of a release of active agents by the cells, that might interact with oxidants or light-emitting molecules in the medium, unless the substances secreted would be rather short-lived.

3.3.3 Interaction with external light sources.

In this experiment, we put between a cuvette with sterile medium with a high emission (e.g. M17) and the photomultiplier a second cuvette with a growing culture in either

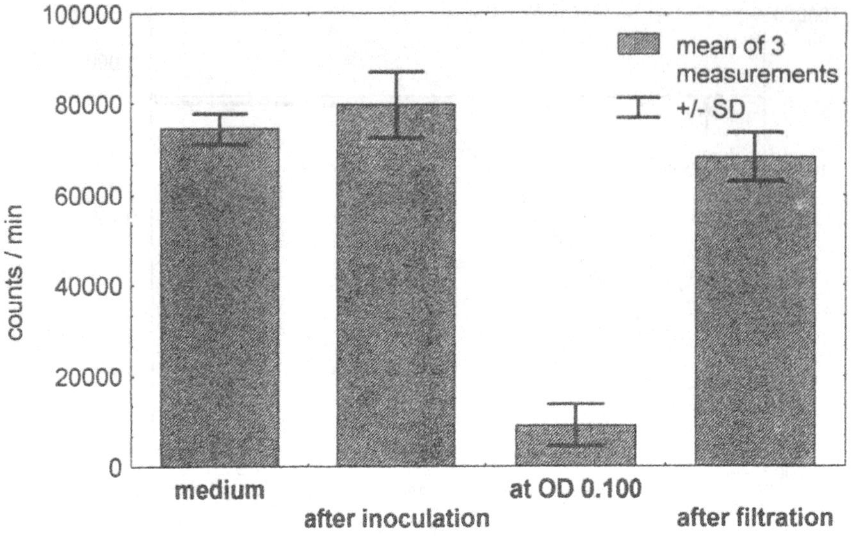

Figure 7. Effect of rapid removal of the cells by sterile filtration of the culture at a turbidity of 0.100 (measured in LSC)

minimal medium with a negligible very low emission (600 cpm) or likewise M17 (25,000 cpm). In this way, the light emission from both cuvettes is measured, but the light from the first cuvette has to pass through the culture in the second cuvette in order to be registered by the detector.

As it can be seen in Fig. 8 (upper), in the first case (culture in minimal medium) the decrease of the intensity does not significantly differ from calculations that take into account only increasing scattering of the light coming from the first cuvette due to increasing turbidity of the culture in the second one.

In the second case (culture growing in M17, Fig. 8, lower), there is a significant deviation from the calculated curve. This might be explained if one considers, that in biological systems photochemical-like processes occur even in the absence of light[8, 9]. Electronically excited species in the medium as triplet carbonyl groups, that are comparably long-lived, may transfer their excitation energy to appropriate acceptors on the cell surface before they fall back to their groundstate under emission of a photon. Thus photochemical reactions might be induced and go along with a quenching of the weak photon emission of the medium.

Thus, the decrease of the emission in the second case, additional to the decrease due to the increasing turbidity of the culture, may be explained if one considers that M17 shows fluorescence. While in the first case (minimal medium) the light may pass almost unhindered through the medium, it is permanently absorbed and re-emitted in the second case (M17) by fluorescent compounds in the medium. In a sterile medium this multiple absorption / re-emission is in dynamic equilibrium and does not change the emission intensity over time. This may be different, however, in the presence of multiplying cells,

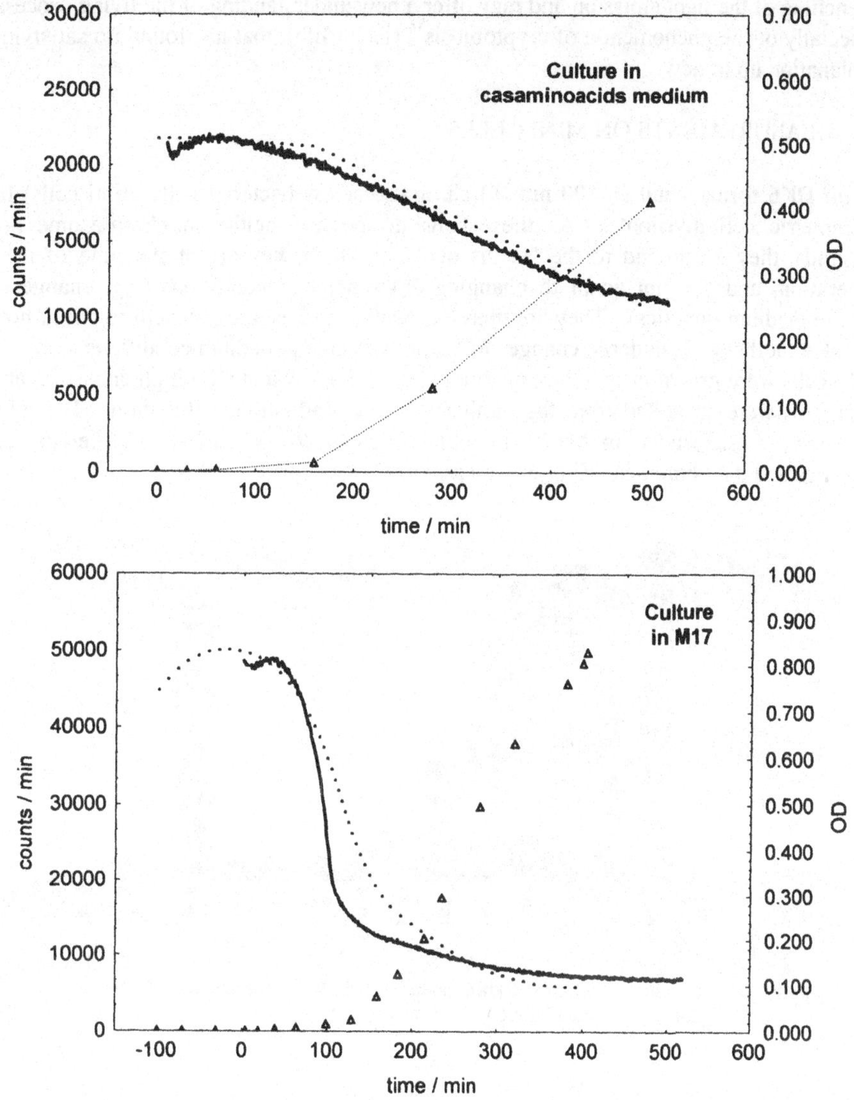

Figure 8. Emission of sterile M17 passing through a growing culture of E.coli B (solid line) in minimal medium (upper) and M17 (lower) and theoretical calculations (dotted line) that consider only scattering due to the turbidity of the growing culture (points)

as they may interact with excited states as described above and therefore lower the weak emission intensity in the course of growth of the culture additionally to the absorption due to its increasing turbidity.

This mechanism is still speculative, but it can be at least partially responsible for the quenching of the light emission and may offer a new understanding of the living process, especially of the phenomenon of cryptobiosis[14] (latent life), that has found no satisfying explanation up to now.

3.4 MEASUREMENTS ON MINI CELLS

E.coli DK6 forms small (≈ 500 nm ∅) chromosomeless bacterial cells (mini cells) by asymmetric cell division[23]. As these cells do possess neither a chromosome nor plasmids, they are bound to the "status quo", in which they are at the time of their generation, and can not adapt to changing environmental conditions (e.g. change of culture medium qualities). They are therefore well suited as a test system to check how far DK6 mother cells undergo changes to adapt to changing medium conditions.
DK6 cells were grown in specified medium (M17 with low and with high emission) and mini cells were separated from the mother cells by gradient centrifugation. The mini cells were re-suspended in M17 with either high or low emission to different cell densities and the photon emission was measured.

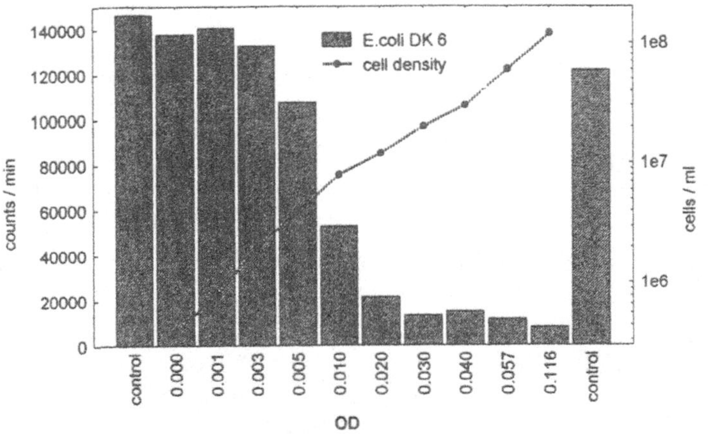

Figure 9. Emission of cultures of DK6 mother cells in M17 as a function of cell density (measured in LSC),

As shown in Fig. 9, DK6 mother cells show the same behavior as the other strains examined, with the cross-over to the stage of low emission at a turbidity of about 0.020. Mini cells of DK6, grown in medium with low emission, parallel this behavior when re-suspended again in medium with low emission. If re-suspended in medium with high emission, the quenching is rather incomplete, even 18 hours after re-suspension, and can at least partially be ascribed to the turbidity of the samples and the decay of the emission of the medium. However, if the mini cells were grown in medium with high emission, the quenching in medium with high emission was well measurable, although it was not as effective as with the mother cells.

These results suggest that the quenching ability is not a static characteristic of the

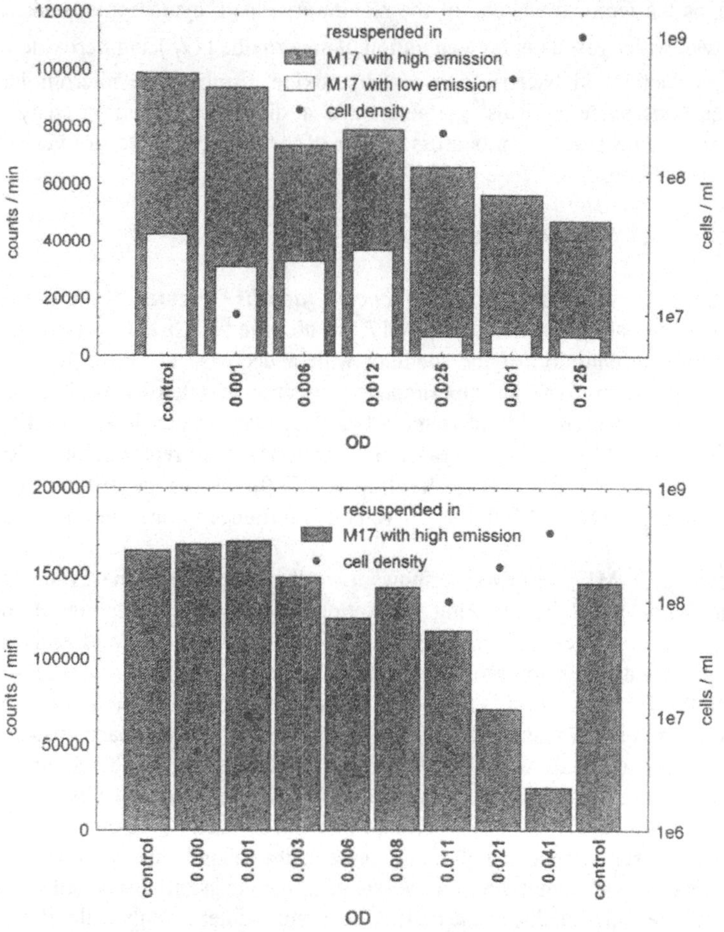

Figure 10. Emission of cultures of DK6 mini cells harvested after growth in medium with low (upper graph) or high (lower graph) emission and re-suspended in medium with low or high emission (measured in LSC)

cells, but that it is in response to the medium conditions and involves some kind of adaptation to it, starting from the genetic level.

4. Measurements on bacteria of the lactic acid group

Bacteria of the lactic acid group (all gram-positive) grow as well under aerobic as under anaerobic conditions. They lack cyanide sensitive respiration and do not use molecular oxygen as the primary electron acceptor. They share several properties as the lack of catalase activity, resulting in a frequent accumulation of H_2O_2, and the presence of flavin-linked oxidases[3]. These flavoenzymes do not transfer 4 electrons to molecular oxygen, as the terminal enzyme of the respiratory chain, cytochrome oxidase, but only one or two, which results in the generation of superoxide (O_2^-) and peroxide ions (O_2^{2-}) [26]. Some lactic acid bacteria possess superoxide dismutase in micromolar levels to scavenge superoxide radicals[15], others have a different defense strategy basing on manganese[2]. Few strains that possess neither of these mechanisms, are very sensitive to oxygen exposition[3].

4.1 *ENTEROCOCCUS FAECALIS*

We will first present results on *Enterococcus faecalis*. Figure 11 shows the emission curves in different culture media. In M17, we observe two distinct periods of emission: Once again the emission of the medium with a decrease at OD 0.010 as described before, now followed by an superimposed increase in intensity leading to a second period of maximum emission at roughly OD 0.250, apparently linked to the growth of the cells, followed by a slow decrease. This pattern is well reproducible. Some details as the little peak at OD 0.050 or the drop at OD 0.800 may be differently marked as medium preparations vary, but they are reliably reproduced if medium from one batch is used.

Growing in M17, *E.faecalis* shows two distinguishable phases of exponential growth, characterized by differing generation times. The beginning of the second, slower phase coincides with the increase of emission of the second emission period, which might indicate some growth inhibition linked to the photon emission. In MRS medium, the second period of emission is more stretched, the maximum emission seems to be lower and is not reached before the culture enters the stationary growth phase. In HNB medium, which shows only a weak own light emission, the behavior is similar as in MRS, but here the emission of the second phase exceeds the emission of the medium. The distinct peak at OD 0.020 can probably be traced back to the quenching at low cell densities, now superimposed with the increase of the second emission phase. Obviously the two mechanisms, underlying on the one hand the emission of the medium and on the other hand the emission due to the cells itself, seem to interact only little, if at all.

In order to examine the involvement of peroxide and superoxide ion in the *E.faecalis* emission, we employed the scavenger enzymes catalase and superoxide dismutase (SOD):

$$2\,H_2O_2 \xrightarrow{\text{catalase}} 2\,H_2O + O_2$$

$$2\,O_2^- + 2\,H^+ \xrightarrow{\text{SOD}} H_2O_2 + O_2$$

Addition of catalase leads to a distinct decrease of the intensity of the second emission period with increasing concentration of the scavenger enzyme, wiping it almost entirely out at a concentration >1000 units/ml (leaving over a weak emission, that seems not to be affected by catalase and that might be responsible for the small emission peak at OD 0.040). This suggests, that the accumulation of H_2O_2 is involved in the generation of light in this phase of growth. Moreover this explains the growth inhibition in the second exponential growth phase described above and is in accordance with previous experiments[13]. The effects of added catalase on the emission of the medium

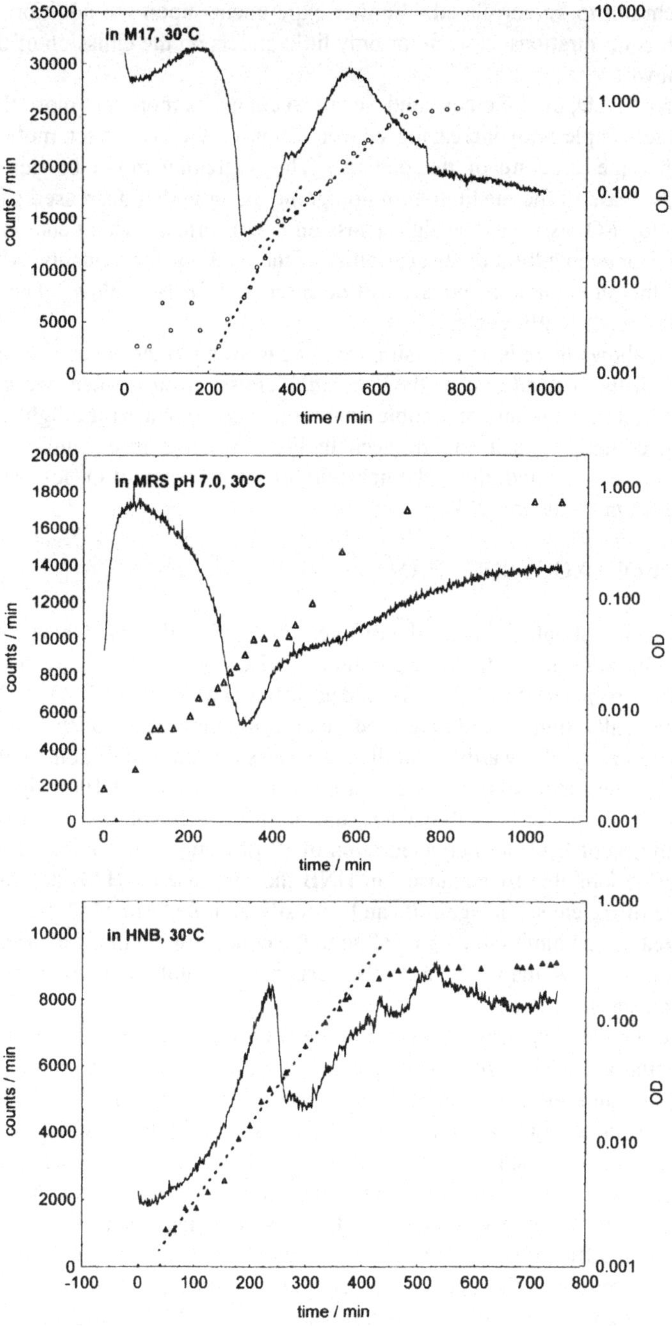

Figure 11. Emission pattern of E.faecalis in different culture media

alone, however, are not clear. Intensity of the medium emission and concentration of catalase seem not to be correlated. Further experiments upon the addition of catalase even in high concentrations show no or only little effects on the emission of the medium (data not shown).

Addition of SOD, on the other hand, seems to compress more and more the emission pattern of the sample with increasing concentration of the scavenger molecule, but it does not alter the structure of the pattern. This flattening might be due to a SOD dependent decrease of the medium emission going along with a decreased sensitivity of the medium to H_2O_2 as far as the light emission is concerned. This behavior, that may constitute a change in a kind of susceptibility of the medium for oxidants, with the light emission as the measurable response, will be discussed further below. The underlying chemistry, however, is still unclear.

As shown above, there is no emission from the medium in the absence of oxygen. In order to check the dependence of the *E.faecalis* emission on oxygen, we cultured the strain under both aerobic and anaerobic conditions and compared the light emission of the growing cultures. As it can be seen in Fig. 13, there is no emission from the anaerobic culture, what underlines the necessity of the presence of O_2 for the cells to be chemiluminescent (compare [28]).

4.2 EFFECT OF EXOGENOUS H_2O_2

In order to investigate, whether the different intensities of the *E.faecalis* emission in different media are due to different amounts of H_2O_2 accumulated, we tested several media for their response on H_2O_2. We added different amounts of H_2O_2 to M17, HNB and a minimal salts solution and measured the chemiluminescence of the sample for ten minutes immediately afterwards. Parallel, we measured the inhibition of the induced chemiluminescence upon addition of catalase at a concentration of 1000 sigma units/ml. As it can be seen in Fig. 14, above left, the minimal salts solution does only poorly respond, addition of H_2O_2 to a concentration of 10 µM only doubles the emission. The increase can be inhibited by catalase. In HNB the response on H_2O_2 addition is a 17-fold increase of the emission, again it can be mostly inhibited. In M17 the response is a 14-fold increase, non inhibitable by catalase in the concentration that was used. At H_2O_2 concentrations of less than 100 nM, the response is inhibitable by catalase at the concentration stated.

Thus the relative responses of M17 and HNB on H_2O_2 are similar, but the absolute responses stand about one order of magnitude apart. Therefore the conclusion can be drawn, that the emission response of different media on H_2O_2 is not only dependent on the concentration of H_2O_2 accumulated. There is an additional susceptibility factor which is specific for each medium and which shows some proportionality to the emission intensity of the medium without stimulation. The reason for this is probably rather complex, it might be connected to differing reducing potentials of the different media, or due to different concentrations of anti-oxidants being present in the medium.

We also determined the spectral distribution of the response of M17 on H_2O_2 stimulation and compared it to the spectral distribution of M17 without stimulation and of *E.faecalis* emission at the maximum. All three spectra, that have been corrected for the detector's sensitivity, show maximal emission in the 500 to 600 nm range and are,

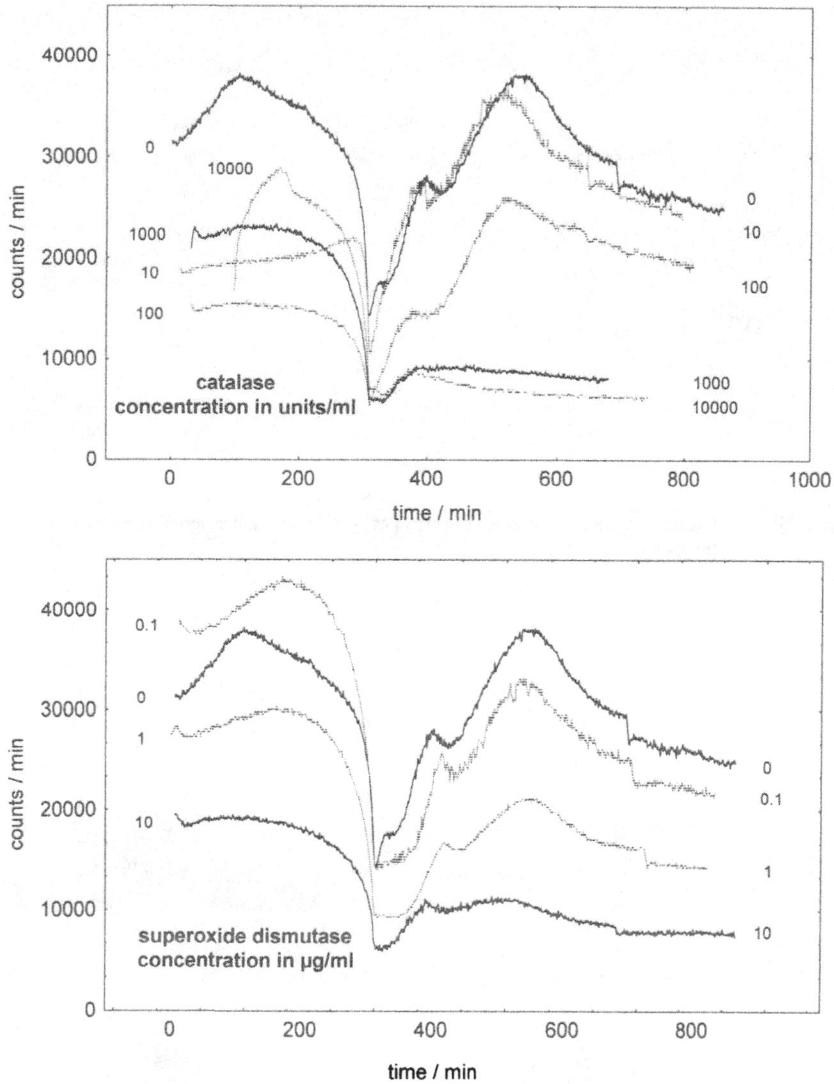

Figure 12. Emission of E.faecalis in M17 after addition of catalase (upper) and superoxide dismutase (lower) in the concentrations stated

within the limits of confidence, not distinguishable. This can be due to being it the same process that leads to the light emission or being it a whole range of different processes, like the generation of a variety of excited carbonyl chromophores as postulated in [1], such that differences are averaged out. Most probably, however, it can be traced back to the presence of fluorescent molecules in the medium that broadens eventually existing narrower emission bands and shifts them to longer wavelengths.

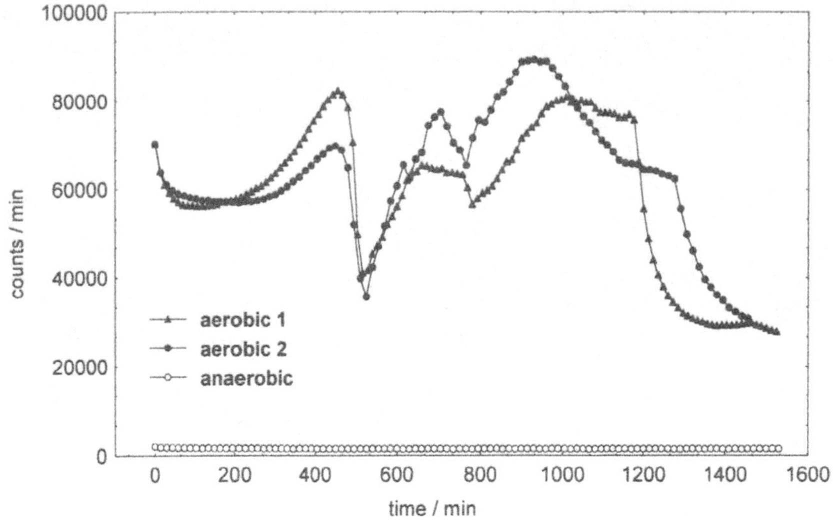

Figure 13. Emission curve of E.faecalis growing in M17 at aerobic and anaerobic conditions (measured in LSC)

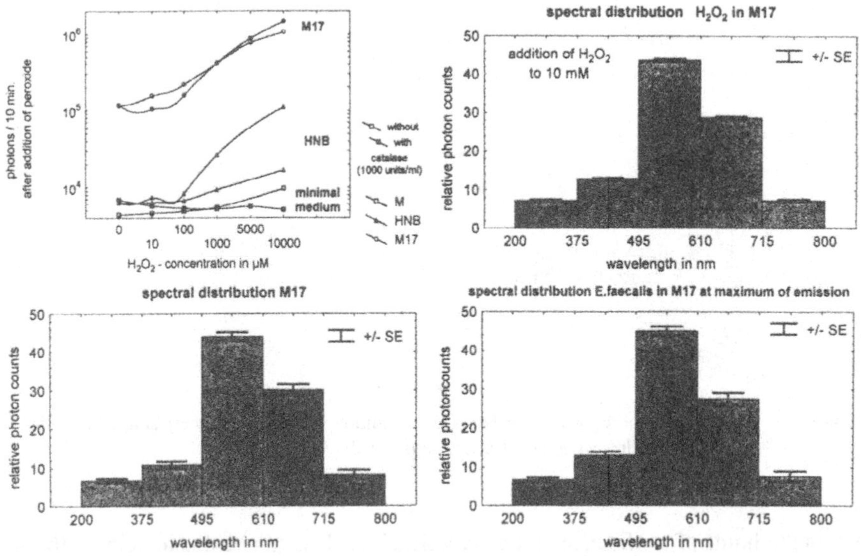

Figure 14. Chemiluminescence response of different media on stimulation with H_2O_2 and comparison of the response spectrum with the spectra of M17 without stimulation and of the E.faecalis emission

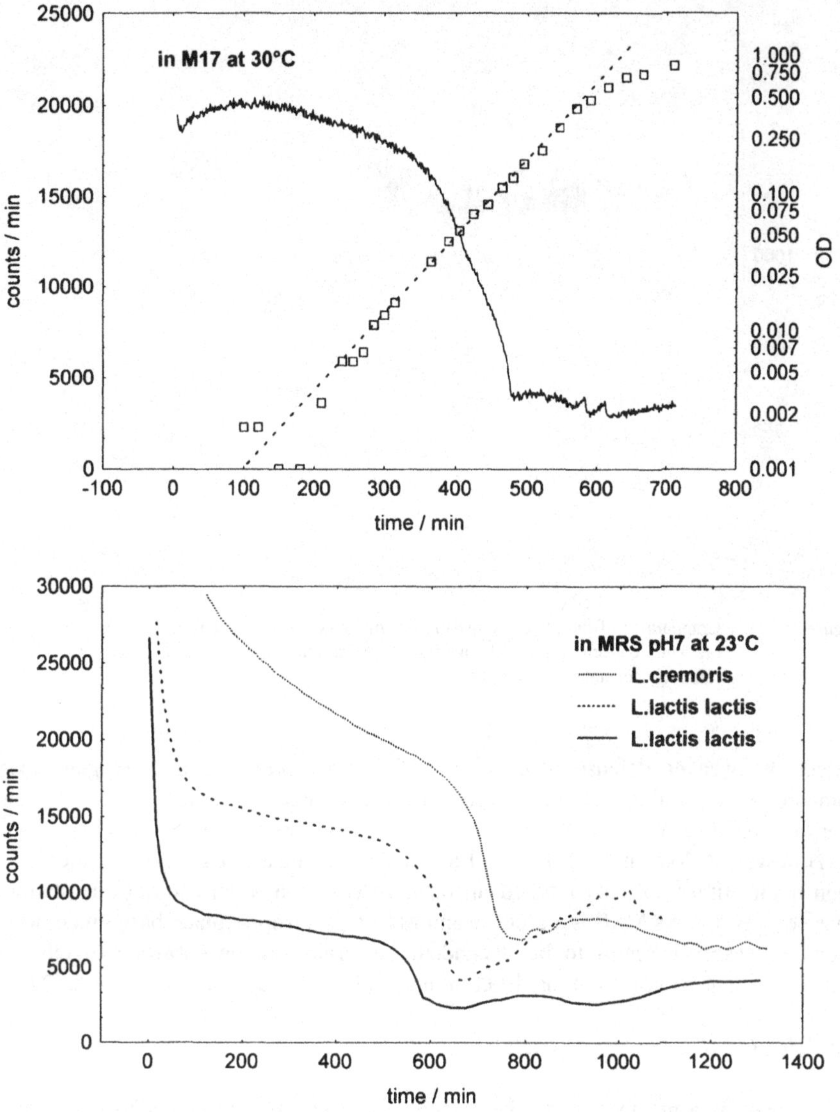

Figure 15. Emission pattern of L.lactis lactis in M17 (upper graph) and MRS (lower graph, here compared with L.cremoris, measured in LSC)

4.3 *LACTOCOCCUS LACTIS*

Fig. 15 shows the emission curve of *L.lactis lactis* in M17 and in MRS. In contrast to the other measurements, the quenching of the medium emission is here not accomplished until an OD of roughly 0.200 is reached (independent of the medium). This might

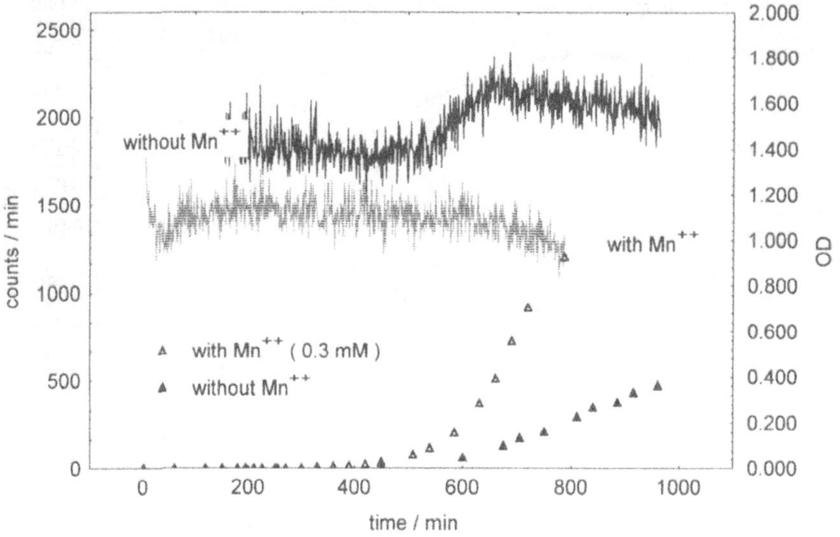

Figure 16. Comparison of the photon emission during growth of L.plantarum depending on Mn^{++} concentration of the medium. The medium emission (not shown) is relatively constant on 1600 cpm.

reflect the weaker defense mechanisms of *L.lactis lactis* against oxygen toxicity compared to the strains examined before. It is remarkable that *L.lactis lactis*, in contrast to *E.faecalis*, does not seem to cause any second light emission period in M17.

However, it does in MRS (as well as *L.cremoris*, data not shown), as there is an, even if sometimes not very marked, increase of emission when entering the stationary phase of growth. As MRS is a glucose and M17 a glycerophosphate based medium, the chemiluminescence seems to be dependent on strain and on substrate together and therefore on particularities of the different metabolic pathways (see also [13] & [24]).

4.4 *LACTOBACILLUS PLANTARUM*

L.plantarum is a member of the lactic acid group and possesses neither catalase nor SOD. While in 1971 it was still claimed that *L.plantarum* does not show any oxygen consumption during aerobic growth[15], it has been shown later that O_2 may be involved in its metabolism[2, 19], resulting in a production of H_2O_2 which is delayed to the start of oxygen exposition for up to several hours. However, this accumulation of H_2O_2 affects the growth of *L.plantarum* only slightly[20]. *L.plantarum* produces a substitute for the lacking enzymatic SOD activity[2], that is basing on free Mn^{++}. Thus, it is possible that H_2O_2 accumulation is the product of the dismutation of superoxide and represents the lesser detriment to the cells compared to the accumulation of superoxide and its other reaction products, like the hydroxyl radical. This hypothesis is underlined by the observation that the growth of *L.plantarum* is only poor in media with low Mn^{++}-

concentrations.

Therefore we examined the photon emission of *L.plantarum* in MRS, which is rich in Mn^{++} (0.3 mM) and in MRS with no Mn^{++} added. As described above, the growth rates in the two different media are differing considerably (Fig. 16). It is striking, that the quenching of the medium emission is not very pronounced here (as well as with *Lactobacillus casei*, data not shown). It can be observed in a medium with high emission, however only at very high cell densities (OD > 0.800, data not shown). This parallels the strain's sensitivity to oxygen toxicity due to its lack of efficient defense mechanisms. In Mn^{++}-rich medium, no significant emission pattern can be observed despite the accumulation of peroxide described above. Growth in Mn^{++}-poor medium, however, leads to an increased photon emission at higher cell densities, obviously due to the increased oxidative stress imposed by the lack of SOD-activity.

In summary, the quenching ability seems to correlate to the oxygen tolerance of the different strains in the lactic acid group (*E.coli* in comparison), which decreases in the order[4] *E.faecalis* (0.010) > *E.coli* (0.020) > *Lactococcus lactis* (0.200) > *Lactobacillus plantarum* (0.800) > *Lactobacillus* casei (no data), in brackets the turbidity values, at which the cross-over to the stage of minimal emission takes place.

4. Effects upon phage infection of bacterial cells

Bacteriophages are small (≈200 nm) viral particles with mostly double stranded DNA, that infect bacterial cells for reproduction. In course of the infection, the cell is being lysed and new phage particles are being released. We wanted to investigate, how phage infection and the following cell lysis affect the emission pattern of the infected culture. In a simplified scheme the infection cycle (*lytic cycle*) is generally as follows[26]:

- phage particle attaches to bacterial cell wall and injects its chromosome
- phage enzymes for transcription and replication of the phage chromosome are formed
- the bacterial chromosome is being cut down by phage enzymes and built in into phage DNA, proteins for new phage particles are being synthesized
- new phage DNA is being packed into protein envelope (maturation) and 15 - 30 min. after infection new phage particles (≈ 20 - 200) are released by cell lysis
- new infection cycle starts until all cells will be lysed

Therefore, the last infection cycle ends with a mass lysis of the bacterial culture. The length of the period between infection of the culture with phage suspension and mass lysis is depending on the replication time of the phage, the cell density of the culture and the quotient of the phage density and the cell density before the first infection cycle. We decided to start each measurement with the same cell density and to vary the number of phages added to the culture.

First, we infected *Lactococcus lactis lactis* 530-12 with its homologue bacteriophage. We started the measurements at an OD of 0.100, at which *L.lactis lactis* just starts quenching the medium emission. Figure 17 shows the results of a control measurement without phage addition and of three measurements with a decreasing number of phages added. The emission pattern seems not to be affected by the infection until mass lysis occurs. When mass lysis occurs , the emission of the culture either stays

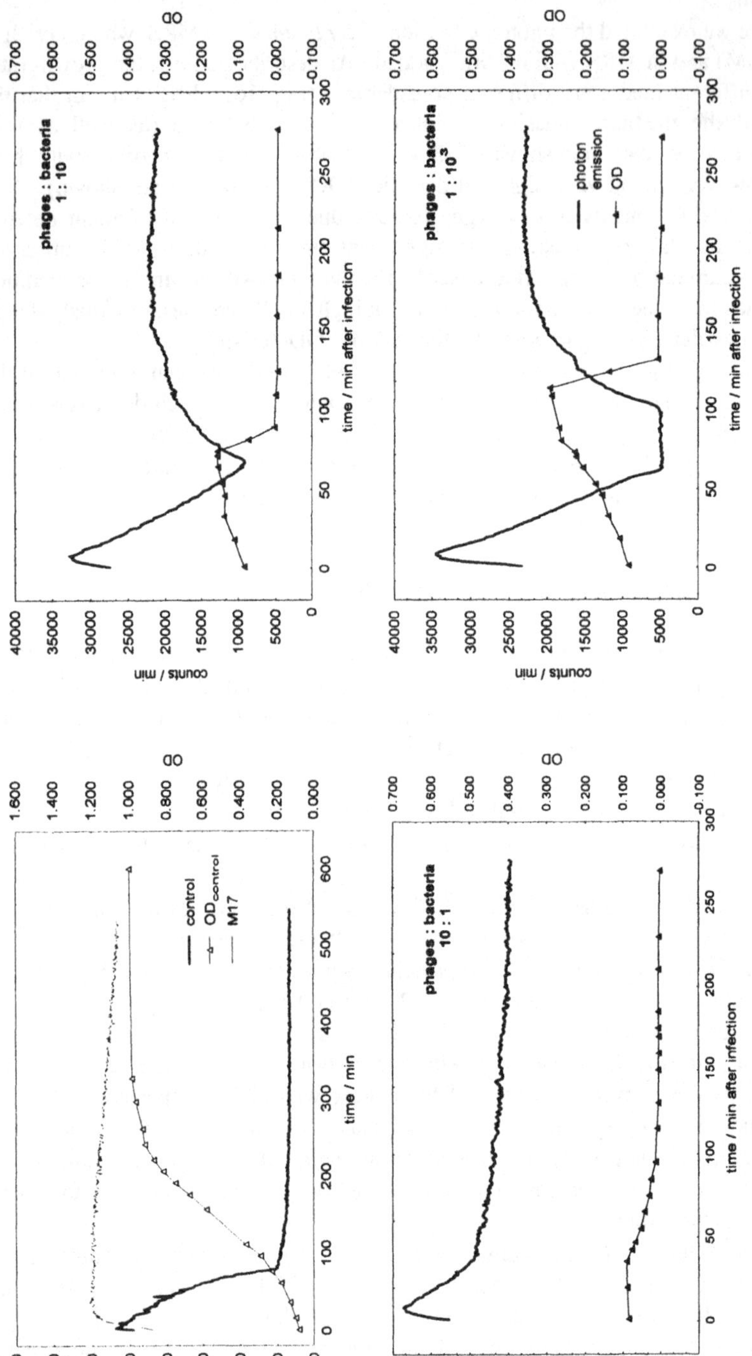

Figure 17. Mass lysis of L.lactis lactis after phage infection with homologue phage in M17 varying the ratio phage density : cell density at t=0, the control measurement is without phage infection

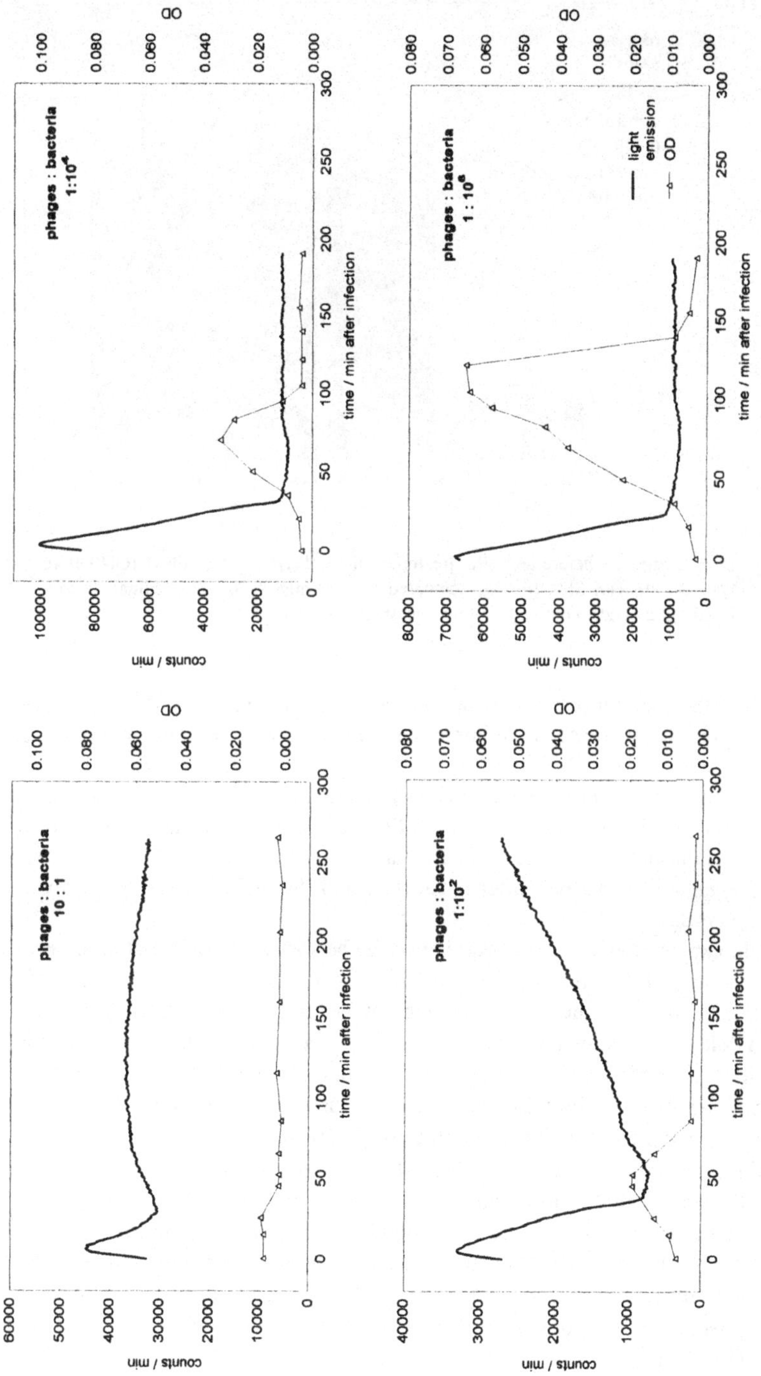

Figure 18. Mass lysis of E.coli B after phage infection with T7 in M17 in dependence of the quotient phages/cells at t=0

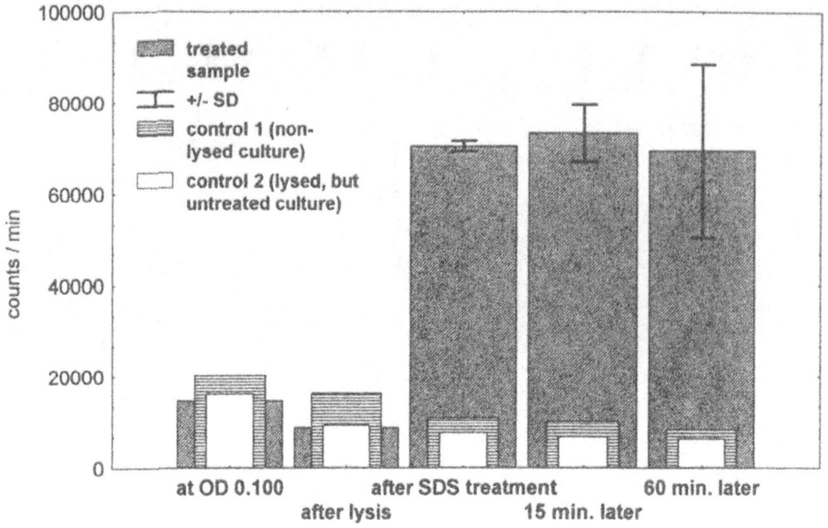

Figure 19. Photon emission before and after treatment of a cell lysate of E.coli B (OD before lysis 0.100) with SDS (0.1%), compared to the emission of the medium, a non-lysed culture and an untreated lysate (measured in LSC)

on the level of the medium emission, or, if quenching already had started, increases again to this level. This increase is not immediate, but a slow process and may take some time.

Repeating this experiment with *E.coli* B and phage T7 yields the same results. But, as it can be seen in Fig. 18, the increase of light emission is not as marked as with *L.lactis lactis*. If the culture reaches turbidity values of more than 0.030 before lysis, the quenching is even still continuing after mass lysis and the emission is rising only very slightly during lysis.

A possible explanation for this phenomenon is the formation of membrane vesicles after lysis, that may still have some activity and can continue quenching the medium emission. In the course of time, these may decompose or loose their activity, leading to the observed delayed and slow increase. The differing behavior of the two strains might just reflect the different abilities of the strains to interact with the medium emission. This is already expressed in the different cross-over turbidity values at which quenching occurs during normal growth (0.020 and 0.200, respectively).

In addition, we examined the response of the photon emission of the lysate on treatment with sodium dodecyle sulfate (SDS). SDS acts as a detergent and is the common agent to dissolve sphaeroplasts or membrane vesicles. A culture of *E.coli* B lysed at an OD of 0.100 by phage infection shows no increase in photon emission for at least two hours after lysis occurred. Addition of SDS to the lysate, however, leads to a rapid decomposition of membrane vesicles and increases the emission of the lysate immediately (Fig. 19).

These experiments point to the cell membrane as the location where the quenching occurs. A participation of the cytoplasm seems to be improbable as the quenching is to effective to involve a crossing of oxidants or excited molecules through the cell membrane(s). Moreover, the increase of the emission to the initial level could hardly be explained.

5. Acknowledgments

We are grateful to *Studienstiftung des deutschen Volkes* for the generous support of this work. Moreover, we wish to thank Priv.-Doz. Dr. Greiner, University Hohenheim, for his help with the fluorescence spectroscopy and to Prof. Bakker, University Osnabrück for advice and supply of strains. Above all, we are greatly indebted to Prof. Popp, IIB Neuss, for his support of this work.

6. References

1. Allen, R.C., (1979), Chemiluminescence from eukaryotic and prokaryotic cells: Reducing potential and oxygen requirements, *Photochem.Photobiol.* **30**, 157-163
2. Archibald, F.S., & I. Fridovich (1981), Manganese and defenses against oxygen toxicity in *Lactobacillus plantarum*, *J.Bact.* **145**, 442-451
3. Archibald, F.S., & I. Fridovich (1981a), Manganese, superoxide dismutase and oxygen tolerance in some lactic acjd bacteria, *J.Bact.* **146**, 928-936
4. Atkins, P.W., 1994, *Physical chemistry*, Oxford University Press
5. Cadenas, E., (1984), Yearly review: Biological Chemiluminescence, *Photochem. Photobiol.* **40**, 823-830
6. Carlsson ,J., G. Nyberg & J. Wrethen (1978), Hydrogen peroxide and superoxide formation in anaerobic broth media exposed to atmospheric oxygen, *Appl.Env.Microbio.* **36**, 223-229
7. Chwirot, B.W., (1992), Ultraweak luminescence studies of microsporogenesis in larch, in *Recent advances in biophoton research and its applications*, eds. Popp, F.A., Li, K.H., & Gu, Q., World Scientific, Singapore, 259-286
8. Cilento, G., & W. Adam, 1988, Photochemistry and photobiology without light, *Photochem.Photobiol.* **48**, 361-368
9. Cilento, G., 1984, Generation of electronically excited triplet species in biochemical systems, *PureAppl.Chem.* **56**, 1179-1190
10. Difco GmbH (1997), Augsburg, Germany, personal communication
11. Friedrich, W., 1987, *Handbuch der Vitamine*, Urban & Schwarzenberg, München
12. Günther, K., (1990), Biochemie der Zellstrahlungsreaktionen: Ein Anzeichen für ablaufende Schutzmechanismen gegen oxidative Zellschädigungen, *Naturwissenschaften* 77, 412-420
13. Hunter, D.J., & R.C. Allen (1981), Oxygen-dependent *Streptococcus faecalis* chemiluminescence: The importance of metabolism and medium composition, in *Bioluminescence and Chemiluminescence*, eds. DeLuca, M.A. & McElroy, W.D., Academic Press, New York
14. Kennedy, M.J., S.L. Reader & L.M. Swierczynski, 1994, Preservative records of microorganisms: Evidence of the tenacity of life, *Microbiology* **140** 2513-2529
15. McCord, J.M., B.B. Keele & I. Fridovich (1971), An enzyme based theory of obligate annaerobiosis: The physiological function of superoxide dismutase, *Proc.Nat.Acad.Sci.USA* **68**, 1024-1027
16. Mei, W.P., (1992), Ultraweak photon emission from synchronized yeast as a function of the cell division cycle, in *Recent advances in biophoton research and its applications*, eds. Popp, F.A., Li, K.H., & Gu, Q., World Scientific, Singapore, 243-258
17. Mendenhall, G.D., (1977), Analytische Anwendungen der Chemilumineszenz, *Angew.Chem.* **89**, 220-

228

18. Mieg, C., W.P. Mei & F.A. Popp, 1992, Technical notes to biophoton emission, in *Recent advances in biophoton research and its applications*, eds. F.A. Popp, K.H. Li & Q. Gu, World Scientific, Singapore, 197-206
19. Murphy, M.G., & S. Condon (1984), Correlation of oxygen utilization and hydrogen peroxide accumulation with oxygen induced enzymes in *Lactobacillus plantarum* cultures, *Arch.Microbiol.* **138**, 44-48
20. Murphy, M.G., & S. Condon (1984a), Comparison of aerobic and anaerobic growth of *Lactobacillus plantarum* in a glucose medium, *Arch.Microbiol.* **138**, 49-53
21. Nagl, W., & F.A. Popp (1983), A physical (electromagnetic) model of differentiation, *Cytobios* **37**, 45-62, 71-83
22. Popp, F.A., Q. Gu & K.H. Li (1994), Biophoton emission: Experimental background and theoretical approaches, *Mod.Phys.Let.B* **8**, 1269-1296
23. Reeve, J.N., (1984), Synthesis of bacteriophage and plasmid encoded polypeptides in minicells, in *Advanced molecular genetics*, eds. Pühler, A., & Timmis, K.N., Springer, New York, 213-223
24. Roth, J.A., & M.L. Kaeberle (1980), Chemiluminescence by *Listeria monocytogenes*, *J.Bact.* **144**, 752-757
25. Ruth, B., 1979, Experimental investigation on ultraweak photon emission, in *Electromagnetic bio-information*, eds. F.A. Popp, G. Becker, H.L. König & W. Peschka, Urban & Schwarzenberg, München, 107-122
26. Schlegel, H.G., (1992), *Allgemeine Mikrobiologie*, Thieme, Stuttgart
27. Slawinski, J., (1988), Luminescence research and its relation to ultraweak cell radiation, *Experientia* **44**, 559-571
28. Tilbury, R.N., & T.I. Quickenden (1988), Spectral and time dependence studies of the ultraweak bioluminescence emitted by the bacterium *Escherichia coli*, *Photochem.Photobiol.* **47**,145-150
29. Tilbury, R.N., (1992), The effect of stress factors on the spontaneous photon emission from microorganisms, *Experientia* **48**, 1030-1040
30. van Wijk, R., & D.H.J. Schamhart (1988), Regulatory aspects of low intensity photon emission, *Experientia* **44**, 586-593

Note:

Some of the figures in this paper have been reproduced from

Journal of Bioelectrochemistry and Bioenergetics, (in press), R. Vogel and R. Süssmuth, Interaction of Bacterial Cells with Weak Light Emission from Culture Media, copyright 1998, with permission from Elsevier Science.
Journal of Bioluminescence and Chemiluminescence, (in press), R. Vogel and R. Süssmuth, Weak Light Emission Patterns from Lactic Acid Bacteria.

BIOPHOTONS AND DEFENSE RESPONSE IN PLANTS

M. HIRAMATSU

Central Research Laboratory, Hamamatsu Photonics K. K.
(Present address: Laboratory of Molecular Biophotonics)
Hirakuchi, Hamakita, 434, Japan

Abstract

Two-dimensional imaging of ultraweak emissions from intact bean sprouts and soybean roots was achieved by a photon counting imager with ultrahigh sensitivity. As an new application of a use of ultraweak luminescence, we have studied plant-fungus interactions to reveal the nature of the defense responses. Ultraweak luminescence generated by sweet potato and nonpathogenic *Fusarium oxysporum* interactions associated with a defense response was detected by a photon counting method using ultrahigh-sensitive photodetectors. The time-dependent intensity variation, the spectrum and the two-dimensional imaging of the ultraweak luminescence are indicative of the defense response of the sweet potato to *F. oxysporum*. The production of ipomeamarone as a phytoalexin means that *F. oxysporum* induced the defense response in the sweet potato.

1. Introduction

Many living organisms are known to emit ultraweak luminescence, referred to as "biophotons"[1]. This is different from the bioluminescence that arises from reactions such as those occurring with luciferin-luciferase systems, which proceed with high quantum efficiencies. The existence of ultraweak luminescence was not known before the 1950s because of its extremely weak intensity (i.e., 10-10^4 photons s^{-1} cm^{-2}), but it has been demonstrated by means of photon counting techniques, first observed from plants by Colli et al.[2] in 1955. The ultraweak luminescence is believed to be strongly related to mechanisms of life phenomena themselves, but at the present time almost all of the phenomena associated with ultraweak luminescence are poorly understood, including their mechanisms.

Ultraweak luminescence is known to be associated with biological defense responses. For example, polymorphonuclear leukocytes emit photons during the process of phagocytosis[3], but to our knowledge there are no reports about plant-fungus systems, such as the one reported here. Plants have various kinds of biological defense strategies in response to infection by microorganisms. One defense mechanism is the synthesis of organic molecules called phytoalexins. The

J. J. Chang et al. (eds.), Biophotons, 45-55.

enzymes that synthesize phytoalexins are induced only after infection or by other stresses. Phytoalexins play a partial role in induced resistance in plants[4]. Though some pathogenic fungi *Fusarium oxysporum* strains cause considerable damage to various agricultural plants, some *F. oxysporum* strains are non-pathogenic and are useful for biological control in place of the conventional agricultural chemicals. The nonpathogenic *Fusarium oxysporum*, which provides protection by pre-inoculation against *Fusarium* wilt of the sweet potato, has been used to control the disease[5].

As an application of the use of ultraweak luminescence, we have studied plant-fungus interactions to reveal the nature of the defense responses. We did not use a pathogenic, but nonpathogenic *Fusarium oxysporum*, which infection result in responses being met by defenses, in order to elucidate the relationship between ultraweak luminescence and the defense response. We were able to obtain information about the defense response of the sweet potato to infection by the nonpathogenic fungus *Fusarium oxysporum* by measuring the ultraweak luminescence that results.

Here we demonstrate two-dimensional imaging of photons emitted from bean sprouts and soybean roots, revealing the evidence that an intact root itself emits very weak light. We report the time-dependent intensity variation, the spectrum and the two-dimensional imaging of the ultraweak luminescence generated by the plant-fungus interactions.

2. Materials and Methods

Bean sprouts and soybean seeds obtained from Sakata Seed Co. were incubated on a wet filter in a Petri dish at 24. The root grew to be about 20 mm in a few days. The root was used to detect photons emitted by intact tissues. The sample of the root in a Petri dish with water was set in a completely dark box and the detector was put above the sample.

F. oxysporum SK102 was isolated from lettuce (*Lactuca sativa* L. var. *capitata* L.), confirmed that it was nonpathogenic to the sweet potato by inoculation test, and cultured in potato dextrose broth by shaking at 28 for 4 days. Conidia were collected by filtration through gauze, washed by centrifugation and resuspension in 1 mL of sterilized distilled water five times, and adjusted to appropriate concentrations for the experiments (see below). Conidia were inactivated by boiling a portion of the suspension for 5 min, and then washed as with living conidia.

Samples of storage roots of sweet potato were obtained by cutting cylindrical sections (40-50 mm in diameter and 8 mm in thickness) around the roots. In order to suppress photon emission reported to occur from injured tissue by Colli et al.[2], the sweet potato sections were placed in plastic Petri dishes (60 mm in diameter) and incubated at 20 for 12 h. Inactivated sweet potato was prepared by boiling sections of sweet potato for 10 min.

It has been reported that the sweet potato responds to infection with black rot fungus, *Ceratocystis fimbriata*, by synthesizing ipomeamarone, a phytoalexin of the sweet potato[6]. Lipid fraction including ipomeamarone was extracted as described by Oguni and Uritani[6-8]. Sweet potato roots (variety Benikomachi) were cut into about 5 mm thick and 70 g weighed sections. The causal fungus of sweet potato rot disease, *Ceratocystis fimbriata* (IFO 30501), was purchased from Institute for Fermentation, Osaka (IFO). The *Ceratocystis fimbriata* was cultured in potato-dextrose broth by shaking at 28C for 4 days. The upper surface of the sections was inoculated with conidial suspensions of non-pathogenic *F. oxysporum*, *Ceratocystis fimbriata* or sterile water. The sections were incubated at 20 under humid and dark conditions for 48 h. After incubation, each group of the sections was homogenized with 100 mL of chloroform-methanol (1:1 vol/vol). The homogenate was filtrated through a glass fiber filter (Advantec GA200) and the tissue residue was washed with 50 mL of extracting solvent. The filtrate was shaken with 50 mL of deionized water, and the suspension was centrifuged to separate it into two layers. The lower chloroform layer obtained was evaporated. The concentrate was dissolved in 2 mL of extraction solvent. Two hundred microliter portions of each extract were applied onto a silicagel plate (Whatman PK5) and developed with n-hexane-ethyl acetate (4:1 vol/vol) for 1 h. After developing, the silicagel plate was sprayed with Ehrlich's reagent (10 % *p*-dimethylaminobenzaldehyde in 95 % ethanol-concentrated HCl, 1:1 vol/vol). The presence of ipomeamarone was identified by the method previously described[9].

A 0.5 mL portion of the conidial suspension was uniformly inoculated on the surface of sweet potato sections in the states described in Table 1. To observe the time-dependent intensity variation in the photon emissions, a C1230 photon counter (Hamamatsu Photonics K.K.) with a built-in, high-voltage, stabilized, direct-current power source was used. The device counts the number of photons detected by an R208 photomultiplier tube (Hamamatsu Photonics K.K.) with bialkali photocathode, providing a spectral response from 185 to 650 nm. A special dark-box system with a rotating disk for sixteen samples was used to measure multiple samples under the same conditions. Each sample was measured every 42.4 s at 20 for 24 h. The time interval used for the photon counting was 1 s (gate time).

An ARGUS-50/VIM photon counting camera (Hamamatsu Photonics K.K.) was used to detect individual photons[10]. Two-dimensional imaging of the ultraweak luminescence was done by accumulating the photons over a given period of time. A conidia suspension with a concentration of 10^7 conidia/ml was inoculated in a pattern on the cut sweet potato by drawing the Japanese character "イ"(i) on the surface. Six hours after the inoculation, a two-dimensional image was accumulated over a 5 min period using the photon counting camera.

The ultraweak luminescence spectrum was measured by using a method with 20-40 nm wavelength resolution developed by Inaba[11]. Soybean seeds (Sakata Seed Co.) were incubated on a wet filter in a Petri dish at 24C for a period of 5 days. The ultraweak luminescence from the intact soybean roots that there is no infection was also measured by the method. Twenty-eight panes of cut-off color filter glass (5050 mm, Toshiba Glass Co. Ltd.) were used to obtain photon count measurements at different wavelengths (280-640 nm). The luminescence intensity at the various wavelengths was corrected because each depends on transmittance of the respective cut-off filters and the quantum efficiency of the photomultiplier tube.

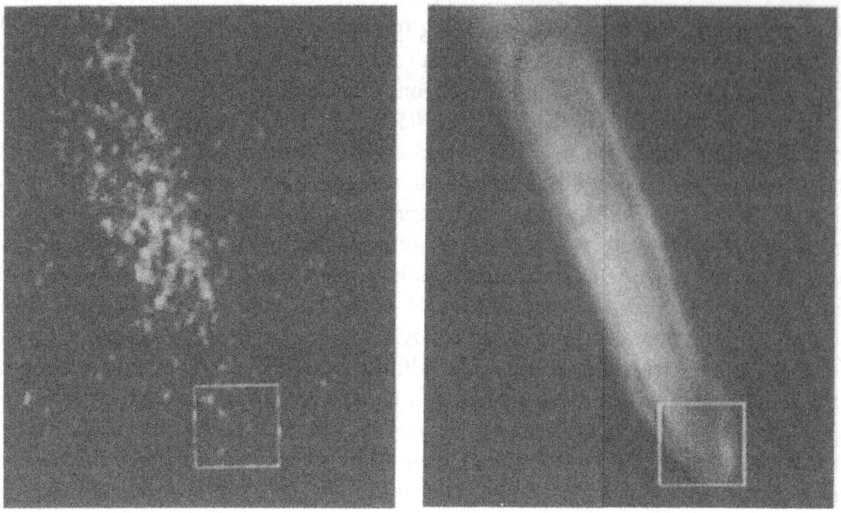

Figure 1. The photon-counting image of the ultraweak luminescence emitted from an intact bean sprout root (left) and the common image of the bean sprout root (right).

The spectrum was obtained by computing the count rate for the individual spectral window defined by the subtraction of two transmission curves with different sharp short-wavelength cutoffs corresponding to the successive color filter glass, followed by the calibration of the quantum efficiency of the photocathode.

3. Results and Discussion

3.1. PHOTONS EMITTED FROM BEAN SPROUTS AND SOYBEAN ROOTS

It was confirmed that seedlings spontaneously emitted photons. The intensity depended on the kind of seedlings. The intensity of the ultraweak luminescence was approximately 210^3 photons/sec/cm^2 for bean sprouts(*Phaseolus mungo*) and 310^2 for both *Vicia fava* and *Raphanus sativus*. It was found that the intensity of the ultraweak luminescence depended on the concentration of oxygen dissolved in the water.

In order to obtain direct evidence of spontaneous ultraweak luminescence, two-dimensional images of photons emitted from bean sprout roots were taken with an ultrahigh sensitivity camera(Hamamatsu Photonics K.K. C1966-20). Fig. 1 (left) shows the first photon counting image of the ultraweak luminescence emitted from the bean sprout root. This figure and the ultrahigh sensitivity camera used were first presented at the Third International Congress on Cell Biology in 1984[12].

With the development of the ultrahigh sensitivity camera, the quality of the photon counting image has been improved. We reported two-dimensional imaging of ultraweak emission from intact soybean roots by means of the Photon Counting

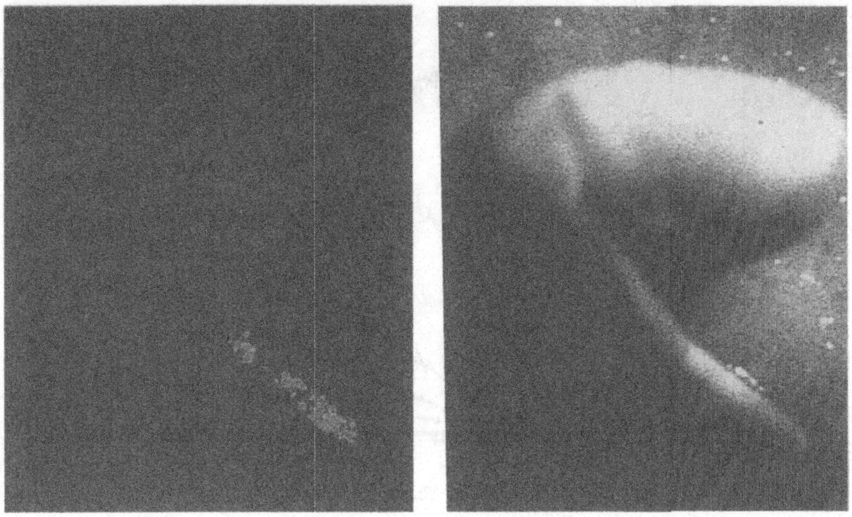

Figure 2. The photon-counting image of the ultraweak luminescence emitted from an intact soybean root (left) and the common image of the soybean and the root (right).

Image Acquisition System(Hamamatsu Photonics K.K. PIAS) together with Dr. Teijiro Ichimura in 1989[13]. Now we can obtain the photon counting image of ultraweak luminescence emitted from the soybean root as shown in Fig.2 (left). The figure indicates that photon flux from the root is more intense than that from the seed itself. From the finding we may conclude that the observable photon emission is related to the enhanced metabolic activity occurring in these regions.

3.2. PHOTONS GENERATED BY THE PLANT-FUNGUS INTERACTIONS

Through collaboration with Dr. Takahiro Makino, Dr. Kimihiko Kato and Mr. Hiroyuki Iyozumi, we found a new application for the use of ultraweak luminescence[14].

3.2.1. *Time-dependent Intensity Variation of The Ultraweak Luminescence*

It was found that the sweet potato inoculated with *F. oxysporum* emitted photons as a result. As shown in Fig. 3 and Table 1, the intensity of the ultra-weak luminescence was high only for the living sweet potato inoculated with living *F. oxysporum*, whereas ultraweak luminescence was scarcely detected from the dead sweet potato inoculated with either dead or living *F. oxysporum*. Furthermore, the integrated intensity of the ultraweak luminescence, (i.e., the total number of photons emitted in a 24 h period) increased as the conidia inoculation concentration was increased. There was a log/linear relationship between the concentration of the conidia and the integrated photon counts(Fig.4). The time when the maximum

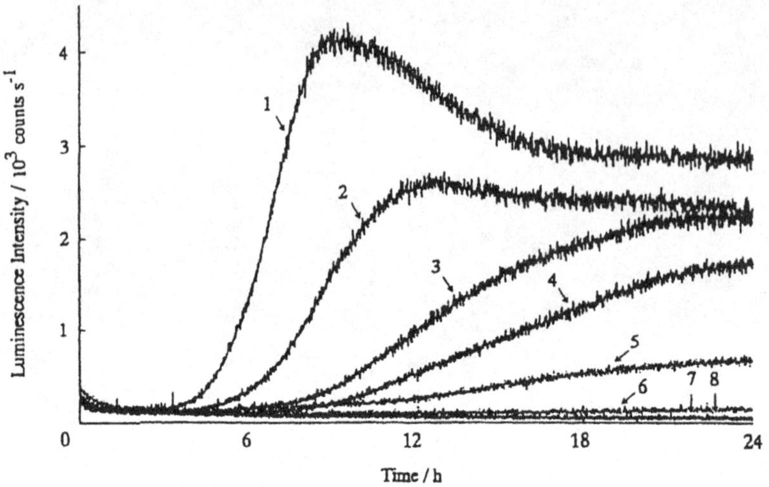

Figure 3. Ultraweak luminescence profile generated by sweet potato and F.oxysporum. The traces 1-8 correspond to the experiment numbers 1-8 in TABLE 1. The measurement temperature was 20°C.

TABLE 1. Effect of conidia concentration (0.5 mL added) and conditions of organisms on the intensity of the ultraweak luminescence generated from the sweet potato by *Fusarium oxysporum.*

		F. oxysporum (SK102)		Ultraweak luminescence		
Exp. no.	Sweet potato condition	Condition	Concentration (conidia $mL^{-1} \times 10^{-5}$)	Maximum (counts $s^{-1} \times 10^{-3}$)	Integrated (counts $\times 10^{-5}$ for 24 h)	Time (h at peak counts)
1	Living	Living	100	4.3	49	10
2	Living	Living	20	2.7	33	13
3	Living	Living	5	2.4	21	22
4	Living	Living	1	1.8	14	≥24
5	Living	Dead	100	0.7	6.6	≧24
6	Living	Distd. H_2O	0	0.3	2.3	0*
7	Dead	Dead	100	0.4	1.7	0*
8	Dead	Living	100	0.3	1.5	0*

*Experiment nos. 6–8 had no peaks in the ultraweak luminescence.

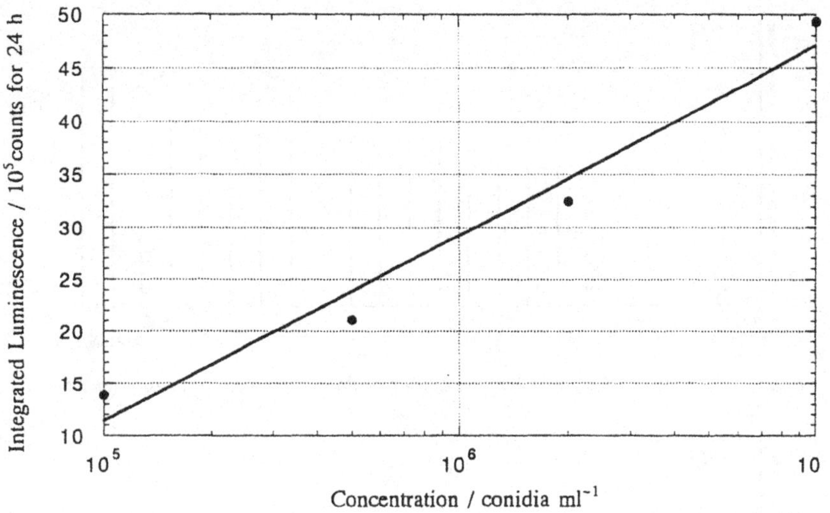

Figure 4. Relationship between the conidia concentration of *F. oxysporum* and the integrated ultra-weak luminescence intensity.

intensity of the ultraweak luminescence was reached also occurred earlier with higher conidia inoculation concentrations. These results indicate that the ultraweak luminescence was generated as a result of the interaction between the sweet potato and *F. oxysporum*.

To obtain spectral information, we measured the emission using a combination of filters. The wavelength range of the ultraweak luminescence from the sweet potato-*F. oxysporum* system was 450-630 nm (Fig.5), whereas that of the ultraweak luminescence from intact soybean roots that there is no infection was 400-600 nm (Fig.6). Both spectra are similar to the emission spectrum (broken lines in Fig.5 and Fig.6)[15] that arises from the process of phagocytosis. These three spectra show that a common species may exist in these systems. The fine structure of these three ultraweak luminescence spectra has not been clarified yet and more detailed work is necessary to elucidate the emitting species[16].

3.2.2. *Detection of Defense Response in Sweet Potato against Non-Pathogenic F. Oxysporum*

Some of the substances produced by plants in response to the stimulation caused by infection from a microorganism are called phytoalexin. It has been reported that the *Ceratocystis fimbriata*-sweet potato system produced ipomeamarone as a phytoalexin[6]. Phytoalexin was analyzed to determine whether the non-pathogenic *F. oxysporum*-sweet potato system produced ipomeamarone or not. Here the phytoalexin ipomeamarone was detected by TLC on the sweet potato inoculated with *F. oxysporum* (Fig.7). The production of phytoalexin indicated that *F. oxysporum* induced a defense response in the sweet potato. The dynamic defense

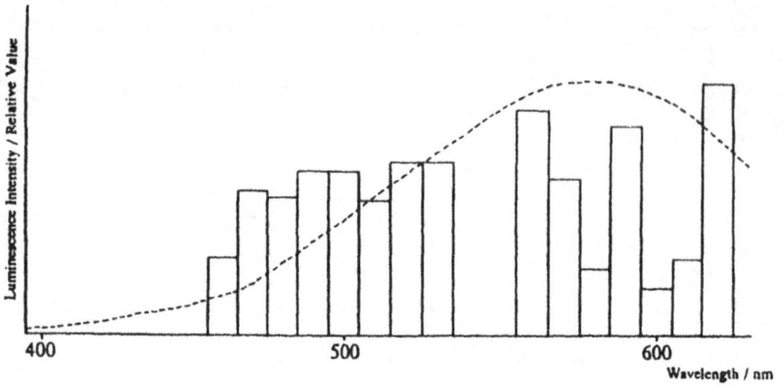

Figure 5. The ultraweak luminescence spectrum of the sweet potato-*F. oxysporum* system

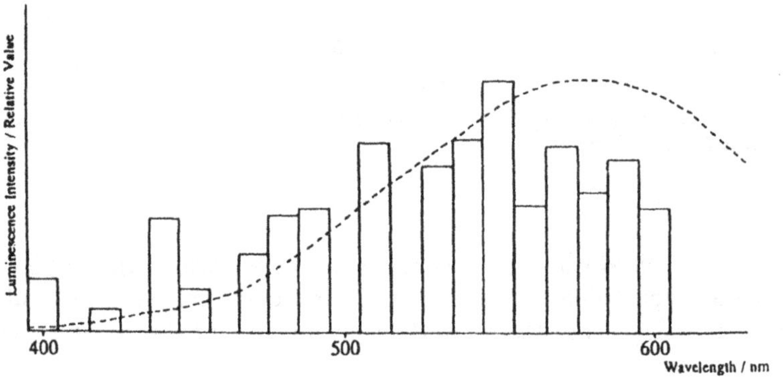

Figure 6. The ultraweak luminescence spectrum of the intact soybean roots.

response of the sweet potato probably prevents *F. oxysporum* from being pathogenic to it.

3.2.3. *Two-dimensional Imaging of the Ultraweak Luminescence*

Two-dimensional imaging was done to obtain direct visualization of the generation of the ultraweak luminescence due to the interaction between the sweet potato and the nonpathogenic *F. oxysporum*. The strongest intensity for the ultraweak luminescence was observed only from the part of the sweet potato that was inoculated with the *F. oxysporum* in the shape of the Japanese character "イ"(i). As shown in Fig. 8, the two-dimensional imaging of the ultraweak luminescence revealed this pattern.

The photon counting image obtained indicates that the ultraweak luminescence arising from the sweet potato due to treatment with the *F. oxysporum* was observed only from cells directly in contact with the non-pathogenic *F. oxysporum*. The intensity of the ultraweak luminescence depended on the concentration of the conidia inoculation solution as shown in Table 1. Furthermore, the inoculation of

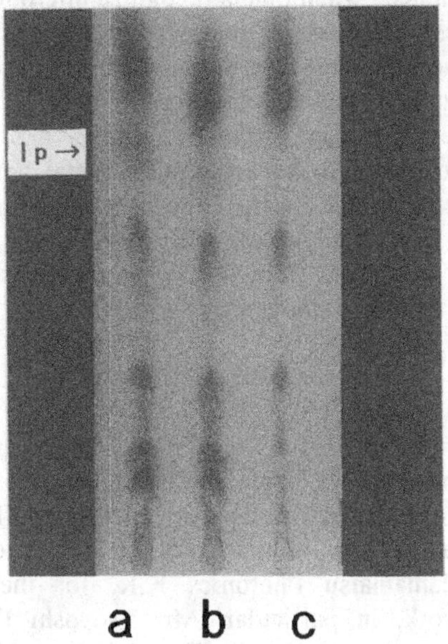

Figure 7. Ipomeamarone production in sweet potato tissue treated with *Ceratocystis fimbriata* (a), nonpathogegenic *F. oxysporum* (b) or sterile water (c). Ipomeamarone is indicated as Ip.

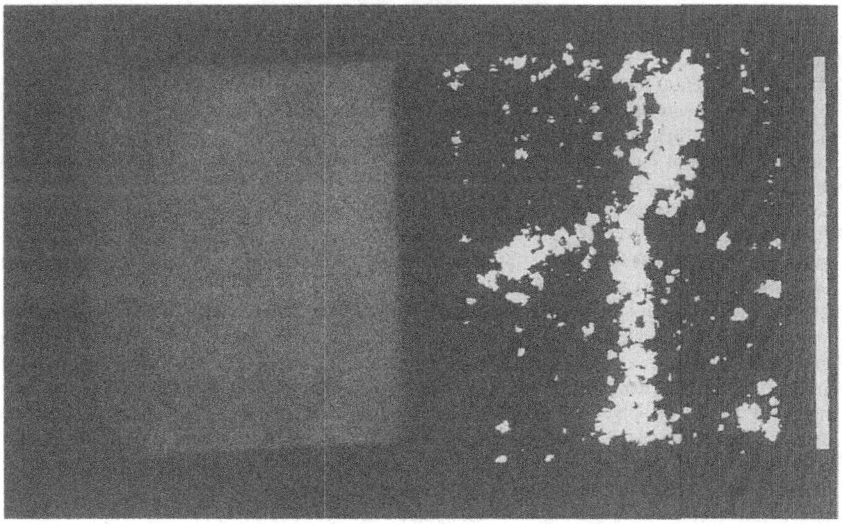

Figure 8. The common image of a sweet potato section (left) and the photon-counting image of the ultraweak luminescence emitted from the sweet potato and *F. oxysporum* (right). Photon signals were accumulated for 5 min.

the conidia induced a defense response in the sweet potato as shown in Fig. 7. Defense responses are induced when plants recognize microorganisms, so the sweet potato appears to express a defense response when the conidia of *F. oxysporum* germinate and invade the tissue because the conidia were used as the inoculum in our system. Because more than half of the conidia of *F. oxysporum* that were dusted onto a potato dextrose agar germinated after 7 h of incubation at 20, *F. oxysporum* is thought to begin to invade within several hours. This time delay corresponds approximately to the onset of the ultraweak luminescence generated from the sweet potato (Fig. 3). From these results, it is concluded that the ultraweak luminescence generated from the sweet potato is related to a defense response to infection by the non-pathogenic *F. oxysporum*.

4. Acknowledgments

I am grateful to Dr. Masahiro R. Ishida for giving me the opportunity to study biophotons. I would like to thank Mrs. Hiroe Honzawa and Mr. Yoshiaki Tachiiri for excellent technical assistance and many helpful discussions. Thanks are also due to everyone in Hamamatsu Photonics K.K. for their many and varied contributions to this work, in particular, Mr. Tsuyoshi Hayakawa, Mr. Isuke Hirano, Mr. Makoto Suzuki, Mr. Nobuyuki Hirai, Mr. Hiroshi Tsuchiya, Mr. Shinji Ohsuka and Dr. Hideshi Fujiwake, without whose significant suggestions and useful discussions, this work would have been impossible.

5. References

1. Slawinska, D. and J, Slawinski (1983) Biological chemiluminescence. *Photochem. Photobiol.*, **37**, 709-715.
2. Colli, L., U. Facchini, G. Guidotti, R. Dugnani Lonati, M. Orsenigo and O. Sommariva (1955) Further measurements on the bioluminescence of the seedlings. *Experientia*, **11**, 479-481.
3. Allen, R.C., R. L. Stjernholm and R. H. Steele (1972) Evidence for the generation of an electronic excitation state(s) in human polymorphonuclear leukocytes and its participation in bactericidal activity. *Biochem. Biophys. Res. Comm.*, **47**, 679-684.
4. Ebel, J. (1986) Phytoalexin synthesis: the biochemical analysis of the induction process. *Annu. Rev. Phytopathol.*, **24**, 235-264.
5. Ogawa, K. and T. Komada (1984) Biological control of Fusarium wilt of sweet potato by non-pathogenic Fusarium oxysporum. *Ann. Phytopath. Soc. Japan*, **50**, 1-9.
6. Oguni, I. and I. Uritani (1974) Dehydroipomeamarone as an intermediate in the biosynthesis of ipomeamarone, a phytoalexin from sweet potato root infected with ceratocystis fimbriata. *Plant Physiol*, **53**, 649-652.
7. Imaseki, H. and I. Uritani (1964) Ipomeamarone accumulation and lipid metabolism in sweet potato infected by the black rot fungus II. Accumulation mechanism of ipomeamarone in the infected region with special regard to contribution of the non-infected tissue. *Plant Cell Physiol*, **5**, 133-143.
8. Uritani, I. and T. Akazawa (1955) Antibiotic effect on Ceratostomella fimbriata of ipomeamarone, and abnormal metabolite in black rot of sweet potato. *Science*, **121**,216-217.
9. Akazawa, T. (1960) Chromatographic isolation of pure ipomeamarone and reinvestigation on its chemical properties. Arch. Biochem. Biophys., **90**, 82-89.
10. Hayakawa, T., K. Kinoshita, S. Miyaki, H.Fujiwake and S. Ohsuka (1986) Ultra-low-light-level camera for photon counting imaging. *Photochem. Photobiol.* **43,** 95.

11. Inaba, H., Y. Shimizu, Y. Tsuji and A. Yamagishi (1979) Photon counting spectral analyzing system of extra-weak chemi- and bioluminescence for biochemical applications. *Photochem. Photobiol.*, **30**, 169-175.
12. Fujiwake, H., Hiramatsu, M., Suzuki, M., Hayakawa, T. and Ishida, M. R. (1984) Direct evidence for plant root chemiluminescence, *Abstracts of Third International Congress on Cell Biology*, 561.
13. Ichimura, T., M. Hiramatsu, N. Hirai and T. Hayakawa (1989) Two-dimensional imaging of ultraweak emission from intact soybean roots. *Photochem. Photobiol.*, **50**, 283-286.
14. Makino, T., Kato, K., Iyozumi, H., Honzawa, H., Tachiiri, Y. and Hiramatsu, M., Ultraweak luminescence generated by sweet potato and Fusarium oxysporum interactions associated with a defense response, *Photochem. Photobiol.*, **64**, 953-956.
15. Andersen, B.R., A.M.Brendzel and T.F.Lint (1977) Chemiluminescence spectra of human myeloperoxidase and polymorphonuclear leukocytes. *Infect. Immun.*, **17**, 62-66.
16. Cilento, G. (1974) Exited electronic states in dark biological processes. *Quarterly Rev. Biophys.*, **6**, 485-501.

EXPERIMENTAL EXAMINATION ON THE POSSIBLE OPTICAL INTERACTION BETWEEN TWO SEPARATE CELL POPULATIONS

L. BEI, T-H. HU AND X. SHEN*

Institute of Biophysics, Chinese Academy of Sciences, Beijing 100101, China

1. Introduction

The cell to cell "communication" is commonly believed to be mediated by some special "messenger" molecules such as hormones, antibodies, growth factors and neurotransmitters[1]. However, whether some kind of physical communication exist between cells has been a question for long time. The so-called "mitogenetic radiation" reported by A.G.Gurwitsch in 1920's was the first indication that a physical interaction between cells might exist[2]. In 1992, G. Albrecht-Buehler reported that the cells were able to detect the orientation of others by signals that penetrated glass but not thin metallic films, therefore, appeared to be carried by electromagnetic radiation[3]. In 1993, Moltchanov et.al. in St.Petersberg University reported that a mammary explant of lactating mice stimulated with some secretion-regulative hormones such as oxytocin, acetylcholine, epinephrine and norepinephrine can induce protein secretion in the other mammary explant of the mice even when separated by a quartz glass wall[4]. Shen et.al. found that the neutrophils stimulated to undergo respiratory burst can activate a second, chemically separated, but optically coupled population of neutrophil[5]. The response of the latter was visualized as a temporary rising of their low-level chemiluminescence and enhanced generation of superoxide radicals detected by both the reduction of ferricytochrome c and spin trapping. In an International conference dedicated to the 120th birthday of A.G.Gurwitsch (Moscow,1994) Kuzin reported that seeds (Raphanus sativus) acquired a new property after they were γ-irradiated at low dose[6]: some hours after irradiation the seeds exert distant influence on the native seeds used as detectors. The distant influence is to accelerate germination and development of the native seeds(160-180% of the control samples). Such a distant influence can go through a quartz layer (0.2 mm), but not a glass one. All above mentioned evidence suggest that the intercellular "communication" may to some extent be mediated by optical interaction. Using neutrophils isolated from pig blood as a model system, a further examination on the possible optical interaction between two separated cell population was conducted. Instead of looking on the influence of a stimulated neutrophil population on a separated resting neutrophil population [5], it was carefully investigated whether

* The principal investigator and correspondence author.

J. J. Chang et al. (eds.), Biophotons, 57-64.

the chemiluminescence burst of the neutrophils stimulated by phorbol myristate acetate(PMA) could be modulated by the presence of a separated neutrophil population in a close vicinity. Instead of detecting the overall effect on the reductive capacity of the neutrophil suspension by an optically coupled separated neutrophil population, which were stimulated to undergo respiratory burst, flow cytometry was used to gain an insight into the reduction ability of each cells in the suspension. In order to minimize the error introduced by the time-dependent variation in cell viability, kinetics and intensity of respiratory burst in the neutrophil suspension in measuring the difference between the samples for seeking intercellular communication and the controls, a special chemiluminescence detecting system was designed. In the system, both the sample for seeking intercellular interaction and control were placed in a rotatable sample holder, the chemiluminescence burst of the cell suspension in both sample and control were detected by a photomultiplier by turns. Therefore, the kinetic curves of the respiratory burst from the cell suspension for seeking intercellular interaction and the control were measured almost simultaneously. The results are reported in this article.

2. Material And Method

2.1 NEUTROPHILS

The neutrophils were isolated from pig blood (4,000 I.U. heparin sodium salt in 200 ml blood) according to the following procedure: in a beaker, 200 ml blood was mixed with 50 ml 4.5% Dextran (MW=515,000, Fluka) in 0.9% solution. The red cells were allowed to settle for 40 min at 4°C (or at room temperature) and the leukocyte-rich plasma (4.5 ml per tube) was layered on top of 1.5 ml lymphocytes separating solution (with a density of 1.007±0.002, Institute of Hematology, Chinese Academy of Medical Sciences) in sterile tubes. The neutrophils were obtained as a pellet after centrifugation of the tubes at 500g for 15 min. Contaminating erythrocytes were removed by hypotonic lysis and centrifugation. The cells were finally washed with saline and resuspended in Hank's balanced salt solution (HBSS) containing 5 mM D-glucose. Neutrophil viability was checked by trypen blue exclusion and always found to be greater than 90%. All the tubes, beakers and pipes used for isolation of the cells were sterilized, and great caution was taken to avoid any stimulation of neutrophils during the isolation procedure.

2.2. SIMULTANEOUS DETECTION OF THE CHEMILUMINESCENCE BURSTS OF THE NEUTROPHIL SUSPENSION IN TESTING CUVETTES AND CONTROL CUVETTES

The experimental setup is shown in Fig.1. The same volume of neutrophil suspension (1×10^6 cells/ml, 6 ml) was added in two identical large quartz cuvettes ($24\times24\times40$ mm^3) labeled as L_A and L_B. Two identical small quartz cuvettes ($10\times10\times45$ mm^3) labeled as S_A and S_B were immersed in the large cuvettes respectively with a fixed combination of L_A-S_A and L_B-S_B. Special caps were used to position the small cuvettes in the center of the large cuvettes. One of the small cuvettes was filled with the neutrophil suspension (3

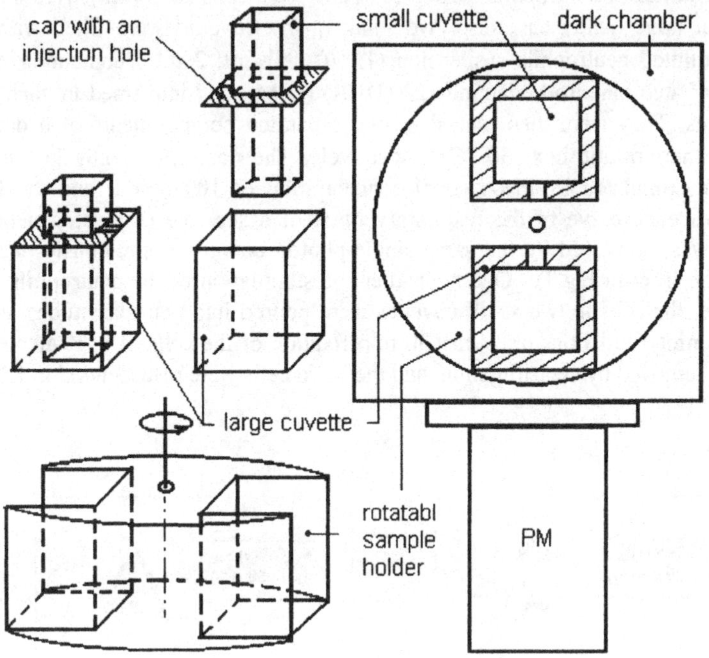

Figure 1. Scheme of the experimental setup for the investigation of optical interaction between the neutrophil suspension in a large cuvette and the neutrophil suspension in a small cuvette immersed in the former. The neutrophil suspension in another identical large cuvette embedded with a small cuvette containing HBSS was used as control. Both testing complex cuvette and control were placed on an rotatable sample holder. The photomultiplier (PM) periodically detected the chemiluminescence bursts of the PMA-stimulated neutrophils in the two complex cuvettes by turns under the control of a computer program.

ml) and the other with HBSS. The two complex cuvettes were placed in the rotatable sample holder of a laboratory-made single-photon counter.

The neutrophils in both large cuvettes were stimulated with 0.5 ml PMA (1 μg/ml in 10% dimethyl sulfoxide and 90% HBSS) and their chemiluminescence bursts were measured by a cooled photomultiplier (at -15°C) periodically by turns. The complex cuvettes L_A-S_A were always placed in position A and the L_B-S_B in position B on the ratable sample holder in all measurements. But, S_A and S_B were in turn filled with the neutrophil suspension for influencing the respiratory burst of the leukocytes in the large cuvette in consecutive tests.

2.3. FLOW CYTOMETRIC MEASUREMENT OF THE ACTIVITY OF THE NEUTROPHILS INFLUENCED BY A SEPARATED STIMULATED NEUTROPHILS SUSPENSION

The same volumes of neutrophils suspension (5×10^6 cells/ml , 6 ml) were added in two identical large quartz cuvettes. Two small quartz cuvettes were filled with same volume of diluted neutrophils suspension (1×10^6 cells/ml, 2 ml) containing exactly same concentrated dihydrorhodamine-123 (DHR) (1μM), and immersed in each of the large cuvettes. They were then placed in two separated compartments of a dark and heparinized thermostatic box (30 °C) respectively. The neutrophils only in one large cuvette were stimulated by PMA (final concentration of 100 ng/ml) and the kinetic chemiluminescence curve of the respiratory burst of a separate stimulated neutrophil suspension was monitored by the same single-photon counter as previously described (shown as the inset in Fig.3). One hour after the stimulation of the neutrophils in one large cuvette, the cells in two small cuvettes were poured into two glass tubes, each of which containing 6 ml Ethanol . After 30 min fixation of the cells in 75% alcohol, the alcohol was removed by centrifugation, and the fixed cells were resuspended in HBSS.

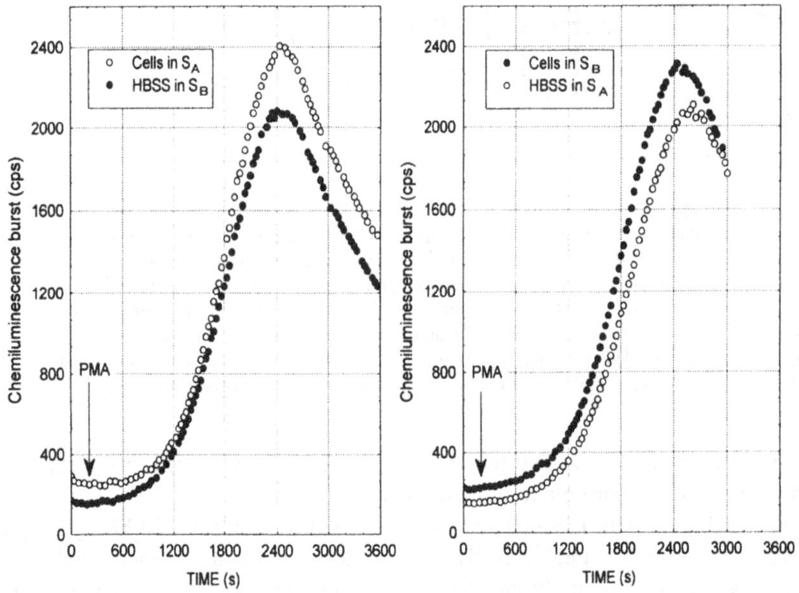

Figure 2. The chemilumenescence bursts from PMA-stimulated neutrophil suspensions in two large cuvettes with two immersed small cuvettes, one of which was filled with cells and the other with HBSS respectively, measured simultaneously in two typical tests. The left part of the figure shows the result when the small cuvette S_A contained cells but S_B contained HBSS. The right part of the figure shows the other alternative in which S_A contained HBSS but S_B contained cells. However, the complex cuvettes L_A-S_A were always placed at the position A on the sample holder.

TABLE 1. Summary of the twelve independent tests for the enhancement of the respiratory burst concomitant chemiluminescence of the neutrophils in a large cuvette by a separate neutrophil suspension in the immersed small cuvette.

No. of test	Date	Cells in immersed cuvette				HBSS in immersed cuvette				Enhancement of the CL burst of the cells in large cuvette by immersed cells
		Position on holder	CL before stimulation	Peak CL after stimulation	Net CL	Position holder	CL before stimulation	Peak CL after stimulation	Net CL on	(unit in σ)
1	26/12/96	A	114 ± 1.4	346 ± 1.5	232 ± 2.0	B	198 ± 1.4	413 ± 2.3	215 ± 2.7	6.4
2	26/12/96	B	94 ± 1.8	509 ± 3.5	415 ± 3.9	A	153 ± 1.9	479 ± 3.9	326 ± 4.4	20.2
3	26/12/96	A	121 ± 1.6	524 ± 3.3	403 ± 3.7	B	179 ± 2.2	506 ± 2.1	327 ± 3.0	20.5
4	26/12/96	B	224 ± 2.3	480 ± 3.3	256 ± 4.0	A	130 ± 1.1	494 ± 3.4	364 ± 3.6	– 27.2
5	26/12/96	A	127 ± 1.0	492 ± 3.9	365 ± 4.0	B	224 ± 2.3	491 ± 1.9	267 ± 3.0	24.4
6	04/06/97	A	267 ± 3.6	2074 ±13.7	1807 ±14.1	B	384 ± 2.9	2146 ±17.5	1762 ±17.7	2.5
7	04/06/97	B	310 ± 3.1	2358 ± 7.6	2048 ± 8.2	A	200 ± 3.8	2321 ± 7.7	2121 ± 8.6	– 8.5
8	04/06/97	A	263 ± 4.7	2388 ± 9.2	2125 ±10.3	B	159 ± 2.1	2063 ± 6.2	1904 ± 6.5	21.5
9	04/06/97	B	224 ± 3.1	2296 ± 8.9	2072 ± 9.4	A	150 ± 1.2	2035 ± 8.2	1885 ± 8.3	19.9
10	04/06/97	A	196 ± 1.7	2114 ± 3.8	1918 ± 4.2	B	157 ± 1.6	1960 ± 4.8	1803 ± 5.0	22.9
11	04/06/97	B	231 ± 9.0	2183 ± 8.3	1952 ±12.3	A	212 ± 5.2	2072 ±10.3	1860 ±11.5	7.5
12	04/06/97	A	186 ± 2.6	1978 ± 7.1	1792 ± 7.6	B	211 ± 3.1	2206 ± 5.7	1995 ± 6.5	– 26.8

The green fluorescence (535 ± 15 nm) intensities of oxidized DHR in 10^4 neutrophils from the cuvette immersed in the PMA-stimulated neutrophil suspension and the one immersed in the resting neutrophil suspension were assayed using a 488 nm argon laser equipped FACScan flow cytometer (Facs 420, Becton Dickinson San Jose, CA).

3. Results And Discussion

3.1. INFLUENCE OF AN IMMERSED SEPARATED NEUTROPHIL SUSPENSION ON THE RESPIRATORY BURST OF THE PMA-STIMULATED NEUTROPHILS

As a test for the possible interaction between two separated but optically coupled cell population, it was carefully investigated whether the respiratory burst of the PMA-stimulated neutrophils in a large cuvette could be influenced by the presence of a second neutrophil population in a immersed small cuvette. As shown in Fig.1, two identical large cuvettes were filled with exactly the same isolated pig neutrophil suspension (identical volume and concentration), and two identical small quartz cuvettes were immersed in each of them. However, one small cuvette was filled with the same neutrophil suspension, but the other with same volume of cell-free HBSS. The neutrophils in two large cuvettes were then stimulated with PMA at almost the same time (with a time interval of only 20 s) and the concomitant chemiluminescence from the two PMA-stimulated neutrophil population were simultaneously recorded by a photomultiplier in terms of a rotatable sample holder. A typical results of the measurement is shown in Fig.2. Twelve independent tests were conducted.

In order to know whether such a enhancement could be due to more favorable scattering of the photons originated in the large cuvette by the neutrophil suspension in the immersed small cuvette, the neutrophil suspension in two large cuvettes were replaced by same concentrated luminol solution and the chemiluminescence from the autoxidation of luminol were measured simultaneously. No difference between the chemiluminescence from the large cuvette with immersed cell suspension and that with immersed HBSS was detected. Even the measured luminol CL from the cuvette immersed with cell suspension was frequently lower than that from the cuvette with cell-free HBSS. This may due to the fact that cell suspension absorbs more photons coming from the area behind the small cuvette than more transparent HBSS does. Although luminol or lucigenin can amplify the chemiluminescence from neutrophils by several orders of magnitude, they were not used in this investigation. Since they are also a stimuli for neutrophil and can contribute photons by their autoxidation, some considerable error could be introduced in the experiment described above. Thus, throughout this investigation, the chemiluminescence burst of the neutrophils was detected without using chemical amplifier.

3.2. INFLUENCE OF THE SURROUNDING PMA-STIMULATED NEUTROPHILS ON THE NEUTROPHIL SUSPENSION IN A SMALL CUVETTE

The influence on the respiratory burst of the PMA-stimulated neutrophils by an immersed separated neutrophil suspension has been demonstrated. Since any interaction would be two ways in nature, would the neutrophils in a small cuvette be also affected

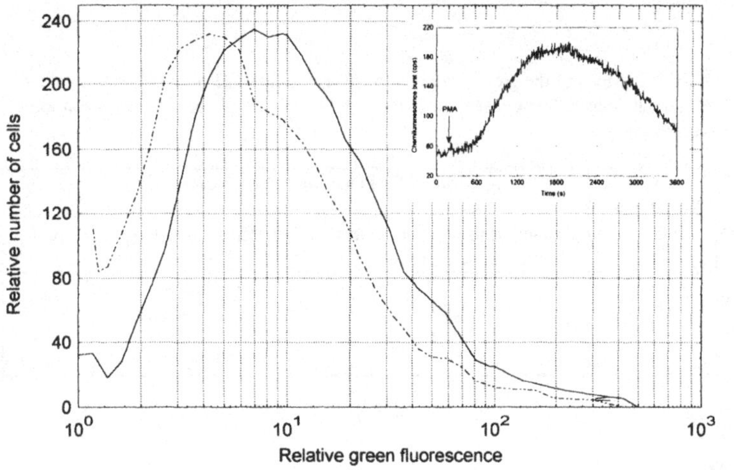

Figure 3. The flow cytometric measurements of the redox activities of the neutrphils in either the small cuvette surrounded by PMA-stimulated neutrophils (dotted line) or in the cuvette surrounded resting cells (solid line). The redox activity of the cell is expressed as the green fluorescence of intracellular rhodamine 123. The measurement were taken 1 h after the cells in small cuvettes containing 1 μM dihydrorhodamine 123 had been optically coupled with the stimulated and resting cells in large cuvettes. The insert: the chemiluminescence burst of the PMA-stimulated neutrophils in the large cuvette.

by surrounding stimulated neutrophils on its outskirts ? Flow cytometry was used to measure the single-cell redox activities of the neutrophil suspension containing nonfluorescent DHR in two small cuvettes, which were immersed in two identical large cuvettes filled with same concentrated neutrophil suspension. The neutrophils in only one large cuvette was stimulated with PMA. Oxidation of nonfluorescent DHR to the green fluorescent rhodamine 123 (R123) was highly specific for the redox activity of neutrophils[7]. If the redox activities of the cells in the small cuvette immersed in the stimulated neutrophil suspension were affected, the distribution of the relative number of neutrophils against the intensity of the green fluorescence from single cell would differ from the distribution obtained from the cells in the small cuvette immersed in the resting neutrophil suspension. Four tests were made according to the experimental procedure and a typical result is shown in Fig.3. All 4 tests showed that the distribution of relative cell number Vs fluorescence from neutrophil population surrounded by the stimulated cells is shifted to the left (towards lower fluorescence) in comparison with the distribution from the cell population in the small cuvette surrounded by resting neutrophils. Table 2 shows how much the means of green fluorescence over 10^4 cells in the cell population optically coupled to the stimulated neutrophils are lowered from the values obtained from controls. The reduction of the green fluorescence of intracellular R123 in the neutrophils surrounded by stimulated cells suggests that the neutrophil suspension in a small cuvette was really affected by the surrounded activated leukocytes. However, the explanation for the less oxidation of DHR in the affected cells has not been well established. It may be suggested that the DHR inside the affected cells were oxidized through a normal redox metabolic process, but reduced by the superoxide anions, O_2^-, which may be generated more in the affected cells.

TABLE 2. Mean of the green fluorescence of rhodamine 123 over 10^4 neutrophils in the small cuvette immersed in either PMA-stimulated or resting neutrophil suspension

No. of test	Mean of the green fluorescence over cells	
	Immersed in PMA-stimulated neutrophil suspension	Immersed in resting neutrophil suspension
1	18.7	25.3
2	28.6	34.7
3	25.8	32.6
4	31.8	38.6

4. Summary

Since the present investigation shows an enhancement of the respiratory burst of the PMA-stimulated neutrophils by the presence of a second, chemically separated but optically coupled neutrophil population in a close vicinity and the less oxidation of dihydrorhodamine 123 in the neutrophils surrounded by PMA-stimulated neutrophils, we may concluded that an optical interaction between two separated cell populations possibly exists. If this is true, it would open a new horizon for understanding cell to cell communication.

5. Acknowledgment

This investigation was supported by the National Natural Science Foundation of China.

6. References

1. S.H.Synder, The molecular basis of communication between cells, *Scient. Am.*, **253**(1985),144
2. A.G.Gurwitsch and L.D.Gurwitsch, *Die mitogeneticshe Strahlung, ihr physikalisch-chemischen Grundlagen und ihr Andwendung auf die Biologie und Medizin*, G.Fisher, Jena, 1959
3. G.Albrecht-Buehler, Rudimentary form of cellular "vision", *Proc.Natl.Acad.Sci.USA*, **89**(1992),8288-8292.
4. V.P.Golantsev, S.G.Koralenko, A.A.Moltchanov and V.I.Prutskov, Lipid peroxidation, Low-level chemiluminescence and regulation of secretion in the mammary gland, *Experientia*, **49**(1993), 870-875
5. X.Shen, W.Mei and X.Xu, Activation of neutrophils by a chemically separated but optically coupled neutrophil population undergoing respiratory burst, *Experientia*, **50**(1994), 963-968.
6. A.M.Kuzin and G.N.Surbenova, Secondary biogenic irradiation of plant structures after γ-irradiation at low dose, in *Biophotonics* (L.V.Beloussov and F.A.Popp. eds.), P.257-265, Bioinform Services Co., Moscow, 1995.
7. G.Rothe, A.Emmendörffer, A.Oser, J.Roesler and G.Valet, Flow cytometric measurement of the respiratory burst activity of phagocytes using dihydrorhodamine 123, *J.Immun.Methods*, **138** (1991), 133-1135.

LUMINOMETRY IN CELLULAR STRESS RESEARCH

Continuous detection of firefly luciferase activity in intact mammalian cells during heat stress

J.E.M SOUREN AND R. VAN WIJK

Department of Molecular Cell Biology, Utrecht University
Padualaan 8, 3584 CH Utrecht, The Netherlands.

1. Introduction

1.1. STRESS AND DEFENCE MECHANISMS

Life and environment are in a constant interaction, and temperature plays in this a crucial role. It affects the rate of all cellular reactions and it determines the survival of the organisms. Notoriously sensitive to temperature is mammalian cell survival which is mainly limited to a narrow range of temperatures, from a few degrees to approximately 40°C. Cells differ in the range of temperatures that they can tolerate. This may change in time as a certain degree of adaptation is possible. It has been known that exposure to a sub-lethal temperature often leads to adaptations so that previously lethal temperatures are now tolerated. This response to heat shock has attracted considerable attention from molecular biologists over the last decade, which has resulted in a rapid accumulation of data providing considerable insights not only into the molecular basis of thermotolerance, but also into stress physiology in general. The heat shock response is now known to occur from bacteria to man (Schlesinger *et al.* 1982). It is accompanied by a transient reprogramming of cellular activities to ensure the cell's survival by stimulating defence mechanisms and by protecting essential cell components against the heat damage.

A main feature of a heat shock response is clearly the vigorous but transient activation of a small number of specific genes previously either silent or only active at low levels. New mRNAs are actively transcribed from these genes and translated into proteins which are collectively referred to as heat shock proteins, or hsps. The exact number of different types of hsp varies considerably in different organisms and cell types, but it became clear in mammalian cells that particular hsps such as Hsp28, Hsp60, Hsp70, Hsp84, Hsp100 were all transiently synthesized at an elevated rate after heat stress. Usually, proteins of approximately 70 kDa are among the most prominent. In this class a distinction can be drawn between the 'constitutive' Hsp70 (Hsc70) and the 'inducible' Hsp70 (Hsp70). These proteins are similar but not identical. The Hsp70 function in cell protection to stress has been supported by several lines of evidence:
(a) microinjection of antibodies against Hsp70 suppresses survival of fibroblasts at 45°C

J. J. Chang et al. (eds.), Biophotons, 65-78.

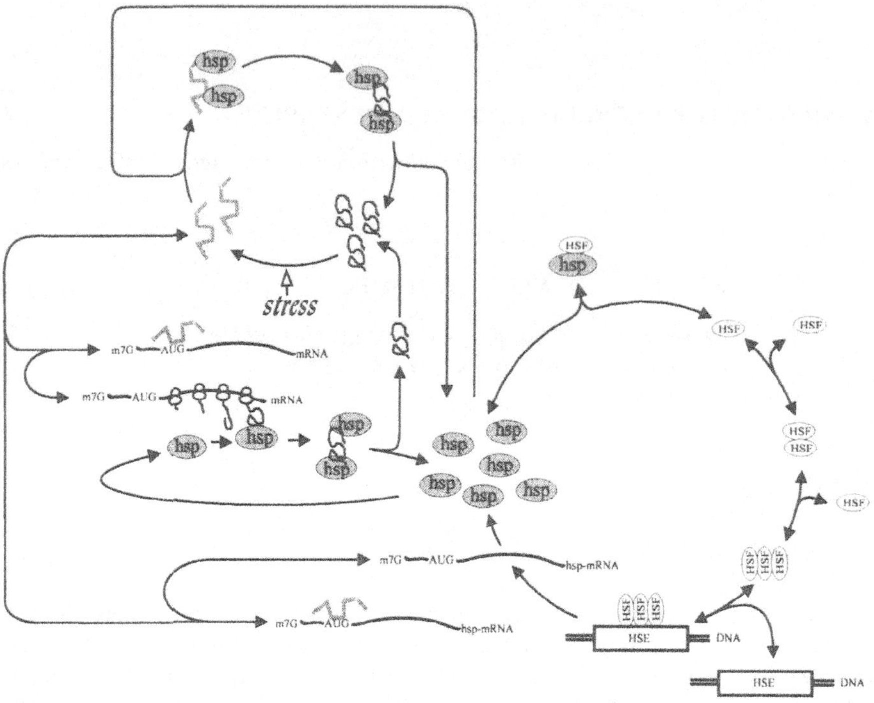

Figure 1. A diagram of the functions of Hsp70 and autoregulation of its levels in the cell.

(Riabowol *et al.* 1988), (b) transformation of rat cells (Li *et al.* 1991) with constitutively expressed human Hsp70 genes dramatically increases thermotolerance. In the last case, a careful two-dimensional analysis of total cellular proteins showed that increased basal expression of Hsp70 is the only substantial alteration in the protein profile, making this approach the most convincing demonstration of the importance of Hsp70 in mammalian thermoprotection.

1.2. REGULATION OF HSP70 GENE EXPRESSION

The model presented in figure 1 summarizes the cell's response to changing temperature conditions depicting the functioning of Hsp70 and the central role of free Hsp70 in the regulation of its own synthesis (for reviews: Parsell *et al.* 1993; Lindquist, 1993; Wu, 1995). For the induction of heat shock proteins after a temperature challenge a short special DNA sequence adjacent to the hsp genes is required. This DNA sequence has been called the heat shock element, or HSE to which a regulatory protein can bind. Transcription of hsp genes appears to correlate with the presence of this HSE-binding protein. This protein is called heat-shock transcription factor, or HSF. Having identified HSFs and their binding to HSEs, a question is how the gene activation by heat shock is regulated. Most of the present data suggest that the family of Hsp70 proteins plays an important role in this regulation.

One of the most exciting concepts that emerged from the study of the Hsp70 family members is their role as molecular chaperones. Molecular chaperones are proteins that mediate in the folding of other polypeptides. The tracing of the chaperoning pathways in eukaryotic cells started with the observation that the constitutively synthesized member of the Hsp70 family, Hsc70, transiently associates with many nascent polypeptide chains in non-stressed mammalian cells. Hsc70 is also thought to deliver polypeptides synthesized in the cytoplasm to the rough endoplasmic reticulum or mitochondria in an unfolded state ready for membrane translocation.

During stress, the structural integrity of part of the proteins in the cell become compromised. The binding of the Hsp70 to these damaged proteins prevents them from aggregating and provide for them the opportunity to refold. For the refolding reaction itself, Hsp70 probably functions in concert with other chaperones. The binding of Hsp70 to partially unfolded proteins after a stress condition depletes the pool of free Hsp70 proteins and triggers the binding of HSF to HSE.

1.3. PERSPECTIVES

While we have learned a great deal about the general biochemical activities of Hsp70 and its importance in protecting organisms at high temperatures, we still have not defined the critical biological processes that are protected by it at elevated temperatures. In line with the suspected role that Hsp70 and Hsc70 play in chaperoning and salvaging denatured proteins in stressed cells the investigations have to be extended to their role in the processes of denaturation and renaturation of specific thermolabile enzymes.

Up till now several methods have been used to study the protein denaturation in cells. Differential scanning calorimetry is a method by which denaturation of the total of proteins in the cell can be studied. However, a major disadvantage of this method is that the denaturation of the cellular proteins cannot reliably be measured under non-harmful conditions. Acquisition of enzyme activities of specific proteins in the cell have also been used. However, not all proteins are equally sensitive for heat and one has to search for proteins that are thermolabile and can be easily and reproducibly detected. A recent development is the extension of these studies by transformation of cells with reporter enzymes. With the use of the appropriate reporter enzymes the important question can be asked whether Hsp70 protects proteins against denaturation or whether Hsp70 functions in the process of renaturation.

2. Firefly luciferase as reporter protein

Reporter proteins are useful tools in molecular biology for the assessment of gene promoter activity. Some of the most commonly used are chloramphenicol acetyl transferase (cat) (Gorman *et al.* 1982), green fluorescent protein (GFP) (Chalfie *et al.* 1994) and luciferases from bacteria (Meighen, 1991) and from fireflies (De Wet *et al.*

1987). Their usefulness relies not only on the sensitivity, reproducibility and simplicity of the corresponding detection assay, but also on the range of cell types where they can be expressed. Among these reporter genes, firefly luciferase (LUC) is growing in importance. In mammalian cells, firefly luciferase provides one of the best non-toxic and sensitive methods to measure gene expression (De Wet *et al.* 1987).

Luciferase from the firefly Photinus pyralis has an apparent molecular weight of 62 kDa and requires luciferin, ATP and O_2 as substrates. The reactions catalyzed by firefly luciferase are:

$$\text{luciferase} + \text{luciferin} + \text{ATP} \rightleftharpoons \text{luciferase}\cdot\text{luciferin}\cdot\text{ATP} \quad (1)$$

$$\text{luciferase}\cdot\text{luciferin}\cdot\text{ATP} \longrightarrow \text{luciferase}\cdot\text{luciferin}\cdot\text{AMP} + \text{PPi} \quad (2)$$

$$\text{luciferase}\cdot\text{luciferin}\cdot\text{AMP} + O_2 \longrightarrow \text{luciferase}\cdot\text{oxyluciferin}^* + \text{AMP} + CO_2 \quad (3)$$

$$\text{luciferase}\cdot\text{oxyluciferin}^* \longrightarrow \text{luciferase}\cdot\text{oxyluciferin} + \text{light} \quad (4)$$

$$\text{luciferase}\cdot\text{oxyluciferin} \longrightarrow \text{luciferase} + \text{oxyluciferin} \quad (5)$$

The first reaction is the formation of a luciferase-luciferin-ATP complex followed by an enzyme-bound luciferyl-adenylate and PP_i. During the next reaction, the luciferyl-adenylate undergoes an oxidative decarboxylation which results in the production of CO_2, AMP and oxyluciferin in an excited state. With the return of the excited oxyluciferin to the ground state the light is emitted. In the last step oxyluciferin has to dissociate from the enzyme for a new reaction cycle can begin. The production of light by firefly luciferase is very efficient; the reaction yield is at least 0.88 with respect to luciferin. The reaction catalyzed by firefly luciferase emits at 25°C at pH 7.5 to 8.5 a yellow green light with the peak emission at 560 nm.

2.1. FIREFLY LUCIFERASE AS REPORTER PROTEIN FOR HEAT SHOCK INDUCED PROTEIN DENATURATION

Particularly interesting is the observation that luciferase is a thermolabile protein and therefore also can be used as a reporter of the effect of the temperature on protein denaturation in a cellular environment. Even at 37°C some denaturation of this protein has been reported but the rate of denaturation is enormously amplified at higher temperatures. Generally the inactivation and renaturation of this enzyme has been studied by measuring the enzymes activity in the cell lysates. A disadvantage of this method is that for each time point separate samples must be used. This leads to high variability caused by the variation in transfection efficiency between the samples. Another disadvantage are the artifacts due to the lysis procedure of the cell.

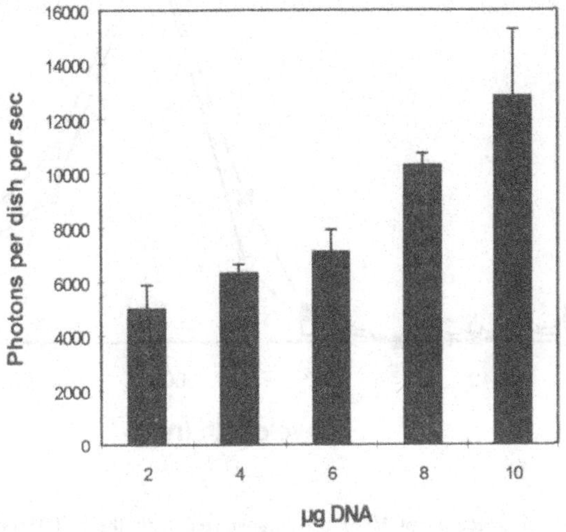

Figure 2. In vivo measurements of luciferase activity at 37°C of H9c2 cells transfected with various amounts of pGL3 plasmid coding for luciferase with an extracellular luciferin concentration of 0.1 mM, three days after transfection.

In this chapter we demonstrate that luciferase is one of the enzymes which can be detected intracellularly, non-invasively and with high sensitivity

2.2. MEASUREMENT OF *IN VIVO* LUMINESCENCE

The utility of luciferase as a reporter of *in vivo* regulation has been controversial due to confusion about whether luciferin can permeate cell membranes. At neutral pH, luciferin carries a single negative charge, and it was presumed a priori that the molecule could not cross the hydrophobic lipid bilayer. Several reports, however, describe measurements of *in vivo* luminescence in animal, plant, and bacterial cells and tissues with luciferin (Gould and Subramani, 1988).

In figure 2 we demonstrate the *in vivo* detection of luciferase in cultured mammalian cells. For this purpose we first used the established H9c2 cell line which is a subclone of an original cell line derived by selection from embryonic rat heart tissue. These cells still do exhibit cardiac-specific characteristics (Hescheler *et al.* 1991) but also show several features of skeletal muscles (Kimes and Brandt, 1976). We transfected the H9c2 rat myocyte cells in monolayer culture with the pGL3 luciferase expression vector, having a

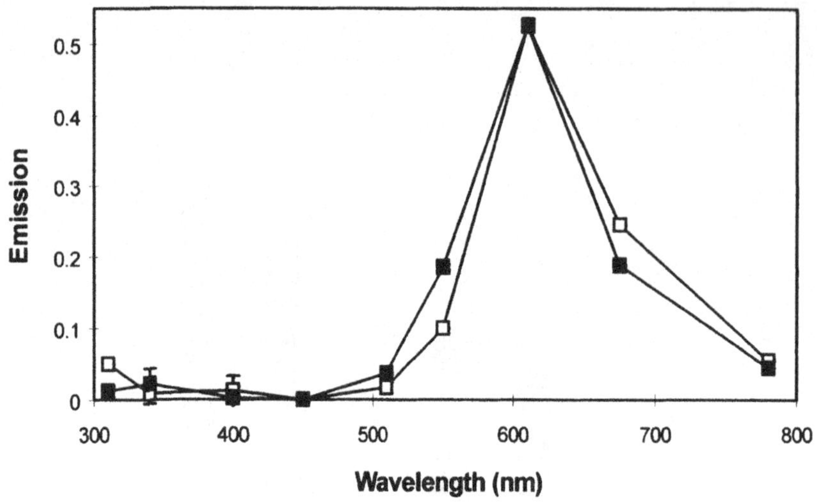

Figure 3. Emission spectrum of H9c2 cells transfected with the pGL3 control plasmid coding for luciferase after addition of 0.1 mM luciferin at 25°C (closed squares) or 37°C (open squares).

modified luciferase gene integrated under the control of the SV40 promoter with the SV40 late poly(A) signal and the SV40 enhancer sequence. The luciferase gene has a deletion in the coding sequence for the C-terminal tripeptide in order to express the enzyme activity in the cytoplasm constitutively and to eliminate peroxisome targeting of the expressed protein. Using a simple calcium phosphate precipitation method for transfection, we could detect the first luciferase activity about seven hours after plasmid addition. The signal increases for about two days after which a plateau was reachedwhich after that remained constant for about a week. Monolayers of pGL3 transfected H9c2 rat heart cells at a density of $5x10^5$ per petri dish with 1 ml culture medium supplemented with 0.1 mM luciferin resulted in a signal which could vary but was generally in the order of about 2000 photons per sec. These measurements were performed with a temperature controlled photo detection system. The bottom of the dish was approximately 7 cm below the photomultiplier tube window of a photon counter, in a single photon counting mode, equipped with a Hamamatsu R550 photomultiplier tube (spectral response 280-850 nm, 1.5 kVolt) kept at -20°C. Standard high performance photon counting electronics consisting of a low-noise preamplifier, amplifier, discriminator and ratemeter were used. Under these conditions, background or dark current of the photomultiplier amounted to 60-90 counts per second (Souren *et al.* 1996).

Besides the time after transfection the amount of added plasmid-DNA determines the amount of luciferase present. In figure 2 is shown the photon emission of cell transfected with different amounts of plasmid, and measured three days after transfection.

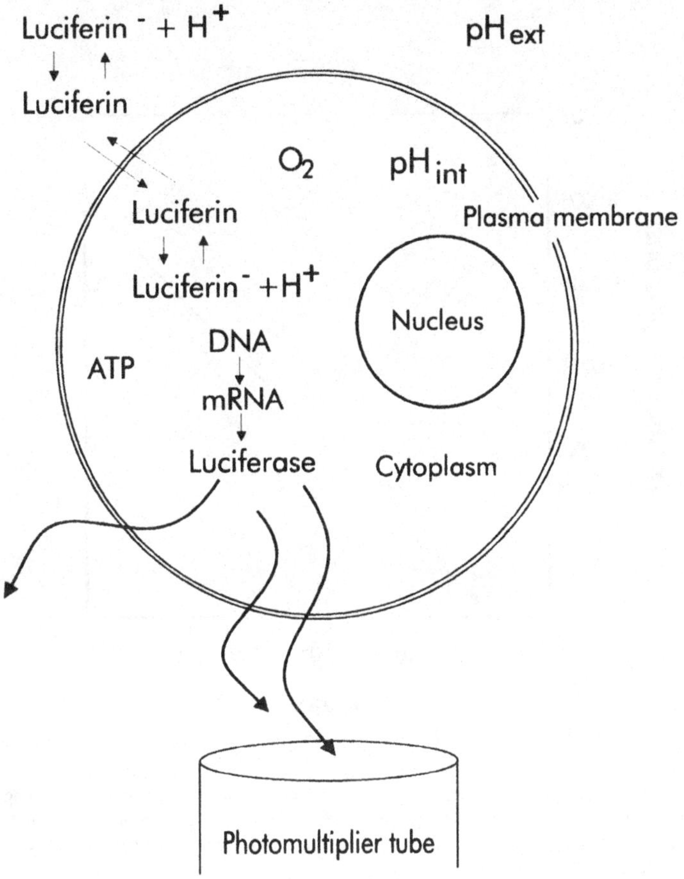

Figure 4. Luciferase activity measurements of intact cells.

In figure 3 the emission spectrum of H9c2 cells transfected with a luciferase-coding plasmid is shown at the culturing temperature of the cells (37°C) and at room temperature (25°C). In the firefly Photinus pyralis the emitted light is yellow-green and has an emission peak at 560 nm. From the data it is obvious that at both temperatures the peak value of the emission spectra of the luciferase measured in the H9c2 cells is shifted to the higher wavelengths. This shift is even further to the red part of the spectrum at the culturing temperature of the cells. Red shifts of the emission spectrum of luciferase have also been reported in studies with luciferase in solution. This is the case under various experimental conditions as at acid pH, at higher temperatures, in the presence of cations as Zn^{2+}, Cd^{2+}, Hg^{2+} and/or anions as in buffers with a high phosphate concentration (Seliger and McElroy, 1964).

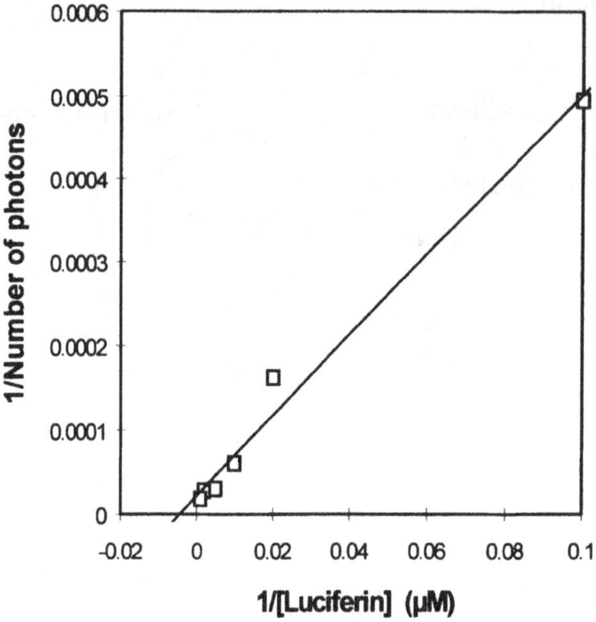

Figure 5. Luciferase activity of H9c2 cells transfected with pGL3 plasmid coding for luciferase at various luciferin concentrations 15 min after substrate addition. The data are given in a Lineweaver-Burk plot. The intercept value with the x-axis provides a K_m value of 0.22 mM.

3. Conditions regulating the luciferase activity *in vivo*

The measured luminescence as a result of the luciferase activity *in vivo* depends on various variables as outlined in figure 4.

3.1. LUCIFERIN

In figure 5 the dependence of the luciferase activity of the external luciferin concentration is given in a Lineweaver-Burk plot. The data points can be fitted to a straight line suggesting simple Michaelis-Menten enzyme kinetics with a K_m value of 0.22 mM. This value is high compared with the Km value found in cell lysates (73 mM at pH 7.8 in a tricine buffer). It is not known whether this is due to the fact that the internal luciferin concentration does not correspond with the external luciferin concentration or whether this is due to other different environmental conditions of the enzyme.

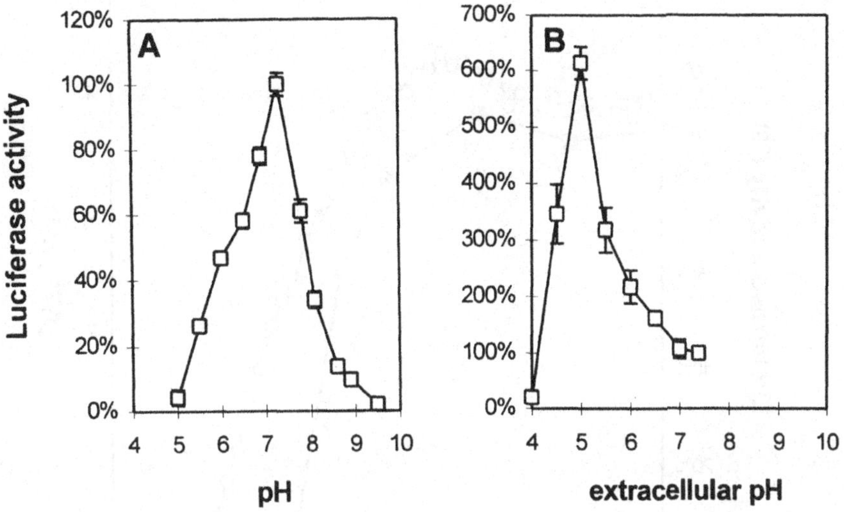

Figure 6. (A) Luciferase activity of lysed H9c2 cells transfected with pGL3 plasmid coding for luciferase at various pH (B) Luciferase activity of intact H9c2 cells transfected with pGL3 plasmid coding for luciferase at various extracellular pH.

3.2. pH

In figure 6A the dependence of the luciferase activity of lysed H9c2 cells on the pH is given. From this figure it can be derived that the optimum pH for this enzyme is about pH 7.5. In figure 6B the intracellular luciferase activity is given as a function of the extracellular pH. A large difference exists between the pH optimum of the enzyme in lysates and the apparent pH optimum of the enzyme measured in intact cells. Several factors could be responsible for the change in the enzyme activity at changing pH. As already mentioned before, at neutral pH, luciferin carries a single negative charge. Luciferin is an acid which at neutral pH is dissociated to a high degree. However, even at neutral pH a small part of the luciferin is not negatively charged and could therefore permeate the cell membrane. As it is this uncharged luciferin for which an equilibrium will be reached, the concentration of luciferin in the cell will also be determined by the difference between intracellular and extracellular pH. As the cell will try to maintain its intracellular pH, a decrease in the extracellular pH will lead to a lesser decrease in the intracellular pH. A decrease in extracellular pH will therefore result in a higher luciferin concentration in the cell than extracellarly present. Furthermore, as a larger fraction of the luciferin molecules will be in their uncharged state, luciferin will reach faster its equilibrium concentration. Contrary to these stimulating conditions the decrease in the intracellular pH could lead to protein inactivation ultimately resulting in a net reduction of the luciferase activity at low pH.

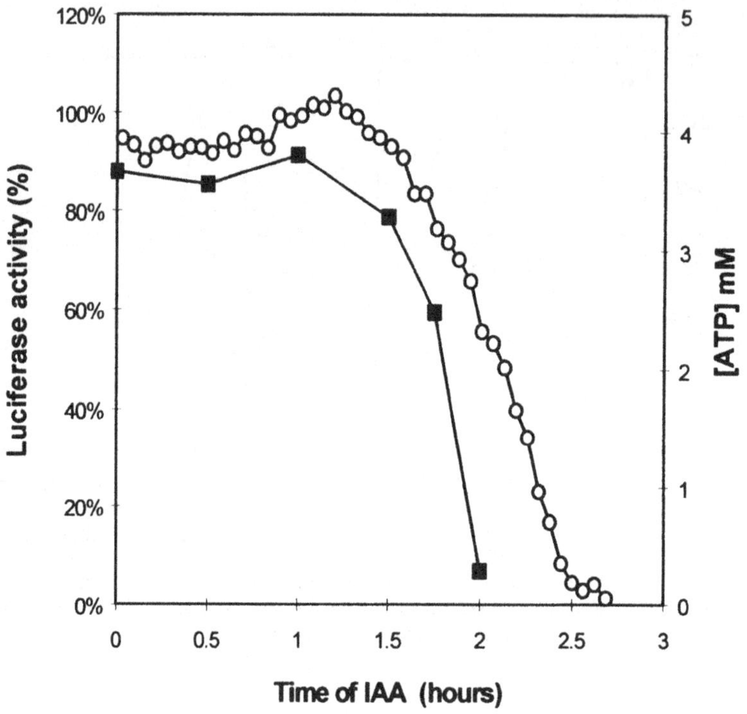

Figure 7. Luciferase activity (circles, left Y-axis) and the ATP concentration (squares, right Y-axis) of pGL3 plasmid (coding for luciferase) transfected H9c2 cells in time after addition of iodoacetic acid, an effective inhibitor of glyceraldehyde-3-phosphate dehydrogenase, the key regulatory enzyme of the glycolytic pathway.

3.3. ATP

The ATP concentration in mammalian cells is normally in the range of 2-4 mM. To study the effect of intracellular ATP on the *in vivo* luciferase activity the intracellular ATP concentrations were influenced by using iodoacetic acid, an inhibitor of the glycolytic pathway. Iodoacetic acid inactivates glyceraldehyde-6-phosphate dehydrogenase, the rate limiting enzyme in this pathway. After addition of iodoacetic acid to the cells the ATP concentration in the cell decreases after about one hour and is less than 10% of the original value after about two hours (figure 7). The luciferase activity declines after 1.5 hours steadily. These results suggest that intracellular luciferase activity is not limited by ATP under normal culturing conditions. The results further suggest that the luminescence from the enzyme is proportional with the

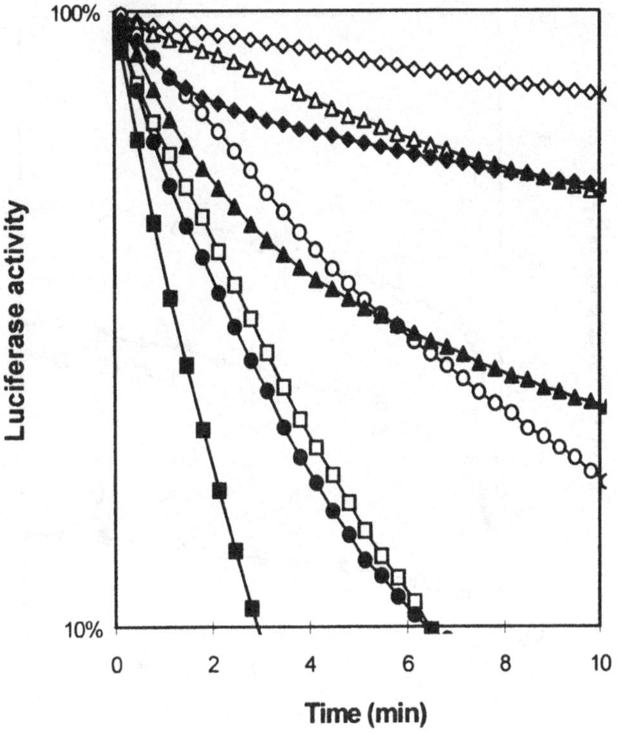

Figure 8. Luciferase activity in time of pGL3 transfected Rat-1 cells (closed symbols) and (Hsp70 overexpressing) HR24 cells (open symbols) incubated at 41°C (diamonds), 42°C (triangles), 43°C (circles) and 44°C (squares).

intracellular ATP levels below 1 mM and can therefore also be used to monitor these levels.

4. Effect of temperature on protein denaturation and the role of Hsp70

Heat shock proteins are believed to play an important role in stabilising defolded proteins and assisting these proteins in regaining their native structure. In this paragraph we report our studies on the denaturation effect of heat shock on luciferase activity and its subsequent recovery at 37°C in Rat-1 cells, and Rat-1 cells transfected with human Hsp70 gene (HR24). Specifically, we ask whether overexpression of Hsp70 protects

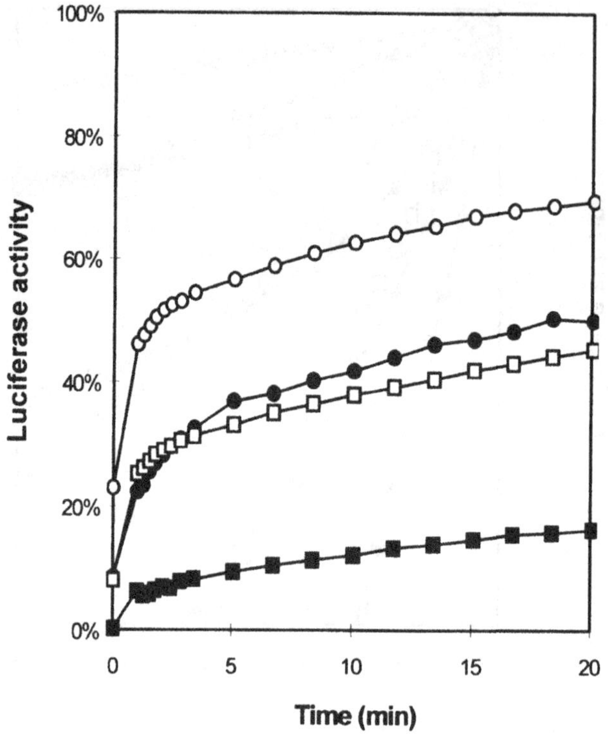

Figure 9. Recovery of luciferase activity in time of pGL3 transfected Rat-1 cells (closed symbols) and (Hsp70 overexpressing) HR24 cells (open symbols) incubated at 37°C after a heat shock of 10 min at 43°C (circles) or 44°C (squares).

cells from heat-induced protein denaturation and/or facilitates renaturation of proteins after the heat shock treatment.

Successful transfection and expression of the human Hsp70 was characterized by Li *et al.* (1991). A 2.3 kb DNA fragment of the human Hsp70 was excised from pHHsp70, a plasmid containing the entire portion of the Hsp70 locus plus 5'-sequences required for the heat-inducible expression (Wu *et al.* 1985; Hunt and Morimoto, 1985). In the resulting plasmid, pSV-Hsp70, transcription of the Hsp70 gene is driven by the SV40 early promoter and enhancer, and ends with termination signals in the 3'-region of the Hsp70 locus. The Hsp70 overexpressing cells were derived by Li *et al.* (1991) from Rat-1 cells which were transfected with pSV-Hsp70 as well as with a plasmid containing the gene conferring neomycin resistance. Neomycin-resistant cells were selected, grown to monolayers and subjected to heating cycles. HR24 was obtained from an individual colony and the expression of Hsp70 has been verified by Northern blot analysis. The identity and integrity of the human Hsp70 expressed in HR24 cells were assessed by two-dimensional gel electrophoresis of proteins extracted from these cells (Li *et al.* 1991).

In our studies exponential growing Rat-1 and HR24 cells were transfected with pGL3 plasmid using the calcium phosphate precipitation method as mentioned before. The expression of luciferase in the two cell lines was about the same. The effect of the temperature on luciferase activity in Rat-1 and HR24 cells is shown in figure 8. The exponential decay of the luciferase activity at the higher temperatures suggests a first order reaction kinetics of the denaturation process. Obviously, at the higher temperatures the inactivation rate of luciferase is also higher. When comparing the denaturation of luciferase in Rat-1 cells with the denaturation in the Hsp70 overexpressing cell line HR24 the higher level of Hsp70 clearly has a protecting effect of about 1°C on the inactivation rate of luciferase activity.

Whether this difference is due to a decreased denaturation of the enzyme or due to a faster renaturation has now to be distinguished. In figure 9 the recovery of the luciferase activity is shown when after the heat shock the cells are returned to their culturing temperature of 37°C. As can be seen from this figure a part of the luciferase activity is recovered very fast followed by a much slower recovery of enzyme activity. This biphasic recovery kinetic may be the result of the different degree of denaturation of the enzyme after the heat shock. Denaturation of a protein is a multi step process in which loss of a part of the tertiary structure often facilitates the further loss of structure. Minor losses in the tertiary structure can be repaired fast and easily while extended loss of structure is a time-consuming process, possibly amplified by interaction of the denatured proteins in larger complexes. After a heat shock at the same temperature the luciferase activity in the Hsp70 overexpressing HR24 cells recovers to a higher extent than in their parent cells, the Rat-1 cells. However, the recovery of the luciferase activity in the Rat-1 and the HR24 cells is almost identical when comparing the recovery of heat shocks which invoke an equal loss of luciferase activity.

So, considering the available data we conclude that the main action of Hsp70 in mammalian cells is the prevention of the protein inactivation.

5. Discussion

An exciting and unique application of firefly luciferase is the use as an *in vivo* reporter of denaturation processes. It is possible to treat living cells with bio-active agents (e.g., environmental contaminants, potential chemotherapeutic drugs or antibiotics) or altered environmental conditions (e.g., temperature or osmotic shock, circadian effectors, oxygen deprivation, or electrical fields) and assess their effects on physiological response by monitoring real-time changes in luciferase activity. With the appropriate equipment, adherent cells cultured in flat-bottom dishes can be easily and rapidly analyzed for luciferase expression. These luciferase assays do not destroy the cells and can be performed aseptically in culture vials, allowing for repeated analysis on the same culture over time. This provides great flexibility in experimental design and eliminates the problem of compensating for variable transfection efficiencies between cell populations within an experimental group. Furthermore, luciferase can be used as an *in vivo* reporter of gene expression. For this purpose the luciferase gene has to be brought under control of the promoter region of the gene of interest. Firefly luciferase should be of enormous potential in the quantification of very weak promoters due to its null background. Studies using the luciferase gene under control of the inducible Hsp70 promoter region are under way.

6. References

Chalfie M., Tu Y., Euskirchen G., Ward W.W. and Prasher D.C. (1994). Green fluorescent protein as a marker for gene expression. *Science* **263**: 802-805.

De Wet J.R., Wood K., DeLuca M., Helsinki D.R. and Subramani S. (1987). Firefly luciferase gene: structure and expression in mammalian cells. *Mol Cell Biol* 7: 725-737.

Gorman C.M., Moffat L.F. and Howard B. (1982). Recombinant genomes which express chloramphenicol acetyl transferase in mammalian cells. *Mol.Cell.Biol.* **2**: 1044-1051.

Gould S.J. and Subramani S. (1988). Firefly luciferase as a tool in molecular and cell biology. *Anal.Biochem.* **175**: 5-13.

Hescheler J., Meyer R., Plant S., Krautwurst D., Rosenthal W. and Schulz G. (1991). Morphological, biochemical, and electrophysiological characterization of a clonal cell (H9c2) line from rat heart. *Circ.Res.* **69**: 1476-1486.

Hunt C.R. and Morimoto R.I. (1985). Conserved features of eukaryotic hsp70 genes revealed by comparison with the nucleotide sequence of human hsp70. *Proc.Natl.Acad.Sci.U.S.A.* **82**: 6455-6459.

Kimes B.W. and Brandt B.L. (1976). Properties of a clonal muscle cell line from rat heart. *Exp.Cell Res.* **98**: 367-381.

Li G.C., Li L., Liu Y.K., Mak J.Y., Chen L. and Lee W.M.F. (1991). Thermal response of rat fibroblasts stably transfected with the human 70-kDa heat shock protein-encoding gene. *Proc.Natl.Acad.Sci.U.S.A.* **88**(5): 1681-1685.

Lindquist S. (1993). Autoregulation of the Heat-Shock Response. In: *Translational regulation of gene expression 2*, edited by Ilan J. Plenum Press, New York. pp.279-320.

Meighen E.A. (1991). Molecular biology of bacterial luminescense. *Microbiol.Rev.* **55**: 123-142.

Parsell D.A., Taulien J. and Lindquist S. (1993). The role of heat shock proteins in thermotolerance. In: *Molecular chaperones*, edited by Ellis R.J., Laskey R.A. and Lorimer G.H. Chapman & Hall, London. pp.23-30.

Riabowol K.T., Mizzen L.A. and Welch W.J. (1988). Heat Shock is lethal to fibroblasts microinjected with antibodies against hsp70. *Science* **242**: 433-436.

Schlesinger M.J., Aliperti G. and Kelley P.M. (1982). The response of cells to heat shock. *Trends Biochem.Sci.* 7: 222-225.

Seliger H.H. and McElroy W.D. (1964). The colors of firefly bioluminescence: enzyme configuration and species specificity. *Proc.Natl.Acad.Sci.U.S.A.* **52**: 75-81.

Souren J.E.M., van Aken H. and Van Wijk R. (1996). Enhancement of superoxide production and protection against heat shock by hsp27 in fibroblasts. *Biochem.Biophys.Res.Comm.* **227**: 816-821.

Wu B.J., Hunt C. and Morimoto R.I. (1985). Structure and expression of the human gene encoding major heat shock protein HSP70. *Mol.Cell.Biol.* **5**: 330-341.

Wu C. (1995). Heat shock transcription factors: structure and regulation. *Annu.Rev.Cell Biol.* **11**: 441-469.

UV-INDUCED DNA DAMAGE AND REPAIR: A POWERFUL LIGHT TRAPPING SYSTEM IN DNA IN ORDER TO CONVERT LIGHT ENERGY INTO BIOCHEMICAL SIGNALS

H.J. NIGGLI

Biophoton AG
P.O. Box 31, CH-1731 Ependes, Switzerland
Tel. / Fax. 0041-264131445

1. Summary

Photons participate in many atomic and molecular interactions and changes. Recent biophysical research has shown the existence of photons in biological tissue and plants, animal and human cells emit a very weak radiation which can be readily detected with an appropriate photomultiplier system. Although the emission of this radiation is extremely low in mammalian cells, it can be efficiently induced by ultraviolet light. Ultraweak radiation in human fibroblasts can also be enhanced by a brief illumination with an ordinary fluorescent light source. The nature of the emitted light can then be examined as rescattered emission, an experimental approach which allows the investigation of ultraweak re-emission patterns in the differentiation system of normal- and DNA repair-deficient Xeroderma Pigmentosum (XP) fibroblasts. We have reported that postmitotic XP-fibroblasts lose the storing capacity of ultraweak photons which are efficiently trapped in normal cells. Thus, it is evident that there exists an important difference between normal and XP cells and this suggests that there is an effective intracellular mechanism of photon trapping in normal human cells. It is proposed that nucleic acids may play via pyrimidine dimer formation, a similiar role of light trapping as is known for vitamin A in the retinal isomerization reaction found in the light driven photon conversion process of the eye. This light-trapping system can influence metabolic and cellular events by triggering amplification mechanisms and promoting photochemical processes long after exposure to light via excision repair as shown recently by Gilchrest and co-workers (1994) for induced melanogenesis via excised UV-induced pyrimidine dimer molecules.

J. J. Chang et al. (eds.), Biophotons, 79-86.

2. Introduction

Spontaneous ultraweak photon emission (PE) has been extensively described in yeast, plant and animal cells [1-7]. In a recent report, experiments with cultured human cells in which normal and DNA-excision-repair-deficient XP cells were UV-irradiated in medium and balanced salt solution (BSS) were assessed for ultraweak photon emission [4]. There was evidence of induced-photon emission from normal cells in BSS but clear evidence of a UV fluence-dependent emission in XP cells in medium and in BSS. Overall, these results revealed an important difference between normal and XP cells and it was proposed that XP cells are unable to store ultraweak photons which are efficently trapped in normal cells and perhaps used to regulate metabolic activity [4]. In the same study, we found in defined stages of the fibroblast differentiation system, which has been recently described in detail by Bayreuther and co-workers [8,9], that UV-light elevates photon emission in MMC-induced postmitotic XP-fibroblasts at least by a factor of 2 compared to mitotic XP-cells. In order to investigate this finding further, in this report light-induced ultraweak photon re-emission (IPE) studies similar to those described previously [5-7] are performed in human skin cells and IPE processes are studied in mitotic and postmitotic stages of normal and DNA-repair-deficient XP fibroblasts [10].

3. Results and Discussion

In normal cells, we have found that white light-induced photon re-emission relaxation dynamics are identical after successive irradiations of approximated 1 min intervals, and even several cycles of illumination and measurement do not quantitatively change the re-emission intensity for several hours . The re-emission curves are hyperbolic as we have reported most recently with mouse melanoma cells [7]. These results confirms several previous investigations in plant and mammalian cells [1,3,5-7]. As interpreted by Li and Popp [11], the hyperbolic decay kinetics found in living systems after pre-illumination with white light may indicate coherent re-scattering of ultraweak photons due to collective excitation of nucleic acids within the DNA of the investigated fibroblasts.

Bayreuther and co-workers showed biochemical and morphological evidence for the fibroblast differentiation system in vitro. They showed that normal human skin fibroblasts in culture spontaneously differentiate along the cell lineage of mitotic (MF) and postmitotic fibroblasts (PMF). Additionally, they developed methods to shorten the transition period and to increase the frequency of distinct postmitotic cell types using physical agents such as ultraviolet light (UV) and mitomycin C (MMC). Figure 1 depicts total light-induced photon emission in distinct differentiation stages of normal fibroblasts (GM38). There is no discernible difference between untreated mitotic fibroblasts, MMC treated mitotic cells and MMC-induced postmitotic cultures three weeks following MMC-treatment.

Similiar results (not shown) were obtained for three other cell lines. However, a marked increase in IPE is detected in postmitotic XP-fibroblasts (CRL 1223) following the third exposure of white light as shown in Fig. 2.

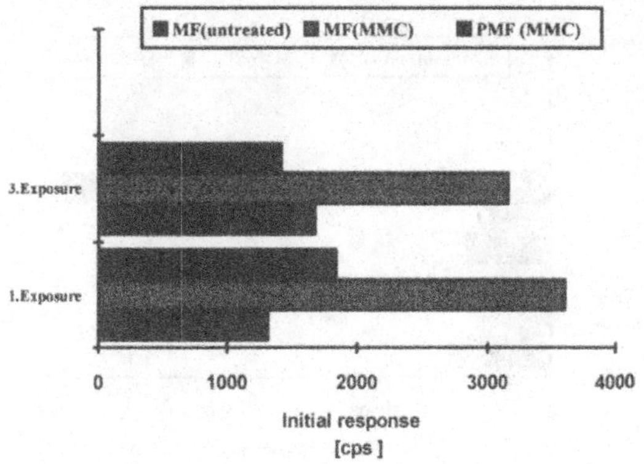

Figure 1. Total counts of photoinduced ultraweak photon emission registrated in the first second after illumination with white light during 30 s of normal human skin fibroblasts (GM38). Background values obtained with medium are subtracted.
For the experiments, medium was removed from the tissue culture flasks prior to irradiation, the cells were trypsinized and the action of trypsin was stopped by the addition of 50 ml of a 5% aqueous solution of soy-bean trypsin inhibitor to each flask. For the IPE determinations in mitotic (MF) and postmitotic fibroblasts (PMF), the contents of 20 and 100 flasks(surface, 75 cm^2; Gibco, Basel, Switzerland), respectively, were pooled ($3x10^7$ cells), centrifuged at approximately $1x10^3$ rpm for 3 min and resuspended in 10ml Earle's balanced salt solution (EBSS).

It has to be noted that the total light induced photon emission after the second light exposure is the same as shown for the first light treatment in Fig. 2. The increase in IPE obtained following the third exposure was even more pronounced after successive irradiation. To confirm these experiments further, other XP-cells (GM05509 A) were used in a subsequent study, showing similiar results although the IPE-burst was one illumination cycle retarded in these cells (data not shown). Therefore, it seems evident that MMC-induced postmitotic XP-fibroblasts cells tend to lose the capacity to store photons efficiently, confirming our recent report [4].

Because the previous published experiments were performed with UV-light, however, it was of interest to determine if this response could also be reproduced by the white-light source used in the present investigation. Tilbury discussed in the multi-author review of van Wijk [3] that the observation of ultraweak photon emission is for biochemists and biophysicist still on a very phenomenological level.

As he pointed out, most of these investigations were undertaken in microorganisms which have been shown to emit ultraweak photons and this very weak radiation has been detected in both the visible and ultraviolet region. Radiation in the visible region appears to be due to excited carbonyl groups and/or excited singlet oxygen dimers arising from lipid peroxidation, which in turn are associated with an increase in various

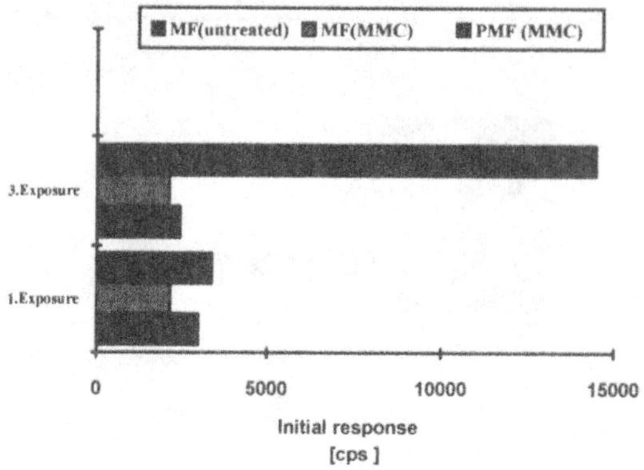

Figure 2. Total counts of photoinduced ultraweak photon emission registrated in the first second after illumination with whitelight during 30 s of repair deficient Xeroderma Pigmentosum human skin fibroblasts (CRL 1223). Background values obtained with medium are subtracted. Error bars shown represent standard deviations of two independent experiments with different cells. Experiments are peformed as described in Fig. 1.

reactive oxygen species such as the superoxide anion, hydrogen peroxide, hydroxyl radical and singlet oxygen.There is also substantial evidence for DNA playing a key role in these emissions [2,4,7,12]. This macromolecule may especially be involved in the emission of ultraweak photons by the cell in the UV-region of the spectrum. Very little work has been done with cells obtained from human sources [4]. The results presented herein show that human cells have an effective intracellular mechanism of photon trapping which MMC-induced postmitotic XP-cells do not possess. These postmitotic repair-deficient XP-cells are not capable of storing photons efficiently if they are subjected to repeated light exposure. In order to examine this phenomenon further, we extended our experiments to irradiation of cells with UV-light, similiar to studies previously reported, where the largest increase in ultraweak photon emission was measured in MMC-induced post-mitotic Xeroderma pigmentosum cells [4]. As depicted in Fig. 3, we were indeed able to increase this white-light induced burst of photons in MMC-induced postmitotic cells by irradiation with a fluence of 800 J/m^2 of artificial sunlight by a factor of at least 10^2.

The maximum photon release was reached at 5.5 h post UV-irradiation. Since this marked photon burst was observed by illuminating the UV-irradiated postmitotic XP-cells with white light, we looked for the wavelength region responsible for the induction. We found that the peak wavelengths responsible for this effect are between 360-420nm. It is well known that this region of UVA light induces oxidative stress in human skin and

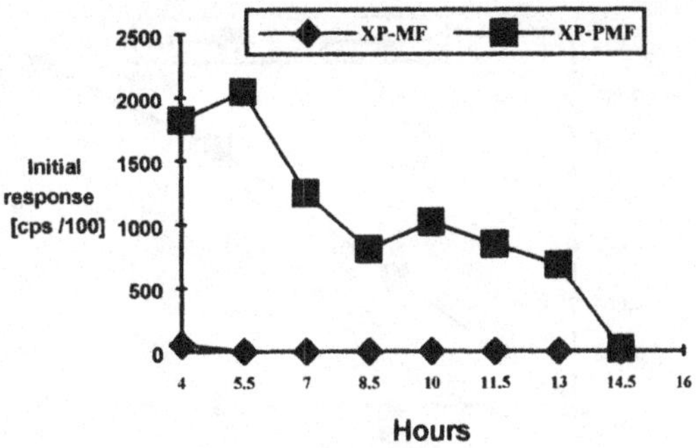

Figure 3. Initial counts in the measured time range of 10 ms of photoinduced ultraweak photon emission in repair deficient Xeroderma Pigmentosum human skin fibroblasts (CRL 1223) after 30s pre-illumination with white light. Experiments are performed at a cell densitiy of $8x10^6$ cells/ml and are described in detail by Niggli [4]. The high sensitivity ultraweakphoton counting system used is described in detail by Popp and co-workers [12].

it has been recently shown that heme oxygenase is a stress protein activated by these wavelengths [13]. The biological effectiveness of UVA radiation to induce erythema in skin is 1000 times less than that of an UVB-source. The same wavelenths of UVA-light also efficiently induces thymine photodimers as depicted in Fig. 4 by fluence response curves in the DNA of normal fibroblasts (CRL 1221). It is evident that thymine-thymine (T-T) dimers are produced in significant yields in the UVA region,although low fluences of this radiation do not produce measurable amounts of cytosine-thymine (C-T) dimers nor cell killing.

The long irradiation times and the need for irradiation of the cells in suspension does not allow us to determine the exact ratio of for C-T/T-T, which can be assumed from Tyrrell's work in bacteria of approximately 1:5 [14]. A fluence of $8x10^5$ Jm^{-2} shown in Fig. 4 is necessary to produce the same amount of T-T dimers as 1 Jm^{-2} of 254 nm light.

This same fluence of UVA radiation would produce approximately 60-90% cell killing in normal human fibroblasts whereas only 10-20% cell killing for this fluence of UVC radiation is found. Because UV-light induces ultraweak photons in repair deficient XP-cells more efficiently than in normal cells, pyrimidine photodimers may play an important role in the emission of ultraweak photons. Birks [15] suggested 20 years ago that the precursors of pyrimidine photodimers are the excimers of nucleic acids. As reported by Vigny and Duquesne in 1976 [16] such excimers represent an efficient excitation energy trap. From this latter observation, Popp and co-workers developed the genetic photon storage concept, showing indirect evidence for DNA playing a key role in the emission of ultraweak photons [12]. It is conceivable, therefore, that nucleic acids act as chromophores in the genetic material and one part may be involved via the UV

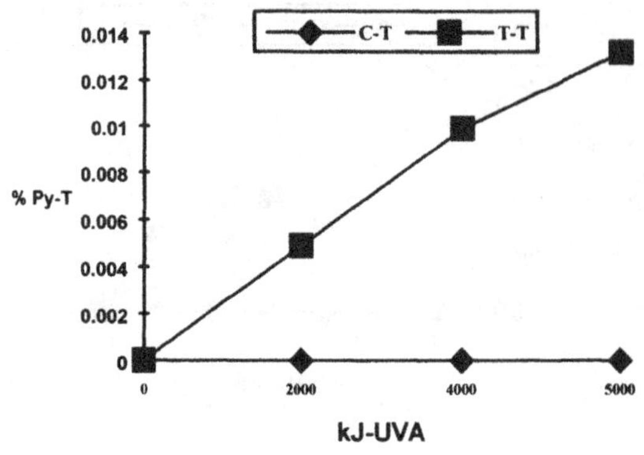

Figure 4. Formation of thymine-thymine (T-T)(I) and cytosine-thymine dimers (C-T) (m) in DNA of normal cells (CRL 1221) upon irradiation with UVA-light (365 nm) from a monochromator.
At 365 nm the slits were set at 5 mm and the light was filtered by a Corning 0-52 filter in order to remove scattered light of shorter wavelength. The fluence rate was 453 Wm^{-2} determined by a radiometer IL 770 A with a SEE 400 photodetector. The half band width at these settings is 17 nm. The cells were irradiated in 2 ml suspensions in PBS using a water-cooled quartz cuvette at approximately 4° C. Mean values of 2 experiments with standard deviation are given. Symbols with no error bars represent duplicate points with values within the dimensions of the symbol.
Determination of photodimers are performed by HPLC as described in detail by Niggli and Cerutti [20].

light-induced photochemical reaction of dimer formation in a storage process of light energy into DNA.
This mechanism would be similiar to the light-driven photoisomerization of the chromophore retinal discovered 25 years ago in bacteriorhodopsin [17] and subsequently confirmed in the visualisation pathway of the eye. This light-trapping system may influence metabolic and cellular events by triggering amplification mechanisms and promote photochemical processes even in the dark as proposed by Cilento in Popp's multi-author review [1]. Since the time that Cleaver found in 1968 [18] that Xeroderma pigmentosum patients which are extremly sensitive to sunlight, showed a defect in pyrimidine dimer excision, this observation has led to the present concept that the subsequent conversion of light energy into biochemical processes may be blocked in such cells. This speculation is supported by the finding that the rate of thymine dimer formation after UV-irradiation of MMC-induced postmitotic XP cells is slightly higher compared to mitotic XP-fibroblasts [9], thus making it possible, that successive irradiations of MMC-induced postmitotic XP-fibroblasts with white light, would lead finally to the limit of the light storing capacity of the DNA. In this respect, a

part of this energy may be released subsequently by these repair deficient cells as enhanced light emission where as normal cells are able to convert this light energy into biochemical processes via excision repair. As recently shown by Gilchrest and co-workers [19], UV-induced pyrimidine photodimers are stimulating melanogenesis confirming therefore that photodimers initiate biochemical processes and serve as molecular signals. In conclusion, our data show experimental evidence that the genetic material is involved in the ultraweak photon emission process. Furthermore, our results imply that the complex mechanism of UV-induced DNA damage and repair is a powerful light trapping system in DNA in order to convert light energy into biochemical signals.

4. References and Notes

1. Chwirot,W.B., Cilento, G., Gurwitsch, A.A., Inaba, H., Nagl, W., Popp, F.A., Li, K. H., Mei, W.P., Galle, M., Neurohr, R., Slawinski, J., Van Wijk, R., and Schamhart, D.H.J.: Multi-author review on Biophoton emission, *Experientia* **44** (1988), 543 -600.
2. Niggli, H.J.: Ultraweak photons emitted by cells: biophotons, *J. Photochem. Photobiol. B: Biol.* **14** (1992a), 144 -146.
3. Tilbury, H.J., Slawinski, J., Ezzahir, A., Godlewski, M., Kwiecinska, T., Rajfur, Z., Sitko, D., Wienzuchowska, D., Kochel, B., Gu, Q., Popp, F.A., Lilius, E.M., Marnila, P., Van Wijk, R., and Van Aken, J.M.: Multi-author review on biophoton emission, stress and disease, *Experientia* **48** (1992), 1029 -1102.
4. Niggli, H.J.: Artificial sunlight irradiation induces ultraweak photon emission in human skin fibroblasts, *J. Photochem. Photobiol. B: Biol.* **18** (1993), 281-285.
5. Scholz, W., Staszkiewicz, U., Popp, F.A., and Nagl, W.: Light-stimulated ultra-weak photon re-emission of human amnion and wish cells, *Cell Biophys.* **13** (1988), 55-63.
6. Van Wijk, R.V.,. Mei, W.P., Popp, F.A., and Van Aken, H.: Light-induced photon emission by mammalian cells, *Photochem. Photochem.B: Biol.* **18** (1993), 75-79.
7. Niggli H.J.: The cell nucleus of cultured melanoma cells as a source of ultraweak photon emission, *Naturwissenschaften* **83** (1996), 41-44.
8. Bayreuther, K., Rodemann, H.P., Hommel, R., Dittman, K., ,Albiez, M., and Francz, P.I.: Human skin fibroblasts in vitro diferentiate along a terminal cell lineage, *Proc. Natl. Acad. Sci. (U.S.A.)* **85** (1988), 5112-5116.
9. Niggli, H.J., Bayreuther, K.,. Rodemann, H.P., Röthlisberger, R., and Francz, P.I.: Mitomycin C-induced postmitotic fibroblasts retain the capacity to repair pyrimidine photodimers formed after UV-irradiation, *Mut. Res.* **219** (1989), 231-240.
10. Skin fibroblasts from normal individuals (GM38, GM 1717) and from a xeroderma pigmentosum patient of complementation group A (XPA), XP12BE (GM05509A) were obtained from the Human Genetic Mutant Cell Resository (Camden,NJ, USA). Normal cells (CRL 1221) and XPA cells derived from a 10-year old female (CRL 1223) were purchased from the American Type Culture Collec tion (Rockville, Maryland). The cells were grown as monolayers fetal calf serum and 100 U/ml penicillin/100 Eg/ml streptomycin as described in [4].
11. Li, K.H., and Popp, F.A.: non-exponential decay law of radiation systems with coherent rescattering, *Phys. Lett.* **93 A** (1989), 262 -266.
12. Popp, F.A., Li, K.H., and Gu, Q.: *Recent advances in Biophoton Research and its Applications*, World Scientific, Singapore,1992.
13. Niggli, H.J., and Applegate, L.A.: Glutathione response after UVA irradiation in mitotic and postmitotic human skin fibrolbasts and keratinocytes, *Photochem. Photbiol.* **65** (1997), 680-684.
14. Tyrrell, R.M.: Induction of pyrimidine dimers in bacterial DNA by 365 nm radiation, *Photochem. Photobiol.* **17** (1973), 69-73 .
15. Birks, J.B.: Excimer, *Rep. progr. Phys.* **38** (1975), 903 -974.

16. Vigny, P., and Duquesne, M.: On the fluorescence properties of nucleotides and polynucleotides at room temperature, in J., Birks (ed.) *Excited States of Biological Macromolecules* ,Wiley, London, 1976, pp. 167-177.
17. Oesterhelt, D., and Stoeckenius, W.: Rhodopsin-like protein from the purple membrane of Halobacterium halobium, *Nature* **233** (1971), 149-151.
18. Eller, M.S., Yaar, M., and Gilchrest, B.A.: DNA damage and melanogenesis, *Nature* **372** (1994), 413 - 414.
19. Cleaver: Defective repair of DNA in xeroderma pigmentosum, J.E , *Nature* **218** (1968), 652.-656.
20. Niggli, H.J., and Cerutti, P.A.: Cyclobutane-type pyrimidine photodimer formation and excision in human skin fibroblasts after irradiation with 313-nm ultraviolet light, *Biochemistry* **22** (1983), 1290-1295.
21. I would like to thank Elena Alborghetti, Corina Benkel, Anne-Lise Etter, Beate Hemmer and Wei-Ping Mei for excellent technical assistance. I am grateful to Urs Pfefferli for help with the graphics. Especially, I am indebted to Drs. Thomas Brown and Lee Laurent-Applegate and Prof. Martin Spiess for critical reading of the manuscript. This work was supported by Cosmital SA, a research company of Wella AG, Darmstadt (Germany).

THE PHOTON COUNT STATISTIC STUDY ON THE PHOTON EMISSION FROM BIOLOGICAL SYSTEM USING A NEW COINCIDENCE COUNTING SYSTEM

F. A. POPP[1] AND X. SHEN[2]

1.International Institute of Biophysics, D-41472 Neuss, Germany
2.Institute for Biophysics, Chinese Academy of Sciences, Bejing 100101, China

1. Introduction

Theoretical work [1,2] and some experimental evidences[3,4] suggest that either spontaneous or photon-induced photon emission from biological system is at least partial coherent. This was verified by statistical distribution of photocounts measured in a short time interval such as 50 to several hundred milliseconds. It has been firmly established by quantum optics that the photons from a fully coherent field obeys Poissonian distribution, but the photons from a mono-mode chaotic field are subjected to geometrical distribution [5,6.] In the photocounts statistics experiment performed with conventional counting system, two experimental conditions must be met in order to obtain the real statistic distribution of the photon counts from the source under investigation. First, a stationary or a quasi-stationary photon field is required. However, this requirement is hardly satisfied in most circumstances due to the continuos metabolic changes in living system and slowly decayed fluorescence. Second, the noise of the counting equipment should be much lower than the net photon counts from living organism in order to prevent the measured photon counts statistics from distortion caused by the noise. This requirement is even more difficult to be fulfilled in the experimental study on living systems, since the spontaneous photon emission from living organisms, such as cell suspension and tissues, is rather weak and is often not one order of magnitude higher than noise. In order to overcome the above difficulties, a new coincidence counting method has been developed. The new system and the coincidence counting method can be used to determine the photon counts statistics of a non-stationary photon field with an ever wanted high accuracy.

2. Principle of the Coincidence Counting Method in Determining the Photon Counts Statistics

J. J. Chang et al. (eds.), Biophotons, 87-92.

Consider two photon sources X_1 and X_2. Channel 1 measures the photons from X_1, and channel 2 those from X_2. The coincidence counting system (CCS) is constructed in such a way that a photon, which is counted in channel, is registered as a coincident one as soon as at least one other photon has been counted in channel 2 in a time interval Δt after the photon counting happens in channel 1(see Fig.1).

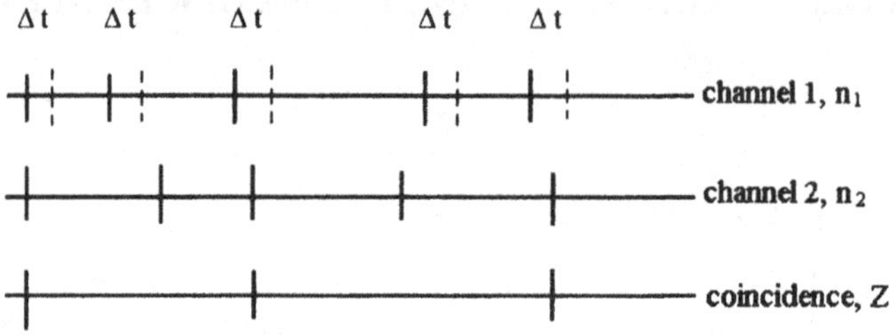

Fig.1. The correlation of coincident counts with the counts in two channels.

The number of random coincidences, Z_j, in the j-th time interval ΔT_j is then:

$$Z_j = n_{1j}\, P_2(\Delta t,\, n_{2,\Delta t} \geq 1) \tag{1}$$

where n_{1j} is the number of counts in channel 1 within the j-th time interval ΔT_j, and $P_2(\Delta t, n_{2,\Delta t} \geq 1)$ is the probability of counting at least one photon in channel 2 in the time interval (or coincidence time) Δt.

$P_2(\Delta t, n_{2,\Delta t} \geq 1) = 1 - P_2(\Delta t, 0)$, where $P_2(\Delta t, 0)$ is the probability of counting no photon in the time interval Δt in the channel 2. From (1) we get:

$$Z_j = n_{1j}\,[\,1 - P_2(\Delta t,\, 0)] \tag{2}$$

Since the CCS allows to measure as well Z_j as n_{1j}, we can determine $P_2(\Delta t, 0)$ of the photon source X_2.

Let us compare a fully coherent and a completely chaotic photon source. The photon count statistics of a fully coherent source follows a Poissonian distribution, while that of a completely chaotic field is subject of a geometrical distribution. This means that the probability $P(\Delta t, n)$ of counting n photons in a time interval Δt is for a coherent field:

$$P(\Delta t,\, n)_{coherent} = \frac{<n>^n}{n!} e^{-<n>} \tag{3}$$

and for a chaotic field:

$$P(\Delta t,\, n)_{chaotic} = \frac{<n>^n}{(1+<n>)^{n+1}} \tag{4}$$

It is obvious that we can fairly well distinguish a coherent from a chaotic field, since two completely different probabilities for n≥0 are obtained:

$$P(\Delta t, 0) = e^{-<\dot{n}>\cdot\Delta t}$$

$$P(\Delta t, 0) = \frac{1}{1+<\dot{n}>\cdot\Delta t}$$

where $<\dot{n}>$ is the expected value of the photon count rate. This leads obviously to different values of the coincidence, Z_j:

$$Z_{j\,coherent} = n_{1j}\left(1-e^{-<\dot{n}_2>\Delta t}\right) \tag{5}$$

$$Z_{j\,chaotic} = \frac{n_{1j} <\dot{n}_2> \Delta t}{1+<\dot{n}_2>\Delta t} \tag{6}$$

where $<\dot{n}_2>$ is the expected value of the photocount rate in channel 2. Equations (5) and (6) can also be rewritten as:

$$\frac{Z_{j\,coherent}}{n_{1j}} = 1-e^{-<\dot{n}_2>\Delta t} \tag{7}$$

$$\frac{Z_{j\,chaotic}}{n_{1j}} = \frac{<\dot{n}_2>\Delta t}{1+<\dot{n}_2>\Delta t} \tag{8}$$

If N measurements on the photon counts in channel 1, channel 2 and corresponding coincident counts were made with time interval ΔT, and the Z_j/n_j (j=1, 2, ⋯,n) was plotted against the $(n_{2j}/\Delta T)\cdot\Delta t$, the data points will be very close to the curve $1-e^{-<\dot{n}_2>\Delta t}$ for the coherent field, but to the curve, $<\dot{n}_2>\Delta t/(1+<\dot{n}_2>\Delta t)$ for the chaotic field.

The data points from a partial coherent photon field will appear between these two curves in such a plot. From the above presented formulas, it should be noticed that there is no limitation regarding how stationary the photon field is. The equation (7) and (8) provide an extremely powerful means to determine the photon count statistics of a photon source by registering the photon counts in two channels and their coincidence.

3. Experimental Measurements on the Photon Count Statistics of the Photon Emission from Mungbean Seedlings and Leaflets of an Elder Bush

The mungbean seedlings with a length of about 2 cm (two days after germination) and the leaflets from the elder bush were used as the investigated living organisms. 20

mungbean seedlings or one leaflet were placed in each of the two quartz cuvettes (24 x 24 x 40 mm^3), then the two cuvettes were placed in two dark chambers corresponding to the two photon counting channels after exposing to daylight for ten seconds. The counting was immediately started as soon as the chambers were closed in order to get a fast decayed photon emission from the seedlings or the leaflets. The kinetics of the photon counts in channel 1, channel 2 and their coincidence registered for mungbean seedlings and leaflets are shown in Fig.2 and Fig.3 respectively. The time interval for

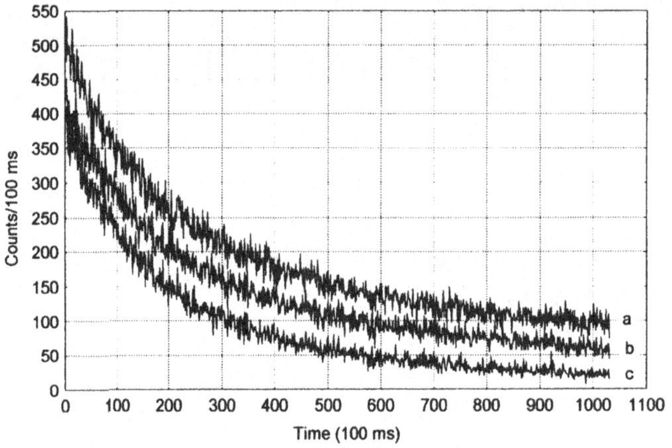

Fig.2. The decay of the photon emission from Mungbean seedlings after exposing to sunlight. a: counts in channel 2; b: counts in channel 1; c: coincident counts.

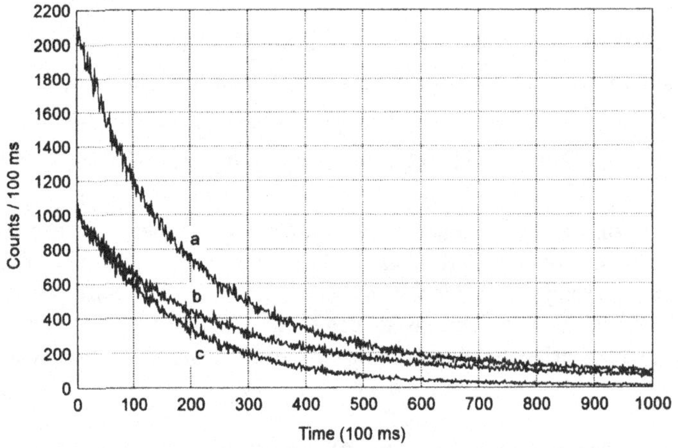

Fig.3. The decay of the photon emission fromleaflet of elder bush after exposing to sunlight. a: counts in channel 2; b: counts in channel 1; c: coincident counts.

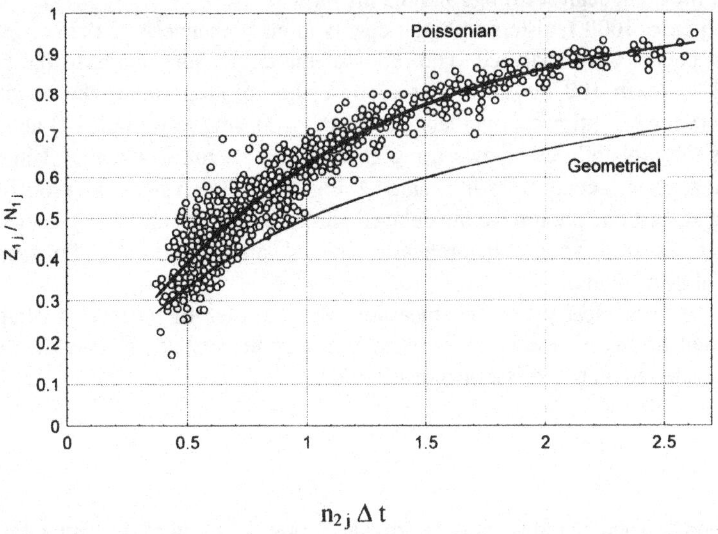

Fig.4. The probabilities of measuring at least one photon from the mungbean seedlings after exposing to sunlight. The time interval for photon counting and the coincidence time are 100 ms and 0,5 ms respectively. 1000 measured counts are analyzed.

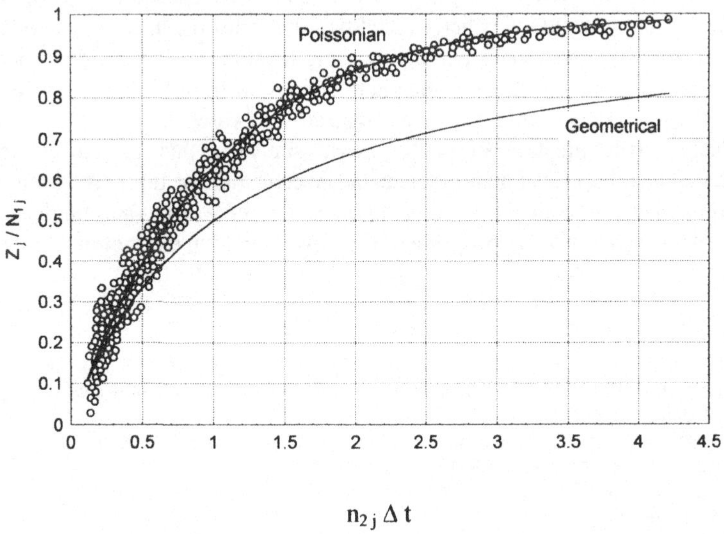

Fig.5. The probabilities of measuring at least one photon from the leaflet of elder bush after exposing to sunlight. The time interval for photon cointing and the coincidence time are 100 ms and 0.2 ms respectivety.

measuring the photon counts in each channel was 100 ms, and the coincidence time was 0.5 ms for the mungbean seedlings and 0.2 ms for leaflets.

In each case, 1000 registered photon counts in each channel and the corresponding coincident counts were analyzed. The ratios of the coincident counts to the counts in channel 1 for each 100 ms time interval were then plotted versus the count rate in channel 2 (i.e. $n_2/\Delta T$) times coincidence time Δt. According to formula (7) and (8), this plot gives the probabilities of measuring at least one photon in channel 2 in the time interval Δt at various count rates in channel 2. Fig. 4 and Fig. 5 show the probabilities of measuring at least one photon for the delayed photon emission from mungbean seedlings and leaflet, as well as the theoretical-calculated probabilities for Poissonian and geometrical distribution.

It can be seen clearly that the measured probabilities of measuring at least one photon from the two investigated living systems are in very good agreement with the probabilities based on the Poissonian distribution.

4. Discussions

The new coincidence counting method provides a powerful tool to determine the photon count statistics for a photon field, in particular for the photon emission from biological systems. Since there is no requirement for a stationary field, this method can be used to determine the photon count statistics for the delayed luminescence or the photon-induced photon emission from living systems. In comparison with the spontaneous photon emission from biological systems, the photon-induced photon emission can be several orders of magnitude higher. It endows this new method with another great advantage over the conventional photon counting method, in that the distortion of the measured photon count statistics for the photon emission from biological system by the unavoidable noise of the counting system can be actually ruled out.

Besides the determination of photon count statistics, this coincidence counting technique can also be used for the systematic investigation of the correlation between two photon sources, in particular the biocommunication between two biological systems. It is our hope to have further application of this new method in biophoton research.

5. References

1. F.A. Popp and K.H. Li, Hyperbolic relaxation as a sufficient condition of a fully coherent ergotic field, *Int. J. Theoretical Phys.* **32** (1993), 1573-1583.
2. F.A. Popp, Q. Gu and K.H. Li, Biophoton emission : Experimental background and theoretical approaches, *Modern Phys. Lett. B* **8** (1994), 1269-1296.
3. F.A. Popp, K.H. Li, W.P. Mei, M. Galle and R. Neurohr, Physical aspects of biophotons. *Experientia* **44** (1988), 576-585.
4. X. Shen, F. Liu and Xy. Liu, Experimental study on photon count statistics of the ultraweak photon emission from some living organisms. *Experientia* **49** (1993), 291-295.
5. F.T. Arecchi, Photon count distribution and field statistics, in R.J. Glauber (ed.) *Quantum Optics*, Academic Press, New York, pp.57-110, 1969.
6. J. Perina, *Coherence of Light*, Reidel Publ. Co, Dortecht, 1985.

WEAK PHOTON EMISSION OF NON-LINEAR CHEMICAL REACTIONS OF AMINO ACIDS AND SUGARS IN AQUEOUS SOLUTIONS

Evidence for Self-Organizing Chain Processes with Delayed Branching

V.L. VOEIKOV[#+] AND V.I. NALETOV[#]

[#]*Bioorganic Chemistry Department, Faculty of Biology, Lomonosov Moscow State University, Moscow, 119899, Russia;*
[+]*International Institute of Biophysics e.V., Neuss, German.*

1. Background

1.1. THE PROBLEM MITOGENETIC RAYS ORIGIN

1.1.1. *On the Reality of Mitogenetic Rays*
Ability of many living systems to emit extra-weak UV-light capable to induce cell divisions ("mitogenetic radiation," MGR) was discovered by A.G. Gurwitsch (1923). In 1920s-1930s the existence of MGR was confirmed by a large number of workers in laboratories throughout the world. There are also reports in recent literature on the ability of living cells to emit UV-photons (Slawinsky, 1988; Quickenden and Tilbury, 1991), on their ability to stimulate cellular divisions (Grasso, *et al.*, 1991; Beloussov, Baskakov, 1995) and to produce other physiological effects (Galantsev, *et al.*, 1993). However, there is no agreement on the mechanisms of processes in living systems leading to high energy photon emission especially in the UV-range (Slawinsky, 1988; Popp, *et al.*, 1994; Beloussov, *et al.*, 1997; Vladimirov, 1997). On the other hand, seemingly simple *in vitro* chemical systems are also producing UV-photons. This phenomenon attracted less attention than low-level luminescence from biological systems, whereas its analysis may be helpful for understanding the mechanisms of electron excited states generation in the latter.

1.1.2. *Glycin Oxidation as a Source of MGR*
Among other chemical systems in which low-level chemiluminescence (CL) is observed, the one discovered by A. and L. Gurwitsch (1938) deserves special attention. They reported that 15-30 min after a brief (few seconds) irradiation of aqueous glycin (Gly) solution with strongly attenuated source of UV-photons (mercury lamp) the solution becomes a long-lasting source of MGR. MGR emission is more intense if the solution is kept in diffuse daylight and completely disappears under anaerobic conditions. It was suggested that UV irradiation induces an autocatalytic reaction of oxidative deamination of an amino acid. Some indications were obtained that the catalysis was performed by a polypeptide-like substance emerging in the solution after its irradiation and staying in it

J. J. Chang et al. (eds.), Biophotons, 93-108.

at a very low stationary level. When an aliquot of the emitting solution was transferred into a test-tube with a fresh portion of Gly solution, the latter turned also into a MGR emitter. This operation could be repeated many times. Thus, the process induced in this system was extremely unusual. On the one hand, free energy of the system was increasing (a reaction system becomes a UV-photon emitter). On the other hand, the process of self-organization, manifested in the emergence of complex structures with catalytic function, took place in this system.

This discovery "passed without comment by other workers in this field" (Gurwitsch, 1946), as it sharply contrasted with theoretical concepts that had been dominating at this time and are still dominating now. It is impossible to explain basing on these concepts how irradiation with several dozens of UV-photons of highly transparent Gly solution could induce in this system such a reaction. Gurwitsch, of course, could not give a comprehensive explanation of mechanisms of such processes. Nevertheless, he was the first to understand that both this reaction and a lot of effects induced in living systems by UV-photons should be considered as specific manifestations of branched chain processes (Gurwitsch, Gurwitsch, 1942).

1.2. PECULIARITIES OF BRANCHED CHAIN PROCESSES

1.2.1. *General Features of Branched Chain Reactions*

Illumination of mixture of H_2 and Cl_2 with a flash of light (λ<478,5 nm) results in HCl synthesis. Max Bodenstein (1913) first noted that the quantity of HCl molecules produced up to 10^5 times exceeded the quantity of photons absorbed. Such a high quantum yield of the reaction is explained by a model according to which photon absorption causes Cl_2 dissociation into atoms. Latter start chains of intermittent reactions: $Cl + H_2 = HCl + H$; $H + Cl_2 = HCl + Cl$; $Cl + H_2 = HCl + H$..., ets. Chain propagates until two homologous atoms recombine or disappear, for example, by interaction with a wall of the reaction vessel. When light is switched off, generation of new chains is halted and the reaction fades. Processes of this type are referred to as linear chain reactions (LCR)

Branched chain reactions (BCR) were discovered at the end of 1920s by Nikolai Semyonov (Semyonov, 1935) and Cyrill Hinshelwood (Hinshelwood, 1940). Both of them became in 1956 Nobel Prize winner for this discovery. BCR fundamentally differ from LCR in that their rate is accelerated after the completion of the initial stimulus action. However, acceleration of the chemical process is also observed in autocatalytic reactions where the reaction rate growth is provided by autocatalysis by stable final products. In BCR the mechanism of acceleration is completely different. BCR may be initiated by an introduction of a vanishingly small number of active species (free radicals, atoms) into a reaction mixture or emergence of such species in it, for example, due to photodissociation. Reaction rate accelerates in these processes according to the exponential law due to multiplication of active centers which may eventually reach high stationary concentrations in totally branched chain reactions (Semyonov, 1986, p. 459). However, these intermediate products completely disappear from the reaction mixture when the process comes to its end. The final result of BCR is the same as for any other chemical and physical process proceeding in closed systems: absolute reduction of free energy and entropy growth in the system. However, the essence of BCR is manifested in its dynamics rather than in its overall balance.

Several specific features distinguish BCR from "normal" chemical reactions (Semyonov, 1986, p.376-381). First, similar to LCR the quantum yield in BCR is very high. Second, BCR development may be often registered only some time after the completion of the initiating stimulus action (induction period). After the end of the induction period initial rate of BCR increases according to the exponential law. Third, critical conditions are characteristic for BCR: temperature threshold, critical volume of the reaction mixture or critical ratio of the reaction mixture volume to the surface of the reaction vessel, critical concentrations of reagents. The reaction proceeds at a very low rate below and above the threshold values of the critical parameters. If critical parameters change in already developing BCR, reaction rate may decrease and the process may be even halted. The fourth specific feature of BCR is a very strong accelerating or rate-retarding effect of minute admixtures in the reaction mixture. Fifth, strong deviation of kinetics from classical laws of chemical kinetics -- Arrhenius temperature law and the law of mass action is observed at certain stages of BCR development.

Totally branched chain reactions in gaseous phase usually develop as explosions. However a lot of reactions with organic compounds, that meet many criteria of BCR take sometimes many minutes or even hours to reach maximal rate. Semyonov suggested that these reactions proceed as LCR, but during the major chain propagation intermediate metastable products emerge. Much lower energy of activation is needed to produce new active centers from these products than from the initial substances. Thus, even if the "parent" chain is eliminated, the new chains arise in the system provided that enough energy for the activation of these metastable products is available. Such reactions are named "degenerate-branched chain reactions," though referring to them as to "Chain Reactions with Delayed Branching" (CRDB) seems to be more appropriate (Semyonov, 1986, p. 488).

1.2.2. *Energy Self-pumping Phenomenon in BCR*

Practically all CRDB are reactions of oxidation of organic compounds. The peculiar property of O_2 is that it is a diradical molecule in its ground state due to the presence of two unpaired electrons each occupying one of the two antibonding π^*-orbitals. Thus, dioxygen in its "usual" state is a triplet, high potential energy molecule, though very high energy of activation is needed to make its energy free. Oxidation of an organic compound RH is initiated with one-electron reduction of dioxygen. Two radicals then appear: $HO_2\bullet$ and a radical of the organic compound, R• . The latter is easily (zero energy of activation) oxidized by dioxygen to a peroxide radical ROO•. The chain propagates due to ROO• interaction with RH again producing R• and hydroperoxide ROOH. Hydroperoxides are metastable compounds and eventually dissociate giving two new active centers, RO• and OH•, that may initiate two new chains (Emanuel, *et al.*, 1985).

Thus, oxygen activation with few or even a single quantum of energy may be enough to overcome the energy barrier of the first reaction. Following it chain branching and elevation of free energy in the system occur. It also follows from this model, that electron excited species should arise in the reaction system. Initial linear chains brake when peroxide radicals ROO• dismutate. Products of their dismutation are highly unstable tetroxides [R-O-O-O-O-R]. They disintegrate producing carbonyl compounds in a triplet state or/and singlet oxygen -- an electron excited molecule (Vassiliev, 1985). Triplet excited carbonyls arise also from the fragmentation of dioxetane/dioxetanon compounds (internal peroxides) (Adam, Cilento, 1983). The latter may be produced from aldehydes that are used as substrates or often emerge in the course of CRDB. Development of CRDB is inevitably followed by increase of "concentration" of species in electron excited triplet states in the system. This effect is not considered as significant for the reaction dynamics in current chemical literature, although progressive generation of such species in the reaction system may be regarded as elevation of the level of its free energy, as self-pumping of the system (Baskakov, Voeikov, 1996b). Emergence and propagation of electron excitation in the course of chain reactions may allow to explain a lot of their "oddities."

BCR and CRDB are usually considered as uncontrolled explosions. However, elements of dynamic orderliness may be observed even in oxidative reactions in gaseous phases. For example, during propane oxidation under the conditions of "cold flame" initial flashing of the reaction mixture is followed with several minor flashes (Asmore, 1966, p. 450). Rate of ClO_2 decomposition in the presence of O_2 also changes in oscillatory manner. Depending upon the *material* of the reaction vessel, its *geometry* the reaction after the initial oscillation may either proceed at a stationary rate for a long time or new oscillations with increasing amplitude emerge (Semyonov, 1986, p. 335).

In CRDB of methane oxidation formaldehyde (H_2CO) is produced as an intermediate product giving start to new chains. This reaction behaves in a very unusual manner. Stationary concentration of H_2CO in its course is kept at surprisingly stable level until 90% of methane is consumed. Only then H_2CO concentration abruptly drops. Besides, if H_2CO is initially added to the reaction mixture in excessive concentrations its concentration rapidly decreases to the same stationary level as it reaches in case when H_2CO is produced only in the course of the reaction. "It is such a striking fact, that its disclosure should undoubtedly give us essential new knowledge in the field of kinetics," -- writes Semyonov in 1969 (1986, p. 496).

Paradoxically, but in 1934 Semyonov had already outlined the way towards understanding a strange behavior of BCR. As noted above, at the stage of multiplication of active centers, spontaneous growth of free energy in the system occurs due to the liberation of the potential energy stored in oxygen in the course of oxidative processes. Experiments demonstrate, that free energy may build up to extremely high levels of electron excited states. For example, when phosphorus is oxidized by oxygen at low partial pressures of both reagents, the temperature of the reaction mixture is only few degrees above the ambient one. Nevertheless, under these conditions "cold flame" is observed, and significant flux of UV-photons ($\lambda < 280$ nm) from it can be registered (Oullet, 1933). Slow electrons passed through the zone of this reaction are accelerated up to 10 eV! (Semyonov, 1986, p.447). As Semyonov points out, these phenomena indicate of the simultaneous existence of two Maxwell-Boltzmann distributions in a system, where BCR takes place. One of them is characteristic for the kinetic energy distribution of molecular species comprising the bulk of the reaction system. Another one is valid for active species, leading the process. The second distribution arise, if "the energy of electronic excitation of particles in the reaction system does not readily dissipate in the vibration and kinetic modes. If constant supply of the energy of excitation such as light quanta is provided, the Maxwell-Boltzmann-like distribution is established between different electron excitation levels with "temperatures" many thousands degrees exceeding the temperature of the gas. The less is fluorescence, the more retarded is the transformation of electron excitation into another forms of energy, the higher is the "temperature" of electron excitation. Naturally, under such conditions electron excitation may reach levels greatly exceeding those corresponding to the energy of the initially absorbed quantum" (citation from: Semyonov, 1986, p. 384). Semyonov predicted also, that negative and positive feedback may take place under such conditions in BCR due to coupling between different elementary stages and chains.

Semyonov stressed out that at a close view the overwhelming majority of chemical and *other* processes, may be regarded to as branched chain processes. He proposed that "the course of *processes* in general, and of chemical processes in particular should obey some most general principles, as fundamental as the second law of thermodynamics to which different *equilibrium states* obey." However, several decades of investigation of BCR and CRDB in gaseous and organic liquid phases did not bring us more close to the comprehension of these principles. Probably studies of "odd" chemical reactions taking place in aqueous media, similar to those that were first described by A. Gurwitsch may shed more light on principles defining the behavior of such processes.

2. Induced Chain Processes with Delayed Branching in Amino Acid Solutions: Electron Excited States and Self-Organization

2.1. CHEMILUMINESCENCE IN AQUEOUS AMINO ACID SOLUTIONS IN THE PRESENCE OF HYDROGEN PEROXIDE

The unusual reaction developing in an amino acid solution after its UV-irradiation discovered by Gurwitsch was neglected probably because all information about it was obtained with a controversial biological test -- MGR-analysis. Much later Anna Gurwitsch *et al.* managed to register UV-photon emission from this reaction system with a photomultiplier. However, light intensity was so low (around 1 photon/cm^2/sec) that it was impossible to study kinetic and other properties of the process (Gurwitsch, A.A., *et al.*, 1987). On the other hand, they demonstrated that CL develops and lasts for several hours in Gly solution after addition of H_2O_2 to it. Contribution of UV-radiation in total luminescence was increasing up to 20-50% in the course of the process (Gurwitsch A.A. *et al.*, 1965).

Photon emission from oxidative reactions is currently explained as an *immediate* result of very rare acts of radical recombination, singlet oxygen and its dimers (excimers) fluorescence and triplet carbonyl phosphorescence (Vladimirov, 1966, Cadenas, 1984, Vassiliev, 1985). This model does not explain situations in which contribution of UV photons in the total light emission may become so high. On the other hand, if self-organization during the development of CRDB takes place in the system, electron excitation of the participating species may indeed reach very high levels . To examine this possibility we performed macrokinetic analysis of the process emerging in solutions of non-aromatic amino acids after H_2O_2 addition using the chemiluminescent method.

To augment signal/noise ratio we added different luminophores (ethidium bromide, Cascade Blue, tyrosine, tryptophane and some others) to the reaction mixture in micromolar range of concentrations. Concentrations of substrate amino acids varied in the range of tens millimilar (Asn, Asp, Gln, Gly, Ala) to decimals molar (Gly). Milli-Q or distilled-deionized water was used as a solvent. All the reagents used were analytical or higher purity grade. CL was registered using a single photon counter with a cooled PMT or liquid scintillation spectrometer Mark II (Nuclear-Chicago) in the coincident count or single photon counting modes depending on the intensity of CL.

Though initial concentration of H_2O_2 was rather high (0.4-0.6 M), initial CL just after addition of H_2O_2 to amino acid solution was low. Duration of the induction period varied from several to tens of minutes depending on minor (submicromolar) admixtures of metal ions in reagents. After the completion of the induction period CL intensity began to increase exponentially (Fig. 1). Count rate accumulation perfectly correlated with accumulation of ammonia in the solution indicating that CL reported of amino acid oxidation. Surprisingly, amino acid oxidation was not followed with H_2O_2 consumption. On the contrary, H_2O_2 concentration increased 5-15% at the stage of exponential CL growth and began to decrease only when photon count rate acceleration slowed (Voeikov, *et al.*, 1996). Thus, amino acid oxidation could not be explained by its trivial "burning" with reactive oxygen species arising from H_2O_2 decomposition. Rather, it resulted from some other process related to the emergence of species excited to such high levels that dissociation of dioxygen into atoms oxidizing water to H_2O_2 could become possible.

Induction of CL growth depended upon initial conditions in a strongly non-linear manner (Voeikov, Baskakov, 1995). At Asn concentration below 25 mM no CL developed in solution even 16 hours after H_2O_2 addition, while at Asn concentrations above it typical CL response was observed (Fig. 1). Similar results were obtained with other amino acids. Dependence of CL kinetics upon the reaction volume at other identical conditions is also strongly non-linear (Fig. 2). First, the duration of the induction period in a 3 ml sample is several times longer than in the samples of larger volumes. Second, integral of counts accumulated in the course of the reaction in a 6 ml sample exceeds that in a 3 ml sample 9-fold rather 2-fold that should be expected from the "normal" chemical reaction.

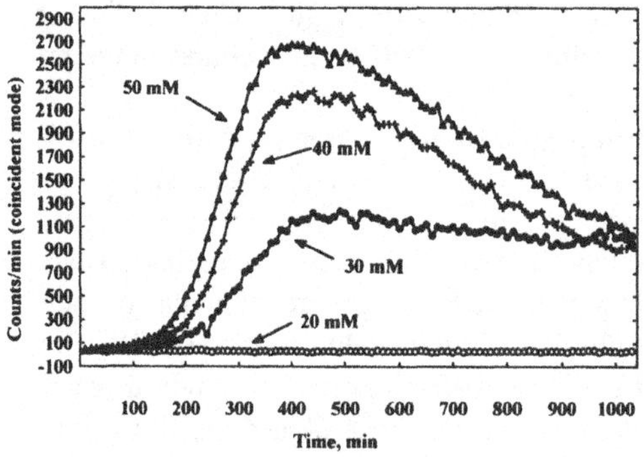

Figure 1. Dependence of the reaction kinetics (0.4 M H_2O_2, 20 mkM ethidium bromide, 6 ml) on Asn concentration.

Dependence of the reaction parameters on the reaction volume (to be more exact, on the surface/volume ratio) had so far been demonstrated for BCR in gaseous phase at low pressures of reagents and in a solid body ("critical mass" for the chain reaction of nuclear disintegration). It is considered to be one of the major arguments in favor of branched chain mechanism of the process. Such dependence is explained by active particles scavenging on the reactor walls or by their missing from the reaction volume, so that the multiplication coefficient can not overcome the threshold value (>1). This mechanism may play some role here, though, as it will be demonstrated in the next section, it alone can not explain the phenomenon of volume dependence of CRDB parameters in amino acid aqueous solutions.

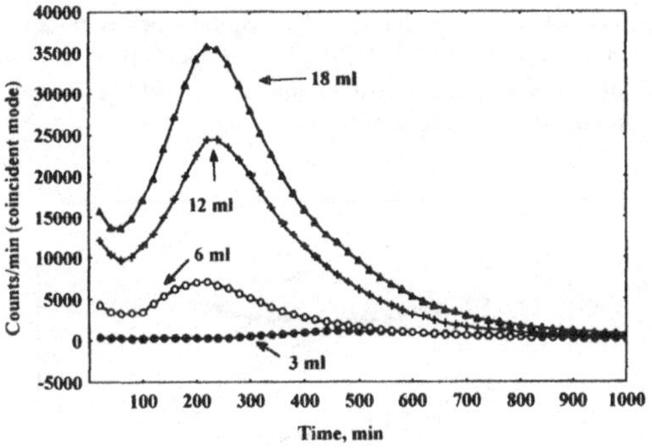

Figure 2. Dependence of the reaction kinetics (100 mM Asn, 0.4 M H_2O_2, 20 mkM ethidium bromide) on the reaction volume.

Deceleration of branched chain reactions rate is usually explained by consuming substrates. In the case of reactions studied here deceleration of count rate after CL has reached maximal intensity can not be explained in such a way. Chemical analysis of the reaction mixture has showm that less then 5% of amino acid and H_2O_2 initially present in the solution was consumed by this time. Neither it could be explained solely by luminophore decomposition, because sometimes secondary "waves" of CL were observed, and, on the other hand, addition of fresh portion of luminophore did not influence CL kinetics. Inhibition and reversal of CL growth could be associated with the arousal of some inhibitory reaction products.

Essential non-linearity of the reaction kinetic parameters on temperature were also revealed (Baskakov, Voeikov, 1995). When temperature was elevated from 20-30 °C to 48 °C at the stage of chains branching apparent energy of activation was increasing from almost zero up to 80-90 kkal/mole. These are enormously high values for reactions of slow oxidation. On cooling the solution the intensity of CL continued to rise for a certain time (negative apparent activation energy!) (Fig. 3). Temperature dependence of the reaction rate was nearly mirror opposite at the stage of its deceleration (Fig. 4).

Resistance of the reaction rate to temperature changes indicates that the process is not primarily driven by the kinetic energy of reactants. Elevation of apparent energy of activation to extremely high values at the initial stage of the process agrees with Semyonov's speculation of the emergence and building up of a stable distribution of electron excited states of species participating in the process.

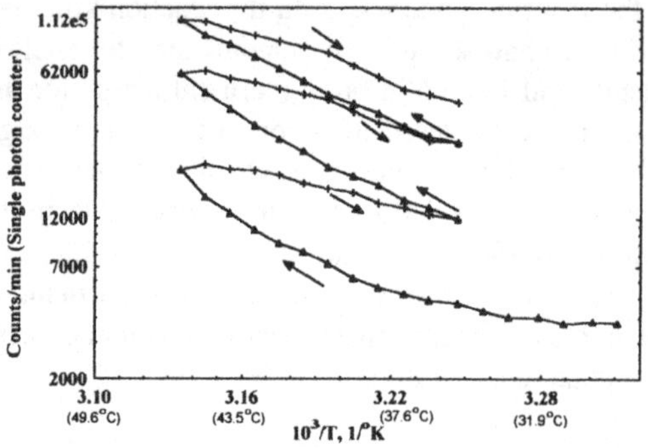

Figure 3. Temperature dependence of the reaction kinetics in Asp solution (50 mM) in the presence of 0.4 M H_2O_2 and 50 mkM ethidium bromide at the stage of rate acceleration. Arrows indicate heating or cooling of the reaction mixture at a rate 1 °C/2.5 min.

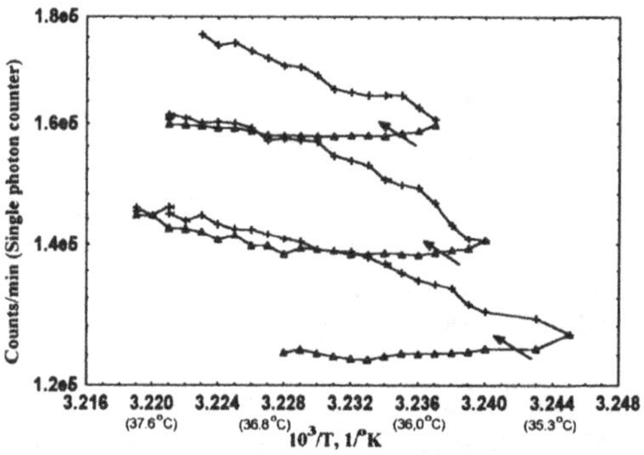

Figure 4. Temperature dependence of the reaction kinetics (same preparation as in Fig. 3) at the stage of rate deceleration.

On the other hand, strong hysteresis at points of temperature reversions indicate of strong cooperative effects emerging in the reaction system. Cooperativity in a seemingly simple solution may be provided only by long-distant interactions among the participating species.

As noted above, A.A. Gurwitsch *et al.* claimed that a significant part of emission from Gly solution in the presence of H_2O_2 was in the UV range. In our experiments practically all light emission was in the visible part of the spectrum, because it was provided by the excitation of luminofores added to the reaction mixture. Among the luminofores used tryptophane and tyrosine could be excited only by UV-light (for example, $\lambda_{excit.}$ for Tyr is 274 nm). Light emission in their presence reached hundreds of thousands CPM (in the mode of single photon counting). Notably, that maxims of emission from reaction systems in the presence of indicative concentrations (10-100 mkM) of Tyr or Trp were close to their spectra of phosphorescence, rather then fluorescence (Voeikov, *et al.* 1996).

Thus, all the features of the process initiated in amino acid solutions by H_2O_2 are characteristic for chain reactions with delayed branching. The results also strongly suggest, that after the reaction initiation products in electronically excited states appear and that non-radiant energy transfer on luminofore takes place. Many features of this model process, in particular, its temperature dependence, imply that dynamic self-ordering occurs in this process that may provide stability of its progress and resistance to external interference. However, this reaction takes place under rather artificial conditions (presence of luminofore, high H_2O_2 concentration) that are far from physiological. Nonetheless, similar regularities were revealed by us in a well-known and biologically significant Maillard reaction.

2.2. NON-LINEAR PARAMETERS OF NON-ENZYMATIC AMINO-CARBONYL REACTIONS BETWEEN GLUCOSE AND GLYCIN

Nonenzymatic glycation of free or peptide bound amino acids, also known as Maillard reaction or browning reaction, plays an important role in aging, pathogenesis of different diseases correlated to hyperglycemia (diabetes, atherosclerosis, Alzheimer disease, uremia, cataractogenesis, etc.), and also in food processing and storage (Namiki, 1988). Products at the early steps of the reaction have been relatively well characterized (Dyer, *et al.*, 1991). Sugars initially react in their open chain aldehydo or keto forms with deprotonated amine group of an amino acid through Shiff base condensation. The resulting aldimine or ketimine products then undergo Amadori rearrangements to give ketoamine Amadori products. The latter stages of Maillard reaction are characterized by a wide spectrum of reactions processing in different combinations resulting in formation of colored and fluorescent products (Yaylayan, *et al.*, 1994; Telegina, Davidiantz, 1995). Many of late products represent high molecular weight substances arising due to reactions of polycondensation, in particular, to polymers cross-linking.

Recently, weak CL has been revealed during the early stages of Maillard reaction (Kurosaki, *et al.*, 1989; Namiki, *et al.*, 1993, Wondrak, *et al.*, 1995). It has been suggested that light emission originates from oxygen-dependent generation of exited states and energy transfer to simultaneously formed fluorescent products of the browning reaction (Wondrak, *et. al.*, 1995). It was observed, that CL was prominent at higher temperatures, and in most cases CL was measured under heating at above 60 °C and was undetectable in the reaction at room temperature (Wondrak, *et al.*, 1995). However, if to speculate, that Maillard reaction belongs to the class of CRDB, heating may play only initiating role and is not needed for further development of the process. Recently we tried to evaluate this hypothesis and obtained evidence in its favor (Voeikov, Naletov, 1997).

Browning (λ_{max} = 305 nm) and blue fluorescence ($\lambda_{exc.}/\lambda_{fl.}$=390/475 nm) was observed in solutions of Gly and D-glucose (Glc) heated at pH≥11 to t≥93 °C for 3-5 minutes. When the solution was cooled down to 20 °C initial CL intensity was low, but later increased in samples exposed to air. CL rising up was followed with decoloration of the solution. In few previous studies it was demonstrated, that formation of colored and fluorescent late Maillard products in Glc/amino acid mixtures is retarded by illumination with white and especially with UV-A (Bohart, Carson, 1955; Sander, Larsen, 1995). CL was negligible in samples kept under anaerobic conditions, and in such solutions more intense browning was observed. These observations indicate, that effects of oxygen and both external and *internal* light on the chemical processes in Maillard reaction may be interrelated.

As we expected, heating of the solution played only initiating role for chemiluminescence development. Most surprising was the existence of *temperature threshold* for the development of CL. Fig. 5 shows, that no CL develops after the reaction mixture heating to 91 °C, and practically full intensity process is observed if mixture has been heated to 93 °C.

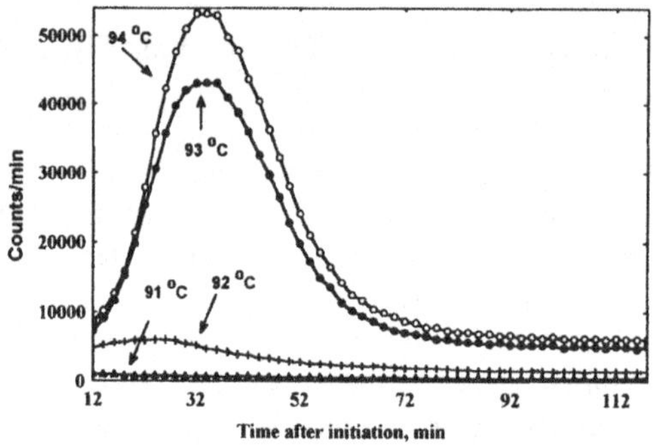

Figure 5. CL development in Gly/Glc solutions (0.14 M/0.12 M, pH 11.3, 5 ml) incubated at indicated temperatures for 5 min.

Development of CL in Gly/Glc solutions depended on pH also in extremely non-linear manner. At pH below 11.0 low CL without characteristic wave-like kinetics was observed. Elevation of pH from 11.2 to 11.4 resulted in a sharp arousal of a wave-like pattern. Amine group of Gly is already completely deprotonated at pH 11.0. However, it is known, that in basic solutions decyclization of Glc is eased, and it may isomerise in endiolic form, which is more active, then cyclic and even linear chain Glc (Morrison, Boyd, 1970). Very sharp dependence of the process initiation on pH indicates of the existence of critical concentration of some intermediate for it.

Non-linear dependence of the reaction kinetics upon Gly and Glc concentrations confirms this proposal. At reagents concentrations below 0.1 M this dependence was similar to that observed for CL from Asn/H_2O_2 mixtures (see Fig. 1). At constant Glc concentration (0.12 M) the first CL maximum was observed at 0.12 M Gly. Elevation of its concentration first was not followed with elevation of CL. At higher Gly concentrations CL again increased and stabilized at 0.18 M Gly. At constant Gly concentration (0.12 M) CL maximum was observed at 0.24 M Glc. At higher Gly concentration CL development was retarded. It worth noting in this connection, that non-linear and non-monotonous dependence of kinetics of ribonuclease glycation by ribose on ribose concentration was recently observed (Khalifah, *et al.*, 1996)

Critical conditions for the development of the process in Gly/Glc solution followed with CL are already indicative of CRDB mechanism. Dependence of CL on the sample volume also turned out to be highly unusual. Fig. 6 shows existence of several optima and pessima in this dependence (note that "specific" light accumulation, that is counts collected for 120 min divided by the volume of each particular sample, are presented in this figure). Such "oscillation"-like pattern of CL kinetics upon the reaction volume implies that some macroscopic spatial orderliness gradually develops in the reaction system (note, that maxima and minima of light emission become evident rather late after the beginning of CL growth).

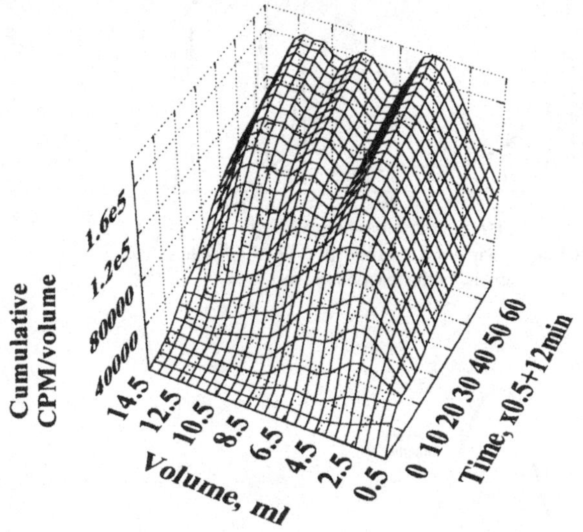

Figure Dependence of Maillard reaction kinetics on the reaction volume. Identical glass vials filled with different volumes of standard Gly/Glc reaction mixture were heated (3 min at 94 °C). CL was registered in vials in turn. Data are presented as cumulative counts divided by sample volume.

CL decay after reaching maximum (cf. Fig. 5) was not due to substrates consumption. CL "waves" could be reinitiated by brief heating of "faded" solution to the critical temperature (Fig. 7). Note, that the maximum CL intensity after the second initiation is much higher than that after the first one. Still higher CL responses follow next initiations. High CL responses may be induced even 20 hours after the first initiation. To explain why successive "waves" of CL are higher then the previous ones the following hypothesis may be suggested. In the course of the process some metastable products capable to quench CL accumulate in the reaction system. They decompose due to brief heating to the "permissive" temperature, and their "splinters" may serve as new active centers giving birth to new chains.

The general shape of kinetic curves of CL in Maillard reaction, as well, as those in reactions induced by H_2O_2 in amino acid solutions (cf. Figs. 1 and 2) is rather uniform. We tried to fit CL kinetic curves to different functions of distributions, and the best fit found so far was to the function of lognormal distribution. As it is shown at Fig. 9, two representative CL curves of two consecutive CL-waves excellently fit to this model (correlation coefficients are 0.9989 and 0.9983 for the data representing the first and second initiations, respectively). Note also, that deviation of experimental data from lognormal distribution is observed only for the very initial parts of the kinetic curves.

Lognormal distribution is characteristic of many physiological parameters. It has been interpreted in terms of a "multiplicative Gestalungs-principle of nature" (Gebelein, Heite, 1950). Zhang and Popp (1994) has traced this principle to coherent states in living organisms. Here the evolution of CL intensity during *the whole* process obeys a lognormal distribution. It may indicate, that the process has "history", in other words,

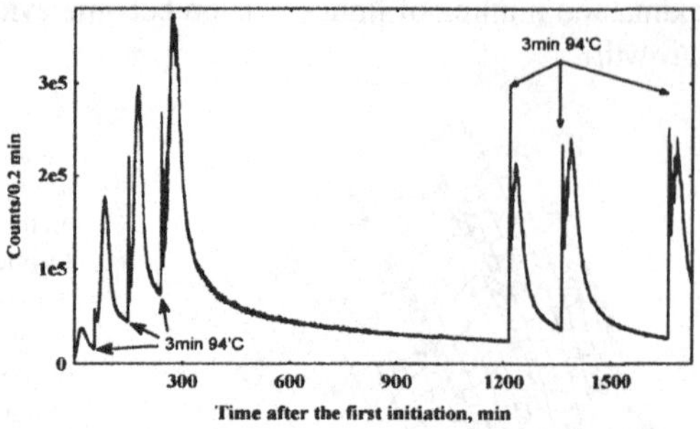

Figure 7. Repeated initiations of CL "waves" in 0.12 M Gly/Glc solution (pH 11,3) with repeated heating of the reaction mixture.

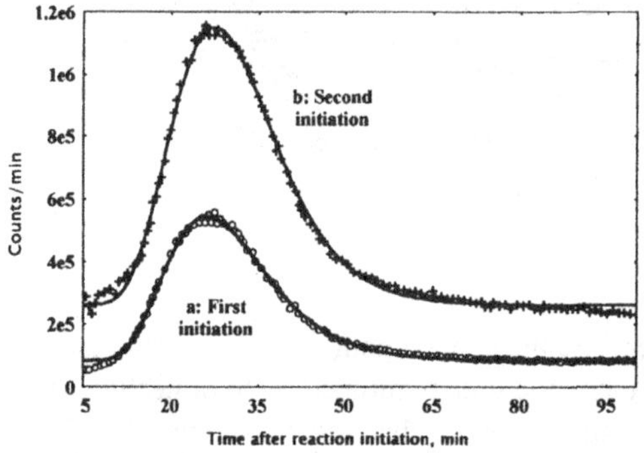

Figure 8. Fitting of CL kinetics in Gly/Glc solutions to lognormal model (solid curves):

$CL(cpm)=A\bullet e[-B\bullet ln^2((t+C)/D)]+E,$

where A, B, C, D, E - constants, CL -chemiluminescence intensity at actual t. Second initiation (94 °C the 3 min) was at t=100 min after the first one. Open circles and crosses -- experimental points.

actual events resulting in photon emission depend not only on the immediately preceding ones, but on the whole preceding network of them. On the other hand, actual events determine not only the immediately following ones, but a succession of events that follow (if no external factors interfere with the trend of the process).

3. General Conclusions

Two reaction systems described in this paper possess all the features characteristic of chain reactions with delayed branching. A lot of "oddities" in these reactions were observed. These "oddities" may be explained basing on slightly modified Semyonov's hypothesis of the two distributions coexisting in those reaction systems, where chain processes develop. The first one, Maxwell-Boltzmann distribution, describes the kinetic energy distribution of the overwhelming majority of solvent and solute molecules. Another distribution is established between different electron excitation levels of species leading the chain process. Electron excitation of Tyr and Trp in H_2O_2/amino acid systems implies, that electron excited levels equivalent to UV-photon energy are reached. This distribution is rather stable, as illustrated by the resistance of the process kinetics to external interference, like temperature changes (Figs. 3 and 4) or mechanical agitation of the solution.

For the first time it was demonstrated here, that in the course of the process evolution this distribution acquires both spatial and temporal stability. Peculiar dependence of CL in Maillard reaction on the reaction system volume (Fig. 6) implies that active species gradually organize in macroscopic domains of certain regular dimensions. Excellent agreement of the experimental data with the lognormal model (Fig. 8) also implies that the evolution of such processes proceeds according to a certain general principle.

Emergence of electron excited states distribution is equivalent to the establishment of "stable non-equilibricity" in a system taken as a whole. According to Erwin Bauer (1935) such a state of stable non-equilibricity is the necessary (though not sufficient) feature of all leaving systems. In the reaction systems studied here energy for non-equilibricity stabilization is provided by oxidation of the substrates. The essential part of this energy is spent, as Gurwitsch noted (1938, 1946, 1959), on synthetic reactions. In their course new substances, in particular, more complex ones and having higher molecular weights than the original substrates emerge. They may be pumped by the energy becoming available due to oxidative reactions occurring in the system and thus they may catalyze further reactions, in particular, oxidation of available substrates providing further energy acquisition. In fact, we had confirmed recently Gurwitsch's claim, that an enzyme-like substance appear in amino acid solutions in which CRDB is initiated with extra-weak UV-light irradiation. We observed formation in such systems of a polymer substances with the activity of glycine deaminase (Baskakov, Voeikov, 1996a).

On the other hand, a lot of evidence has been obtained, that in the course of Maillard reaction a lot of complex compounds, including heterocyclic aromatic molecules and products of their polymerization (melanoidines), arise from simple molecules (Yim, *et al.*, 1995; Telegina, Davidiantz, 1955). It has been demonstrated that melanoidines formed in the heated mixture of amino acids possess a lot of activities, including the synthetic one. For example, when this reaction mixture was irradiated with UV-light condensation of tripeptides added to it into oligopeptides was observed (Telegina, *et al.*, 1990).

During a chain process evolution more and more advanced products may accumulate, which scavenge more and more energy from the system. Fading of CL may be explained by accumulation of such products. However, these products are likely to be metastable enough at least at the early stages of the reactions, and they may be eliminated by an adequate external interference with the course of the process (Fig. 7). Products of their disintegration may serve new active centers for the new series of chain reactions, but as soon as these chains develop already in the new chemical environment, the new branched chain process is in no case just the repetition of the old one.

The results of the present study demonstrate, that in aqueous solutions of amino acids in which chain reactions with delayed branching are initiated, self-organization with far reaching consequences may occur. Such processes, that in some or other form should proceed in living systems, may serve, in particular, the source of pumping a system with energy quanta resulting in the emergence of stationary electron excitation of specific molecular species. Part of this energy may be released in the form of "biophotons" including MGR. *However, it should be emphasized that the chemical-physical nature of emerging excited species, as well as physical foundations for the development of dynamic non-equilibrium stability in reaction systems described here remain to be elucidated.*

4. Acknowledgment

Part of this investigation was performed at the International Institute of Biophysics in the Technology Center Kaiserslautern, FRG. We express our appreciation to Dr. F.-A. Popp for his encouragement. We wish to thank "Center for Analysis of Substance" Moscow for support of this work.

5. References

Adam, W. and Cilento, G. (1983) Four-membered ring peroxides as excited state equivalents: new dimension in bioorganic chemistry, *Ang. Chem. Int. Ed. Engl.* **22,** 529-543.

Ashmore, P.G. (1966) *Catalysis and Inhibition of Chemical Reactions,* Mir, Moscow (Russ. translation of: Buttersworth, London, 1963)

Baskakov, I.V. and Voeikov, V.L. (1995) Temperature dependence of chemiluminescence in aqueous aspartate solutions in the presence of hydrogen peroxide and ethidium bromide, *Biofizika* (Moscow) **40,** 1149-1157.

Baskakov, I. V. and Voeikov V. L. (1996a) Formation of a Polymer with Glycin Deaminase Activity upon UV Irradiation of Amino Acid Solutions, *Russian Journal of Bioorganic Chemistry* **22,** 77-82.

Baskakov, I. V. and Voeikov V. L. (1996b) The role of electron excited states in biochemical processes, *Biochemistry (Moscow).* **61,** 837-844.

Bauer, E.S. (1935) *Theoretical Biology,* VIEM, Moscow-Leningrad.

Beloussov, L.V. and Baskakov, I.V. (1995) A reproduction of the mitogenetic experiments of the Gurwitsch school on frog and fish cleaving eggs, in: L.V. Beloussov and F..A. Popp (eds.), *Biophotonics,* Bioinform Service, Moscow, pp. 447-456

Beloussov, L.V., Voeikov, V.L., and Popp F.-A. (1997) Gurwitsch's mitogenetic rays, *Priroda* (Moscow) N 3, 64-80 (in Russ.).

Bodenstein, M. (1913) *Ztschr. Phys Chem.* **85,** 329-397.

Bohart, G.S., Carson, J.F. (1955) Effect of trace metals, oxygen and light on the glucose-glycine browning reaction, *Nature* **175,** 470-471.

Cadenas, E. (1984) Biological chemiluminescence, *Photochem. Photobiol.* **40,** 823-830.

Dyer, G., Blackledge, J.A., Thorpe, S.R. and Baynes, J.W. (1991) Formation of pentosidine during nonenzymatic browning of proteins by glucose. Identification of glucose and other carbohydrates as possible precursors of pentosidine in vivo, *J. Biol. Chem.* **266,**.11654-11660.

Emanuel, N.M., Zaikov, G.E., and Maizus, Z.K. (1985) *Chain Reactions of Carbohydrates Oxidation in Liquid Phase.* Nauka, Moscow (in Russ.).

Galantsev, V.P., Kovalenko, S.G., Moltchanov, A.A., Prutskov, V.I. (1993) Lipid peroxidation, low-level chemiluminescence and regulation of secretion in the mammary gland, *Experientia* **49,** 870-875.

Gebelein, H. and Heite, H. J. (1950) Uber die Unsymmetrie biologischer Haufigkeitsverteilungen, *Klin. Wschr.* **28.**

Grasso, F., Musumeci, F., Triglia, A., Yanbastiev, M., and Borisova, S. (1991) Self-irradiation effect on yeast cells, *Photochem. Photobiol.* **54,** 147-149.

Gurwitsch A.G. and Gurwitsch L.D. (1938) Uber Anreicherung (Neubildung) von Fermenten auf Kosten von Aminosauren, *Enzymologia* **5**, 17-25.
Gurwitsch A.G., Gurwitsch L.D. (1959) *Die mitogenetische strahlung*, G. Fischer, Iena.
Gurwitsch, A. (1923) Die Natur des spezifischen Erregers der Zellteilung, *Arch. Entwicklungsmech.* **100**, 11-40.
Gurwitsch, A..A., Eremeyev, V..F., and Karabchievsky, Yu. A. (1965) Extra-weak emission in visible and ultra-violet regions in oxidation of solutions of glycin by hydrogen peroxide, *Nature* **206**, 21-22.
Gurwitsch, A.A., Kufal, G.E., Batuanov, A.P., and Lazurkina, N.N. (1987) Physical-chemical processes in photoexcited glycin solution, *Bull. Exp. Biol. Med.* **54**, 683-685 (in Russ.).
Gurwitsch, A.G. (1946) The formation of enzymoids. Review about contributions on fermentology during the last 8 years from the mitogenetic laboratory, *Enzymologia* **12**, 139-141.
Gurwitsch, A.G.,and Gurwitsch, L.D. (1942) Peculiarities of chain reactions and common energy levels in living systems, *Acta Phys. Chem.* **16**, 288.
Hinshelwood, C.N. (1940) *The Kinetics of Chemical Change*, Clarendon Press, Oxford.
Khalifah, R. G., Todd, P. Booth, A. A., Yang Shi X., Mott, J. D., and Hudson, B. G. (1996) Kinetics of nonenzymic glycation of ribonuclease A leading to advanced glycation end products. Paradoxical inhibition by ribose leads to facile isolation of protein intermediate for rapid post-Amadori studies, *Biochemistry* **35**, 4645-4654.
Kurosaki, Y., Sato, H.; and Mizugaki M. (1989) Extra-weak chemiluminescence of drugs. VIII. Extra-weak chemiluminescence arising from amino-carbonyl reaction, *J. Biolumin. Chemilumin.* **3**, 13-19.
Morrison, R.T. and Boyd, R.N. (1970) *Organic Chemistry* Allyn and Bacon, Inc., Boston
Namiki, M. (1988) Chemistry of Maillard reactions: recent studies on the browning reaction mechanism and the development of antioxidants and mutagens, *Adv. Food Res.* **32**, 115-184.
Namiki, M., Oka, M., Otsuka, M., Miyazawa, T., Fujimoto, K., and Namiki K. (1993) Weak chemiluminescence at an early stage of the Maillard reaction, *J. Agri. Food. Chem.* **41**, 1704-1709.
Oullet C. (1933) Trans. Faraday Soc. **29**, 486-494.
Popp, F.-A. Gu, Q., and Li, K.-H. (1994) Biophoton emission: experimental background and theoretical approaches, *Modern Physics Letters B* **8**, 1269-1296.
Quickenden, T.I. and Tilbury, R.N. (1991) Luminescence spectra of exponential and stationary phase cultures of respiratory deficient Saccharomyces cerevisiae, *J. Photochem. Photobiol. B* **8**, 169–174.
Sander, B. and Larsen, M. (1995) Photochemical bleaching of fluorescent glycosylation products, *Int. Ophthalmology* **18**, 195-198.
Semyonov, N.N. (1986) *Chain Reactions*, Nauka, Moscow (in Russ.).
Semyonov, N.N. (1935) *Chemical Kinetics and Chain Reactions*, Oxford University Press.
Slawinsky, J. (1988) Luminescence research and its relation to ultraweak radiation, *Experientia* **44**, 559-571.
Telegina, T.A. and Davidiantz, S.B. (1995) Maillard reaction: amino-carbonyl interactions *in vivo* and melanoidins, *Uspekhi Biol. Chimii (Progress in Biological Chemistry)* **35**, 229-266 (in Russ.).
Telegina, T.A., Masinovsky, Z., Sivash, A.A., Pavlovskaya, T.E., Leibl, V., and Bejsovkova L. (1990) in: *Origins of Life, Evolution and Biosphere* **20**, 269-277.
Vassiliev, R. F. (1985) Ways of chemiluminescence excitation of organic compounds, in: A.I. Zhuravlev (ed.), *Biochemiluminescence*, Nauka, Moscow. pp. 31-55 (in Russ.).
Vladimirov, Yu.A. (1966) *Extraweak Emissions in Biochemical reactions*, Nauka, Moscow (in Russ.).
Vladimirov, Yu.A. (1997) Chemiluminescence in biological systems, *Priroda* (Moscow), N 3, 18-28 (in Russ.).
Voeikov, V.L. and Baskakov, I.V. (1995) Study of kinetics of chemiluminescence in aqueous amino acid solutions in the presence of hydrogen peroxide and ethidium bromide, *Biofizika* (Moscow) **40**, 1159-1166.
Voeikov, V.L. and Naletov, V.I. (1997) Kinetic analysis of chemiluminescent stage of Maillard reaction indicates of the degenerative-branched chain mechanism of the process, *Second Congress of the All-Russia Biochemical Society of Rusiian Academy of Sciences. Abstract Book*, Puschino, part II, 412-413 (in Russ.).
Voeikov, V.L., Baskakov, I.V., Kafkalias K., and Naletov, V.I. (1996) Initiation of degenerate branched chain reaction of glycin deamination with weak UV irradiation or hydrogen peroxide, *Russ. J. Bioorganic Chem.* **22**, 35-42.
Wondrak, G., Pier, T., and Tressl, R. (1995) Light from Maillard reaction: photon counting, emission spectrum, photography and visual perception, *J. Biolumin. Chemilumin.* **10**, 277-284.

Yaylayan, V.A., and Huyghues-Despointes, A. (1994) Chemistry of Amadori rearrangement products: analysis, synthesis, kinetics, reactions, and spectroscopic properties, *Rev Food Sci Nutr.* **34,** 321-369.

Yim, H.S., Kang, S.O., Hah, Y.C., Chock, P.B., and Yim M.B. (1995) Free radicals generated during the glycation reaction of amino acids by methylglyoxal. A model study of protein-cross-linked free radicals, *J. Biol. Chem.* **270**, 28228-28233.

Zhang, C.-L. and Popp, F.-A.(1994) Log-normal distribution of biological parameters and the coherence in biological systems, in M.-W. Ho, F.-A. Popp, and U. Warnke (eds.), *Bioelectrodynamics and Biocommunication,* World Scientific, Singapore, New Jersey, London, Hong Kong, pp. 355-363.

COHERENCE AND BIOPHOTON EMISSION AS INVESTIGATED ON ACETABULARIA ACETABULUM

F.MUSUMECI, A.SCORDINO, AND A.TRIGLIA

Istituto di Fisica della Facoltà d'Ingegneria e sezione INFM, Università di Catania, Viale A. Doria 6, I- 95129 Catania (Italia)

Abstract

In this paper are examined some characteristics of Delayed Luminescence in simple biological systems and in preparations of Cadmium Sulfide which are different only in the dimension of the grains in which the system is aggregated.

The hypothesis suggested is that the D.L. is not connected to the existence of specific fluorophores but to the presence of the band structure created by the short range order conditions present in the samples.

In these conditions the non-linear dependence of the DL on the light dose could reveal the presence of stimulated emissions in the course of this phenomenon.

Key words : Bioluminescence - Biophysical Techniques - Bio-optics

1. Introduction

In the last few years lots of experimental work on several biological systems[1-8] have shown that there is a close connection between the biological state of the system itself and the parameters of Delayed Luminescence (D.L.). Some of these relations were expressed in an analytical form connecting a biological parameter with a parameter of D.L. [9,10].

However notwithstanding the indubitable value of the data reported to show the intrinsic importance of this phenomenon, it hasn't yet been possible to demonstrate conclusively the origin of Delayed Luminescence.

There are two basic hypotheses that seem convincing :

i) the first one, proposed by J. Lavorel [11], gave a biochemical interpretation of the phenomenon considering the luminescence as the visible sign of a minor imperfection in the primary photochemical charge separation

ii) the second, proposed by F.A. Popp [12-14], asserts that hyperbolic relaxation is due to the decaying of an excited full coherent field that plays a role in

J. J. Chang et al. (eds.), Biophotons, 109-120.

promotion and control of living processes and for this reason may become a powerful tool for analysing the state of living systems in terms of coherence.

The experimental evidence showing a close connection between the state of the biological system and the parameters of Delayed Luminescence suggests that Popp' s interpretation is more correct than the biochemical one but it is necessary to take in account certain objections:

a) the supporters of the biochemical hypothesis also sustain [15] that, although D.L. represents an insignificant loss of the total energy stored in one system, it is a sensitive indicator of the many reactions present during interactions between photons and biological matter .
b) it is not possible by means of a mere observation of time dynamic to fully understand the characteristic of a system, and a time trend that appear as hyperbolic inside to the window of measure could be, in principle, described either through Popp' s formalism or through the decay of a continuous set of many equilibrium states with different decay constants exponentially distributed [16,9]. The existence of such a set is consistent with the large number of charge carriers and reaction pathways occurring in biological reactions as well as in solid state systems in which hyperbolic decays of several physical parameters are quite usual.

On the other hand, the hypothesis of the existence of a coherent field requires a description of the basic mechanism which can generate and maintain inside a biological system such a coherent field.

The lack of a full understanding of the origin of the phenomenon is particularly serious, not only from the academic point of view, for the difference between versions of the biology that the two models imply: on the one hand a vision strictly connected to a mechanicistic reductionism that no longer holds in the area of modern physics, on the other a holistic vision of biological systems that considers the complexity and coherence but without having a real predictive capacity for the described phenomena. This lack of understanding infacts also impedes significant steps forward in this sector, which is highly promising from an applicative point of view, limiting the experience realised to a collection of data on the existing correlations between D.L. and the functional state of the system, which, even if useful and important, takes place blindly in the absence of a precise theory.

This article suggests a possible origin of D.L. and is based on a series of observations effectuated essentially on the monocellular alga Acetabularia Acetabulum and on the well known semiconductor Cadmium Sulfide which like other semiconductors, has quite a pronounced delayed luminescence in certain conditions.

2. Material and Methods

The experimental set up, designed to measure photons emitted from biological systems, consists in general of a steel dark chamber where the samples to be analysed can be maintained at a constant temperature.

The radiation emitted from the sample is detected by a low-noise photomultiplier working in single photon counting mode and having a spectral sensitivity ranging from 200 nm to 850 nm. In order to decrease the dark current the photomultiplier is cooled down to -30°C. Due to the very low intensity of the signal it is necessary to increase as much as possible the solid angle of measurements but, because it is necessary to put in shutters and optical filters and to heat insulate the photomultiplier from the surrounding, in a standard experimental set up the solid angle of measurements is of the order of .1 sr.

Several kinds of light emitting devices ranging from high intensity LEDs to halogen lamps can be used as sources of exciting radiation provided that the emission power is stable in time and that the area covered by the light is illuminated with uniform intensity. Moreover , because with rather long periods of illumination the parameters describing time decay are not connected only to the state of the system but also to the temporal duration of its illumination, sources which present extremely short times of illumination were used for almost all the experiments.

Measurements consist in illuminating a biological sample and in counting the number of photons re-emitted from the sample after the light source had been switched off.

During the illumination a light shutter, above which the photomultiplier is fastened, is closed in order to prevent the dimpling of the photomultiplier. After the light source was switched off the shutter is opened by an electromagnetic actuator. Due to this time lag of the experimental set-up, the photon counting starts some tens of ms after the source is switched-off.

The counting of photons emitted by the sample, after each illumination, is stored by a channel scaler using a dwelling time chosen in order to measure the decay dynamics in the best way.

Due to the low level of the signal a spectral analysis of the emitted photons could be performed only by using broadband filters corresponding to rather extended spectral intervals (of the order of tens of nm).

In general it is necessary between two successive illuminations to wait for a time interval long enough to enable the emission from the biological sample to return to, within the experimental error margin, the value prior to the perturbation. This time is at least ten times greater than the time necessary for the signal to reach the background value.

Because it is not possible to neglect the background emission one has to measure the yield of photons emitted from the container of the sample filled only with the culture medium (if it is present) in the same experimental conditions for each set of measurements, in order to obtain the correct background values that will be subtracted from the measurements taken from each sample.

3. Experimental Results and Discussion

First of all we examine the typical time trend of the D.L. shown in fig. 1a : it appears that this trend is hyperbolic and significantly differs from an exponential decay.

It is possible to describe the experimental data with a hyperbolic trend described by

$$I(t) = \frac{I_0}{(t_0 + t)^m} \tag{1}$$

as indicated in figure 1a.

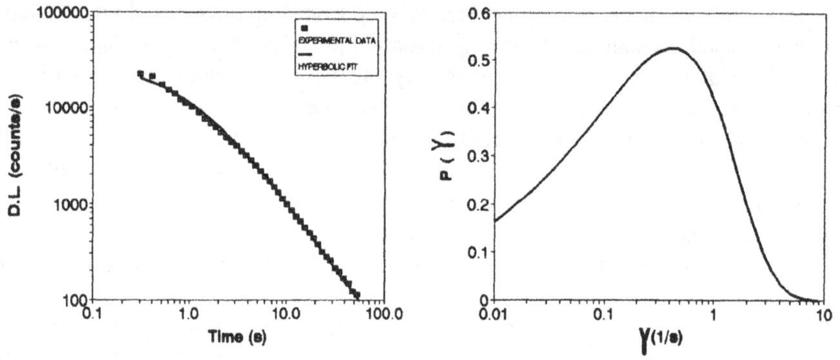

Figure 1 - a) Typical trend of D.L. decay : (▫) experimental data, (——) fit according eq. 1, (- - -) mono-exponential fit .
b) ((——) Laplace transform of the fit of Fig 1a representing the distribution probability of the decay rates γ of single exponential decays whose convolution generate in the time domain the hyperbolic trend of Fig 1a.

The three parameters that characterise the fit are connected to the functional state of the system and to the characteristics of the illuminating source (intensity, time of illumination and wavelength).

It must be pointed out that the hyperbolic fitting of the experimental trend is an extremely practical way for expressing the experimental results independently of the theoretical explanation of the phenomenon. This trend, in fact, is foreseen in Popp's theory but it could be also generated by the weighted sum of a large number of uncorrelated exponential decays characterised by different decay rates $\tau = 1/\gamma$ as indicated in eq. 2.

$$I(t) = \int_0^{\infty} A\, p(\gamma)\, e^{(-\gamma t)}\, d\gamma \tag{2}$$

where A is a normalising factor having the same physical dimensions of I(t). The weights p (γ) to be attributed to the single decay rates are obtainable by the analytical Laplace transform of the hyperbolic fitting function, shown in fig 1b, that represents the distribution probability for the decay rates.

However another model exists which takes into account the behaviour of levels intercorrelated because they are connected to an ordered spatial structure where the molecules are situated. In this case there is no point in talking about transitions between single levels and the number of levels that decay in the unit of time generating photons with energy equal to hν is given by [17]:

$$I(\nu) = \mathbf{B}\int_0^{\infty} n_c(E)\; n_v(E - h\nu)\, F_c(E)\; F_v(E - h\nu)\; dE \tag{3}$$

B being connected to the matrix element that describes the transition and to the volume of crystal, n_c , n_v being the relative density of the states in the conduction and the valence bands, F_c and F_v the distribution function of Fermi-Dirac in the two bands.

It is difficult to evaluate this quantity because it requires the specific knowledge of the structure of the bands of the system that is under analysis. However it certainly does not give origin to an exponential decay because it depends both on the number of excited states and the Fermi quasi-levels E_{Fc} and E_{Fv} vary.

In a simpler case, if there are no rules of selection and there is no degeneration we have, being n and p the total number of electron and holes :

$$I_T = \mathbf{B}\, n\, p = \mathbf{B}\, n^2 \tag{4}$$

which gives origin to a hyperbolic pattern.

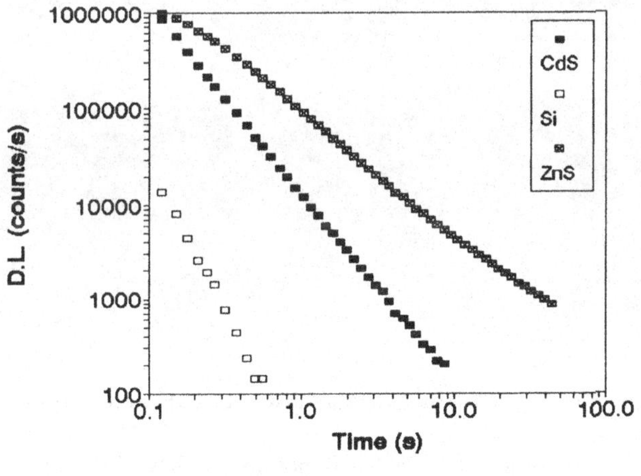

Figure 2 - Trend of D.L. decay for some different semiconductors.

Fig. 2 shows the temporal decay of the case of certain typical semiconductors: can be seen that in all cases the curve is of the hyperbolic kind.

The substantial difference between the models reported above is that, according to the second model, the luminescence is generated by a mixture of fluorophores to which belong the excited levels which are independent and whose time constants are distributed according to equation 2 in such a way as to give origin to a hyperbolic decay, while, according to the first and third model, the luminescence has its origin from the structure of the system itself and can not therefore be attributed to any specific fluorophore.

Generally it is retained that an explanation of the third kind is not possible if not in highly ordered systems. Morever, in the cases reported, it can not be excluded that the radiation is due to some fluoropheres incorporated into the matrix of the semiconductor as impurities.

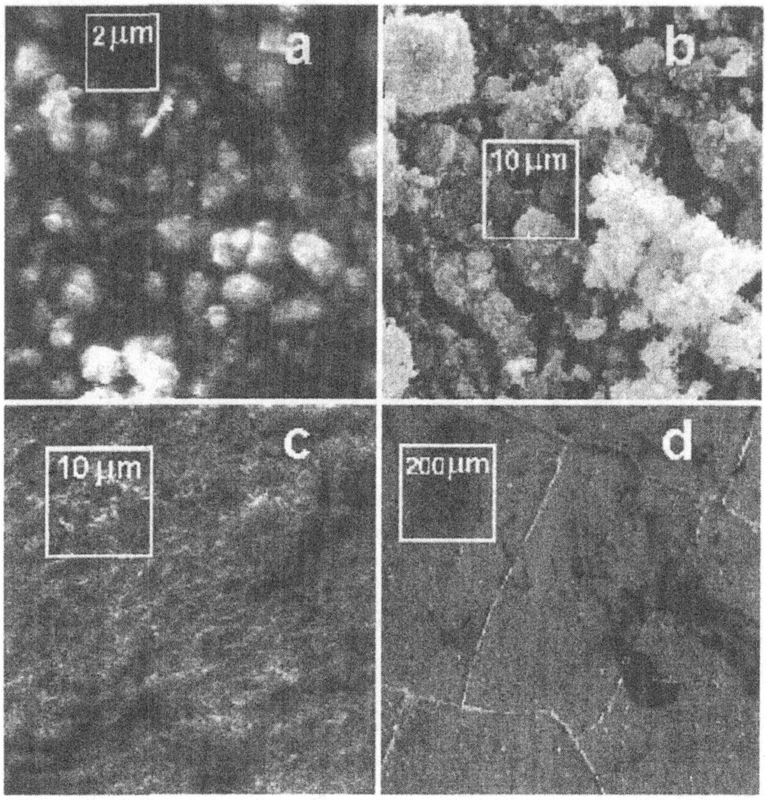

Figure 3a. Images of the four samples of Cadmium Sulfide obtained by an electron microscope

To check this opinion and to see if it is possible to have D.L. in the absence of specific fluorophores even for systems which have only order at short range, a solid state system was studied in which it was possible to control the absence of fluorophores and at the same time possible to check the evolution of the D.L. at different order levels.

With this aim was looked a semiconductor having a gap of the order of a couple of ev in such a way as to wait for eventual photon re-emissions due to the interband recombination in the visible region.

Among the various candidates Cadmium Sulfide was chosen which presents at room temperature a 2.4 eV band and which is very stable in normal conditions.

The emission of this material was measured in various states characterised by different aggregation geometries :

a) powder with very fine granulometry (average dimension < 1 μm)
b) powder with average granulometry (average dimension < 5 μm)
c) material crystallised from the liquid phase (constituting of an almost amorphous stratum characterised by structures of the dimension of a few μm)
d) material obtained by crystallisation at high temperature from a mixture of Cadmium chloride and thiourea (planar crystals with dimensions of several hundred μm).

Fig 3a shows the photographs obtained by electron microscope of the four samples while fig. 3b shows the temporal trends of the D.L. in the four cases.

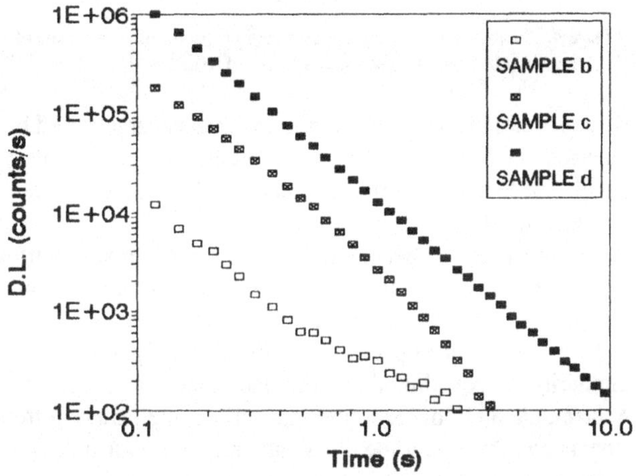

Figure 3b. D.L. from three of the samples of Cadmium Sulfide show in fig. 3a. From the sample "a" no signal has beemeasured

It can be seen that sample "a", having a quantity of Cadmium Sulfide about 100 times more than the others, emits absolutely no measurable signal while the other three samples show higher and higher emissions and increasingly closer to the characteristic hyperbolic trend. The absence of signal in sample "a" testifies the absence of specific fluorophore which, if present, should give rise to a measurable D.L.

It can be observed that samples "b" and "c" present traces of order only at short distances (the grains visible show in fact a maximum dimension of about ten μm) but, notwithstanding that, a distinct signal was produced; given that the chemistry of the system remained unchanged, the D.L. can be attributed to the fact that a band structure has been created and not to the presence of fluorophores immersed in a matrix of Cadmium Sulfide .

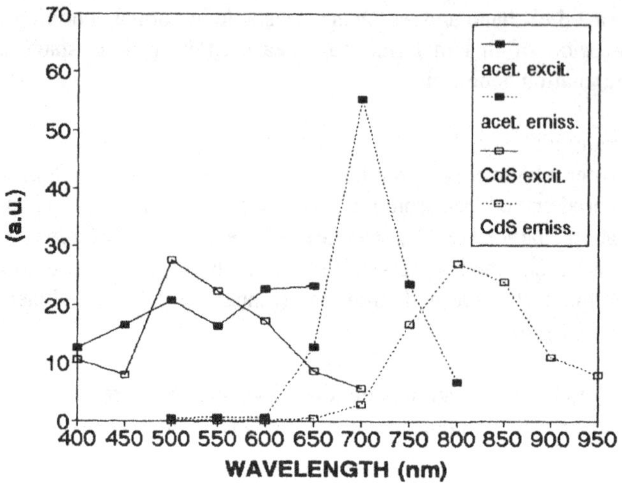

Figure 4. Excitation (——) and emission (- - -) spectra of living cells of Acetabularia (■) and Cds sample "d" (□).

This fact demonstrates clearly that, even in materials characterised by short range orders (on distances in the order of tens of wavelengths), as for example biological systems, it is possible to hypothesise the presence of a D.L. caused by the spatial structure of the system itself.

Significant similarity exists between the D.L. of Cadmium Sulfide and that of biological systems, other than the hyperbolic decay which is common to both; as for istance the excitation and emission spectra can be observed to be similar in that both are wide and without very pronounced peeks as can be seen in fig. 4.

Another similarity is constituted by the fact that, as is shown in fig. 5a, for Acetabularia Acetabulum and 5b, for Cadmium Sulfide, the temporal trend of the single spectral components are absolutely identical showing that, in a first approximation, the probability of decay is independent of the frequency. This phenomenon has been observed in almost all biological systems examined.

One of the characteristics of the D.L. of biological systems difficult to explain is the strong non-linearity of the D.L. with the variation of the illumination intensity present in some cases. On this theme fig. 6 shows the D.L. of Acetabularia Acetabulum doubling the illumination intensity.

As can be seen the trend is completely changes as if the mechanism of emission were mutated. This fact, which is not so evident in all biological systems and that is absent in Cadmium Sulfide, must have an explanation within the scheme proposed to explain the nature of D.L. This can be done by trying to take account of the eventual presence of stimulated emission.

In semiconductor systems the condition of inversion of the population and a consequent positive gain happens when

$$h\nu \leq (E_{Fc} - E_{Fv}) \qquad (4)$$

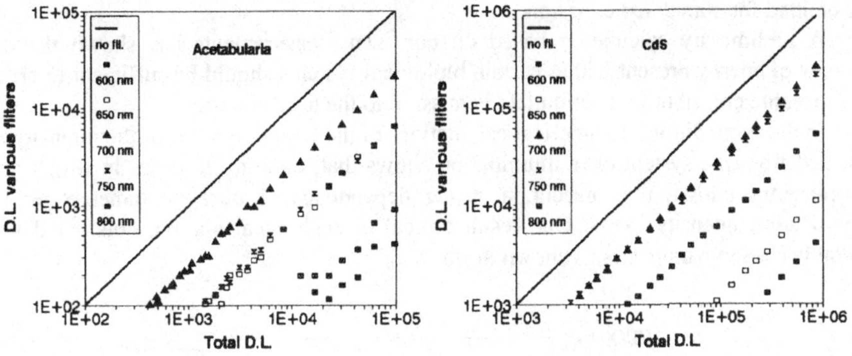

Figure 5. a) D.L. of the various component of the emission spectrum of a living Acetabularia cell versus the total D.L.
b) D.L. of the various component of the emission spectrum of Cds sample "d" versus the total D.L..

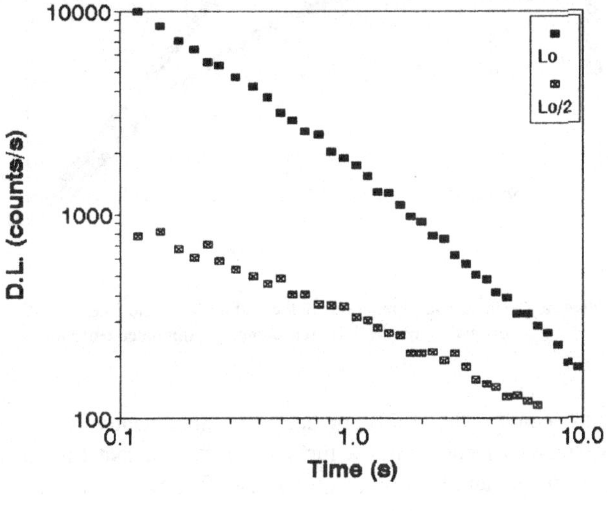

Figure 6. Trend of D.L. from Acetabularia for different intensities (L_0=7 10^{15} phot/s/cm^2) of exciting light at the same light wavelength (450 nm)

and this condition can be verified for the transitions that happen between states close to the base of the conduction band and to the high point of the valence band.

Naturally in our case, the presence of stimulated emission can not mean, due to the absence of continuous pumping and opportune confinement of the irradiated field, that there is a continuous and coherent emission at a determined frequency but that if a photon is spontaneously emitted, there exists a finite probability that other photons will be emitted for stimulated emission.

A preliminary calculation based on our own measurements has shown that the density of energy present within certain biological systems should be sufficient to create a noticeable contribution of stimulated emission to the total emission.

In these conditions a numerical calculation of the temporal trend of the intensity re-emitted from the system after illumination shows that, contrary to cases in which only spontaneous emission is present, a strong dependence of decay parameters on the illumination intensity exists. One result typical of such a calculation, conducted with suitably chosen parameters, is shown in fig. 7.

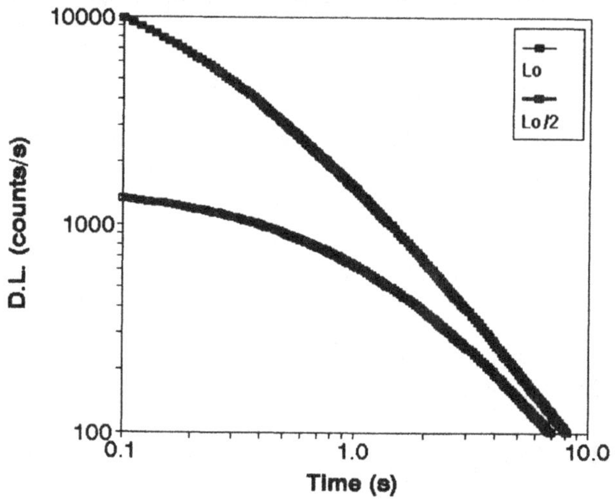

Figure 7. Numerical simulation of the influence of the intensity of excitation on the D.L. in presence of stimulated emission

It can be clearly seen that the trends are similar to those relative to the behaviour of the biological system which constitutes a further clue to the fact that D.L. originates from the structure of the system itself and can not therefore be attributed to any specific fluorophore.

4. Conclusion

The data presented in this article, being still a long way from explaining completely the origin and role of D.L. inside living things, does however allow a beginning in understanding of why such a close connection exists between this phenomenon and the state of such systems.

In fact the data relative to the Cadmium Sulfide show that in the absence of specific fluorophores and for structures having typical dimensions in the order of those present in biological systems, the D.L. is generated as a phenomenon connected to the dimension and the nature of the structure. It is therefore able to give global information about the structure itself even if naturally not in an immediately decipherable way. This vision is encouraged by certain experimental evidence from Acetabularia Acetabulum.

Preliminary studies have shown in fact the existence of a strong correlation between the intensity of the D.L. and the mobility of chloroplasts and, in general, the function of the citoskeleton [18].

The possibility to explain the strong changes in the slope of the time trend with the presence of flashes of stimulated emission constitutes, even in at a rudimentary level, a bridge between Popp's theory and that which explains D.L. by means of the band structure of the system. These flashes of stimulated emission could generate in conditions of continuous illumination a coherent electromagnetic field within the biological structure which could be connected to the processes of control of the whole biological system.

4. References

1. T.V. Veselova, V.A. Veselovsky, A.B. Rubin and V.Z. Bocharov [1985]: Delayed Luminescence of Air-Dry Soybean Seeds as a Measure of their Viability, *Physiol. Plant.* **65**, 493-497.
2. S. Tryka and R. Koper[1986]: Luminescence of Cereal Grain Subjected to the Effect of Mechanical Loads, *Photon Emission from Biological Systems*, B. Jezowska-Trebiatowska, B. Kochel, J.Slawinski and W. Strek eds, World Scientific, Singapore, 248-254.
3. A. Ezzahir, T. Kwiecinska, M. Godlewski, D. Sitko, Z. Rajfur, J. Slawinski, B. Szczesniak-Fabianczyk, A. Laszczka [1992]: *The Influence of White Light on Photo-Induced Luminescence from Bull Spermatozoa*, *ABC* 2/3, 133-137..
4. M.W. Ho, X. Xu, S. Ross, P.T. Saunders [1992]: Light Emission and Rescattering in Syncronously Developing Populations of Early Drosophila Embryos, *Recent Advances in Biophoton Research*, F.A. Popp, K.H. LI and Q. Gu Q. eds, World Scientific, Singapore, 287-306.
5. R. Van Wijk, H. Van Aken, W.P. Mei, F.A. Popp [1993]: Light-Induced Photon Emission by Mamalian Cells, *Journal of Photochemistry and Photobiology B: Biology* **18**, 75-79.
6. B. Chwirot, F.A. Popp [1991]: White Light Induced Luminescence and Mitotic Activity of Yeast Cells, *Folia Histochemica et Cytobiologica*, **29** 155.
7. R. Neurohr [1992]: Non-Linear Optical Properties of Delayed Luminescence from Cress Seeds, *Recent Advances in Biophoton Research*, F.A. Popp, K.H. LI and Q. Gu eds, World Scientific, Singapore, 375-392.
8. H.J. Niggli [1993]: Artificial Sunlight Irradiation Induces Ultraweak Photon Emission in Human Skin Fibroblasts, *Journal of Photochemistry and Photobiology B: Biology*, **18**, 281-285.
9. F. Musumeci, A. Triglia, F. Grasso, A. Scordino, D. Sitko [1994]: Relation Between Delayed Luminescence and Functional State in Soya Seeds, *Il Nuovo Cimento D*, **16**, 65-73.
10. A. Scordino, A. Triglia, F. Musumeci, F. Grasso, Z. Rajfur [1996]: Influence of the Presence of Atrazine in Water on In-Vivo Delayed Luminescence of Acetabularia Acetabulum, *Journal of Photochemistry and Photobiology B*, **32**, 11-17
11. J. Lavorel, J. Breton, M. Lutz [1986] : Methodological principles of measurement of light emitted by photosintetic systems, in *Light Emission by Plant and Bacteria,* J. Govindjee et al. eds , Academic Press, New York, 57.
12. F.A. Popp [1986] : On the Coherence of Ultraweak Photon Emission from Living Tissues. In : *Disequilibrium and Self-Organization*, C.W. Kilmister (Ed), D.Reidel Publishing Company, Dordrecht, 207-230.
13. F.A. Popp [1989] : Coherent Photon Storage of Biological Systems. *In Electromagnetic Bio-Information*, F.A. Popp et al. eds ,Urban and Schwarzenberg Munchen, 144-167.
14. F.A. Popp and K.H.LI [1993]: Hyperbolic Relaxation as a Sufficient Condition of a Fully Coherent Ergodic Field, *International Journal of Theoretical Physics* **32**, 1573-1583.

15. P.A. Jursinic [1986]: Delayed Fluorescence: Current Concepts and Status, in *Light Emission by Plant and Bacteria*, J. Govindjee et al. eds, Academic Press, New York, 291-328.
16. J. Govindjee, M. Van de Ven, C. Preston, M. Seibert and E. Gratton [1990]: Chlorophyll a Fluorescence Lifetime Distributions in Open and Closed Photosystem II Reaction Center Preparations, *Biochimica et Biophysica Acta*,**1015**, 173-179.
17. G.Lasher and F. Stern [1964]: Spontaneous and Stimulated Recombination Radiation in Semiconductors, Phys. Rev. , **113**, A533.
18. A.Scordino, A. Triglia, F. Musumeci, R. Van Wijk [in press] : Delayed Luninescence in Biophysical Investigation , Proceedings of International School of Biophysics on "Chaos and noise in biology and medicine" Island of Ischia 5-10 May 1997

BIOPHOTON EMISSION FROM DEVELOPING EGGS AND EMBRYOS: NON-LINEARITY, WHOLISTIC PROPERTIES AND INDICATIONS OF ENERGY TRANSFER

L.V. BELOUSSOV AND N.N. LOUCHINSKAIA

Laboratory of Developmental Biophysics, Moscow State University, and International Institute of Biophysics, Neuss, Germany

Abstract

Ultraweak photon emission (UWPE) from developing eggs and embryos of sea-urchins, fishes (*Misgurnus fossilus*), amphibians and hens has been measured. Fertilization and (in the case of hen eggs) start of incubation led invariably to the rise of the UWPE level, which decreased again, with some irregular fluctuations, during the subsequent development. In sea urchin and amphibian eggs we detected the UWPE flashes immediately before the cleavage divisions. In *M.fossilus* eggs some more general correlations between the UWPE patterns and first cell cycles could be traced. UWPE from *M. fossilus* eggs was located mostly in the green (400-500 nm) spectral range, that from the early hen embryos (without envelopes) exhibited a substantial UV component (< 300 nm), while the intact hen eggs and their shells emitted exclusively in the red spectral range (≥ 600 nm). UWPE is characterized by a temperature hysteresis and is more sensitive to the temperature gradients than to the absolute temperature values. By hyperbolic decay criteria, the UWPE from the eggs/embryos, as well from their envelopes was highly coherent. In *M. fossilus* embryos we observed extensive UWPE bursts after slight mechanical interventions. We relate this to the existence of some metastable storage of photonic energy. The UWPE rates of the whole hen and amphibian eggs (together with their envelopes) were largely unequal to the sums of the emission rates from the isolated components of the same eggs. These inequalities could be both negative and positive. We regard these inequalities as the manifestations of wholistic properties of the biophotonic fields. They may indicate a radiation-less transfer of photonic energy from the egg's envelopes to the eggs/embryos themselves.

1. Introduction

The capacity of living organisms to emit some small amount of photons (biophotons) in the visible and UV spectral range is now firmly established (reviews: Popp et al eds, 1989, 1992; Slawinsky, 1988). Meanwhile, it remains a matter of strong controversies,

J. J. Chang et al. (eds.), Biophotons, 121-141.

whether the ultraweak photon emission (UWPE) is a mere "metabolic noise", produced in near-equilibrium conditions by some by-products of the oxidative reactions, only slightly correlated with each other and the main biological functions (e.g. Zhuravlev, 1972), or should be treated as a set of highly cooperative, wholistic non-linear phenomena far removed from the thermodynamical equilibrium (Popp, 1993). It is worth mentioning that the latter concept can include the first one as an extreme (border) case.

Remarkably, both concepts have been forwarded long ago by the discoverer of UWPE, Alexander Gurwitsch, and his collaborators (reviews: Gurwitsch & Gurwitsch, 1945; Gurwitsch, 1968, 1988, 1992). On the one hand, these authors showed unambigously that UWPE is fueled by the recombinations of free radicals produced during oxidative and other exergonic reactions, then re-emitted by other substances and hence can be treated, in a number of cases, as a "sensitized fluorescence" with a typical spectral content. On the other hand, they demonstrated a number of collective and relay-like UWPE-associated events and interpreted these properties in terms of a "secondary radiation", branched chain reactions and the formation of common energetic levels embracing large amounts of elementary emitters. Some of the discovered UWPE events (and most of all the so-called degradational radiation) pointed to the existence of molecular photonic stores far removed from thermodynamic equilibrium ("non-equilibrial molecular constellations", Gurwitsch 1945, 1992).

More recently the concept of non-equilibrial photonic storage and non-linear UWPE properties (collective interactions of the photon emitters, memory properties and coherence) became a matter of an extensive elaboration (Popp, 1992; Popp et al., 1994). The main operational criteria for the coherence of biophotons used in these studies was the correspondence of the empirical post-illumination UWPE decays to hyperbolic slopes (Popp & Li, 1993). On the other hand, the more a decay deviates from a hyperbolic slope towards an exponential one, the less coherent should the emission be.

From a mostly general viewpoint the similarity of an empirical decay to a hyperbolic slope means that the elementary photon emitters substantially depend upon each other, while the exponential slopes indicate their independence. Processes in which a great number of elementary components are mutually dependent, often look like avalanches or, in more molecular-oriented terms, branched chain-reactions (see Voeikov, this volume). According to the "self-organized criticality" theory (Bak et al., 1988), these processes also obey the hyperbolic laws. This means, that optical coherence is intimately linked with the dynamics of branched chain reactions.

The main aims of the present investigation were the following ones:

1. To look for the relations between the temporal UWPE patterns, early cleavage cell cycles and developmental dynamics in general.
2. To explore non-linear and wholistic properties of UWPE events, and namely:

2.1. To look whether UWPE responses to temperature shifts include hysteresis loops and a sensitivity to the temperature gradients.

2.2. To induce UWPE flashes by small (very low energy) perturbations.

2.3. To investigate UWPE coherence.

2.4. To compare the UWPE rates from the whole eggs/embryos with those from their constituent parts.

3. We have made also, with the use of the light filters, some rough estimations of the UWPE spectral range in hen and fish eggs.

2. Materials and Methods

All the experiments have been were performed in the laboratories of the International Institute of Biophysics (Kaiserslautern and Neuss). UWPE has been measured by a photoelectronmultiplier (PEM) with the cathode EMI9558 QA, selected type, sensitive within the range of 200 to 800 nm (maximal sensitivity from 200 to 400 nm). For technical details see Mieg et al., 1992. The total duration of the measurments ranged from several minutes to many hours and the dwell time (time of accumulating the photonic signals) from 1s to 5 min. For roughly estimating the UWPE wave-lengths we used the following light filters: BG25 (cutting off frequencies of less than 300 nm; maximal transparency around 400 nm), VG14 (transparency around 500 nm), GG400 (cut-off less than 400nm), RG610 (cut-off less than 600 nm).

The following samples of the eggs and developing embryos have been used for the measurements:

1. Sea urchin (*Psammechinus miliaris*) eggs. Ripe adult *P. miliaris* samples were brought to the lab from Biologische Anstalt Helgoland (courtesy of Dr. Hennemann,) and artificially fertilized. Immediately after fertilization the eggs were placed into a quartz cuvette with a small amount of sea water and shaken before each next measurement which lasted 3 min. Without such a shaking the eggs rapidly sedimented onto the cuvette bottom becoming thus inappropriate for the measurements. In the total, the measurement period covered the time from fertilization to the first cleavage division (about 40 min). 3 separate measurment runs have been performed, all of them giving similar results.
2. Fish (*Misgurnus fossilus*) eggs obtained by artificial fertilization from hormonally treated adult samples. UWPE measurements covered the entire developmental period from fertilization up to a hatching. Eggs kept in a tape water were put into a quartz cuvette and mounted within a narrow slit between a piece of a plankton net and a vertical quvette wall exposed towards PEM cathode. This kind of egg mounting permitted us to produce, in a special set of experiments, the mechanical excitation of the batch by gently pressing (without damaging) the eggs within few seconds by a forceps blade through the plankton net. Since such an intervention is associated with the opening of the PEM chamber and hence with some illumination of the samples, we made the experiments in a semi-dark room, checking each time the effect of the illumination, as well as that of the forceps' immersion without nudging the eggs or by nudging non-developed eggs.
3. Amphibian eggs. They included artificially fertilized *Xenopus laevis* eggs kindly provided by Dr R. Van Wijk and Dr E. Boon (Department of Cell and Molecular Biology, Utrecht University) and *Rana temporaria* eggs obtained from ponds. In the case of *X. laevis* eggs the measurements covered the entire developmental period from fertilization up to the tadpole stage while *R.temporaria* eggs were measured only during the cleavage period. *X. laevis* eggs, similarly to the fish

eggs, were placed into a narrow slit between a vertically oriented coverglass and a cuvette wall, oriented towards the PEM cathode. *R. temporaria* eggs with their extensive jelly coats filled all of the cuvette space. In several experiments we have measured the photon emission from the isolated egg envelopes without eggs.

4. Hen eggs and their constituents. UWPE has been measured from:

4.1. Isolated blastoderms and embryos, cultivated in physiological solution within a quartz cuvette under 37^0 C. Early blastoderms in most experiments were attached by thin parafilm strips to the vertically oriented agarose-coated coverglasses, mounted closely to the PEM-exposed cuvette wall. More advanced embryos (more than 4 incubation days) were simply put into a cuvette.

4.2. Single intact yolks (together with their blastoderms) gently poured out of eggs incubated for 1 - 4 days into empty dry quartz cuvettes. The yolks were oriented towards the PEM cathode either by their embryonic (blastodermal) side, or by the opposite side.

4.3. Similarly treated 33 hr yolks, covered by pieces of a shell taken from the same eggs (yolk + shell combinations).

4.4. Developing, non-fertilized or dead (boiled) eggs; eggs with perforated shells; albumenless eggs; albumen proper; isolated shells.

Each kind of experiment was accompanied by an adequate control (reference) measurement of an emission rate coming from the mounting chamber + medium without a living sample, or from a dead or fixed (killed) sample. We have also made several long-term measurements of a dark (background) PEM emission. In the great majority of cases it was in the range of 10 - 15 counts per second (cps) and never exceeded 20 cps.

3. Results

3.1. GENERAL DEVELOPMENTAL DYNAMICS OF UWPE

Fertilization of sea-urchin, fish and amphibian eggs (and in the case of sea-urchin eggs, also their activation by sea water) invariably led to a significant UWPE increase over that registered in the reference counts, non-fertilized and killed eggs (Table 1, NN 2, 6-8, cf 1; Table 2, NN 2,3, cf 1, 4a, 5a, 14-16). Among the chicken embryos only those renewing their development under the normal incubation temperature rapidly (within about 10 min) increased their UWPE level over that of a background (Table 3, cf 1-1a and 2-4 with 2a-4a). A relatively high UWPE rate have been generally maintained throughout an entire early development, later on gradually declining (Tables 2-4). In amphibian embryos the UWPE level came towards that of a background at a neurula stage. The tadpoles exhibited sudden emission bursts only due to their motile activity. The fish embryos UWPE dynamics was generally the same, although if with some irregular variations.

TABLE 1. UWPE from sea urchin (*Psammechinus miliaris*) eggs. Counts are per 2 s. Duration of measurements is from 3 to 10 min. pf means postfertilization time. In this and other tables densely outlined are the counting data which significantly exceed the background level ($p > 0.95$).

N	Description of cases	Photon counts
1	Control (cuvette+ sea water)	29.6 ± 13.8
2	non-fertilized eggs, 0-3 min in sea water	**120.0 ± 35.0**
3	non-fertilized eggs, 3-6 min sea water	38.2 ± 14.5
4	0-3 min pf	42.3 ± 15.7
5	3-6 min pf	46.6 ± 14.8
6	9-12 min pf	**90.1 ± 18.6**
7	21-34 min pf	**81.4 ± 15.6**
8	34-47 min pf	**102.9 ± 17.1**
9	47-57 min pf (1st cleavage furrow)	39.9 ± 17.0
10	57-67 min pf	40.7 ± 16.4
11	63-73 min pf	43.7 ± 15.7

TABLE 2. UWPE from Xenopus laevis eggs and embryos. Dwell times are 10 s. Duration of measurments is 1.5- 2 h.

N	Description of samples	Photon counts
1	Non-fertilized eggs	574 ± 91
2	Immediately after fertilization, 1	**754 ± 153**
3	Immediately after fertilization, 2	**662 ± 118**
4	8 blastomeres	**704 ± 79**
4a	Same samples, killed by ethanole	542 ± 86
5	Early cleavage	**624 ± 96**
5a	Same samples, killed by ethanole	524 ± 85
6	One gastrula stage embryo, animal pole to PEM	637 ± 80
7	Another gastrula stage embryo, animal pole to PEM	614 ± 75
8	6 gastrula stage embryos, animal pole to PEM	**677 ± 79**
8a	Same 6 embryos, vegetal pole to PEM	**749 ± 97**
9	6 late gastrula embryos without egg envelopes	568 ± 87
10	35 late gastrula embryos with their egg envelopes	**1048 ± 155**
10a	egg envelopes from these embryos	**837 ± 166**
11	35 neurula stage embryos with their egg envelopes	545 ± 68
11a	egg envelopes from these embryos	**643 ± 156**
12	tadpoles	**695 ± 68**
13	another batch of the tadpoles	564 ± 52
14	control (cuvette + water)	581 ± 98
15	control (cuvette + water)	552 ± 69
16	control (cuvette + water)	554 ± 63

TABLE 3. Spontaneous UWPE from the isolated yolks + embryos (dwell time is 5s, measurements durations is10 min). In this and the next table shown are the differences between the cuvettes + samples and empty cuvettes counts. If not mentioned, incubation t^0 is 37C. Samples exposed to PEM cathode by the blastodiscs are defined as "embryo side" and those oppositely oriented as "yolk side". NN 1-1a are non-fertilized sample (nf). All the other samples are fertilized.

N	Incubation time	Photon counts
1	0 - 10 min, nf	31.0 ± 6.9
1a	10 - 20 min, same sample	5.6 ± 4.9
2	0 - 10 min	65.5 ± 7.3
2a	10 - 20 min, same sample	243.6 ± 6.4
3	0 - 10 min	- 45.0 ± 6.4
3a	10 - 20 min, same sample	168.0 ± 9.5
4	0 - 1.5 h, 25°C	70.3 ± 4.6
4a	same sample, 37°C	155.8 ± 4.8
5	12 h	78.5 ± 5.7
6	16 h	55.2 ± 4.8
7	23 h	153.1 ± 2.1
8	30 h	357.8 ± 4.9
9	30 h, yolk side	60.3 ± 5.2
9a	same sample, embryo side	536.9 ± 7.6
10	33 h, yolk side	245.2 ± 4.9
10a	same sample, embryo side	340.3 ± 5.0
10b	same, yolk destructed	159.7 ± 5.8
11	37 h, yolk side	101.8 ± 4.0
11a	same sample, embryo side	231.9 ± 11.2
12	74 h, yolk side	60.9 ± 4.7
12a	same sample, embryo side	111.0 ± 5.8
13	96 h, yolk side	38.9 ± 4.9
13a	same sample, embryo side	34.0 ± 4.5
14	97 h, yolk side	34.1 ± 4.1
15-1	5-6 days, four samples	≈ 0

3.2. UWPE PATTERNS AND EARLY CLEAVAGE CELL CYCLES

The first evidences for the existence of premitotic UWPE bursts (located in UV spectral range) have been obtained long ago with the use of the biodetector (yeast) technique in sea-urchin eggs (Frank & Zalkind, cited from Gurwitsch 1945, 1968). Later on similar effects have been registered by physical methods in budding yeasts (Konev et al., 1966; Quickenden & Tilbury, 1983; Wei-ping Mei, 1992), larch microspores (Chwirot, 1992) and during the first cleavage division in frog eggs (Oonuki & Ryuzaki, 1992). Although,

TABLE 4. Spontaneous UWPE from the isolated (removed from yolk) chicken blastoderms and embryos, maintained in physiological solution. Dwell time is 5s, duration of measurements is 10 min.

N	Incubation time	Photon counts
1	0 - 10 min	**57.5 ± 6.1**
2	3 h	7.4 ± 5.8
3	5 h	**175.1 ± 11.6**
4	6.5 h	**101.1 ± 6.3**
5	7 h	**136.9 ± 9.2**
6	8 h	**65.4 ± 7.4**
7	10 h	**66.2 ± 10.8**
8	3 days	**161.8 ± 6.0**
8a	same embryo, side view	**96.5 ±3.8**
8b	same embryo, scrambled onto a cuvette bottom	**112.2 ± 5.3**
9	3 days	**53.6 ± 4.7**
10	3 days	**47.8 ± 6.3**
11	3 days	**59.5 ± 7.1**
12	5 days	11.5 ± 6.3
13	5 days	**27.0 ± 7.2**
14a	5 days	**25.2 ± 5.9**
15	6 days	5.3 ± 6.0
16	6 days	6.7 ± 5.9
17	6 days	-0.6 ± 6.8
18	8 days	-19.1 ± 7.0

however, these data met sometimes a skeptical attitude, we decided to repeat the similar measurements, although if without making any spectral discriminations. Quite definite results have been obtained on sea-urchin eggs where, in a complete accordance with Frank & Zalkind data, a powerful UWPE burst (in addition to that associated with the eggs activation) have been registered immediately before the 1st cleavage division (Table 1). In 3 experimental runs performed in amphibian eggs we could also detect obvious premitotic UWPE bursts, lasting no more than few seconds each (Fig. 1 A, B). In fish eggs these effects have been shadowed to a great extent by other fluctuations in the emission intensity. However, even in these cases we were able to show some overall correlations between UWPE patterns and cell cycles. For doing this, we brought together the UWPE records in pairs synchronized according to the 1st cleavage divisions and traced the dependence of the correlation coefficients upon the mutual temporal shifts of the records (Fig. 2). In 9 out of 15 such pairs definite correlation maximae could be detected just under zero temporal shifts (ZTS), indicating that the UWPE patterns are really related to the cell cycles (Fig. 2 A-C). In 4 cases no correlation at all could be detected and in 2 cases the correlation maximae were far removed from ZTS.

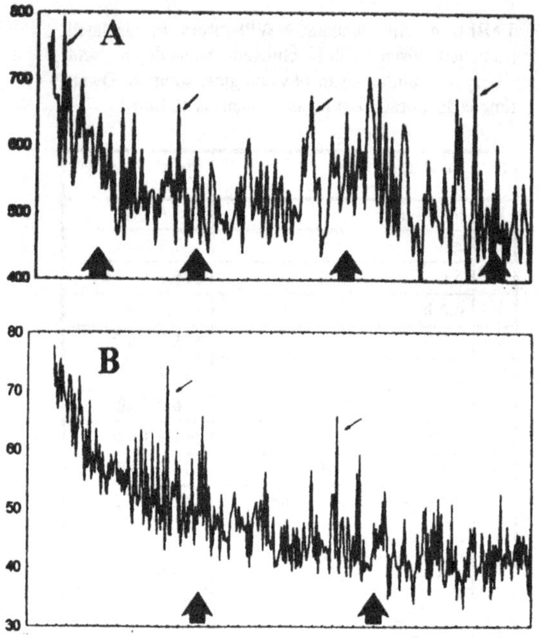

Figure 1. UWPE records of the cleaving eggs batches. Solid upwards arrows indicate the times of the cleavage furrowing. Fine oblique arrows point to pre-cleavage UWPE bursts. A: *Xenopus laevis* eggs, 2nd - 5th cleavage divisions. Dwell time is 10s, no averaging. B: *Rana temporaria* eggs, 2nd - 3rd cleavage divisions. In B dwell time is 2s, 20s averaging. The total measurment time in A was 110 min and in B 1.5 h.

Meanwhile, if one of the paired records corresponded to the reference measurments, a definite correlation maximum between the records have been observed only in 1 case out of 13. In the other cases the correlation patterns were largely smoothed or fragmented (Fig. 2D).

3.3. UWPE DEPENDENCE UPON TEMPERATURE

This was highly non-linear. Both *M. fossilus* and *X. laevis* batches showed quite obvious hysteresis loops (Fig. 3 A, B). They were much more sensitive to the temperature time derivates (rapid temperature uprises and drops) than to the absolute temperature values.

In *M. fossilus* samples the following UWPE properties were of a special interest: (1) The UWPE rate registered at the upper extreme of an optimal temperature range (26^0C) did not exceed that checked under the optimal temperature value ($18-19^0$C), if the latter one have been kept constant for no less than 2 h; (2) The UWPE rate registered during the rapid temperature shifts did not exceed that of the reference level; (3)worth mentioning, the UWPE detected at the lowest temperatures (about 14^0C) was even below the reference level.

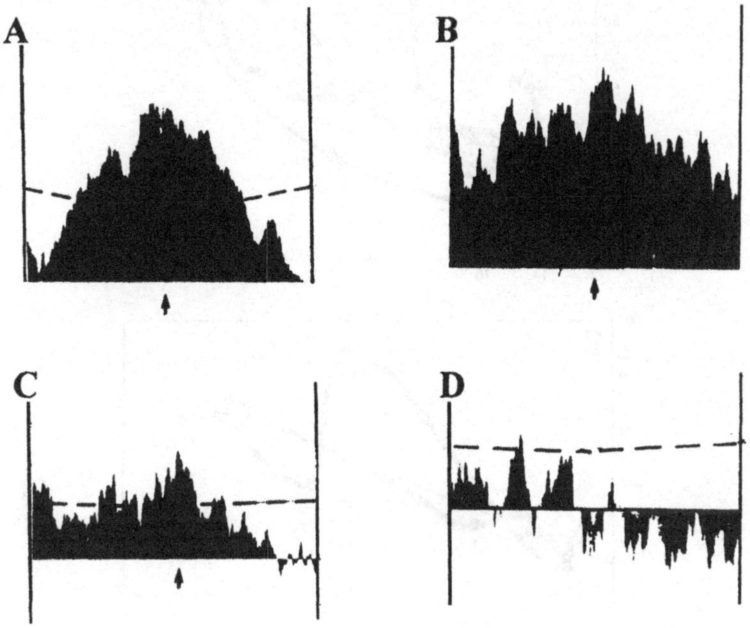

Figure 2. Correlation patterns between the pairs of the UWPE records from the different *M.fossilus* eggs (A-C), as compared with an egg/reference UWPE pair (D). The initial records point was always taken in 20 min before the 1st cleavage division. Vertical axis: correlation coefficients, dotted is 0.95 significance limit. Horizontal axis: temporal shifts. Pointers indicate zero temporal shifts (ZTS) of the records. Maximal temporal shifts are ±10 min.

The UWPE from the developing hen eggs (contrary to that from non-fertilized ones) were also much more sensitive to the temperature gradients, than to the absolute temperature values (Fig. 4 A). Another peculiar property of the developing eggs and even their isolated shells (rather than the dead eggs) was their ability to specifically react to the normal incubation temperature (36-37°C). This is illustrated by Fig. 4 B, C and by the following experiment.

If taking a fertilized non-incubated egg from 5°C temperature and put it into the measuring chamber at 27°C, its emission level will firstly go up, but then rapidly declines up to a very low level, inspite of the temperature remaining constant. Meanwhile, after raising the external temperature to another 9 degrees (that is, up to 36°C) the UWPE rate immediately goes up again, becoming now stabilized at the achieved level. Therefore, in the developing eggs the UWPE rate is stable at 36°C only, and not at 27°C.

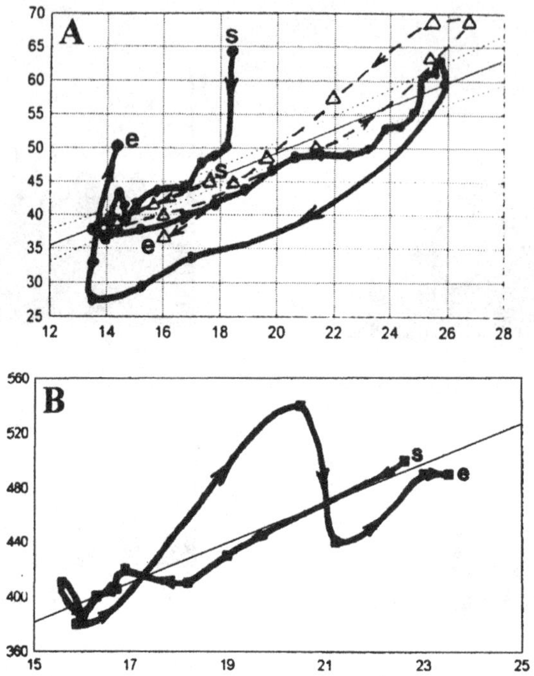

Figure 3. UWPE temperature hysteresis recorded from *M. fossilus* (A) and *X. laevis* (B) cleaving eggs. *s* indicates a starting point of a record while *e* its end. Dotted line in A is the record of a cuvette+solution temperature dependence. Horizontal axis: temperature, C^0. Vertical axis: photon counts per 2 s (A) and 10 s (B).

3.4. MECHANOEMISSION

Starting from a mid-epiboly stage (24 h of development) *M. fossilus* embryos exhibited very pronounced UWPE responses to a slight nudging (15 independent measurments). In few of them the response looked like a slight increase of the UWPE level (Fig. 5 A), but more often it took a shape of a sudden burst with a rapid decay (Fig. 5 B, C). A scatterplot representation of the records (Fig. 5 D) makes it visible the "splashes" and "affiliations" of the main emission trends, indicating the branching chain structure of these reactions and the existence of the metastable photonic storages. Sometimes the bursts led to a stable increase of UWPE level, lasting for at least several dozens minutes (Fig. 5 C). Neither dead nor abnormally arrested embryos revealed any of such responses (Fig. 5 C).

3.5. EVALUATIONS OF THE UWPE COHERENCY FROM THE WHOLE EGGS AND THEIR ENVELOPES

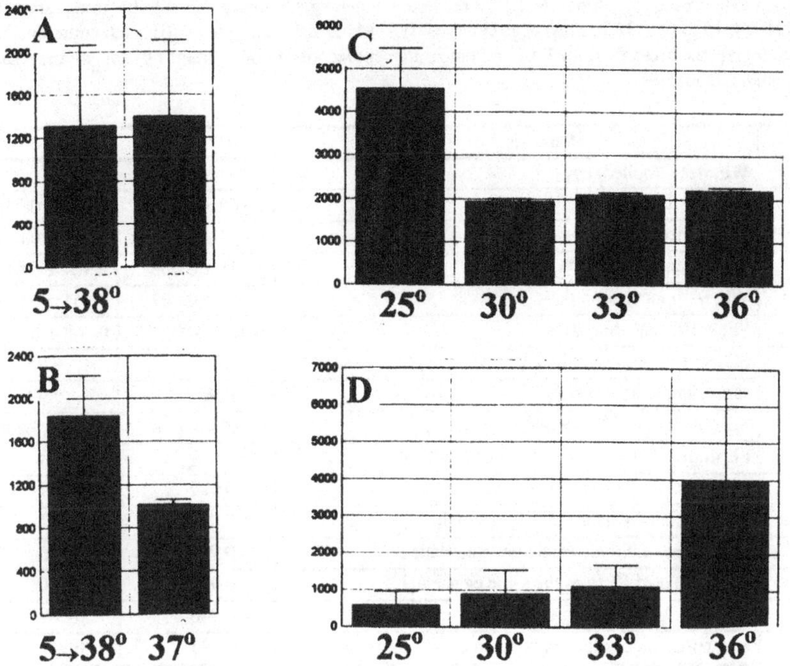

Figure 4. A sensitivity of the developing (as opposed to non-fertilized) hen eggs to the temperature gradients and to the normal incubation temperature. A: the average amount of photons emitted within 40 min by a non-fertilized egg under the temperature shift from 5 to 38^{0}C and under constant 37^{0}C is practically the same. B: a fertilized egg at the start of incubation emits under the temperature shift much more photons than at the constant incubation temperature. C: the UWPE from a dead (boiled) egg is almost indifferent to the temperature changes. Its greatest value under 25^{0}C is due to a post-illumination reaction. D: on the contrary, even a shell taken from a 2 incubation days egg shows a pronounced UWPE maximum just under 36^{0}C. Ordinates: photon counts per 5 min.

In almost all the samples the post-illumination UWPE decays showed a better correspondence with the hyperbolic, rather than the exponential slopes (Table 5, Fig. 6 A-D). Among the whole hen eggs the correspondence with the hyperbolic slope was the lowest in the dead and non-fertilized samples and increased with the age of embryos. Isolated intact shells (with a small perforation only) demonstrated more or less the same dynamics (Table 5, NN 2, 4, 6, 7). Shell emission (SE) directed inside the egg was no less coherent than that directed outside (cf NN 7 and 8). However, any serious damage of a shell geometry led to a considerable decrease of its emission coherency: thus, even in the large shell fragments and, moreover, in the finely dispersed shells the emission decays were closer to the exponential, rather than the hyperbolic slopes (NN 9, 10). Hence, the retention of the ovoidal shell shape seems to be indispensable for a collective interaction of its elementary photon emittors, this very interaction providing the emission coherency.

TABLE 5. Correlation coefficients of the UWPE post-illumination decays with either hyperbolic (H_c) or exponential (E_c) slopes. NN 1-10 relate to the hen eggs samples, NN 11, 12 to the amphibians and NN14, 15 to fish *(M.fossilus)* eggs/embryos. WE: whole eggs emission. SE: shell emission. NN 1-6 and 9 give the averages from 7 different measurement records. In other frames shown are the results of the individual records.

N	Samples	H_c	E_c
1	WE, dead (boiled) eggs	0.951 ± 0.052	0.861 ± 0.070
2	SE, non-fertilized eggs	0.965 ± 0.078	0.905 ± 0.052
3	WE, 2 incubation days	0.977 ± 0.040	0.947 ± 0.021
4	SE, 2 incubation days	0.988 ± 0.005	0.964 ± 0.040
5	WE, 9-10 incubation days	0.990 ± 0.007	0.941 ± 0.011
6	SE, 9-10 incubation days	0.978 ± 0.054	0.968 ± 0.073
7	SE from 10 days sample, 2 days days kept in isolation	0.98	0.95
8	same sample from inside	0.99	0.90
9	SE from large shell pieces	0.960 ± 0.008	0.976 ± 0.006
10	SE from a portion of a finely scattered shell	0.919	0.984
11	100 *R.temporaria* eggs, late cleavage	0.929	0.747
11a	newly isolated jelly coats (JC) from the same eggs batch	0.904	0.772
12	35 *X. laevis* embryos at late gastrula stage	0.963	0.716
12a	newly isolated JC from the same eggs batch	0.908	0.760
13	Cleaving *M. fossilus* eggs, postillumination decay	0.948	0.868
14	*M.fossilus* midblastula embryos, mechanoemission decay	0.985	0.941

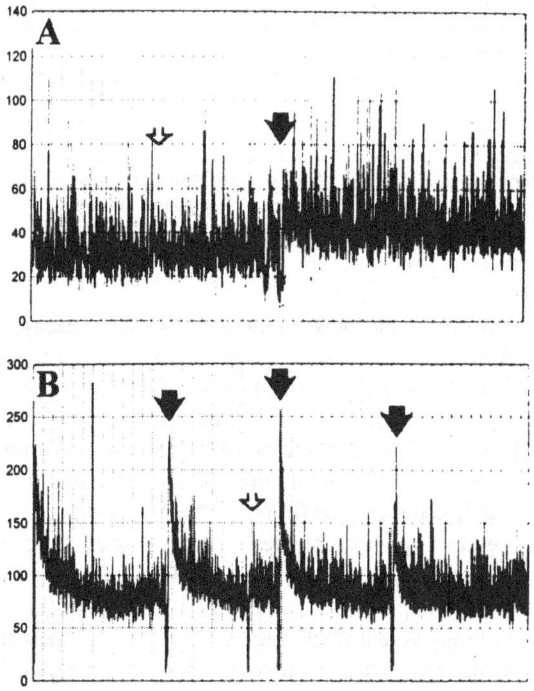

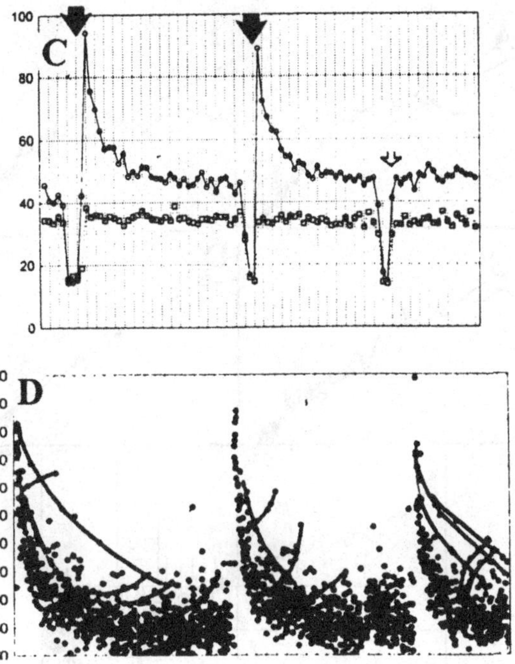

Figure 5. Changes of UWPE rate after nudging *M. fossilus* embryos (mechanoemission). Filled arrows indicate nudgings, empty arrows are PEM door openings without nudging. A: 10 h of development, reaction to the nudging still slight. B-D: 24 h of development (mid-epiboly stage), mechanoemission bursts highly pronounced. In C the rectangles show a simultaneously recorded reference level. Note that already after the first nudging a steady state UWPE value in the embryonic batch shows a stable increase from 40 cps (a brief left part of the record) up to about 50 cps. D is a scatterplot version of some part of B record. Connected are the mostly visible "affiliations" of the main emission trend. In A, B the total length of the record is 80 min and in C it is 50 min.

The UWPE decays from the amphibian and fish eggs, as well as from their isolated envelopes also showed a better correspondence with the hyperbolic, rather than exponential slopes (NN 11-14). Interestingly, in *M. fossilus* eggs the mechanoemission decay was much closer to a hyperbolic slope than the ordinary post-illumination decay (cf NN 13 and 14). Spontaneous UWPE uprises in *M. fossilus* eggs also obeyed a (reverse) hyperbolic law (Fig. 6 E).

3.6. NON-ADDITIVITY OF UWPE AND THE INDICATIONS OF A RADIATIONLESS ENERGY TRANSFER FROM THE EGG ENVELOPES TO THE DEVELOPING EGGS/EMBRYOS PROPER

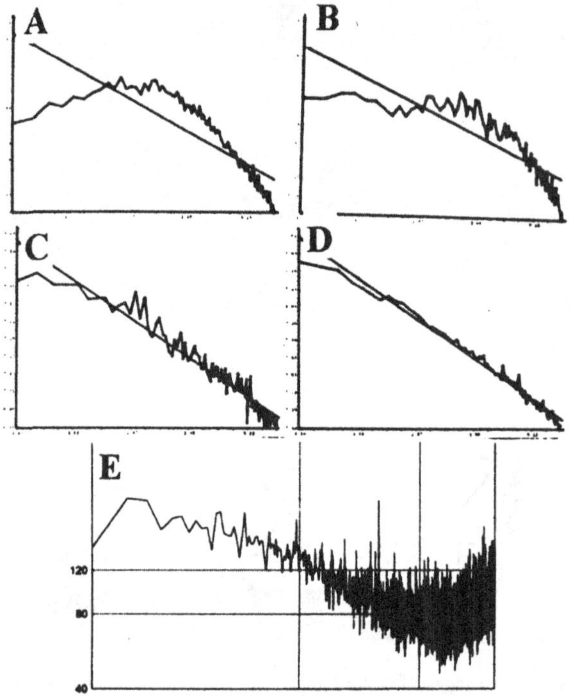

Figure 6. Some of UWPE records plotted in log-log coordinates (horizontal axis: time, vertical axis: photon counts). A-D: post-illumination decays in hen eggs. Straight lines correspond to hyperbolic slopes. A: boiled egg. B: non-fertilized fresh egg. C: fertilized egg, start of incubation. D: 3 incubation days egg. From A to D the empirical slope is definitely approached towards a hyperbolic one. E: a record from a batch of the cleaving *M. fossilus* eggs. Not only a post-illumination decay (left part) but also a spontaneous emission uprise (right part) are close to straight lines (that is obey the hyperbolic laws).

3.6.1. *Hen eggs*

Whole hen eggs, whether they are developing, non-fertilized and even boiled always emit, after being taken from a room light, a great amount of photons (about several thousands cps). Same emission range is typical for the newly isolated egg shells (however, while kept alone, a shell reduces irreversibly its emission level and loses a capacity to be restimulated by light). Following a linear (additive) logic, one should ascribe practically all of the whole egg emission (WE) to its shell, since an egg has no other photon emittors of a comparable range: as mentioned above, an embryo + yolk emits no more than few dozens cps and the albumen (egg's white) do not emit, by our data, any photons at all. It was found, however, that only in the non-fertilized eggs the WE - SE differences were close to zero value, while in all of the developing eggs they were significant (Fig. 7). In most of 2 days eggs, as well as in the yolks+shell

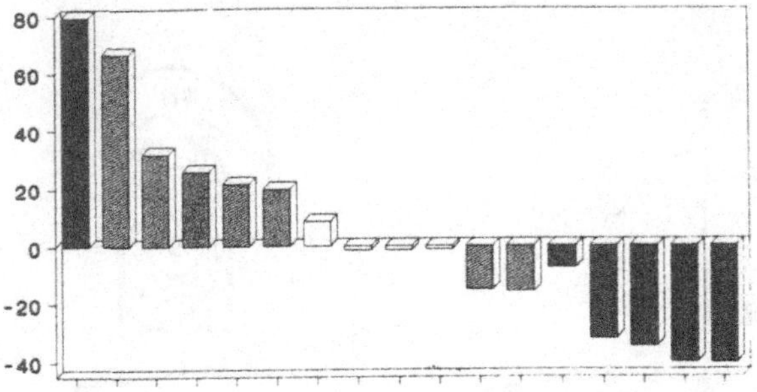

Figure 7. WE-SE differences in the non-fertilized hen eggs (empty boxes), 2 incubation days eggs (hatched boxes) and 9 incubation days eggs (filled boxes). The differences for non-fertilized eggs are insignificant, those of 2 days eggs are mostly positive while those of 9 days eggs are mostly negative. For more comments see text.

preparations from 33 h eggs the differences were positive (WE > SE), while in 9 days eggs most of them were negative (SE > WE). Among the yolk+shell combinations, in 6 cases out of 8 the emission of shell pieces which covered the yolks largely exceeded the emission of the same isolated shell pieces (Fig. 8A). We consider this as an indication of SE stimulation by a yolk.

Among those 2 days eggs which exhibited positive inequalities (WE>SE) in 3 cases out of 7 each next intervention, namely a perforation of a shell, removal of an albumen and removal of a yolk led to a progressive reduction of the emission rate (Fig. 8B). This confirms our suggestion on the stimulation of SE by yolk, looking now to be supported also by the albumen, as well by the very intactness of the shell. As estimated by the first measured post-illumination values, a stimulated SE portion is around 4000 cps, exceeding thus 5-fold an isolated SE rate. About 0.25 of this stimulated value still retains in the albumen-less eggs, indicating again a direct SE stimulation by a yolk. This value is of the same order as obtained in the artificial yolk+shell combinations.

Two next cases have been characterized by reverse inequalities (SE>WE), showing that in the average some part of a shell's photonic energy (corresponding to about 3300 cps) is consumed by an egg's interior (Fig. 8C). Meanwhile, in these cases the removal of an albumen led to SE increase exceeding that of the isolated shell in 3000 cps. The only possible explanation is would be that the remaining yolk stimulates indeed SE, but only in the absence of an albumen. We have to conclude, that the consumption of a shell energy by an egg's interior may go on hand by hand with the oppositely directed SE stimulation.

In other 2 cases WE exceeded SE, whereas the photon emission of the albumen-less eggs was lower than SE. While the first inequality points to some stimulation of SE by an egg's interior, the second one argues that in the albumen-less eggs at least some part of a shell's energy is consumed by the remaining yolk. Again we are dealing with the coexistence of a consumption and a stimulation, the albumen participating now obviously in transmitting towards the shell some of the yolk-derived stimulation

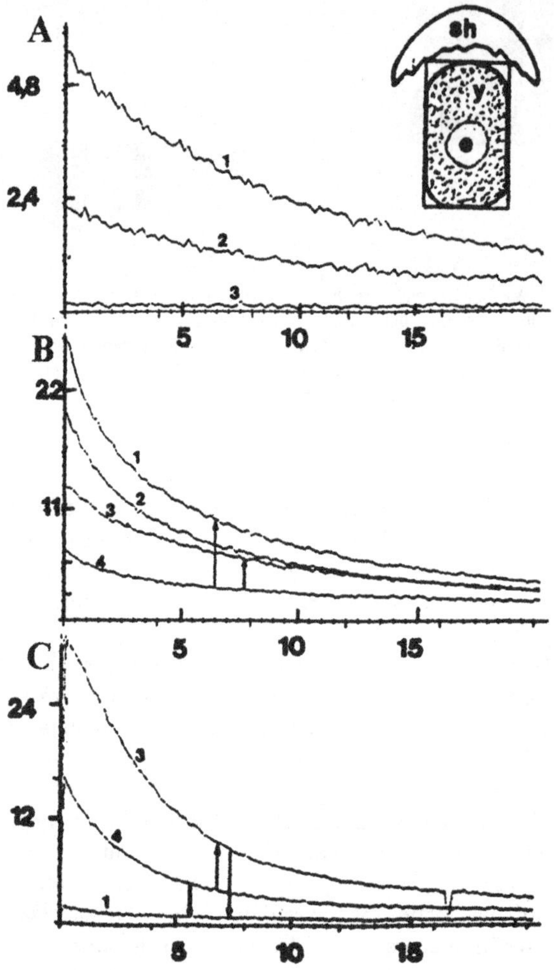

Figure 8. UWPE records from the different components of hen eggs. A: records from the yolk+shell combinations.

1: emission from yolk+shell together. 2: emission from the same shell piece without yolk. 3: emission of yolk alone.

1 is largely greater than 2+3. To the upper left is the components arrangement. Sh - a shell piece, y - yolk within a cuvette, e - embryonic blastodem upon yolk. B, C: records from the different parts of the eggs of 2 incubation days. 1: whole intact egg, 2: egg with a perforated shell (a perforation non-exposed to a PEM cathode), 3: albumen-less egg (yolk+shell), 4: newly isolated shell. Upward arrows in B, C indicate a suggested stimulation of SE by a yolk in the whole egg (1) and in the albumen-less egg (3). Arrows 4→1, 3→1 indicate a suggested SE consumption by an intact and an albumen-less egg correspondingly. Horizontal axis: time, min. Vertical axis: photon counts × 10^3/5 s.

(similaly to the case shown in Fig. 8 B). Such mixed cases seem to be most of all ubiquitous.

3.6.2. *Amphibian eggs*

Similarly to the hen eggs, the UWPE rate of the amphibian eggs/embryos together with their envelopes was largely unequal to the sums of the same components emission. Thus, in *X.laevis* embryos at the gastrula stage the embryos+ envelopes UWPE was substantially greater than that of the envelopes alone (Fig. 9 A), while for *X.laevis* neurula stage embryos (Fig. 9 B) as well as for the cleaving *R. temporaria* eggs the reverse was true. In the latter case the envelopes, while being kept in isolation for 2 days, increased their postillumination emission level in about 4-fold. Meanwhile, after ethanole fixation the emission came down to a background level.

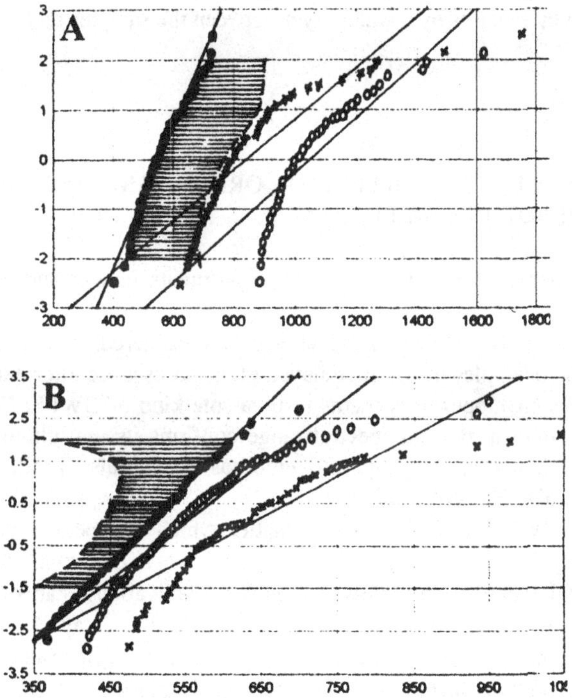

Figure 9. UWPE records from *X. laevis* embryos+envelopes (empty circles) and from the same batches newly isolated envelopes (crosses) as referred to the background photon emission (filled circles). A: 40 midgastrula stage embryos, B: 100 neurula stage embryos. Hatched areas show the (embryos+envelopes) - (isolated envelopes) UWPE differences. In A the difference is positive while in B it is negative. The records are given in the form of the normal distribution plots.

Following the same reasoning as before, we conclude that in those cases when the embryos+envelopes emission exceeded that of the isolated envelopes the embryos stimulated somehow the envelopes emission while in the reverse cases a photonic energy which was produced/accumulated within the envelopes have been consumed by the eggs/embryos. By rough estimations, in *R.temporaria* batch including 90 cleaving eggs the amount of the consumed photon counts per 2-3 postillumination minutes was about 77000, what corresponds to about 10 counts per embryo per second (if such a linear approach is here permitted).

3.7. SPECTRAL ESTIMATIONS

UWPE of early (1-2 incubation days) chicken embryos was almost completely absorbed by all the kinds of light filters used. From this we conclude that it was located mostly in the UV range (<300 nm). The UV component could be also detected from the finely dispersed shell pieces, but not from the intact shells which emitted almost exclusively in the red spectral range (> 600 nm). *M.fossilus* eggs at cleavage stages emitted almost exclusively in green (400-500 nm). We cannot exclude however that a UV component was absorbed in these cases by a water layer between the samples and a PEM cathode.

4. Discussion

4.1.EMISSIONS OF THE DIFFERENT ORIGIN ARE INTEGRATED INTO COMMON BIOPHOTONIC FIELDS

The UWPE phenomenae registered in our experiments consist indeed of at least 3 different groups:

(1) post-illumination UWPE decays, which can be considered, in a first approximation at least, as the examples of a superdelayed phosphorescence after external lightning. As to the eggs envelopes, this seems to be a sole kind of UWPE. For the hen eggs shells, its location in the red spectral range confirms its phosphorescence (Stocks-like) origin. Certainly, the eggs and embryos proper also produce this kind of emission (see also Ho et al., 1992).

(2) spontaneous UWPE from the living matter which can be maintained and even spontaneously increased without any external light. It is correlated with the main developmental events, such as fertilization (eggs activation), start of incubation and cleavage divisions. This kind of UWPE well may be treated as a chemiluminescence, originated from the recombinations of free radicals.

(3) brief UWPE bursts induced by slight mechanical interventions (mechanoemission). We consider these bursts as the discharges of certain pre-existed highly metastable photon storages (their metastability foolows also from a relatively slow development of the "affiliations" of the main burst, see Fig. 5 D). In general, the mechanoemission can be attributed to the "degradational radiation" (Gurwitsch, 1945, 1968, 1992; Popp, 1992), which is known to be inducible by several kinds of non-specific slight interventions. It is noteworthy, that in *M. fossilus* embryos the mechanoemission is firstly indicates only at a mid-blastula stage, that is when the mechanical tensions on

the embryo surface are increased and some crucial events in embryonic determination are taking place.

There are good evidences for suggesting, that all of the above described kinds of UWPE, inspite of their different physico-chemical origin, are integrated, within the whole embryos together with their envelopes and even within the entire egg batches, into common biophotonic fields with a collective non-linear dynamics. Such a view is supported, first of all, by a high degree of UWPE coherency. As shown above (Table 5), this relates not only to the post-illumination decays of both the eggs and their envelopes, but also to the mechanoemission decays and to spontaneous UWPE increases in *M. fossilus* embryos. The next evidence is the UWPE temperature dynamics as characterized by hysteresis loops and a sensitivity to the temperature gradients. And the main argument favouring the existence of the common macroscopic biophotonic fields is based upon a *UWPE non-additivity*, that is a substantial inequality of the UWPE-s from the whole samples to the sums of their components UWPE-s. That makes it possible to extend the fundamental biological conclusion about the wholistic character of an embryonic development also towards those egg parts (envelopes) which were attributed usually to a non-living substance. “Alles Lebendige verbreitet eine Atmosphaere um sich her” (W. Goethe).

4.2. EVIDENCES FOR A RADIATION-LESS TRANSFER OF ENERGY WITHIN A BIOPHOTONIC FIELD

While that kind of non-additivity which correspond to the inequality WE>SE can be interpreted, more or less formally, as a stimulation of shells (envelopes) emission by an eggs interior, the explanation of the reverse inequality (WE<SE) is more intriguing. Three main versions can be suggested in this respect:

(1) The latter inequality can be explained by a pure (“passive”) optical absorbtion of the envelopes emission by the eggs/embryos.
(2) An egg’s interior somehow inhibits the envelopes emission, that is prevents the production of the excited electronic states within the latter.
(3) A photonic excitation energy produced within the envelopes is transferred in a radiation-less manner towards the eggs/embryos interior.

By our view, the first version can be immediately rejected since the optical properties (as well as the positions in relation to the envelopes and the PEM cathode) of the eggs/embryos which *stimulate* the envelopes emission (such as 2 days hen eggs or the gastrula stage *Xenopus* embryos) are practically the same as those of the samples which either *decrease* the envelopes emission (9 days hen eggs, neurula stage *Xenopus* embryos) or do not affect it at all (non-fertilized hen eggs). We have noted also, that among the same stage hen eggs (that is, those sharing the same optical properties) some samples increase the shell emission while others, on the contrary, decrease it.

The second hypothesis also do not seems to be quite plausible, but the arguments against it are more or less indirect. The main one is a difficulty to imagine such a kind of inhibition which, on one hand, is cancelled *immediately* after the removal either of an entire egg’s interior or one of its components and, on the other hand, can be immediately eliminated and overwhelmed by the UWPE-stimulating light exposures. As concerning hen eggs, an additional difficulty is in explaining the inhibition of the photon emittors resided within a solid calcite matrix of an egg shell.

Let us address now to the last possibility, which is at the same time most of all intriguing. It means, that a developing egg/embryo, may be from a definite stage on, directly consumes some part of the excited electronic states energy, which is produced and accumulated in the egg envelopes at the expense of an external light bath. Such an interpretation is in line with Cilento's (1988) "photobiochemistry without light" concept and even earlier Gurwitsch (1945, 1992) ideas about the common energetical levels, embracing the entire macroscopical organs.

At the moment only few suggestions can be made as concerning the biological role of such an energy transportation. One of its main advantages seems to be associated with the above demonstrated high coherency of the envelopes emission: they can be regarded indeed as very weak lasers. Numerous stimulating effects of the laser irradiation upon living tissues, including developing embryos, are now well established. Averyanova et al (1991) and Burlakov et al. (1997) gave a large nnumber of exciting examples of the reparating, stimulating and synchronized effects of the short exposures to a weak infrared laser radiation upon the development of fish eggs. A weak coherent 900 nm radiation was also shown to increase ATP output during mitochondrial respiration (Warnke, 1989). The latter result correlates with our registration of a stable SE>WE difference in hen eggs just at the 9th incubation day, that is when the embryonic respiration pattern, as compared with 2nd days samples, becomes drastically reconstructed and a direct blood vessels pathway is established between a shell and an embryo proper. One may speculate, that a part of a coherent shell's photonic energy is transported towards embryo via blood vessels and is used, in particular, for increasing the effectiveness of an intracellular respiration.

Certainly, the embryonic blood is not the only one suitable substrate for transporting a photonic ebergy in a radiation-less way. We have already presented the evidences indicating that such a transportation may proceed via hen eggs albumen both towards a yolk and in the reverse direction. As concerning amphibian and fish samples, the energy transportation from the envelopes to the eggs proper may go via perivitellinic space which is rich of carbohydrates. Noteworthy, long ago a transfer of a photonic energy via tubes filled with glucose or with the protein solutions have been reported (Gurwitsch & Gurwitsch, 1945). In the mentioned experiments, the transfer distances reached 1 m for a glucose solution and were no less than 20 sm for the protein solutions, while the transfer velocity was about 1 m/s, pointing to a non-photonic transport mechanisms.

Another group of evidences for the transfer of a UWPE energy, now from an external medium towards the developing embryos comes from the cases of the negative (that is, lower than that of a sole medium) UWPE of the living samples, maintained under suboptimal conditions (for example under low temperatures, see 3.4, point (3)). This is in line with Vogel data obtained on bacterial cultures (this volume). Further and more direct experiments are required for approving and exploring in more details a remarkable phenomenon of the photonic energy transfer throughout the living tissues.

5. References

Averyanova, O.V., Burlakov, A.B., Paschenko, V.Z., Sleptzova, L.A., Tusov, V.B., Turovetsky, V.B. (1991) Studies of the influence of laser irradiation upon the early development of *Misgurnus fossilus*, *Vestnik Mosk. Universiteta Ser. Biol.* **1**, 34-40.(in Russian)

Bak, P., Tang, C., Wiesenfeld, K. (1988) Self-organized criticality, *Phys. Rev. A*, **38**, 364-374.

Burlakov, A.B., Averyanova, O.V., Paschenko, V.Z., Tusov, V.B., Golitchenkov, V.A. (1997) Laser correction of the embryonic development in *Misgurnus fossilus*, *Vestnik Mosk. Universiteta Ser. Biol.* **1** (in press, in Russian)

Chwirot, B.W. (1992) Ultraweak luminescense studies of microsporogenesis in Larch, in F.A.Popp, K.H.Li and Q.Gu (eds.), *Recent Advances in Biophoton Research and its Applications*, World Scientific, Singapore etc, pp. 259-286.

Cilento, G. (1988) Photobiochemistry without light, *Experientia* **44**, 572-575.

Gurwitsch, A.A. (1968) *Problems of Mitogenetic Radiation as an Aspect of a Molecular Biology*, Medizina, Leningrad (in Russian).

Gurwitsch, A.A. (1988) A historical review of the problem of mitogenic radiation, *Experientia* **44**, 545-549.

Gurwitsch, A.A. (1992) Mitogenetic radiation as an evidence of nonequilibrium properties of living matter, in F.A.Popp, K.H.Li and Q.Gu (eds.), *Recent Advances in Biophoton Research and its Applications*, World Scientific, Singapore etc, pp. 457-468.

Gurwitsch, A.G. & Gurwitsch, L.D. (1945) *Mitogenetic Radiation, its Physico-Chemical Foundations and the Applications to Biology and Medicine*, Medgiz, Moskva (in Russian).

Ho, M.W.,.Xu, X, Ross, S., & Saunders, P.T.(1992) Light emission and rescattering in synchonously developing populations of early drosophila embryos, in F.A.Popp, K.H.Li and Q.Gu (eds.), *Recent Advances in Biophoton Research and its Applications*, World Scientific, Singapore etc, pp. 287-306.

Konev, S.V., Lyskova, T.I. & G.D. Niesenbaum (1992), Very weak bioluminescence of cells in the ultraviolet region of the spectrum and its biological role. *Biophiysika* **11**, 410-413 (in Russian).

Mei, W.-P. (1992) Ultraweak photon emission from synchronized yeast (Sacharomyces cerevisiae) as a function of the cell division cycle, in F.A.Popp, K.H.Li and Q.Gu (eds.), *Recent Advances in Biophoton Research and its Applications*, World Scientific, Singapore etc, pp. 243-258.

Mieg, C.,.Mei, W.P &. Popp, F.-A.(1992). Technical notes to biophoton emission, in F.A.Popp, K.H.Li and Q.Gu (eds.), *Recent Advances in Biophoton Research and its Applications*, World Scientific, Singapore etc, pp. 197-206.

Oonuki, M. & Ryuzaki, M.(1992) Cortical photons in early development of frog eggs: comparative study of photon emission between nuclear division and cytoplasmic fission, *J.Cell.Physiol* **153**, 518-522.

Popp, F.-A. (1992) Some essential questions of biophoton research and probable answers, in F.A.Popp, K.H.Li and Q.Gu (eds.), *Recent Advances in Biophoton Research and its Applications*, World Scientific, Singapore etc, pp. 1-46.

Popp, F.-A., Gu, Q., & Li, K.H.(1994) Biophoton emission: experimental background and theoretical approaches, *Modern Physic Letters B*, **8**, 1269-1296.

Popp, F.-A., and Li, K.H.(1993) Hyperbolic relaxation as a sufficient condition of a fully coherent ergodic field, *Int. J. Theor. Phys.* **32**, 1573-1583.

Popp, F.A., Li, K.H.and Gu, Q.(eds.), *Recent Advances in Biophoton Research and its Applications*, World Scientific, Singapore, New Jersey, London, Hong Kong.

Popp, F.-A., Warnke, U., Koenig, H.L., Peschka, W.(eds.), *Electromagnetic Bio-Information*, Urban & Schwarzenberg, Munchen, Wien, Baltimore.

Quickenden, T.I. & Tilbury, R.N. (1983), Growth dependent luminescence from cultures of normal and respiratory deficient Sacharomyces cerevisiae, *Photochem. & Photobiol.* **37**, 337-344.

Slawinski, J. (1988) Luminescence research and its relation to ultraweak cell radiation, *Experientia* **44**, 559-571.

Warnke, U. (1989) Influence of light on cellular respiration, in F.-A. Popp, U. Warnke, H.L.Koenig, W. Peschka (eds.), *Electromagnetic Bio-Information*, Urban & Schwarzenberg, Munchen etc, pp. 213-220.

Zhuravlev, A.I. (1972), Ultraweak Luminescence in Biology, *Trudy Mosk. Obsch. Yestestvoispit. (Moscow Society of Naturalists)* **39** (in Russian).

MEASURING WEAK LIGHT SIGNALS NOT FAR FROM NOISE-LEVEL

W. HEERING

Lichttechnisches Institut
Kaiserstraße 12
76131 Karlsruhe

1. Fundamentals

Detecting extremely weak light emission means to measure photon fluxes down to a few photons per second or radiation power of the order of 10^{-19} W in the VIS. Though this is technically possible to do, one needs special detectors and methods of signal processing and it can be performed mostly only on very slowly changing light signals over long integrating or counting times.

It is noise being superimposed over the signal, that limits detection of small light signals. Noise signals are of random nature so that the average value will tend to zero with detection time. The mean square deviation from the average is however different from zero for noise signals.

The ultimate performance of optical detection is reached when there is no noise generated within the detector itself, no amplifier noise and there is is no electromagnetic interference and no radiating background against which the light signal has to be detected. Under these conditions, the electric pulses generated by the light detector are due only to signal photons. As the arrival of photons is randomly distributed over time, shot noise is introduced into the light signal. So, the quantum nature of light truely determines the detection limit, the so called signal fluctuation limit SFL.

Often, the detection limit is not the signal fluctuation limit but the background fluctuation limit BFL. This is why the emitting area is surrounded by a radiating background. The radiation incoming onto the detector is composed of that from the emitter and of that from the background. As the background radiation, normally a thermal radiation from objects at an ambient temperature of about 300 K, is also fluctuating, the signal due to it contains a noisy component which cannot be eliminated by subtraction. Under most operating conditions the background fluctuation limit is the ruling one when infrared radiation is detected, whereas the signal fluctuation limit is operative for UV and VIS detection.

Really, signal detection is not only limited by shot noise of the light signal, but also by additional noise as thermal noise (Johnson noise) found in all resistive materials, including semiconductors, as generation-recombination noise caused by fluctuations in the rates of thermal generation and recombination of free carriers in semiconductors, as flicker (1/f) noise being associated with the presence of potential barriers at the contacts

J. J. Chang et al. (eds.), Biophotons, 143-157.

or surface of semiconductors and as shot noise of the dark current of light sensors. So, the real detection limit is often considerably higher than the signal fluctuation limit. While flicker noise and generation-recombination noise have a power spectrum which falls to higher frequencies as $1/f^n$ with $0.8 < n < 1.5$ respectively $n = 2$, shot noise and thermal noise exhibit a white noise spectrum [1], i.e. its spectral power is independent on frequency, as it is schematically illustrated in Fig. 1 for a photoconductor in the absence of radiation. As a guide, noise becomes nearly white at frequencies beyond 1 kHz. So, there are two ways to reduce noise in the detected photo-signal.

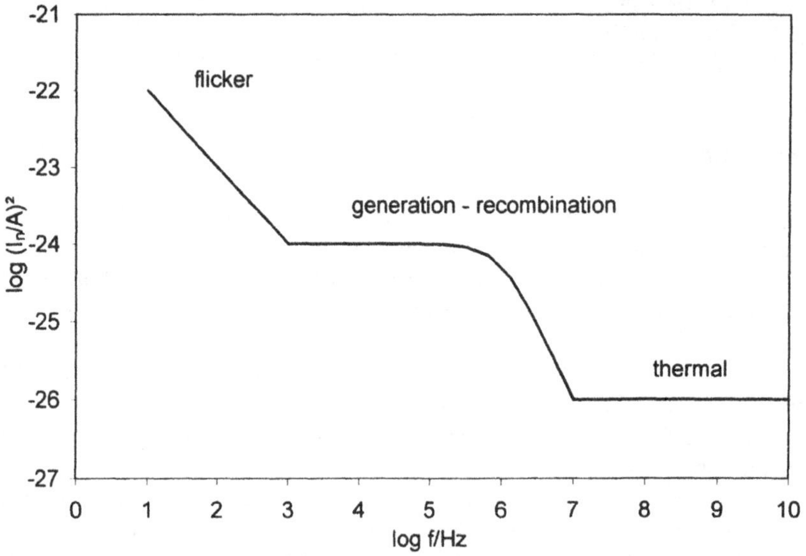

Figure 1. Mean square noise current in a photoconductor as a function of frequency for unit bandwidth.

First, the light to be measured should be modulated and detected at high enough frequencies so that flicker and g-r noise are not involved. Secondly, the electrical bandwidth of the detection system has to be reduced by using active or passive narrow-band filters. However, filtering needs a higher frequency stability of the signal and of the center frequency of the filter. Furthermore, the time of response to an input signal increases as the bandwidth decreases. Thus, the narrower the filter, the slower the signal may be allowed to change if it is not to be distorted, i.e. you buy an improvement in signal-to-noise ratio with (measuring) time.

2. Signal and Background Fluctuation Limits for Ideal and Real Detectors

For any real situation, the observer can wait only a finite length of time for a photon to be detected. So, there must be a reasonably high probability that at least one photon will be detected during the observation time τ_0 for a given average photon flux onto the detector. This is equivalent to say that the probability $p(0) = \exp(-\overline{N})$ that no photon will be detected during τ_0 should be at most 0.368. Here, Poisson statistics can be applied, because the average number of photons $\overline{N}$ being detected within the observation interval τ_0 is assumed to be 1, i.e. it is small. In order to detect one photon during τ_0, a minimum radiation power SFL falling onto the detector with quantum efficiency η is required so that $\eta \tau_0 SFL / (hc/\lambda) = 1$.

Radiation measurements are usually analogue and not digital. In order to reduce amplifier drift and to raise signal-to-noise ratio, the analogue detector signal is modulated and received within the modulation bandwidth, the bandwidth of an electrical filter. It can be shown that the power bandwidth B of an integrating filter is given by the reciprocal of twice the time constant corresponding to the measuring time τ_0. So, the signal fluctuation limit SFL is for a monochromatic light source

$$SFL = 2\frac{hcB}{\eta\lambda}. \tag{1}$$

The background fluctuation BFL limit arises from fluctuations in the background radiation which is incident on the detector with sensing area A. Simplifying, the background is assumed to emit blackbody radiation of temperature T into the viewed solid angle $\pi \sin^2(\theta/2)$ of the detector. Here, θ is the full cone angle. The noise equivalent power is per definition that radiation power which produces the same output signal as the noise power so that the signal-to-noise ratio is unity. The detected noise power is here due to the mean square fluctuation in photons emitted by a blackbody and detected by the optical sensor with quantum efficiency η. If the detector is not selective and only responsive in the wavelength range from 0 to λ_c and receives signal radiation from a monochromatic source, one derives the following expression for BFL, the background fluctuation limit [2]:

$$BFL = 2\frac{hc}{\lambda}\sqrt{\frac{BA\pi kT}{h\eta}}\exp\left(-\frac{hc}{2\lambda_c kT}\right)\left[2\left(\frac{kT}{hc}\right)^2 + \frac{2kT}{hc\lambda_c} + \frac{1}{\lambda_c^2}\right]^{1/2}\sin(\theta/2) \tag{2}$$

Here, the Wien approximation has been appplied. Really, both signal fluctuation as well as background fluctuation are generally active at the same time. The minimum detectable monochromatic power as a function of wavelength for a composite of signal and background fluctuation is shown in Fig. 2 for detector area of 1 cm^2, for bandwidth of 1 Hz, a 2π steradian field of view, background temperature of 300 K and unit quantum efficiency of the detector. The detector long wavelength limit is assumed to be equal to the wavelength of the monochromatic signal source. As an example, for an area of 1 cm^2 and bandwidth of 1 Hz the signal fluctuation limit dominates below 1.0 μm while the background fluctuation limit is operative above 1.0 μm with a steep dependence on wavelength. Here, at 1.0 μm the minimum detectable monochromatic radiant power is 6.8 10^{-19} W corresponding to about 3 photons per second.
In fact, detection limits are higher because the quantum efficiences of the real detectors are smaller than one. The signal detection limit is further increased by the shot

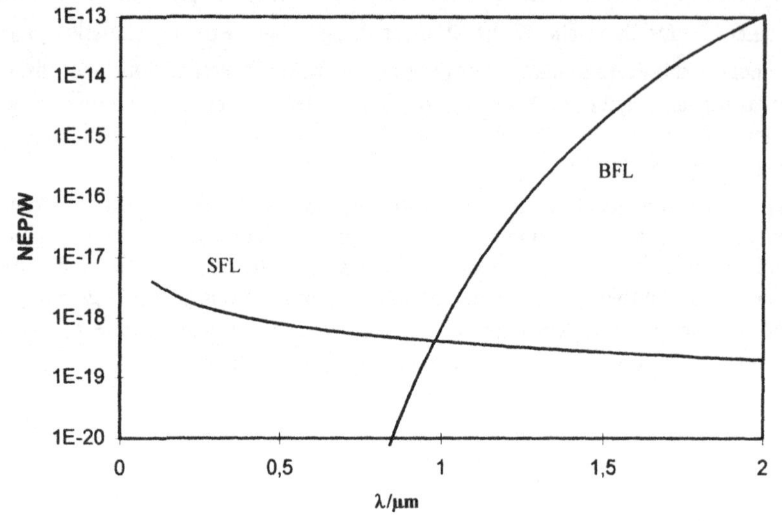

Figure 2. Minimum detectable monochromatic power NEP versus wavelength λ for composed signal and background fluctuations under conditions described in the following text.

noise of the dark current i_d of detectors and the thermal noise of the detector load resistance R_L and the equivalent noise resistance R_{eq} of the signal amplifier. This effects an enhanced noise equivalent power NEP [3]:

$$NEP = \frac{hcB}{\eta\lambda}\left[1+\sqrt{1+\frac{2\gamma q i_d + 4kT/(R_L + R_{eq})}{(\gamma q)^2 B}}\right] \tag{3}$$

Expression (3) turns to the lower limit according to (1) if temperature T is brought to zero, because dark current normally will vanish at absolute zero. (3) as well as (2) imply a signal-to-noise ratio of one. The reciprocal of NEP is the detectivity. It is presented in Fig. 3 as a function of wavelength for different optical detectors responding in the 100 nm to 1200 nm region [4]. Obviously, the sensors with the highest detectivity are photomultipliers, especially in the UV and short range VIS. They are less detective than an ideal detector at least by a factor of ten. The signal fluctuation limit of an ideal detector can nearly be reached by selected cooled (light sensitive) charge coupled devices, CCDs. These MOS capacities are made of silicon the gate electrodes of which are separated from the semiconductor by a silica layer. Mainly the noise of its dark current effects a slightly higher detection limit than that of an ideal detector. The noise equivalent charge of such a light energy measuring device can be down to a few electron charges.

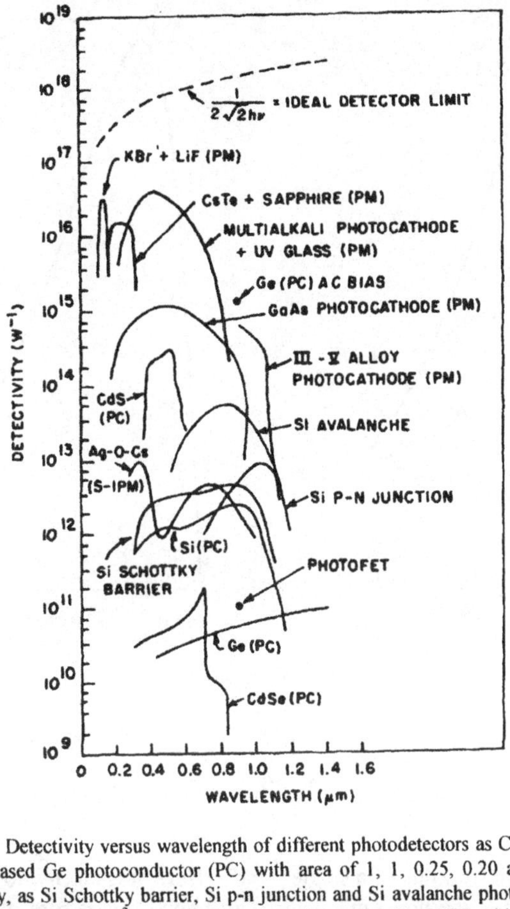

Figure 3. Detectivity versus wavelength of different photodetectors as CdS, CdSe, Si, Ge and AC biased Ge photoconductor (PC) with area of 1, 1, 0.25, 0.20 and 2.4 10^{-5} cm^2 respectively, as Si Schottky barrier, Si p-n junction and Si avalanche photodiode with area of 0.03, 0.25 and 0.07 cm^2 respectively and as photomultipliers (PM) with area of 1.0 cm^2. Bandwidth is 1 Hz. After Seib and Aukerman [4].

3. Methods for Enhancing Signal-to-Noise Ratio

3.1. SIGNAL AVERAGING - BOXCAR INTEGRATOR AND MULTICHANNEL AVERAGER

If we have a signal obscured by noise, then by examining it many times and averaging the results, we would expect random noise to average to zero while the signal tends to its true value. This procedure is only successive if the signal is constant over time or strongly repetitive. Consider a signal u(t) in Fig. 4 composed of the required signal s(t) and random noise n(t) which is sampled at times $t_k + pT$, $t_{k+1} + pT$,... m times, where k = 1,...,m, T the duration of the repetitive signal, $0 < p < 1$, i.e. sampling is performed for equal signal amplitudes $s(t_k + pT) = s(pT)$. So,

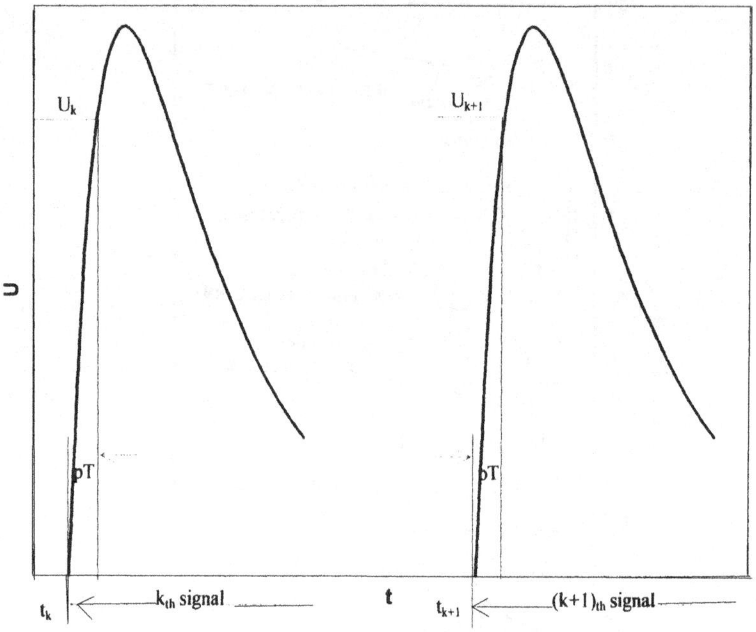

Figure 4. Repetitive sampling of corresponding signal amplitudes.

$$\sum_{k=1}^{m} u(t_k + pT) = \sum_{k=1}^{m} s(pT) + \sum_{k=1}^{m} n(t_k + pT) = ms(pT) + (m\rho^2)^{1/2} ,$$

where ρ is the r.m.s. noise value of a single sample. Here, it is taken into account that for m random noise signals the nett r.m.s. value will be the square root of the sum of the squares. Thus, the signal-to-noise ratio improves as the square root of the number of repetitive measurements:

$$(\frac{S}{N})_m = \frac{ms(pT)}{(m\rho^2)^{1/2}} = m^{1/2}(\frac{S}{N})_1 \tag{4}$$

For such measurements, a synchronizing trigger to which the signal is accurately related in time is needed. The boxcar integrator switches the signal on receipt of a trigger at time t_k and some selectable delay pT m times to a low pass filter. Each sampling is for a relatively short preset time, the gatewidth T_g, which is chosen approximately equal to the RC time constant of the filter, i.e. a weighted average is performed m times to obtain the arithmetic average of the true signal amplitude s(pT). So, signal-to-noise ratio is raised by a factor $(mRC/T_g)^{1/2}$. In order to measure all the points of the signal, the delay pT has to be shifted after every m repetitions. Often, this

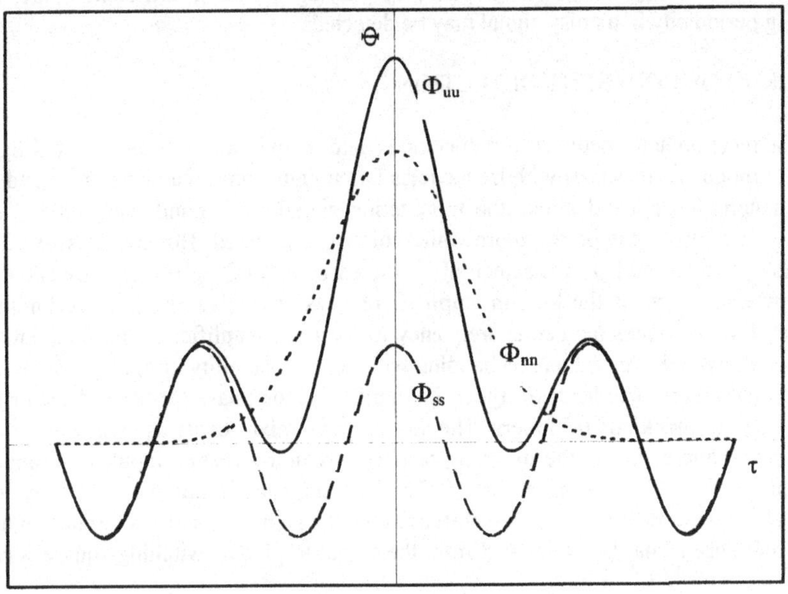

Figure 5. Autocorrelation functions of signal s, noise n and a composite u of both.

takes too much time. Then, the better choice is a multichannel averager, which looks for n signal amplitudes m times. Such an instrument mostly uses digital techniques, because these allow long time drift-free storage and the ready application of operations, such as deviding, over a wide dynamic range.

3.2. CORRELATION

Two signals $u_1(t)$ and $u_2(t+\tau)$ measured with some time delay τ are crosscorrelated by

$$\Phi_{12}(\tau) = \frac{1}{2T}\int_{-T}^{+T} u_1(t)u_2(t+\tau)dt\,, \tag{5}$$

where T depends on the type of functions involved. If the correlated signals are identical, the expression (6) is the autocorrelation function of the signal. For a periodic signal function, the autocorrelation function $\Phi_{11}(t)$ will have a peak whenever τ is a multiple of the period. Pure random noise will have an autocorrelation Φ_{nn} equal zero for an infinite bandwidth unless $\tau = 0$, but it will have finite bandwidth, if the noise bandwidth is limited, as is illustrated in Fig. 5. The narrower the bandwidth the wider Φ_{nn} will be. On the other hand, a periodic signal s(t) will give a periodic autocorrelation function $\Phi_{ss}(\tau)$ of the same period, though the form of $\Phi_{ss}(\tau)$ may be quite different from s(t). For the simplest case of sinusoidal s(t), $\Phi_{ss}(\tau)$ is actually of the same form. A noisy sinusoid will have $\Phi_{uu} = \Phi_{ss} + \Phi_{nn}$. Thus, by varying τ over a sufficient range, an underlying periodicity in a noisy signal may be detected.

3.3. LOCK-IN OR COHERENT DETECTION

One of the most popular methods to reduce noise and recover a low level detected light signal is to modulate the signal with frequencies far enough from 1/f and g-r noise and to provide a narrow passband about the modulation signal by a band pass filter. The smaller the bandwidth can be the more white noise is eliminated. However, usual filter amplifiers become instable as the selectivity is increased up to Q = 100. An extension of the narrowband system is the lock-in amplifier or phase sensitive amplifier technique. This type of system locks the center frequency of the filter amplifier to the modulation frequency, allowing increased narrowbanding without the instability normally associated with tuned amplifiers. The heart of a lock-in amplifier is the phase sensitive detector or mixer. There are two kinds of mixers. The square-wave mixer contains switches which feed the signal alternating to the inverting respectively non-inverting input of an unity-gain amplifier, as it is shown in Fig. 6. This effects multiplying the signal by +1 respectively -1 dependent on the phase shift between the signal U_S and a reference signal U_R. The reference signal U_R is derived from the signal to give a switching square-wave signal

$$U_R(t) = \frac{4}{\pi}\sum_{n=0}^{\infty}\frac{1}{2n+1}\sin\left[(2n+1)\omega_R t\right]$$

with a frequency which is equal to the signal frequency or the harmonics of the signal have frequencies which are multiples of the reference frequency ω_R:

$$U_S = \sum_{m=1}^{\infty} U_{S,m}\cos(m\omega_R + \varphi_m)$$

The output of the phase-sensitive rectifier is then smoothed by a low-pass filter to give the arithmetic average of the „rectified" signal:

$$U_A = \sum_{m=1}^{\infty}\frac{2}{\pi m}U_{S,m}\cos\varphi_m, \qquad \text{if m is odd}$$

$$U_A = 0, \qquad \text{if m is even.} \tag{6}$$

Output signal is maximum when signal and reference signal are in phase, and it is zero, when they have phase differences equal to 90° or 270°. The low-pass filter determines the overall bandwidth, and hence signal-to-noise improvement of this detection system. The noise equivalent bandwidth NEBW is here given by the power transfer function of a RC low-pass filter:

$$NEBW = \int_0^{\infty}\frac{1}{1+(2\pi fRC)^2}df = \frac{1}{4RC} \tag{7}$$

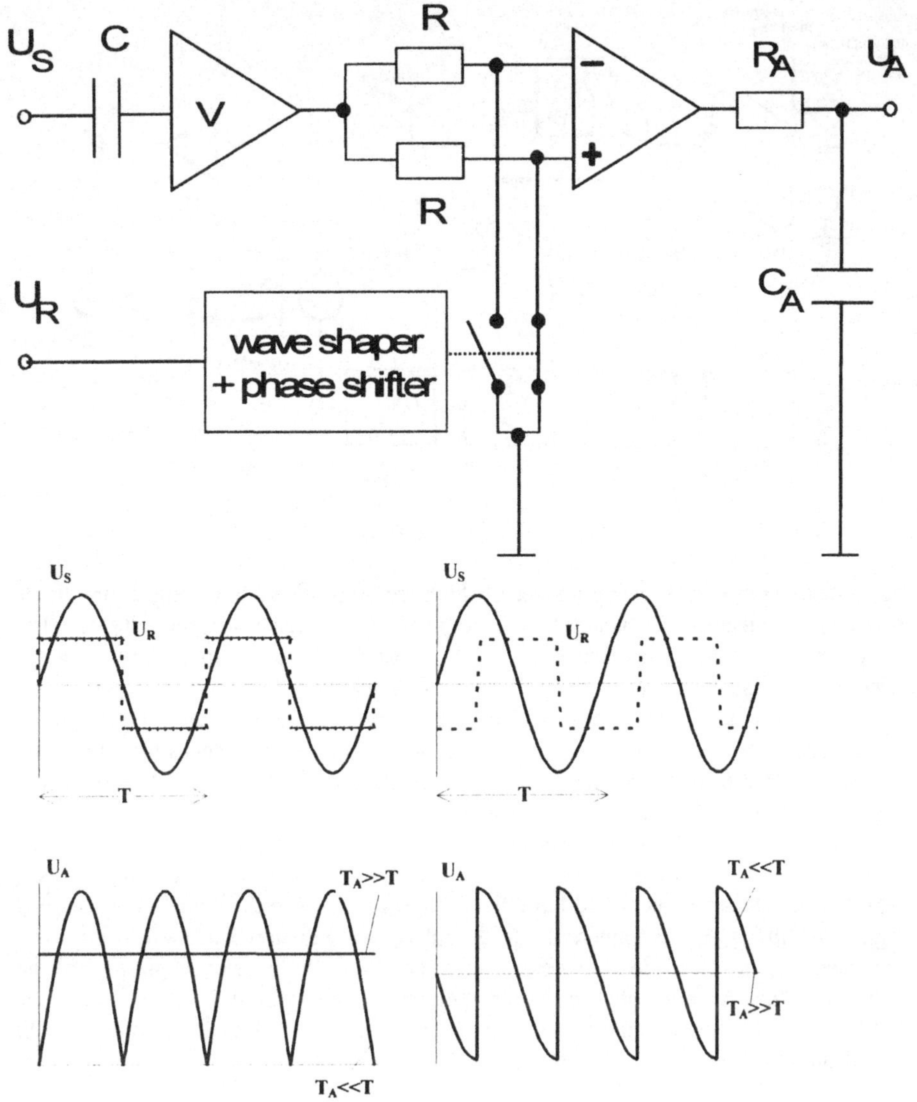

Figure 6. Lock-in amplifier with square-wave mixer and output signals for different phase relations between signal and reference signal, i.e. 0° and 90° phase shift.

The square-wave mixer responses to all odd harmonics of the signal at frequencies f_R, 3 f_R, 5 f_R .... This is why the sine-wave mixer is often prefered. As it is shown in Fig. 7, by means of phase-locked loop (PLL) a reference signal which has nearly the same frequency as the signal is generated. It is then phase-adjusted relative to the signal, converted into a precise sinusoidal waveform and multiplied by the signal.

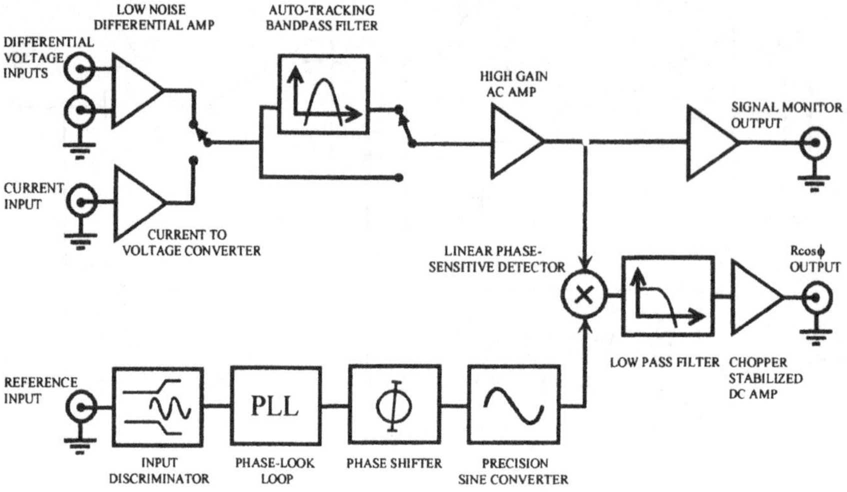

Figure 7. Lock-in amplifier with sine-wave mixer.

Multiplication produces two components which are proportional to the signal amplitude and vary with the sum frequency $f_S + f_R$ respectively difference frequency $f_S - f_R$. The component with the sum frequency is then separated by a low-pass filter. Such a heterodyning system is more diffucult to design than the lock-in with square-wave mixing. Lock-in amplifiers need a high dynamic reserve to reject noisy signals that are much larger than the signal. Typically, 60 dB of dynamic reserve means that if you are measuring 1 μV of real signal at 1 kHz, you can tolerate 1 mV of other frequencies at the input without overloading the instrument.

A kind of digital lock-in measuring system [5] is presented in Fig. 8. Also in this system a PLL is implemented in the reference channel to provide sampling times t_i which are at equal distances and have fixed phases relative to the signal. The sampled signal amplitudes A_i are converted into digital signals and stored in the memory of a computer. By means of discrete Fourier transformation, the Fourier amplitudes of the signal can then be computed from the measured amplitudes A_i. As it is done by multichannel averagers, signal-to-noise ratio is enhanced by multiple measuring of each signal amplitude.

3.4. PHOTON COUNTING

As it can be seen from Fig. 9, the output signal of PMT changes if the light intensity is lowered. At higher light levels the output signal is a superposition of analogue pulses which cannot be separated per time. As light intensity becomes weak one can observe discrete pulses corresponding to single incident photons. The measurement of these dicrete pulses is called the photon counting mode of light measurement. Photon counting is possible with sufficient accuracy approximately up to 10^7 photons per second. This limit arises from the finite width τ_a of a pulse at the output of the PMT which is the response to a single photon. τ_a is of the order of 10 ns and is due to the transit time

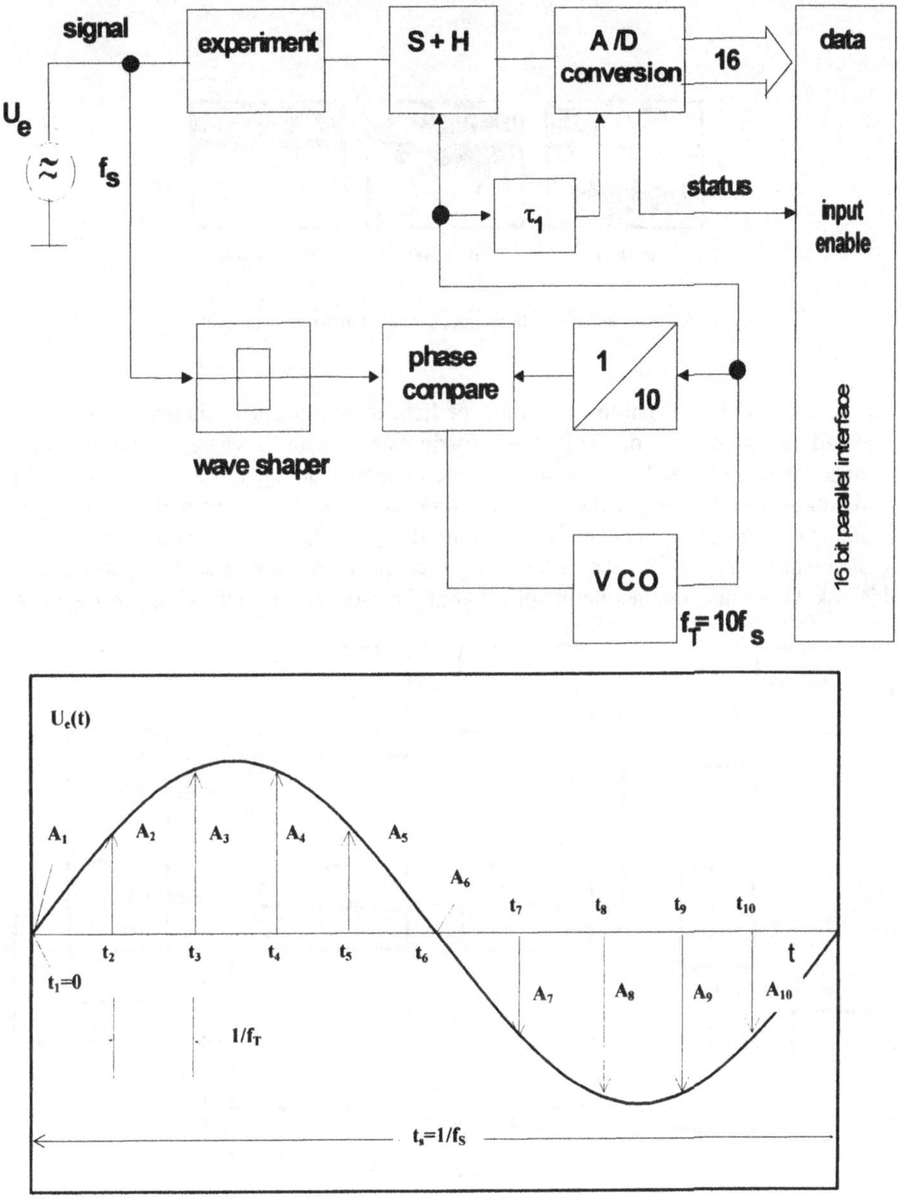

Figure 8. Digital lock-in detection by means of PLL technique [5].

spread. The resulting pulse pile-up error is 1 - exp($-r\tau_a$), i.e. 1 % respectively 9,5 % for τ_a = 10 ns and counting rate r of 10^6 s^{-1} respectively 10^7 s^{-1}. For an ideal photon counting system with unity detector quantum efficiency, vanishing dead-times, no pulse pile-up

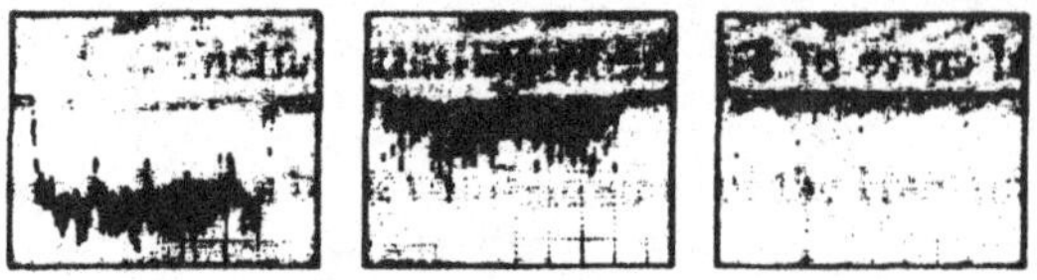

Figure 9. Photomultiplier output signals at different light levels.

and no dark counts, signal-to-noise ratio of light measurement is simply given by the standard deviation $(rt)^{1/2}$ of the Poisson distribution of counts, where r is the expected average photon rate and t the counting time.Longer counting times will improve the signal-to-noise ratio. When the dark counting rate is a significant part of the signal counting rate, the background must be measured and the dark counts subtracted from the total counting rate. Fig. 10 shows how a light chopper can be introduced to accomplish this task. Depending on the rotating position of the chopper wheel the chopper triggers a

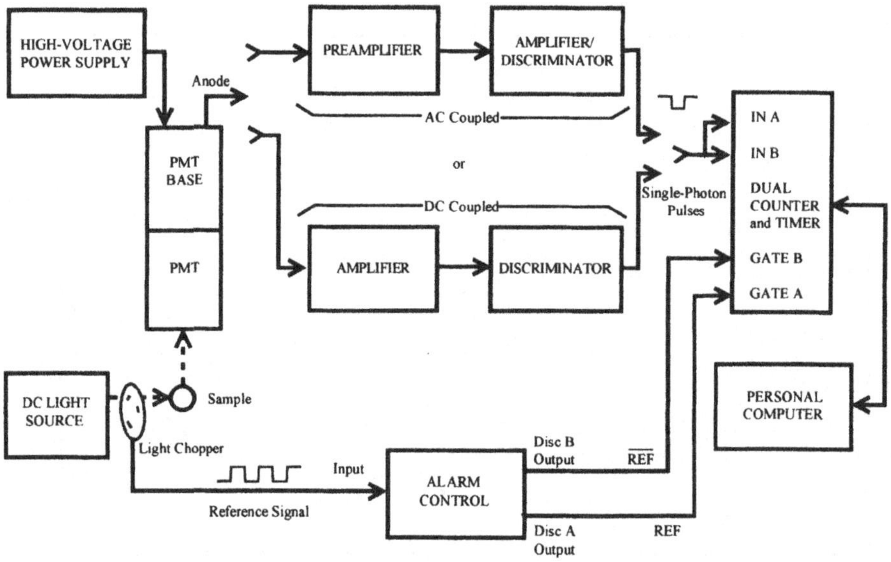

Figure 10. Photon counter with dark signal subtraction.

light or a dark signal measurement. Then signal-to-noise ratio is [6]

$$\frac{S}{N} = \sqrt{\frac{rt}{1+2d/r}}, \tag{8}$$

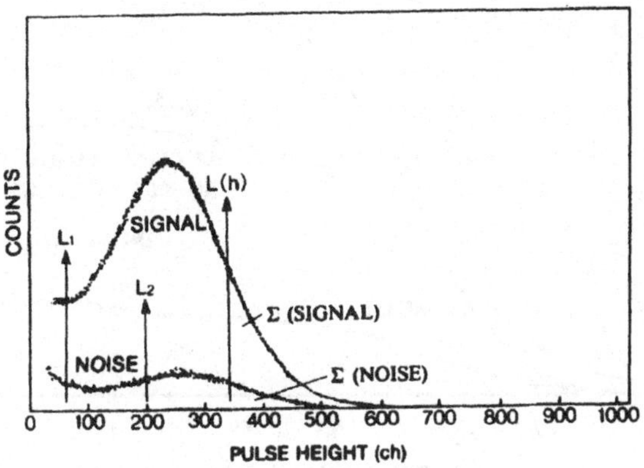

Figure 11. Counting rate versus charge pulse height [7].

where d is the average dark counting rate. To achieve 10 % precision in a measurement with 5 signal counts per second and 50 dark counts per second a counting time of at least 420 seconds would be required.

There is another way to reduce the dark counting rate and its associated noise. As it is demonstrated in Fig. 11 [7], a relatively high dark signal is by thermal emission of electrons from dynodes producing charge pulses with low height and also some pulse rate by glass emission etc. giving large pulse heights. This is why a window dicriminator is used that eliminates all pulses lower than a distinct threshold and also those higher than another higher treshold. This procedure cannot separate pulses due to the thermal emission of the photocathode because they have the same pulse height distribution as the photo pulses. Thermal emission can be reduced by photomultiplier cooling.

The photomultiplier tube used for photon counting should have a high responsivity to the wavelengths of the light signal, high gain, especially at the first dynode, fast rise time followed ba a rapid and smooth return to baseline, low dark counting rate and low transit-time spread. Moreover, the photomultiplier is to be operated in the plateau region shown in Fig. 12. The plateau is reached when the counting rate does not increase with increasing supply voltage any further. AC coupling of the PM output signal to the counting system provides satisfactory performance for average counting rates up to 10^6 s^{-1}. In excess of such counting rate the baseline between pulses at the amplifier output begins to shift so that the area circumscribed by the signal and the baseline is equal on both the positive and the negative sides of ground potential. As this reduces pulse heights, DC coupling is prefered a higher counting rates.

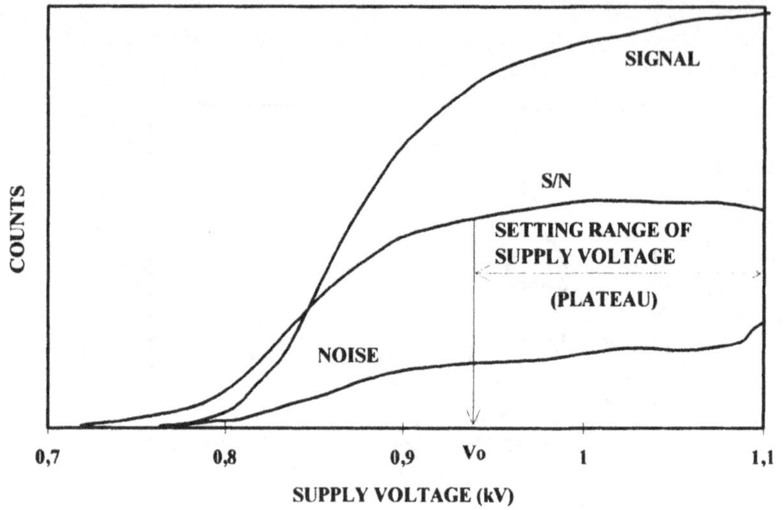

Figure 12. Counting rate versus PM supply voltage.

4. Conclusions

Measurements of weak light signals not far from noise-level are possible only at a limited speed or with limited signal-to-noise ratio. Best results can be obtained if the method of detection is adapted to the waveform of the light signal. Noise in the measured signal cannot be reduced only by filtering respectively averaging but also by methods which separate signal from noise by selection of frequency, pulse height, etc.

Recent advances in image intensifier technology have significantly increased detection capabilities. GenIV image intensifiers based on GaAs photocathode technology with quantum efficiency of about 40 % and equipped with small-diameter microchannel plates have been positioned in front of CCDs having a read-out noise level of as little as 2 e. Such an imaging detector system in the focal plane of a polychromator has been used by X. Sunny Xie and Peter Lu at the Environmental Molecular Science Laboratory of the Pacific Northwest National Laboratory [8] in order to monitor the fluorescence emission spectra of a single immobilized sulphorhodamine 101 molecule at room temperature every 170 ms (Fig. 13). Imaged by means of an inverted microscope, significant spectral shifts have been become evident before the molecule is finally photo-bleached. This impressively demonstrates the innovations in low-level light detection systems.

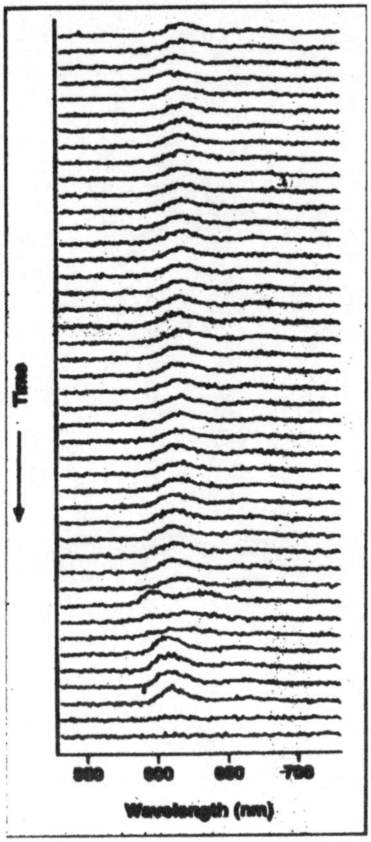

Figure 13. Fluorescence emission spectra of a single immobilized sulphorhodamine 101 molecule monitored every 170 ms by an intensified CCD array in the focal plane of a polychromator. After Lu et al. [8].

5. References

1. Van der Ziel, A. (1968) *Solid State Physical Electronics*, Prentice-Hall INC., Englewood Cliffs, N.J..
2. Limperis, T. (1965) Detectors, in W. L. Wolfe (ed.), *Handbook of Military Infrared Technology*, Office of Naval Research Department of the Navy, Washington D.C., pp. 457-517.
3. Heering, W. (1972) Signal-Rauschverhältnis bei der photoelektrischen Lichtmessung, in *Das Lichttechnische Institut der Universität Karlsruhe 1961-1971*, Monografie, pp. 83-90.
4. Seib, D.H. and Aukerman, L.W. (1973), in L. Marton (ed.), *Advances in Electronics and Electron Physics* 34, Academic Press, New York.
5. Elektronik-Arbeitsblatt Nr. 168 (1985), Lock-In-Meßtechnik mit digitaler Signalauswertung, *Elektronik* **34**, 67-70.
6. EG&G ORTEC (1995) Pulse-Processing Electronics for Single-Photon Counting, *Application Note* AN**51**, Oak Ridge, U.S.A..
7. Hamamatsu (1993) How to Perform Photon Counting Using Photomultiplier Tubes, *Technical Information*, Shimokanzo, Japan.
8. Lu, P. et al. (1997) Single-molecule spectral fluctuations at room temperature, *Nature* **385**, 143-146.

MEASUREMENT OF LOW-LEVEL LIGHT EMISSION UNDER LABCONDITIONS

D. GALL, J. FISCH, R. NOLTE, A. WALKLING
Technical University of Ilmenau
Department for Lighting
PO 10 05 65, D-98684 Ilmenau
Germany

1. Summary

The present paper is based on studies by Ruth /1/, Popp /2/ and by many other authors as well, who were confronted again and again with the problematic nature of the exact laboratory measurement of low-level light emission . The results of our investigations shall show that a variety of Conditions has to be considered with the measurement of the spontaneous emission of very low luminous objects. Special emphasis won't be given to questions which arise in the case of detection by the sensor (e.g. signal-noise-ratio, signal evaluation etc.), but effects will be investigated which have a direct influence on the measuring quantity. These are in particular thermal and stray radiation superimposing on the radiation being measured. It can be indicated by means of theoretical assessments and experimental verification, that a photon flux down to 10 photons per second is undoubtedly detectable with appropriate measurement instruments. The necessary conditions for this outcome shall be demonstrated in this article.

2. Radiometric and photon-technical basis

Concerning the description of the radiation and photon field there exist two systems at present [/3/-/7/], namely
- the radiometric way (indicated with index e) and
- the photon-technical way (indicated with index p).

Regarding the radiometric way the power parameter X_e and regarding the photon-technical way the photon density (photons / time) X_p will be used. An interrelation between both ways of looking at things can be described using the spectral radiometric quantity X_λ (equation 1)

J. J. Chang et al. (eds.), Biophotons, 159-181.

$$X_{e\lambda} = \frac{d\,X_e(\lambda)}{d\,\lambda} = X_{p\,\lambda} \cdot \frac{h \cdot c}{\lambda} \tag{1}$$

if then

$$p(\lambda) = \frac{h \cdot c}{\lambda}$$

$$X_{e\,\lambda} = X_{p\,\lambda} \cdot p(\lambda) \tag{2}$$

Thus the expressions for the integral interrelations are as follows:

$$\begin{aligned} X_e &= \int X_{e\,\lambda} \cdot d\,\lambda = \int X_{p\,\lambda} \cdot p(\lambda) \cdot d\lambda \\ X_p &= \int X_{p\,\lambda} \cdot d\,\lambda = \int \frac{X_{e\,\lambda}}{p(\lambda)} \cdot d\lambda \end{aligned} \tag{3}$$

For the calculation of the complete radiation and photon field there has been derived a number of quantities listed on Table 1.

TABLE 1. Nomenclature and units of used quantities

	Conventional Sign (X)	Nomenclature	Unit
ϕ	ϕ_e	radiant flux	W
	ϕ_p	photon flux	Counts / s
I	I_e	radiant intensity	W / sr
	I_p	photon intensity	Counts / s · sr
E	E_e	irradiance	W / m²
	E_p	photon irradiance	Counts / m² · s
L	L_e	radiance	W / m² · sr
	L_p	photon radiance	Counts / m² · s · sr
M	M_e	radiant exitance	W / m²
	M_p	photon exitance	Counts / m² · s

Meaning of indices: e - quantities related to power
p - quantities related to photons

The geometrical interrelation will be described by Table 2 (see also Figure 1)

TABLE 2. Interrelations of used quantities

1.	$\Phi_z = \int \Phi_{z\lambda} \cdot d\lambda$
2.	$I_z(\gamma_1) = \frac{d\Phi_z}{d\Omega_1}$
3.	$E_z = \frac{d\Phi_z}{dA_2}$
4.	$L_z(\gamma_1) = \frac{dI_z(\gamma_1)}{dA_1 \cdot \cos\gamma_1}$
5.	$M_z = \frac{d\Phi_z}{dA_2}$

Meaning of indices:
z - e or p
1 - luminous area
2 - illuminated area
(see Figure 1)

$d\Omega_1$ = solid angle (Steradiant sr)

$$d\Omega_1 = \frac{dA_2 \cdot \cos\gamma_2}{r^2} \cdot \Omega_0$$

$\Omega_0 =$ 1sr (unit solid angle)

Based on the specified formulas in Table 2 the desired radiometric and photon-technical quantities can be calculated or converted into each other, respectively.

For further consideration the following relationships are of paricular interest:

a. Direction-characteristic of the radiator

Because the angular dependence of the radiation is often unknown not measured ye, a so-called Lambert characteristic will be assumed.

Thus the radiant intensity distribution results (Table 2; row 2) in

$$I_z(\gamma_1) = I_z(\gamma_1 = 0°) \cdot \cos\gamma \quad (4)$$

and regarding the radiance distribution (Table 2; row 4) is valid:

$$L_z\,(\gamma_1) \neq f\,(\gamma_1) = const.$$

b. Point-source radiator:

In the case of a point-source radiator within a point P (Figure 1) with the radiance $I_z\,(\gamma_1)$ the irradiance results in

$$E_{zp} = \frac{I_z(\gamma_1)}{r^2} \cdot \cos\gamma_2 \cdot \Omega_0 \tag{5}$$

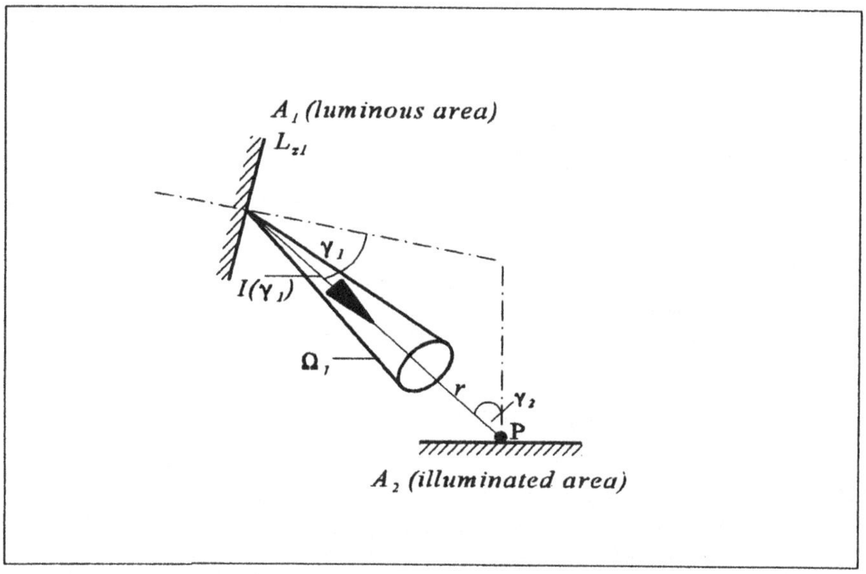

Figure 1. Geometrical relationships
r - distance between radiation source and irradiated area

c. Radiating room:

If radiation from the half-space hits on the point P, then the irradiance (Figure 2) is

$$E_{zp} = \pi \cdot L_{z_1} \cdot \Omega_0 \tag{6}$$

If the radiation will be diffuse reflected from the area into the half-space again, then the radiance of area A_2 has to be taken into consideration

$$L_{z_2} = \frac{\varrho_z \cdot E_{zp}}{\pi \cdot \Omega_0} \tag{7a}$$

The reflected radiant flux from the area dA_2 is

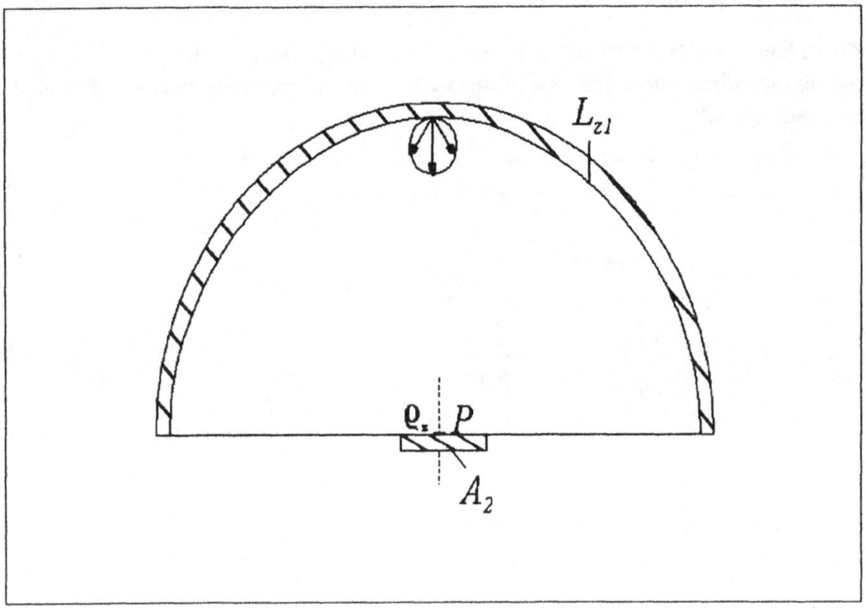

Figure 2. Radiation coming from the half-space
ρ_z - radiometric or photon-technical reflectance

$$d\phi_{z_2} = \pi \cdot L_{z_2} \cdot dA_2 \cdot \Omega_0 \tag{7b}$$

and thus the radiant exitance

$$M_{z_2} = \frac{d\phi_{z_2}}{dA_2} = \pi \cdot L_{z_2} \cdot \Omega_0 \tag{8a}$$

For a self-luminous source is valid:

$$M_{z_1} = \frac{d\Phi_{z_1}}{dA_1} = \pi \cdot L_{z_1} \cdot \Omega_0 \qquad (8b)$$

d. Radiation into a solid angle sector

If the radiation is limited by any aperture, the relationship between the signal determinant quantities (absorbed radiant flux) and geometry has to be described by means of the radiance and radiant intensity:

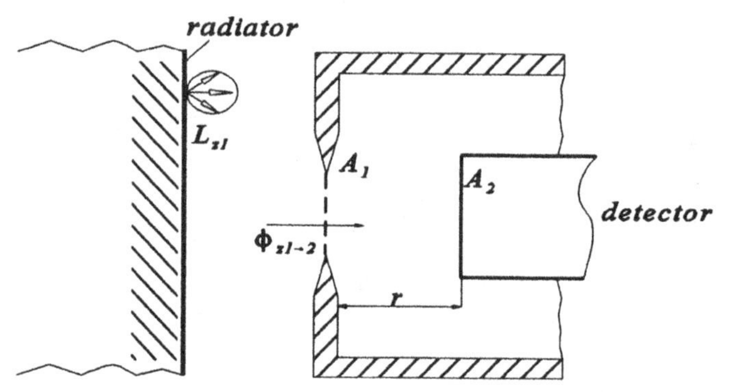

Figure 3. Radiation coming from a solid angle sector

If: A_1 - aperture of the solid angle
A_2 - detector area
R - distance A_1-A_2

$$L_{z1}(\gamma_1) = \frac{d\,I_{z_1}(\gamma_1)}{d\,A_1 \cdot \cos\gamma_1} = \frac{d^2\,\Phi_{z_{1-2}}}{d\,A_1 \cdot \cos\gamma_1 \cdot d\,\Omega_1}$$

respectively (9)

$$\Phi_{z_{1-2}} = \int_{\Omega_1}\int_{A_1} L_{z_1}(\gamma_1) \cdot d\,A_1 \cdot \cos\gamma_1 \cdot d\,\Omega_1$$

Concerning the assembly shown in Figure 3 it can be written approximately

$$\phi_{z_{1-2}} \approx L_{z_1} \cdot \frac{A_1 \cdot A_2}{r^2} \cdot \Omega_0 \tag{10}$$

By means of the photon flux and the quantum efficiency η_q of the detector area the signal quantity (S) can be obtained

$$S = \phi_{p_{1-2}} \cdot \eta_q \quad [cps]$$
$$cps \; - \; measured \; rate \; of \; counts \tag{11}$$

Proceeding from equations 10 and 8b the photon exitance emitted from area A_1 can be determined

$$M_{p_1} = \pi \cdot L_{z_1} \cdot \Omega_0 = \pi \cdot \phi_{z_{1-2}} \cdot \frac{r^2}{A_1 \cdot A_2} \tag{12}$$

$$M_{p_1} = \pi \cdot \frac{cps}{\eta_q} \cdot \frac{r^2}{A_1 \cdot A_2}$$

3. Considerations concerning the radiation and photon field within the laboratory room

3.1. Geometrical conditions of the radiometer head

According to equations 10 and 11 (Figure 3) the radiometer head can be used usually to measure radiances respectively photon radiances. For interpretation of the signal (cps) the specifications of areas A_1 and A_2 and their distance (equation 12) will be needed in order to obtain the photon radiance of the object. The radiance of the object (O) can be composed of a primary (prim.) and a reflected (secondary) part:

$$L_0 = L_{prim} + L_{sec} \tag{13}$$

3.2. Geometrical conditions in the room

In Figure 4 radiation conditions are indicated, as they can occur in a laboratory. Here a human body serves as an object, who is emitting a luminescent and a thermal caused radiation itself and who is receiving stray radiation and thermal radiation as well as reflecting it into the room. (Note: In the following the index marking z will be cut out)

The radiance of the object results from equation 13, while

$$L_{prim} = L_L + L_{TO}$$
$$L_{sec} = L_{\varrho O} + L_{\varrho TO} \tag{14}$$

L_L - luminescence radiation of the object
L_{TO} - thermal radiation of the object within the visible range
$L_{\varrho O}$ - the reflected part of stray light on the object
$L_{\varrho TO}$ - the reflected thermal radiation from the room (visible range) on the object
ϱ_O - reflectance of the object

Using the reflectance of the object (ϱ_O) and the irradiance on the object E_O the secondary part of the radiance (equation 14) can be calculated by means of equations 6 and 7a as follows:

$$L_{sec} = \frac{\varrho_O \cdot E_O}{\pi \cdot \Omega_O} = \varrho_O \cdot (L_{\varrho R} + L_{\varrho T}) \tag{15}$$

$L_{\varrho R}$ - radiance of the room coming from the stray light ϕ_R (Figure 4)
$L_{\varrho T}$ - radiance (visible range) of the room coming from the thermal radiation of the room

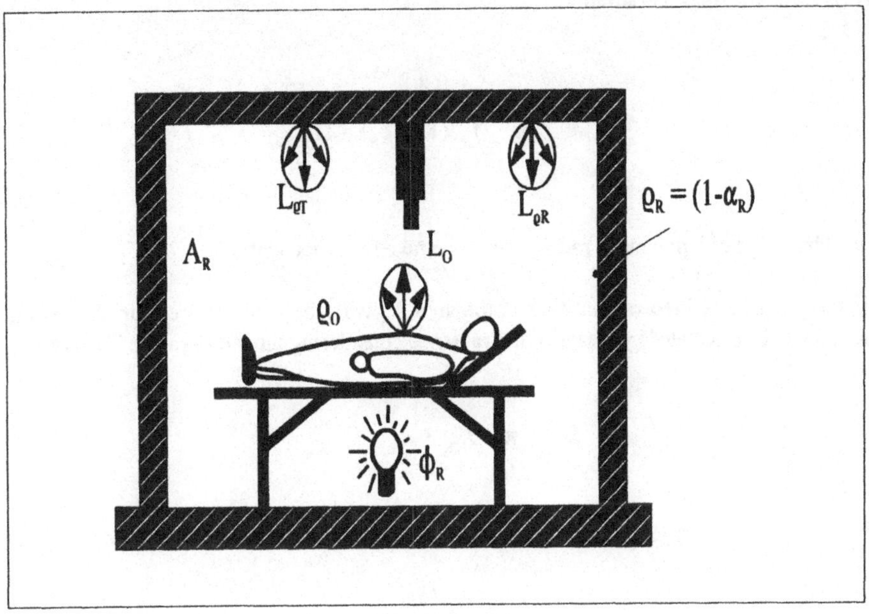

Figure 4. Radiation conditions within the room

Concerning the secondary radiation according to equation 15 it has to be considered that the radiances will be formed as a result of a multiple reflection. If the cubic space will be replaced approximately by a equivalent sphere, then the radiances can be specified as

$$L_{\varrho x} = \frac{\varrho_R \cdot \Phi_x}{\pi \cdot A_R \cdot (1 - \varrho_R) \cdot \Omega_O} \tag{16}$$

A_R - area of the room surface
ϱ_R - reflectance of the room

That means regarding the stray light

$$L_{\varrho R} = \frac{\varrho_R \cdot \Phi_R}{\pi \cdot A_R \cdot (1 - \varrho_R) \cdot \Omega_O} \tag{16a}$$

(Note: The direct component of radiation won't be taken into account) and regarding the thermal radiation

$$L_{\varrho T} = \frac{\Phi_T}{\pi \cdot A_R (1 - \varrho_R) \cdot \Omega_0} \tag{16b}$$

(Note: The direct component of radiation will be taken into account!)

Φ_T is the primary emitted radiant flux of the area A_R with the temperature T in the visible range and will be calculated by means of the radiance considering the temperature T according to the equation 7b

$$\Phi_T = \pi \cdot L_{TR} \cdot A_R \cdot \Omega_O \tag{17}$$

Thus

$$L_{\varrho T} = \frac{L_{TR}}{(1 - \varrho_R)} \tag{18}$$

According to Kirchhoff's law:

$$L_{TR} = \alpha_{TR} \cdot L_{TSK}(T_R) = (1 - \rho_R) \cdot L_{TSK}(T_R) \tag{19}$$

$L_{TSK}(T_R)$ - radiance of the black body with the temperature T_R
α_{TR} - emissivity of the room

Equation 18 can be expressed as

$$L_{\varrho T} = \frac{L_{TR}}{1 - \varrho_R} = \frac{L_{TR}}{\alpha_{TR}} = L_{TSK}(T_R) \tag{20}$$

That means, the radiance of the black body concerning the room surface can be assumed to be the effective radiance of the room in the case of multiple reflection.
The multiple reflections don't have to be considered for the object and thus equation 19 becomes

$$L_{TO} = \alpha_O \cdot L_{TSK}(T_O)$$
$$\alpha_O = 1 - \varrho_0 \qquad (21)$$
$$\varrho_{skin} \approx 0{,}37 \Rightarrow \alpha_O = \underline{0{,}63}$$

It shall be noticed here that the above mentioned considerations regarding the radiation of the room also can be transferred in this case, if the cabinet of the detector has a noticeable influence.

3.3 Specifications concerning the used measurement instruments and laboratory conditions

Laboratory room and measurement object:

The Low Level Emission measurements were carried out on the surface of the human body within a laboratory room (for dimensions see Figure 5), which was blacked out against stray light and where the indoor room surfaces were painted black completely.

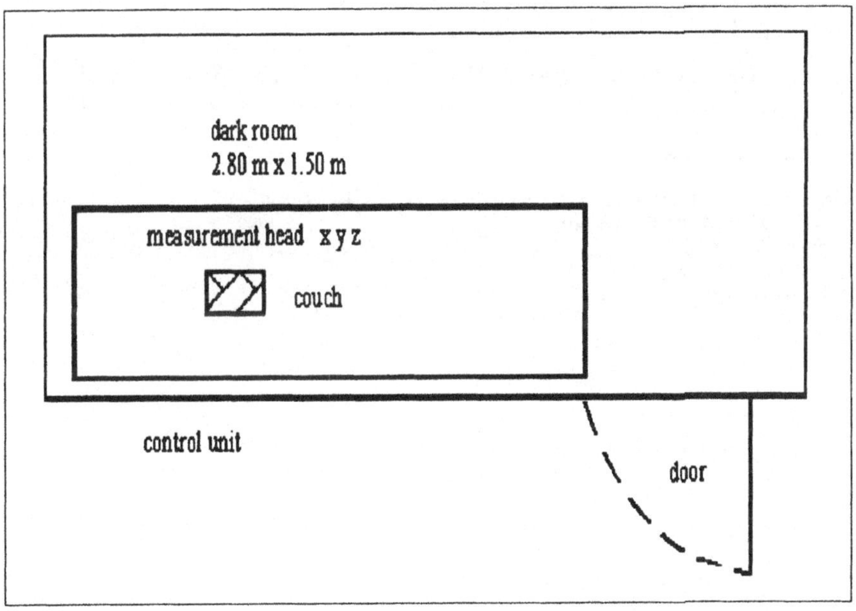

Figure 5. Geometry of the measurement room (height = 2.20 m)

Measurement instrument:

The sensor used was a photomultiplier (9558QA) with a S20-cathode. The spectral sensitivity as well as the quantum efficiency (η_q; see equation 11) is shown on Figure 6.

The photomultiplier was put into a detector head which was equipped with a cooling system and diaphragms (Figure 7). With this construction the area amounted to $A_1 = A_2 = 16\ cm^2$ where the distance of the areas is $r = 15$ cm (see Figure 3 and equation 10).

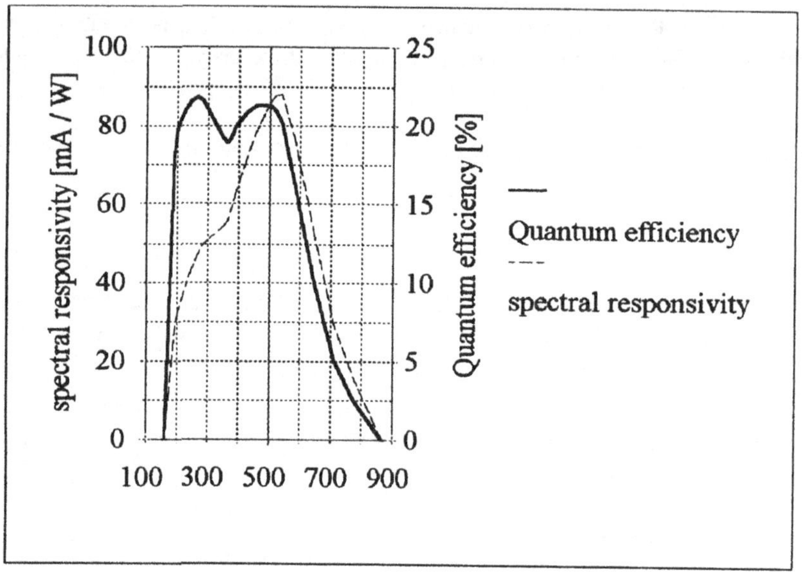

Figure 6. Quantum efficiency and spectral sensitivity of the used photomuliplier 9558QA with S20-cathode

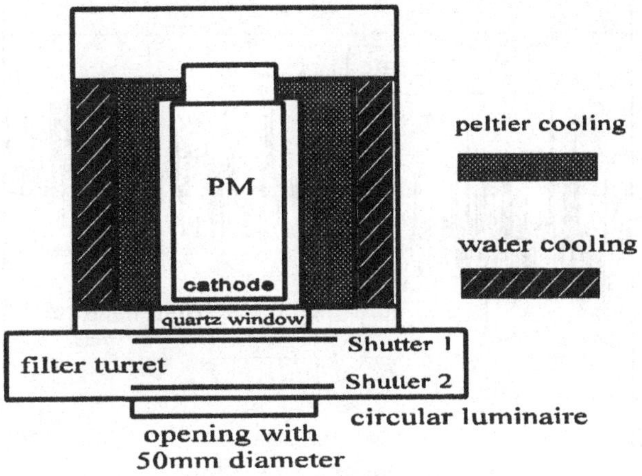

Figure 7. Used detector head with diaphragms and cooling system

4. Assessments of radiation conditions

Following paragraph 3 the photon flux detected by the measurement instrument consists of two different components (equation 13 and following) and therefore it is necessary to estimate how strong the disturbing radiation quantities are as against the relevant quantity L_L (luminescent radiation) of the object (equation 14). This can be done by calculations as well as by experiments. The first measurements of the hand surface resulted in signals of 8 cps above the noise signal (Figure 8). In the case 260 measurements performed it is possible to state a mean value difference of 0.61 with a statistical reliability of 95 %.

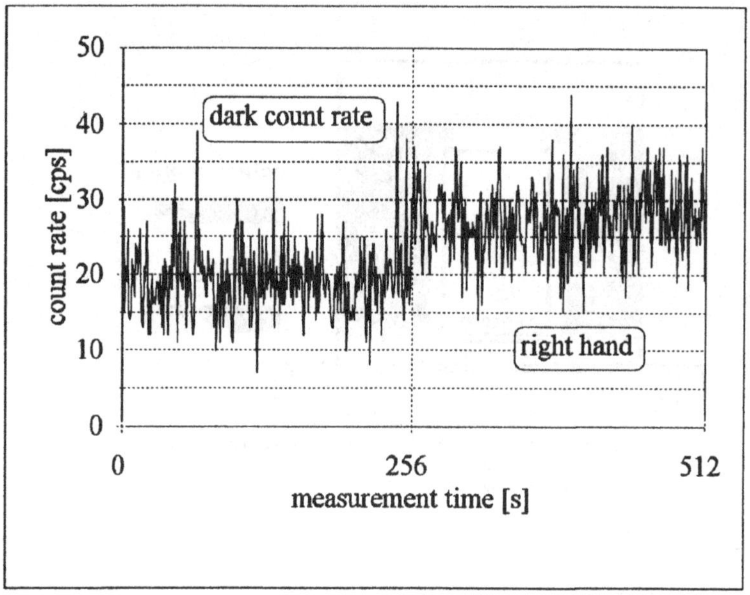

Figure 8. Typical example of a dark-measurement and a measurement of the spontaneous emission of a human

	min.	max.	mean value	standard deviation	median
dark signal	7	43	19.37	4.98	19
hand plate	14	44	27.37	5.17	27

4.1. COMPONENT PART OF THERMAL RADIATION WITHIN THE VISIBLE RANGE

The component of radiation caused by the hand (L_{TO} - equation 14)

By means of Kirchhoff's law (equation 21) as well as Planck's radiation formula law and the geometrical relationships between the radiation quantities themselves (Figure 10), the cps (counts per second) can be calculated for different assumed hand temperatures. From Figure 9 it follows that the count rates found in this way rank far below the those at the measurement of L_O (Figure 8) and therefore can be ruled out as the cause of signal.

But the total thermal radiation of the human hand could also influence the temperature of the multiplier cathode. In order to investigate this influence, the object was replaced by a metal plate instead of the hand surface and its temperature was varied. It can be seen from Figure 10, that heat-up effects caused by the radiation are noticable only if the radiation temperature is above the value of 42°C.

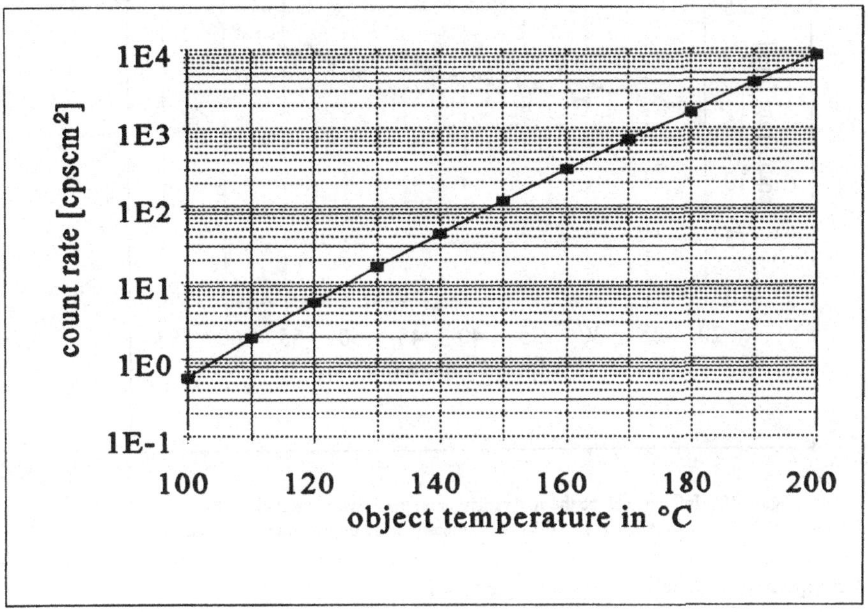

Figure 9. Number of countable photons (cps) as dependent upon the temperature of the hand surface (absorbance of the hand: α_0=0.63)

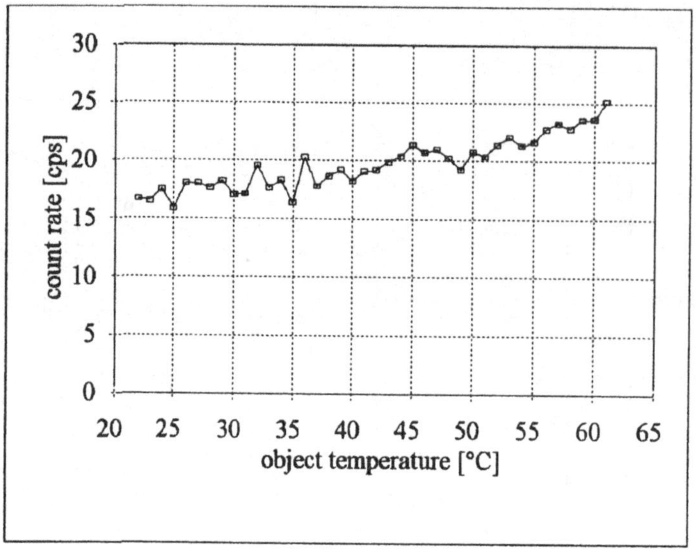

Figure 10. **Influence of the thermal radiation on the measured signal at low temperatures (20°C to 60°C, measurement object: aluminum plate)**

The component of radiation caused by the room

Because the room temperature (22°C) still ranks far below the object temperature (36°C), it can be expected that the thermal radiation within the visible range caused by the room is excluded from the measured signals as well. If heat sources occur with a temperature above 100°C an influence has to be taken into consideration.

Spectral distribution of thermal radiation

A calculated relative spectral radiance ($L_{e,\lambda}$) is shown in Figure 11. It can be identified as extraordinary decrease of radiation in the direction of the short-wave range conditional on the low object temperature. Taking into account the course of the quantum efficiency (Figure 6) me can proceed from the assumption that if thermal photons are really detectable it is only possible within a small spectral range below 800 nm.

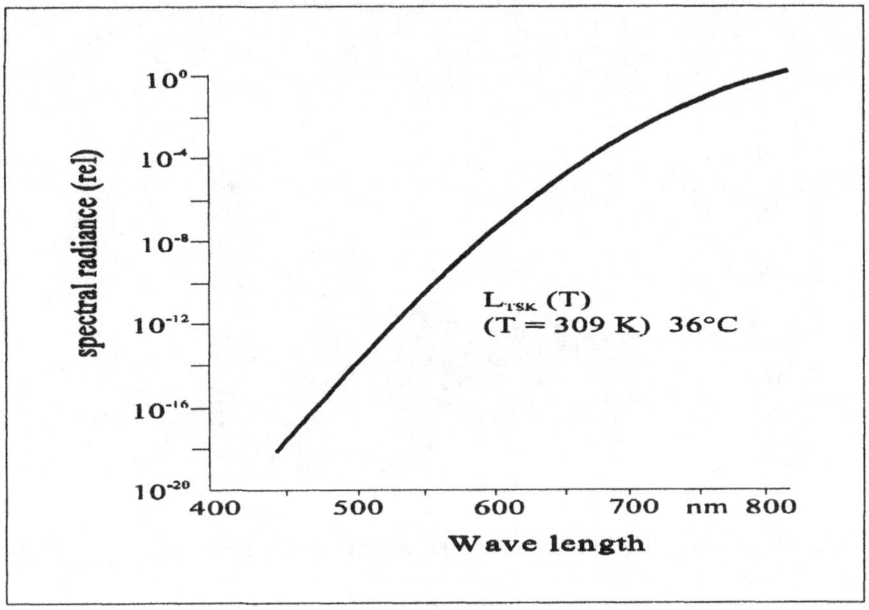

Figure 11. **Relative spectral radiance of the black body at T=309 K**

4.2. STRAY LIGHT COMPONENT WITHIN THE ROOM

If there is any stray light radiator in the room or if disturbing radiation can reach the room from outside (radiant flux ϕ_R - Figure 4), then the signal can change clear (equation 15). This influence can be best determined experimentally best in this way, that non-self luminous materials with different reflectances will be used instead of the human hand. Figure 12 shows a not carefully blacked out room where the straylight indoors is different and noticeable one in various places. With the help of a powerful luminaire, several points transparent to light can be found by illuminating the room from the outside.

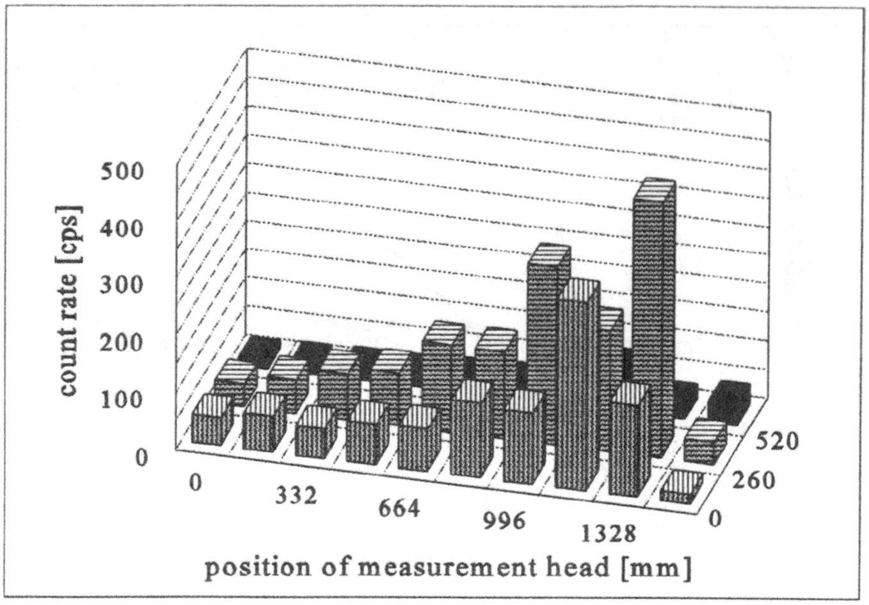

Figure 12. Influence of scattered light on the dark signal
(material of the object: ρ_0=0.99)

If we achieve no signal within the order of the expected object signal by illuminating with an external luminaire, as then it can be considered reliable that photons will be measured by the facility instead of any other parameters (e.g. α, β, γ-radiation; ionisation effects).

5. Experimental findings

Regarding the measurements of the Low Level Emission on hand surfaces (Figures 13a, b, c) a signal course dependent on time of day has been indicated.

A variation of mean values of 9 test persons over 26 days was also observed (Figure 14).

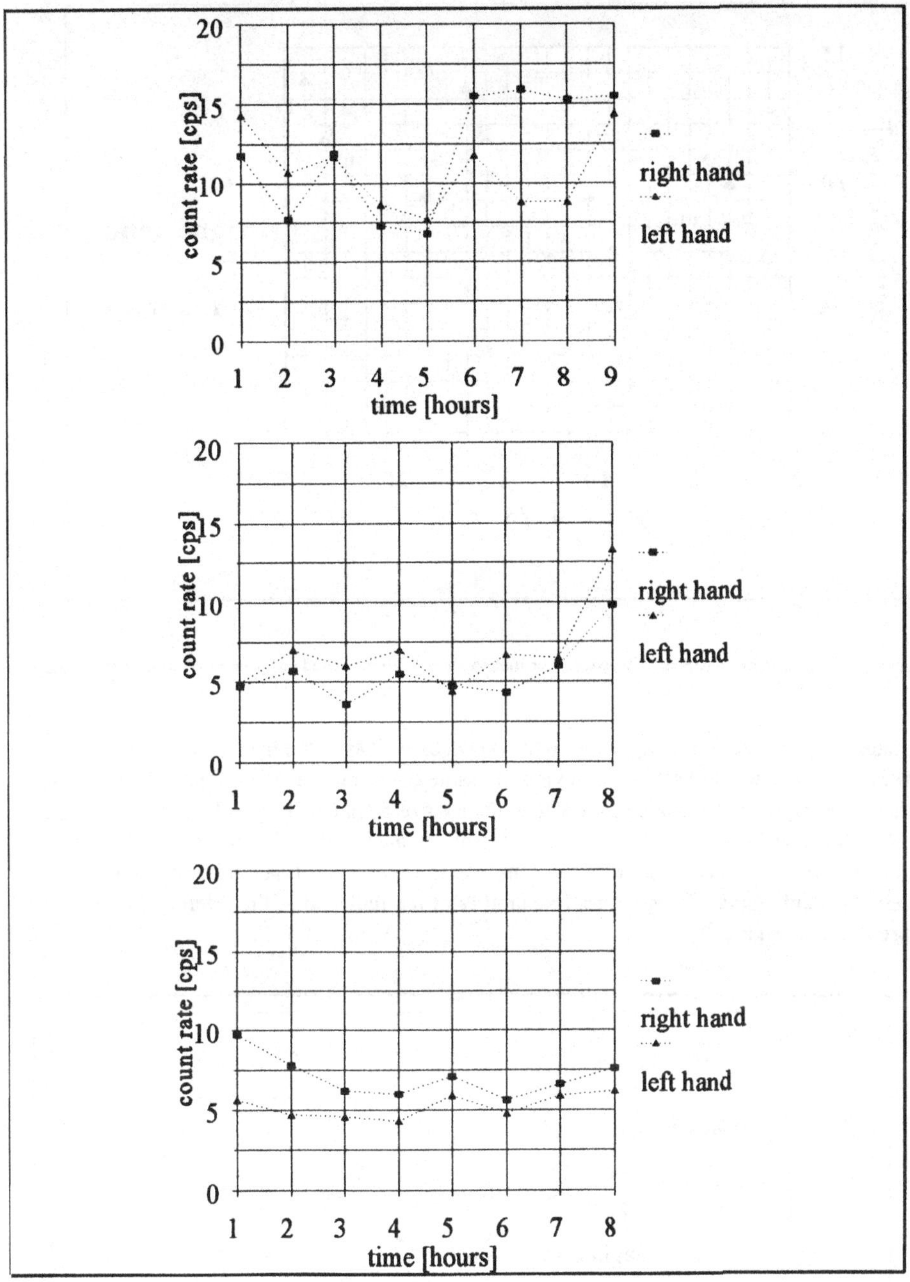

Figure 13a-c. Spontaneous emission of the right and left hand plate - person 1, 2, 3 (course over the day)

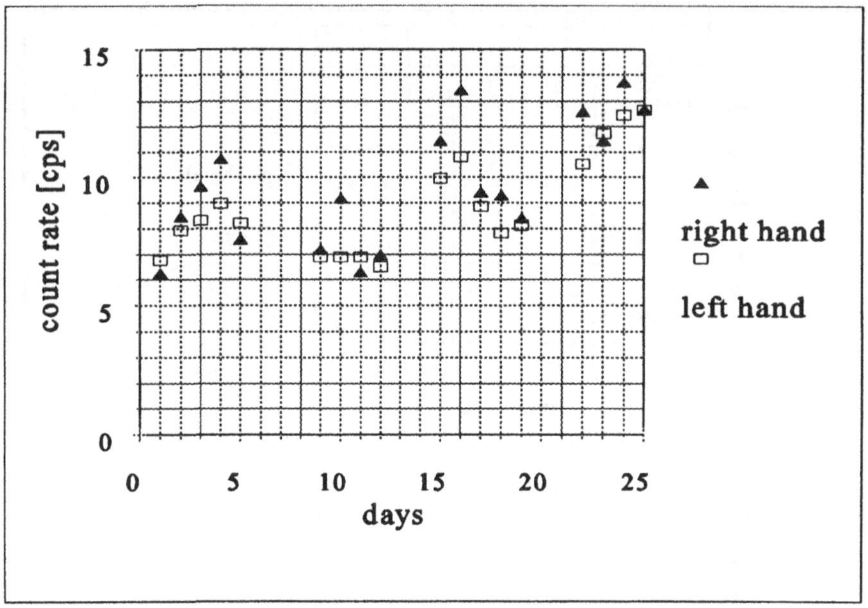

Figure 14. Long-term measurements of 9 persons (the mean values of all measured test persons relating to the day are represented in the diagram)

Because these variations could be due to the object as well as to the measurement instrument, a reference radiator was set up achieving the same count rates as the object. The reference radiator has an uniform radiating area with a diameter of 5 cm (see Figure 15) whose radiance (that is photon radiance) can be made reproducible by means of the lamp current of an incandescent lamp. Because the light of the source is too powerful, it has to be weaken by non-selective diaphragms. The good reproducibility of the radiance of the reference radiator is represented in Figure 16.

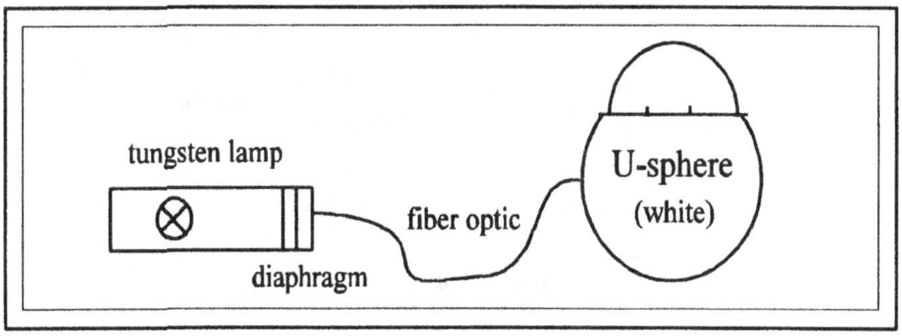

Figure 15. Assembly sketch of the reference radiator 02/ TU Ilmenau

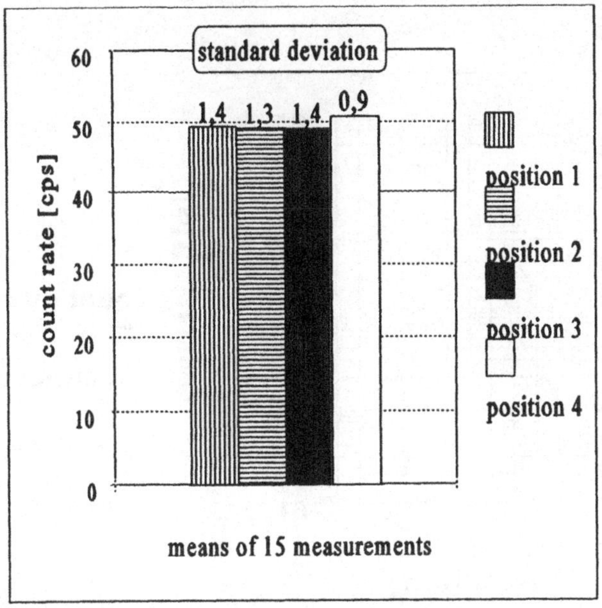

Figure 16. Reproducibility of the measurements using the reference radiator 02

measurement 1-3: position of the standard was not changed, detector head was adjusted to the same position in each case,
measurement 4: standard was new placed under the detector

In order to obtain reproducible signals of enclosed object radiances (L_L - Gl. 14) it is advisable therefore to calibrate the instrument related to the reference radiator.
For the purpose of accurate results this should be done before each measurement!

With the reference radiator the linearity of the measurement instrument was tested. For this purpose first the diaphragms (Figure 15) were removed and the illuminance caused by a light source behind an interference filter (λ = 520 nm) was measured using a linearized luxmeter. (The interference filter was used to leave the different spectral sensitivities of the luxmeter and the measurement instrument out of account). The illuminances were realized by variation of the lamp current (0.01 to 100 lx). Now the complete reference radiator (supplementary with interference filter) was positioned in front of the measurement instrument and the count rate was determined. From Figure 17 it can be seen that similar curves result if the lamp current is be varied.

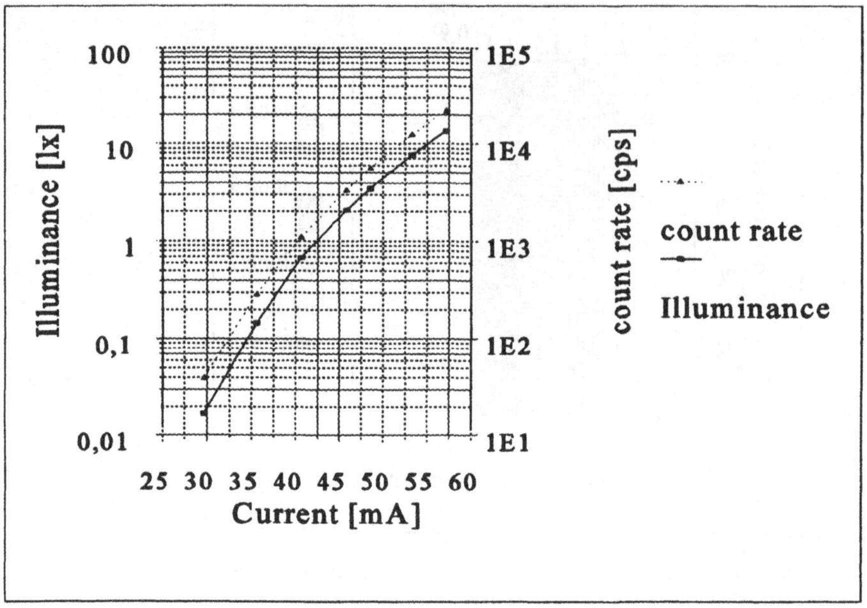

Figure 17. Dependence of illuminance and count rate (cps) on the current intensity of the reference radiator

That means:

a. The photon measurement instrument works linearly as well.

b. The detected count rate really comes from photons. In this case other influences can be excluded, otherwise another shape of the curve would have resulted

Conclusions:

It can be concluded that photon fluxes of about 10 cps can be established with reliability, if other influences can be eliminated.

All examinations were carried out by means of a Biophotonic-Measurement-Device (DGS G 94178453 from 20.4.1995) from the International Institut of Biophysics Kaiserslautern. The authors express thanks especially Prof. F.-A. Popp for the manyfold support at the research work.

6. References

1. Ruth, B. (1977) *Dissertation*, Philipps-Universität Marburg/Lahn, Fachgebiet Physik

2. Popp, F. A. (1994), *Biophoton Emission: Experimental Background and Theoretical Approaches*, Modern Physics Letters B8, 1269-1296

3. *CIE International Electrotechnical Vocabulary* Chapter 845: Lighting (1987) Bureau CIE, Genf

4. Kiefer, J. (1977) *Ultraviolette Strahlen*, de Gruyter Berlin-New York

5. Dereniak, E. L., Boreman, G. D. (1996) *Infrared Detectors and Systems*, John Wiley & Sons, Inc. New York

6. Jezowska, B. *et al.* (1986) Photon Emission from Biological Systems, World Scientific, Singapore

7. Beier, W. *(et al.)* (1980) *Erzeugung , Messung und Anwendung ultra violetter Strahlen*, Georg Thieme Verlag, Leipzig

WHOLE-BODY COUNTING OF BIOPHOTONS AND ITS RELATION TO BIOLOGICAL RHYTHMS

S. COHEN AND F.A. POPP

International Institute of Biophysics, IIB,
Station Hombroich, Vockrather Straße, 41472 Neuss, Germany

1. Introduction

As early as 1979, the first measurements of biophotons from the human body were performed by an American research group [1]. After investigation of only a few persons the experiments were stopped. Later, in about 1990, an English group repeated these first trials [2]. Since these results were as confusing as the former ones of the American group, further investigations of the complicated and time-consuming measurements were stopped again. It turned out that the biophoton intensities, ranging from about 10 to 100 counts per second (cps) , were different on different parts of the body, e.g. hands displayed higher photon emission than forehead and trunk. However, on the same points of the body the intensity remained rather stable. Only the tendency for a slow increase from morning to evening could be registered. The observations of the spectral distribution were different. Whereas the American group [1] stated that significant components have been found in the violet part (at about 400 nm), the English group [2] observed the highest emission in the red part of the spectrum (580 to 650 nm). Correlations between photon emission and physiological or psychological functions could not be found. Only a few persons with individually different intensities have been investigated in both of these research projects. A further attempt of measuring in vivo chemiluminescence of human skin was published by G. Sauermann and coworkers [3].

We started in 1993 to build up a lighttight dark room (4mx3mx2.50m) with black interior walls. Based on the single photon counting technique of Ruth and one of the authors [4], we installed a detector head that by hanging on a runner can be moved over the whole body of a person who is lying on a bed below. The intensity of the stray light in the dark room was minimized to about 5 cps which is lower than the noise of the cooled photomultiplier (EMI 9558 QA, selected type) of about 20 cps. In stationary conditions the detector can register a real count rate of 3 cps in a measurement time of 30 min on a significance level of 95%. Round skin areas of 7 cm diameter were always exposed to the photomultiplier which was placed at a distance of about 5 cm. 256 values were registered in preset time intervals of 100 ms or, alternatively, 1 second. As a measure of the „spontaneous" biophoton emission (bpe), the mean value and variance as well as the normalized factorial moments up to the 7th order were calculated from the

J. J. Chang et al. (eds.), Biophotons, 183-191.

original data (counts/100ms or, alternatively counts/s). In addition to the bpe the delayed luminescence (d.l.), i.e. rescattered light after illumination with an external lamp, was measured. The irradiation time of the 150 W-tungsten lamp was always 5s. Hundred ms after switching off the external lamp, the first measurement value of „delayed luminescence" was then recorded. 256 values of the delayed luminescence were always taken up within preset time intervals of 100 ms. The relaxation function in this measurement time of 25.6 s was recorded and subjected to further analysis [5, 6].

2. Primary Investigations

During the period July 1994 to November 1995 we measured the whole-body biophoton emission of 80 persons, where forehead, ears, hands, cheek, throat, breasts, sternum, belly, hips, knees, calf, feet, bottom, shoulder, arm, regions on the skin neighbouring the stomach, prostate, lung, kidney were subjected to investigation. Not all of these locations have been studied, but the photon emission of at least three of these positions were subjects of measurements at least one time to several times over different periods, sometimes even under changed external conditions as, for instance, before and after a definite treatment. In this way a lot of data has been recorded and some valuable experience was gained about biophoton emission and delayed luminescence of the whole body.

We could not follow a systematic measurement schedule, since the performance was dependent on whether people were available and, in addition, patient enough to serve as subjects of investigation. Our goal was concerned with the question of finding features which are characterizing states of health and disease.

For all people under investigation it turned out that independent of their state of health the increase of the normalized factorial moments with increasing order was significantly lower than that of the dark count rate (Fig.1), which has to be interpreted as proof of a Poissonian distribution of the photocount statistics (PCS). This means that the character of biophotons in the human body is the same as for all other living systems which display, as one knows nowadays [7], a Poissonian PCS. As has been shown in several papers, this means that the photons from the human body are either chaotic with a coherence time much smaller than the preset time interval (of 100 ms), or it means that the photons originate from a fully coherent field. The bpe-intensity values of all persons range from a few up to some hundred cps, the delayed luminescence (d.l.) takes relaxation times of some seconds, where the intensity after excitation may be up to 1000 times higher than the stationary bpe-values. There is no doubt that the d.l. of the human body follows in accordance with that of every living tissue an hyperbolic relaxation function (Fig.2) which indicates again that there is in principle no difference between the biophoton field of the human body and that of all the other living subjects [7]. Moreover, Poissonian PCS and hyperbolic relaxation of d.l. show together evidence of a fully coherent ergodic field. However, up to now we have not been successful in demonstrating a clear connection between deviations from Poissonian distribution and/or hyperbolic relaxation and the kind and the degree of a disease. Rather, it seems that healthy people show moderate variations as well of bpe and of d.l. on different parts

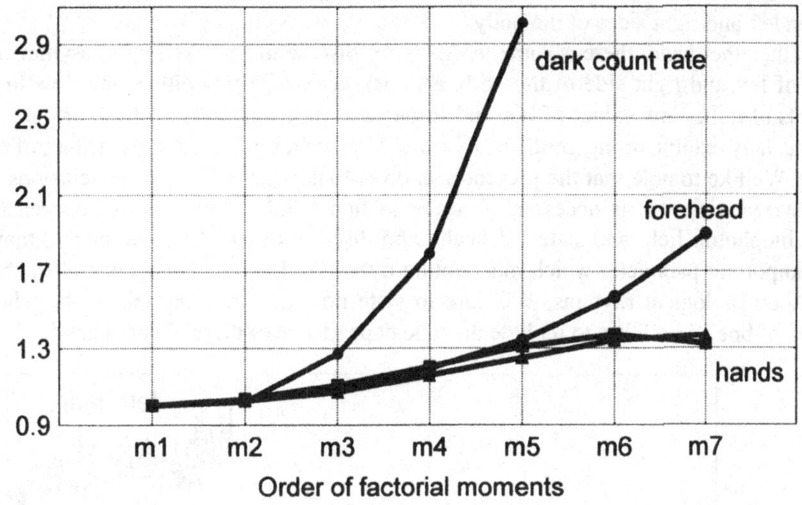

Figure 1. Normalized factorial moments up to the 7th order of the dark count rate and the biophoton emission (+ dark count rate) for different parts of the body.

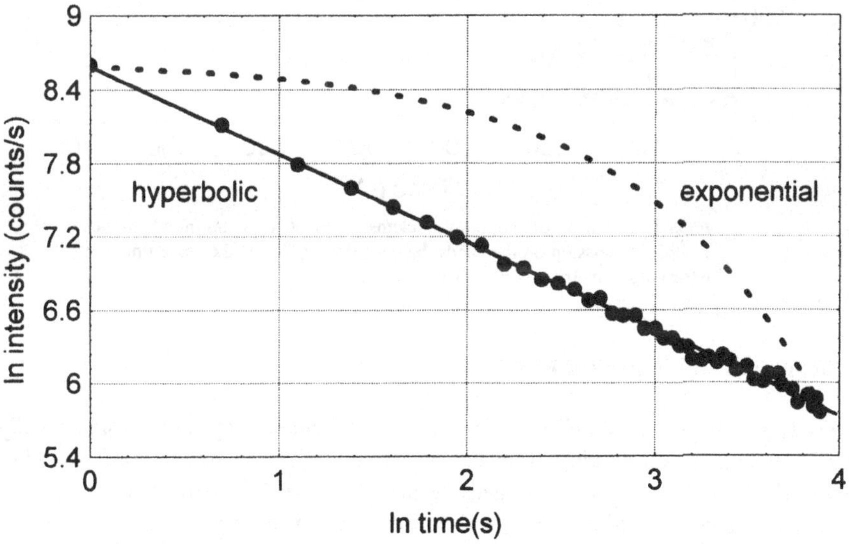

Figure 2. Delayed luminescence of a human hand after white-light illumination. A hyperbolic (1/t) law describes the relaxation rather accurately, whereas an exponential function (dashed line) can be excluded. The points on the theoreticalhyperbolic function represent measured values.

of the body, anti-correlations between bpe and d.l.- values, and rather symmetric values between left and right sides of the body.

On the other hand, there is some tendency for disease to be linked up to asymmetric values of left and right side of the body and sick people display either only low local gradients of rather low values of bpe and at the same time high values of d.l. or, just the opposite, they exhibit strong gradients of rather high values of bpe and low values of d.l. (Fig.3). We like to note that the present data do not allow us to jump to conclusions. A lot of systematic work is necessary in order to find reliable connections between the human biophoton field and states of health and disease. In addition, we learned that a rather important parameter which must not be overlooked is the dependence of the bpe and d.l. on biological rhythms. We dare to state now that the solution to the whole problem of bpe and d.l. has to include the time dependence of these phenomena.

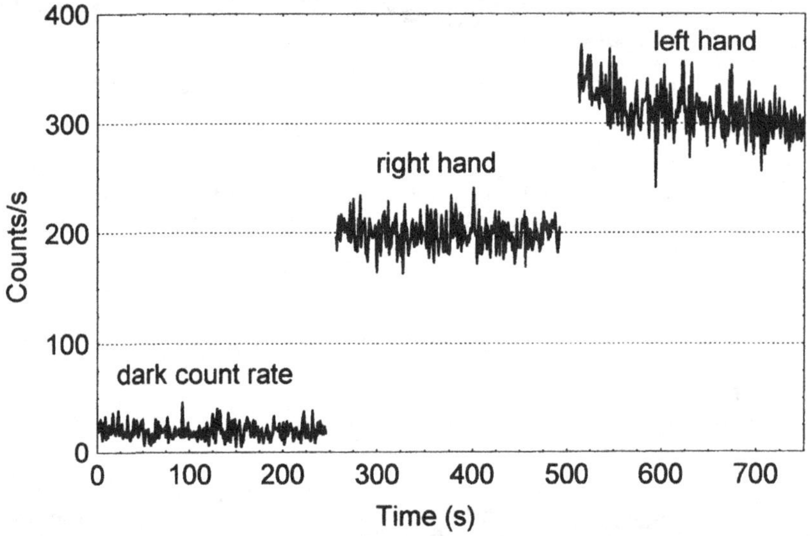

Figure 3. A case of a multiple sclerosis patient who has much higher biophoton emission on the hands than healthy people. At the same time the left-right symmetry is obviously broken.

3. Biophotons and Biological Rhythms

In order to investigate the biophoton emission of the human body more systematically we started a longtime-monitoring of the bpe and the d.l. of the hands and the forehead of a person (healthy woman, 27). The measure of bpe is the the mean value of the 256 measurement values in counts/s, and the first value 100 ms after excitation (counts/100 ms) has been taken as a measure of d.l. These measurements were performed between 8-10 in the morning from June 8, 1995 to March 5, 1996 for the hands and from August 29, 1995 to March 5, 1996 for the forehead.

It turned out that the biophoton intensity of the human skin is measurable at least in the wavelength range from 400 to 800 nm, where the photomultiplier is sensitive. The order of magnitude of the count rate is a few up to some hundred photons /(s cm2).

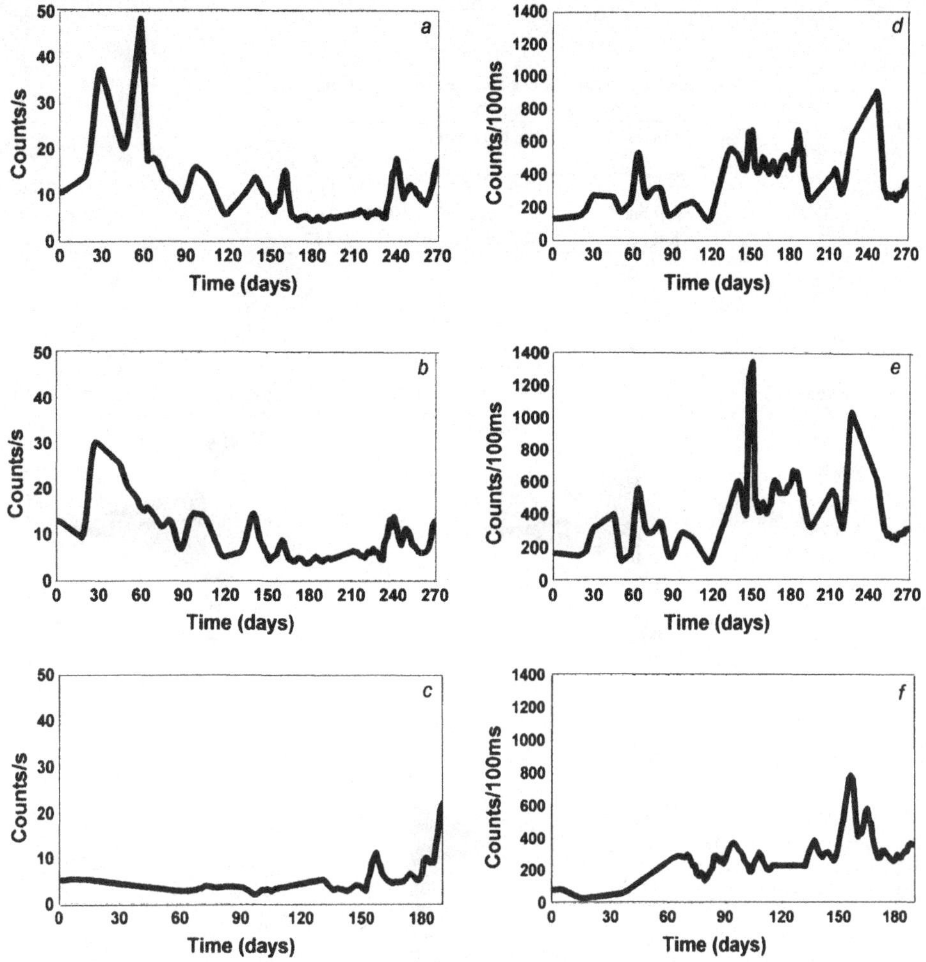

Figure 4. Measurements of the biophoton emission, bpe, (a, b, c) and the delayed luminescence, d.l., (d, e, f) of the right hand (a, d), the left hand (b, e) and the forehead (c, f). These measurements were performed between June 8, 1995 and March 5, 1996 for the hands and between August 29, 1995 and March 5, 1996 for the forehead.

The exact value depends on the individuum, on the time of measurement and on the location on the skin. The biophoton emission is generally rather stable. The mean value does not change considerably during the few minutes of investigation. However, it may vary within a few hours between a few percent of the original value up to ten-times higher or lower values, again depending on the individual situation. Figure 4 shows in the left column the bpe variations and in the right one the d.l. variations of the right and left hand as well as of the forehead. The values display some patterns, where the mean values and the standard deviations over the whole measurement period are shown in the TABLE.

TABLE. Mean values and standard deviations over the whole measurement period

Location	bpe (counts/s)	d.l. (counts/100ms)
right hand	12.9 +/- 8.7	359 +/- 183
left hand	10.7 +/- 6.6	400 +/- 230
forehead	4.7 +/- 2.6	245 +/- 147

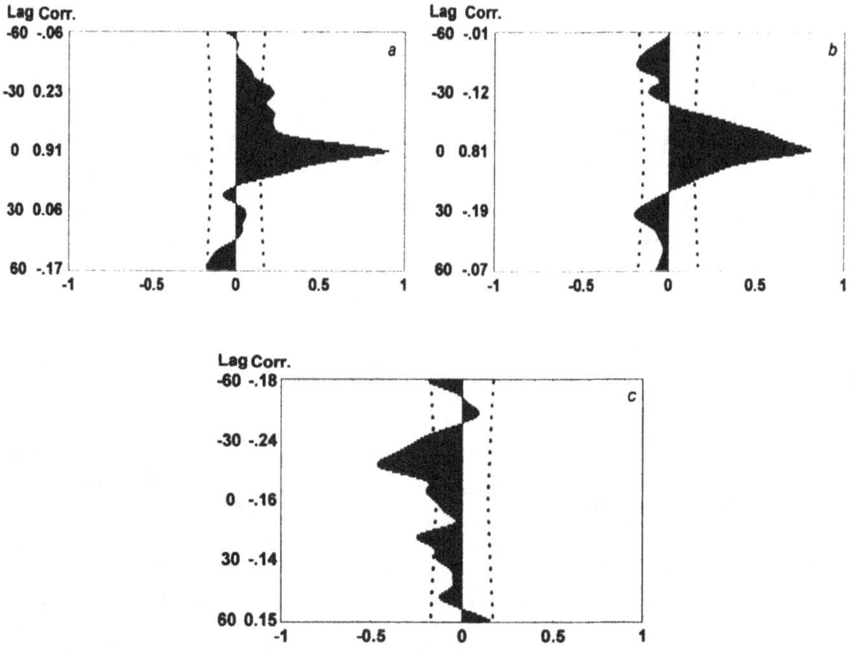

Figure 5. Cross-correlations of the measurements of the hands between August 29, 1995 and March 5, 1996, for a, the bpe variations of left hand (first) and right hand (lagged), b, the corresponding d.l. variations, c, the bpe (first) and the d.l. (lagged) variations of left hand.

The cross-correlation analysis of these temporal bpe-and d.l.-variations reveals the following characteristics:

- Left hand and right hand emissions are strongly correlated, as well for bpe as for d.l. (Fig. 5a, b respectively).

- Bpe and d.l. are not correlated, but to a significant extent anti-correlated (Fig. 5c).

- The bpe-values of the forehead are not correlated with those of the hands (Fig. 6a), but there is a significant correlation of the d.l.-values of the forehead and the hands (Fig.6b).

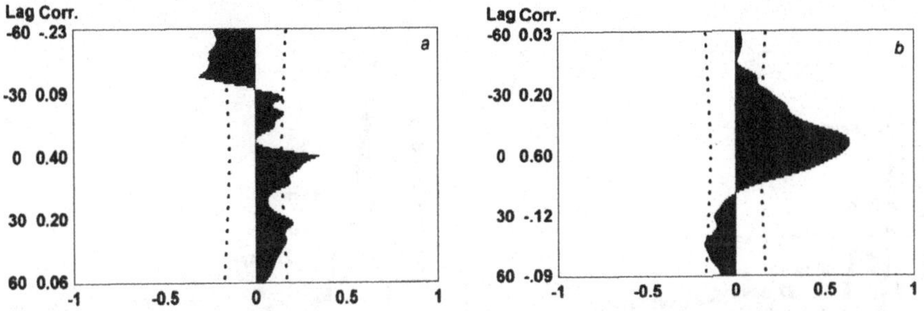

Figure 6: Cross-correlations of the measurements between August 29, 1995and March 5, 1996, of the left hand (first) and the forehead (lagged) for a, the bpe variations and b the corresponding d.l. variations.

The Fourier analysis of the temporal bpe- and d.l.-values of Figure 4 is shown in Figure 7, where the spectral densities of the different measurements of Figure 4 are displayed always on the corresponding positions of Figure 7. It is obvious that all these spectral densitities show a similar pattern of rhythms, including 7 days, 14, 21, 27 and 90 days.

We are not able at this stage to present a complete interpretation of the results. Nevertheless the clear preference of right-and left hand correlation indicates a regulatory role for the biophoton emission, which is confirmed with the Fourier pattern revealing well-known biological rhythms.

We also like to note that we already suggested in 1983 that biophotons trigger the biological rhythm by breathing-modes of the DNA [8].

4. Acknowledgments

We are particularly grateful to the Familie-Ernst-Wendt-Stiftung for its interest and its financial support.

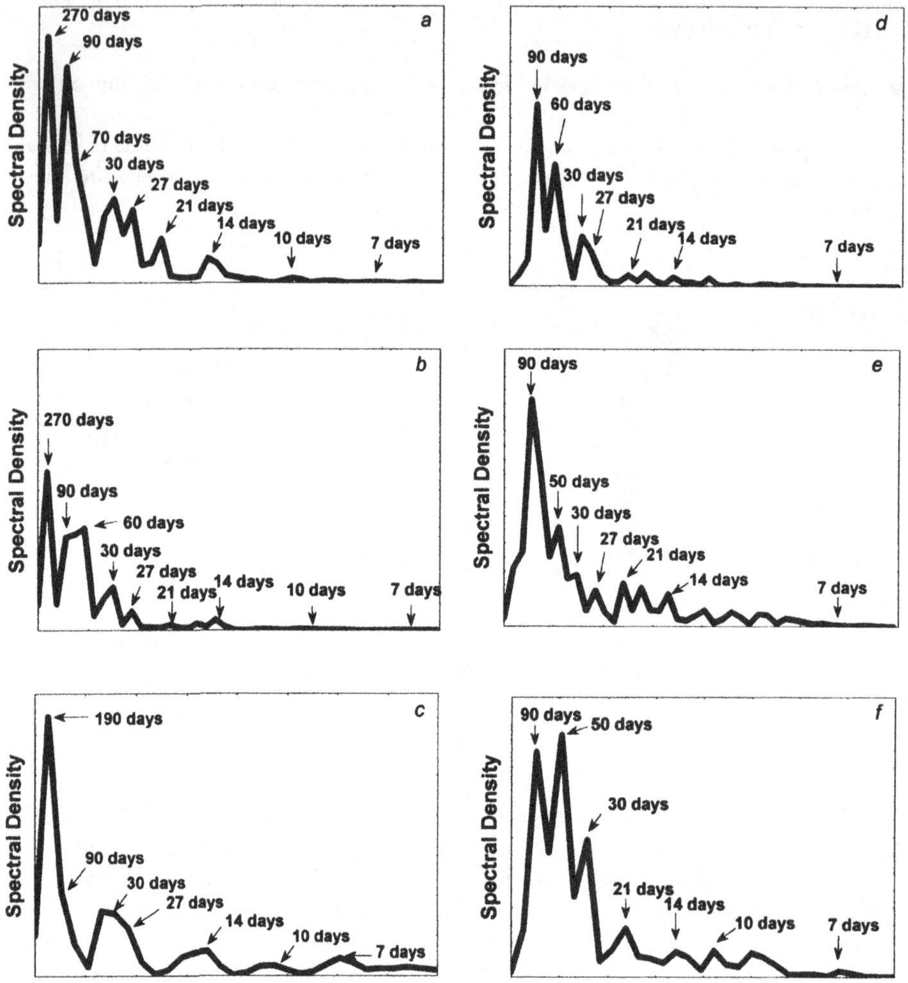

Figure 7. Spectral density of the measurememts of the biophoton emission, bpe, (a, b, c) and the delayed luminescence, d.l., (d, e, f) of the right hand (a, d), the left hand (b, e) and the forehead (c, f).
These measurements were performed between June 8, 1995 and March 5, 1996 for the hands and between August 29, 1995 and March 5, 1996 for the forehead.

5. References

1. Dobrin, R., Kirsch, C.,. Kirsch, S., Pierrakos, J., Schwartz, E., Wolff, T., and Zeira, Y.: Experimental measurements of the human energy field, in S. Krippner (ed.), *Psychoenergetic Systems,* 1979, pp. 227-230.
2. Ibison, M.: Personal communication.

3. Sauermann, G., Hoppe, U., and Stäb, F.: In vivo chemiluminescence of human skin, *Photochem. Photobiol.* (1993), 41S.
4. Ruth, B.:Experimental Investigations on Ultraweak Photon Emission, in F.A. Popp, G. Becker, H.L. König and W. Peschka (eds.), *Electromagnetic Bio-Information*, Urban & Schwarzenberg, München-Wien-Baltimore, 1979, pp. 107-122.
5. Cohen, S., and Popp, F.A.: Biophoton emission of the human body, *J. Photochem. Photobiol. B: Biol.* **40** (1997), 187-189.
6. Cohen, S., and Popp, F.A.: Low-level-luminescence of the human skin, *Skin Research and Technology* **3** (1997), 177-180.
7. Popp, F.A., Gu, Q., and Li, K.H.: *Recent advances in biophoton research and its applications*, World Scientific Publishing, Singapur, 1992.
8. Popp, F.A., and Nagl, W.: A physical (electromagnetic) model of differentiation, *Cytobios* **37** (1983), 71-83.

FLUORESCENCE IMAGING TECHNIQUE FOR DETECTION OF HUMAN MELANOMA

Large Sample Study of Diagnostic Parameters of the Method

B. W. CHWIROT, S. CHWIROT, K. SŁOWIKOWSKA,
J. REDZIŃSKI, J. TAZBIR♣, J. SIR♦, J. NUSSBEUTEL,
AND M. GRADZIEL

Interdisciplinary Group of Optical Methods of Early Detection of Cancer, Institute of Biology and Environment Protection, Nicholas Copernicus University, ul. Gagarina 9, PL 87-100 Toruń, Poland
♣Nicholas Copernicus District Hospital, Łód'z, Poland
♦Regional Centre of Oncology, Bydgoszcz, Poland

Abstract

New fluorescence imaging technique for in situ detection of human melanoma has been tested in a large sample study of 156 skin lesions (57 melanomas, 82 pigmented nevi and 17 cases of verruca seborrhoica). The study showed that the sensitivity of the method was 82% while its specificity was 65%. The results show that the method could be useful for a preliminary selection of potential melanoma cases in screening programmes involving large groups of patients.

1. Introduction

The incidence of melanomas has been increasing world-wide and according to statistical data the number of diagnosed melanomas doubles every 10-15 years. If the trend continues melanoma will become one of the most common malignant cancers after the year 2000 [1] and at that time 50% of all cases will be diagnosed in people less than 40 years old. Our knowledge of factors and molecular mechanisms leading to a formation of melanoma is still very limited and this is the main reason for the lack of a programme of a prevention or at least significant reduction of incidence of this cancer.

Early detection of melanoma is of crucial importance for the success of surgery and for survival of the patient. Unfortunately, in many cases, the clinical diagnosis of pigmented tumours is relatively difficult. According to recent studies on average even 50% of early melanomas may not be recognised in routine clinical diagnosis, with 80-90% being detected by experts [2]. The diagnosis is mainly based on a set of morphological criteria. Thus, the diagnosis is necessarily subjective and depends on the

J. J. Chang et al. (eds.), Biophotons, 193-200.

experience and the oncological awareness of the examiner. An exact diagnosis can be obtained only after histological examination of the removed tissues. Therefore, the availability of a method allowing for a sensitive and objective in situ differentiation between pigmented nevi and melanomas would enable a more rational approach to deciding on a necessity of surgical excision of suspected tissues and would also significantly increase the efficiency of screening programs.

98% of melanomas arise in skin tissues. Such a location makes it possible to use for the melanoma diagnostics a new approach based on methods of so-called tissue spectroscopy. The first attempt in using fluorescence method for diagnosing melanoma in situ was made in 1988 by Lohmann and Paul [3]. The authors excited in vivo autofluorescence of skin tissues with ultraviolet light and recorded the spectra of light emitted by healthy tissues, nevi and melanomas. They found that the fluorescence spectra were similar in all cases and had a simple structure with one broad maximum at approximately 475 nm. In some cases the melanomas yielded more complex fluorescence spectra with additional peaks at 445 and 535 nm. The authors noticed, however, that melanomas generated specific patterns of variations in intensity of the fluorescence. Usually, the intensity was very low for a melanoma and there was a local increase of the intensity in a transition zone between the melanoma and areas of a healthy skin followed by a drop in the intensity measured for the latter. The local maxima of intensity were not found for nevi and the authors concluded that the effect may be of importance for the differentiation between melanomas and nevi. The hypothesis was tested in studies involving a large sample of patients [4].

The results of Lohmann's group were not confirmed by independent study of Sterenborg et al. [5] who measured the excitation-emission matrix of the skin and tried to apply the imaging technique to differentiate between melanomas and benign lesions. Another approach based on studies of reflectance images of the skin lesions has been developed by Bono et al. [6] who found that such images in the infrared band may be useful for the discrimination between melanomas and benign melanocytic lesions.

The aim of the present work was to determine diagnostic parameters of the fluorescence method for diagnosing melanomas in situ developed in our group [7]. Our approach consists in using digital imaging technique for measurements of the intensity of the autofluorescence emitted in a selected band of the spectrum by the cells in areas of interest instead of the point - by - point scanning approach of [3,4] with detecting the whole spectrum and evaluating the intensity from such data. No special qualifications are required to operate our diagnostic apparatus and analysis of a surface distribution of fluorescence intensities can be performed automatically using procedures for digital image analysis.

2. Subjects and Methods

A total of more than 600 skin lesions were tested but only the data obtained for 156 cases with histopathologically confirmed clinical diagnosis were used for estimating the diagnostic efficiency of the method (malignant melanomas - 57, non-malignant pigmented lesions - 99).

The patients from four hospitals and clinics located in four different cities (Toruń, £ód'z, Bydgoszcz, Gdańsk) were examined. The results of histological evaluations performed in those units were used as a reference and a control of the fluorescence diagnosis. All the patients gave their informed consent and the study was approved by the District Committee of Ethics of Scientific Research at School of Medicine, Bydgoszcz, Poland.

The diagnostic method is based on analysis of the spatial distributions of the autofluorescence of the skin excited with the 366 nm line of Hg and recorded in a specially selected band of visible part of the spectrum [7]. The main element of the apparatus comprising the light source and the CCD camera is suspended on a holder with many degrees of freedom allowing for easy illumination and imaging of any place on a patients body. The digital imaging part of the apparatus is based on a CCD camera designed for low light applications (SpectraSource, USA). The CCD element of the camera is cooled to -30° C with resulting low dark count rates allowing for using long exposure times. The images are corrected for dark counts, bias voltages of the pixels and also for variations in a sensitivity of the pixels of the CCD matrix. The optical system of the camera includes a suitable interference filter and objective lens of a typical photographic camera (wide angle, f 37 mm). An AT486 computer is used to control the measurements and to perform digital analysis of the images of the autofluorescence.

Images of the autofluorescence are recorded in a dark room. The autofluorescence is typically excited in areas with a diameter on the order of 3-5 cm. The distance between the area subject to examination and the camera is on the order of 30 cm. Such an arrangement reduces errors of the determination of the fluorescence intensity due to changes of the distance between the often curved regions of the skin and the objective lens.

In our method the exciting light does not need to be focused and its intensity at the skin level is on the order of less than 0.1 mW/cm^2. Typical exposure times range from seconds to one minute depending on the pigmentation of skin and lesions and on the distance from the source of the excitation light. Thus the doses of the UV radiation are much lower than minimal average minimal erythema doses [8].

An attempt has also been made to correlate the results of the fluorescence studies with a level of expression of HMB-45, a cellular antigen considered a marker of melanocytic activation [8,9] and commonly used as a sensitive marker for a differentiation of malignant melanoma from nonmelanocytic malignant neoplasms [10]. In a typical procedure paraffin tissue sections were incubated with antibodies against HMB-45. The reaction product was visualised using alkaline phosphatase system. All chemicals used were supplied by DAKO as a kit and the reactions were carried out according to instructions provided with the kit. The level of expression of HMB-45 was determined by measurements of optical absorption of histochemically stained cells selected in microscopic images. The images were acquired using the Olympus BX-50 microscope equipped with a CCD camera (DEI-470T-CMT, Optronics, USA). The intensity of the transmitted light was normalised to the intensity of the light passing through the cytoplasm of negatively stained cells in the same section.

3. Results

Images of the autofluorescence recorded for lesions under examinations always included also areas of surrounding visually normal skin. For each lesion another fluorescence image, of a normal skin from a place located symmetrically on the patient's body was also acquired. Such a methodology allowed estimations of: (1) natural variability of the level of the fluorescence within the areas subject to investigation for a large group of patients of different skin types and with different locations of the lesions; (2) characteristics of the spatial distributions of the autofluorescence that might have been due to a presence of melanoma.

The variability of the intensity of the fluorescence was measured by a ratio (MM) of the maximum and minimum levels of the emission determined for a given area of the skin. Within the limits of experimental errors the ratio was on the order of 2.0 for a healthy skin for all the localisations with exception of the skin surrounding melanomas. For the latter the mean value of MM was 2.4 (Table 1 and Figure 1). As for the pigmented lesions, the variability was highest for melanomas. The pigmented nevi and verrucas yielded similar values of MM which were on average twice as low as those determined for melanomas (Table 1 and Figure 1).

TABLE 1. Variability of the intensity of the autofluorescence determined as the ratio MM of the maximum and minimum intensity measured for areas subject to examination (The errors are the standard deviations of the mean calculated for each group)

		healthy skin from the areas surrounding the lesions			healthy skin from the areas located symmetrically with respect to the lesions		
	lesion	skin type I	skin type II	skin type III-IV	skin type I	skin type II	skin type III-IV
melanoma	5.7±4.1	2.4±0.9	1.9±0.4		2.2±0.7	1.9±0.5	
nevus	2.6±2.0	1.8±0.4	1.9±0.5	1.8±0.4	1.8±0.6	1.7±0.3	2.0±0.6
verruca	2.8±1.4		2.0±0.7			1.9±0.4	

Typical images of the fluorescence of the lesions and the surrounding skin are characterised by a presence of zones of the fluorescence of different intensities. Often, the zones closest to the lesion outline its shape and with the increasing distance the zones become more diffused. At the same time, however, the zones of similar fluorescence intensities are not homogenous and can be overlapped by local areas of increased or decreased intensity. Nevertheless, we practically never observed in regions between melanomas and normal skin the zones of a maximum fluorescence intensity described by Lohmann et al. [3,4]. From the point of view of detecting melanomas most interesting were the results of a determination of ratios (R) of the maximum intensities of the fluorescence measured in a neighbourhood of pigmented lesions to the minimum of the intensity determined for the lesions as such. The ratio R was found to be the most sensitive parameter for a differentiation between melanomas and other pigmented lesions. Histogram of the values of R determined for 51 cases of melanoma and 99 benign pigmented lesions (nevi and verrucas) is shown in Figure 2. It should be noted that we excluded from our analysis 6 cases of melanoma characterised by a presence of inflammatory reactions, callous skin and other morphological features making them unsuitable for the fluorescence examination.

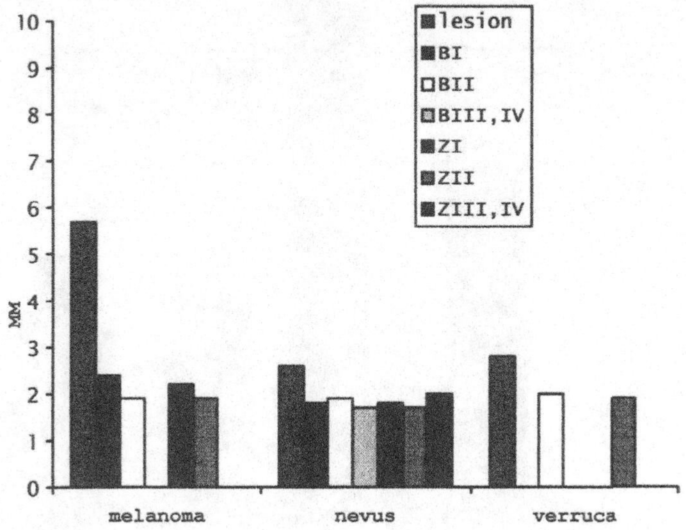

Figure 1. Histogram of the MM values obtained for different skin lesions and areas of the visually healthy skin classified according to skin types and locations with respect to the lesions of different types (skin types according to [11]: I - never tan, always burn, II - always burn but sometimes tan, III - always tan, sometimes burn, IV - always tan, never burn; B - visually healthy skin surrounding the lesion, Z - visually healthy skin located on the body symmetrically to the lesion).

It can be clearly seen from Figure 2 that indeed basing on the values of the parameter R one can with a high probability determine if the pigmented skin lesion subject to fluorescence examination may be melanomatic. Our studies show that for diagnostic purposes one can reasonably assume that lesions yielding $R \geq 7.0$ should be tentatively classified as melanomas. With the threshold value of $R=7.0$ the sensitivity of our method, defined as the ratio of the detected melanomas to all the cases of melanoma found in histopathological examinations, is equal 82%.

Another important parameter of any diagnostic method is its specificity defined as a fraction of really positive cases in the group of those with a positive result of the test. The specificity of our method is equal 65% i.e. 35% of cases yielding $R \geq 7.0$ is benign lesions.

4. Discussion

The most important result of this work is undoubtedly a confirmation of earlier work [7] suggesting that measurements of the spatial distribution of the autofluorescence may

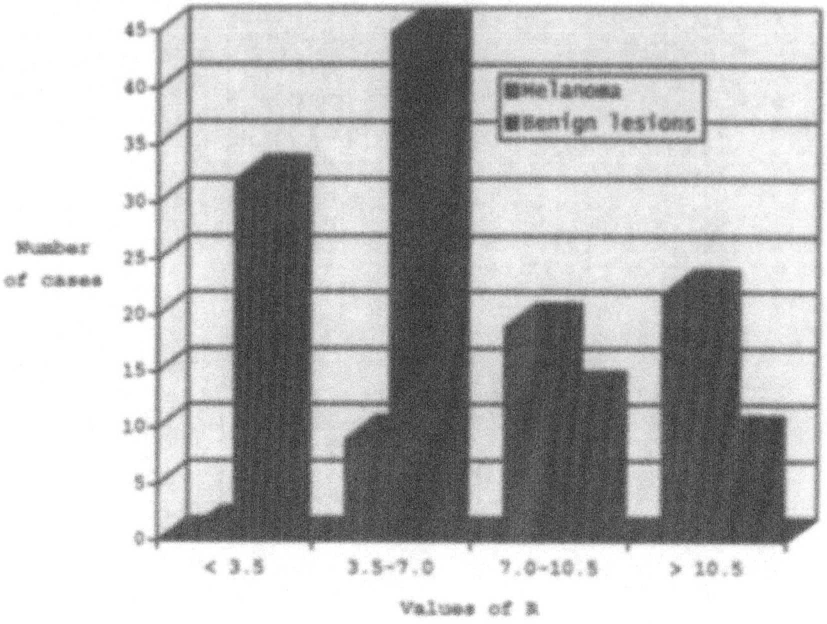

Figure 2. Histogram of the values of the parameter R determined for melanomas and benign pigmented lesions.

allow for a detection of human melanoma in situ. The diagnostic parameters of our method obtained from the study of a large population of patients indicate that with its simplicity and objective character the method constitutes a useful tool for screening of large groups of people and for a selection of those who should seek advice of a specialist. The examination of the patients is absolutely non-invasive. Only a little training is required to learn how to operate the diagnostic unit and calculations of the diagnostic parameter R can be done automatically.

It is surprising that similarly to the report of [5] our study does not confirm the observation of local maxima of the intensity of the autofluorescence in the zone of transition from melanomas to healthy skin as described by Lohmann et al. [3,4]. A possible explanation of this apparent discrepancy may be that the intensity of the exciting light was in our experiments kept at a very safe level and was much lower than used by Lohmann et al. who focused the exciting UV light onto small spots on the skin surface. At the same time, however, using a procedure different from that of [3,4] we are able to detect melanoma in situ with a good sensitivity and reasonably good specificity.

Another interesting result is that the intensities of the autofluorescence measured within areas of a surface of a few centimetres squared may differ by a factor of 2. This result indicates that more careful studies are required to confirm some earlier literature reports on relations between the absolute autofluorescence intensities and processes such as a natural ageing and a photoageing of human skin [11].

At the moment both the origin of the autofluorescence and the biological background of the presented method still require further studies. Several endogenous fluorophores capable of emitting in the visible spectrum have been suggested in the literature but all the suggestions have been so far based mostly on similarities between the observed spectra and spectra measured for chemically pure substances studied in vitro. It is also difficult to explain why the values of the parameter R measured for melanomas are so different from those determined for benign lesions. One of the histopathological criteria for identification of melanomas is a presence of the cellular antigen HMB-45 considered a marker of melanocytic activation. Our quantitative immunocytochemical measurements of levels of the expression of this antigen in the cells of the skin lesions subject to the fluorescence examinations did not reveal any correlations between the intensity of staining and the R values.

An interesting result (not shown) was that in a majority of cases the visually healthy skin near melanoma seemed to emit the fluorescence of a slightly lower intensity than the skin of the same patients from areas located symmetrically to the lesion on the patient's body. The mean value of the appropriate ratio of the intensities calculated for all the cases differs from one within the limits of the standard deviation but the phenomenon was observed for a majority of the patients. The phenomenon, if confirmed by a future studies, might be related to interactions between the tumour and the host cells. Such interaction, probably of a defence character, undoubtedly take place at least at some stages of a tumour formation and progression and most probably it is them that are responsible for the structural and metabolic changes causing the characteristic dependence of the values of the parameter R on a presence of melanoma or other skin lesions of a more benign character.

5. Acknowledgments

This work was supported by a grant No 4 P05B 055 12 (Komitet Badañ Naukowych) and a grant from the Regional Fund of Environment Protection, Toruñ, Poland. The authors thank Prof. Dr. Hab. A. Kopacz, Dr. A. Kucharski and Dr. T. Jastrzêbski of Medical School, Gdañsk and Dr. W. Czechowicz and Dr. J. Œniegocki of Regional Center of Oncology, Bydgoszcz and Dr. Z. Morawiec of Nicholas Copernicus District Hospital of £ód'z for their help in encouraging patients to volunteer to undergo fluorescence examinations. We also thank the staffs of the histopathological units at those clinics for providing histological evaluations of the tissues subject to the fluorescence studies.

6. References

1. Balch, C.M., Houghton, A. and Peters L.: Cutaneus melanoma in V.T. de Vita, S. Hellmann and S.A. Rosenberg (eds.), *Principles and Practice of Oncology*, J.B. Lippincot Comp., Philadelphia, 1989, p. 1499.

2. Grin, C.M., Kopf, A.W., Welkovich B., Bart, R.S. and Levenstein, M.J.: Accuracy in the clinical diagnosis of malignant melanoma, *Arch. Dermatol.* **126** (1990), 763-766.
3. Lohmann, W. and Paul, E.: In situ Detection of Melanomas by Fluorescence Measurements, *Naturwissenschaften* **75** (1988), 201-202.
4. Lohmann, W., Nilles, M. and Boedeker, R.H.: In situ Differentiation Between Nevi and Malignant Melanomas by Fluorescence Measurements, *Naturwissenschaften* **78** (1991), 456-457.
5. Sterenborg, H.J.C.M., Motamedi, M., Wagner R.F., Duvic, M., Thomsen, S. and Jacques S.L.: In Vivo Fluorescence Spectroscopy and Imaging of Human Skin Tumours, *Lasers Med. Sci.* **9** (1994), 191-201.
6. Bono, A., Tomatis, S., Bartoli, C., Cascinelli, N., Clemente, C., Cupeta, C. and Marchesini, R.: The Invisible Colours of Melanoma. A Telespectrophotometric Diagnostic Approach on Pigmented Skin Lesions, *Europ. J. Cancer* **32A** (1996), 727-729.
7. Chwirot, B.W., Chwirot, S., Jêdrzejczyk, W., Redziñski J. and Michniewicz, Z.: Sposób wykrywania znamion czerniaka. *Biuletyn Urzêdu Patentowego* **21** (1996), 8 (Bulletin of Patent Office, patent application).
8. Tronnier M. and Wolff, H.,H.: UV-Irradiated Melanocytic Nevi Simulating Melanoma In Situ, *Am. J. Dermopathol.* **17** (1995), 1-6.
9. Skelton H.G., Smith K.J., Barrett T.L., Lupton G.P. and Graham J.H.: HMB-45 Staining in Benign and Malignant Melanocytic Lesions. A refelction of Cellular Activation. *Am. J. Dermopathol.* **13** (1991), 543-550.
10. Smoller B.R., McNutt N.S. and Hsu A.: HMB-45 recognizes stimulated melanocytes. *J. Cutan. Pathol.* **20** (1989), 49-53.
11. Leffel D.J., Stolz M.L., Milstone L.M. and Deckelbaum L.I.: In vivo Fluorescence of Human Skin. A Potential Marker of Photoaging. *Arch. Dermatol.* **124** (1988), 1514-1518.

ELECTRO-LUMINESCENCE AND ITS APPLICATION

A Sensitive Method for Testing Microbial Contamination

J.J. CHANG*# AND F.A. POPP*

**International Institute of Biophysics, IIB,*
Station Hombroich, Vockrather Straße,
D-41472 Neuss, Germany
 # Institute of Biophysics,
Chinese Academy of Sciences, Beijing 100101, China

Abstract

By use of electro-luminescence at least 10 to 100 bacteria/ml in aqueous solutions can be distinguished. This method is sensitive, fast and can provide a real time monitoring which is most important and useful for industrial application.

keywords: electro-luminescence, photon emission, bacterial contamination.

1. Introduction

There are different methods for estimating microbial contamination in a sample. They can be categorised according to their principle into physical, chemical and biological tests. They have been reviewed [1-2]. Up to now among all these methods the colony counting method is the most reliable one [3]. However it requires a long time and cannot reach the sensitivity of our method which is able to distinguish down to 10 or 100 bacteria /ml. Although some bacteria have special fluorescence emission which can be used for their identification by luminescence, most of bacteria do not emit fluorescence light. Actually, immunoassay [4] can also be used with a high sensitivity for this purpose, but it is rather expensive. Therefore we developed the electro-luminescence method for testing microbial contamination. We will concentrate in this paper onto the description of this method.

Although both luminescence and electric conductivity have been used for estimating bacterial mass, measuring electro-luminescence for testing contamination of bacteria has not been mentioned up to now. Besides high sensitivity, our method is fast and can provide a real time estimation of microbial infection.

* Supported by the National Natural Science Foundation of China.

J. J. Chang et al. (eds.), Biophotons, 201-215.

2. Materials and Methods

TABLE 1 shows the materials used in our experimental studies.

TABLE 1: Materials

Material	Time of experiments	Results	Document
Dinofllagellates P. elegans	July 1993		Ref. (6)
Suspended chicken embryonic brain cells	April 1994		
Suspended chicken embryonic heart cells	April 1994		
Erythrocytes	May 1995		
Soybean plant root, stem and leaves	Aug.1993		
Bryophyllum plant, stem and leaves	Aug.1993		
Bacteria	May 1995		Ref.[7]-[8]

For testing bacterial contamination, cultured soybean root nodule bacteria Rhizobium japonicium (Rh.j.) 1132-2 were used [9]. In order to provide a simple system, the bacteria were cultivated in a modificated medium according to Tsein [10]. The contents of the medium are the following :
Yeast extract 1.5 g, D-Monnitol 10 g, NH_4Cl 3mM contained in 1000ml water. The culture medium was adjusted to pH 6.8 to 6.9. For solid cultures 15g Agar was added. The concentration of Rh.j. was counted under microscope by hematometry.
In some experiments, Rh.j. and other bacteria were suspended and cultivated in beer. Pure water and tap water contaminating Rh.j and cultivated Pseudomonas diecinuta (from Dr. Tomath, Stuttgart University) were also used as tested samples.

2.1. PREPARATION OF MEASURED SAMPLES

For any unknown tested liquid samples or samples which could be put into aqueous solution, the controls were made by filtering the solution by means of a micropore film of pore size 0.3μm or 0.45μm which can also be used as medium for dilution. A series of measuring samples of different concentration of the unknown bacteria were made by using the medium.

2.2. ELECTRODE AND ELECTRIC STIMULATION

In the examples presented here we used two parallel needles made of platinum between which the distance could be changed from 1mm to 3cm according to the samples and the electric generator, but in principle other electrodes, for example plates, can also be used for these purposes. Some experiments were also carried out by using two parallel needles made of silver and other materials. Platinum electrodes can be used several times but before use it is necessary to wash them at first in diluted KOH aqueous solution and then in water.

For commercial purpose cheaper material can also be used for making electrodes, but for each measurements it is then necessary to use new ones. In this case the electrodes can be made in a fixed form like a lamp which can be simply inserted into plugs. The plugs can be included in the photon-measurement equipment. In our experiments different intensity of dc- and ac-current were used for the electric stimulation.

2.3. PHOTON EMISSION MEASUREMENTS

The photon emission of measuring samples were recorded by specific single photon counting systems PMS-1, PMS-4 and the double chamber measuring system from Popp's Laboratory [5]. The devices were equipped with a EMI 9558 QA photomultiplier which is sensitive in the range between 200 nm and 800 nm. Bacteria and suspended cells were measured with PMS-1 and PMS-4, plants were measured with the double chamber system. Intensities of photon emission were measured as the numbers of photons integrated over a time interval by computer.

The sample is put into a photon-measurement cuvette and the electrodes are introduced into the sample so that they are located just underneath the surface of the sample solution. The electrodes are then connected to the generator. In a series of tests one has to take care to keep always the same location for the electrodes and the same pole connections.

Before photon-measurement the sample was kept 5 min in darkness. The measuring parameters were chosen according to the sample and the electric parameters. In our experiments we chose 100 ms for the time interval measurement and a total count of 2048 for each measuring sample. During the 2048 counts one stimulation was given for some samples and for most samples three electric stimulations controlled by computer system and distributed at 500, 1000, and 1500 ms were made. In that case the first stimulation was the most important for analysing the results.

3. Results

3.1. ELECTRO-LUMINESCENCE FROM CULTIVATED RH.J.

Fig.1 and Fig.2 show the spontaneous emission patterns from liquid cultures of soybean root nodule bacteria, Rhizobium japoninum 1132-2. From Fig.1 we see that different cultures but cultivated in the same manner gave similar results which show that the electro-luminescence method is a stable reproducible method. Compared to the medium without bacteria, the total photon counts of the electro-luminescence-counts from the bacteria contaminated medium were lower under certain conditions. Fig.2a (top) shows the results from the medium whereas Fig.2b (middle) and Fig.2c (bottom) show the results from the medium containing respectively about 10^2 and about 10^3 /ml bacteria.

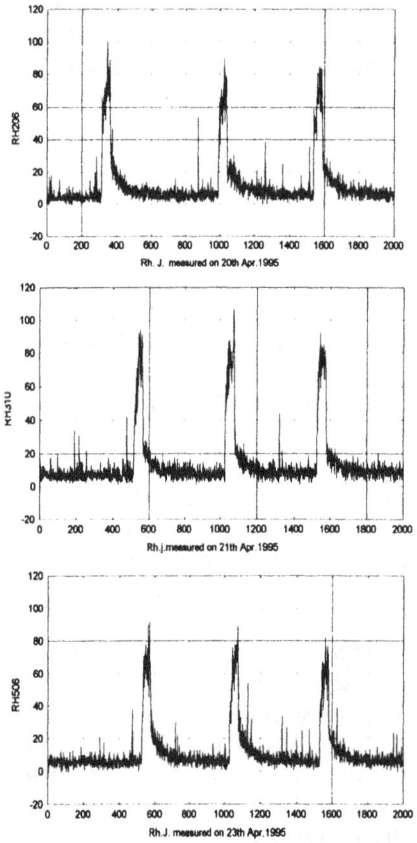

Figure 1. Electro-luminescence from liquid cultured Rh.j. Each sample contains 10^6 bacteria /ml, stimulated by 30V for 5s.

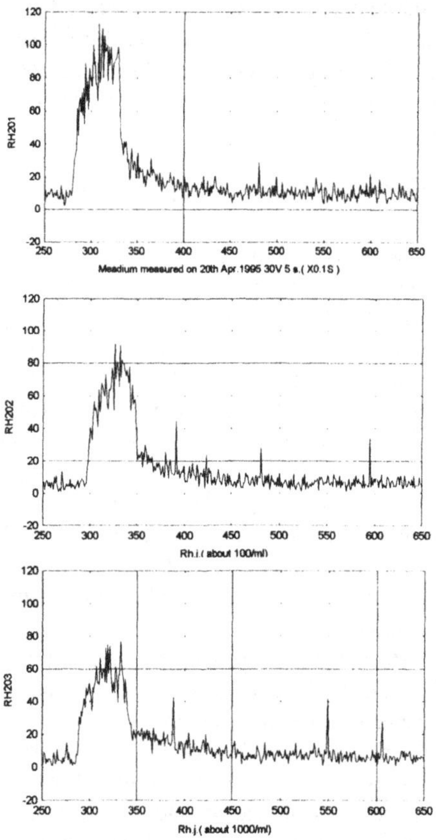

Figure 2. Electro-luminescence from cultured Rh.j. stimulated by 30Vfor 5s. From the top to the bottom the concentration of bacteria is 0, 10^2 and 10^3 /ml.

3.2. ELECTRO-LUMINESCENCE FROM BEER

Similar results were obtained with electro-luminescence from beer. Fig.3 and Fig.4 show photon emission patterns from different beers stimulated by dc current. From these figures, decreases of electro-luminescence can be observed even up to 10^2 beer contaminating bacteria/ml.

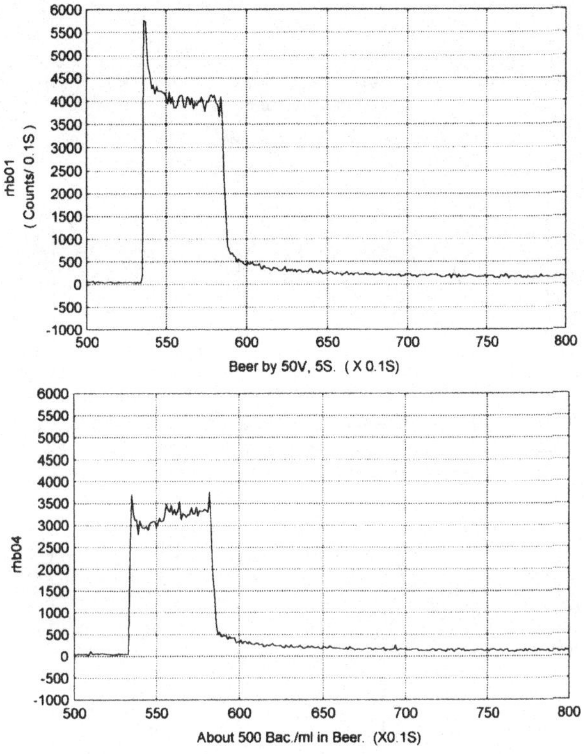

Figure 3. Electro-luminescence from beer (top) and beer containing about 5.10^2 bacteria /ml (bottom).

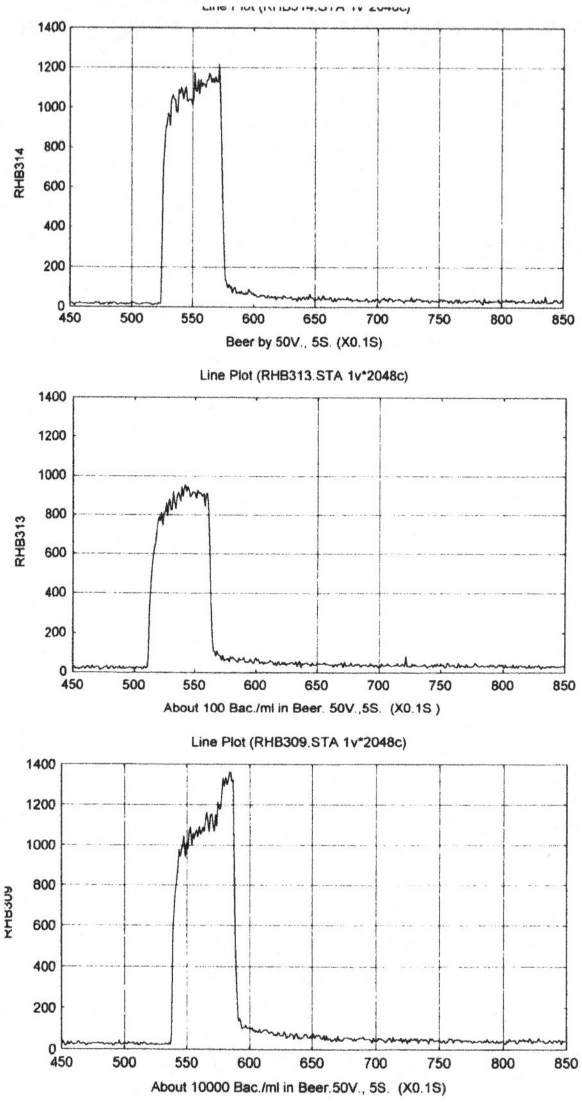

Figure 4. Electro-luminescence from beer (top) and beer containing about 10^2 /ml bacteria (middle) and about 10^4 bacteria /ml (bottom)

3.3. ELECTRO-LUMINESCENCE FROM WATER

Pure water and Rh.j. even in the concentration of 10^6 /ml suspended in pure water without ions do not give measurable electro-luminescence emission pattern (Fig.5a), whereas electro-luminescence can be distinguished from water containing some ions (Fig.5b).

The electro-luminescence method is also a sensitive method for monitoring the tap water quality. As shown in Fig.6a and Fig.6b, within 24 hours the tap water shows some changes in the electro-luminescence which means that there must have been some changes either in the contents of ions and/or in other properties. Under suitable conditions water containing 10 bacteria /ml showed an identical decrease of electro-luminescence photon counts (Fig.7). Cultivated Pseudomonas diecinuta gave also similar results as shown in Fig.8. As the concentration of the bacteria increases from o/ml to 10^3 /ml,. the total photon counts of electro-luminescence of the cultures decreased.

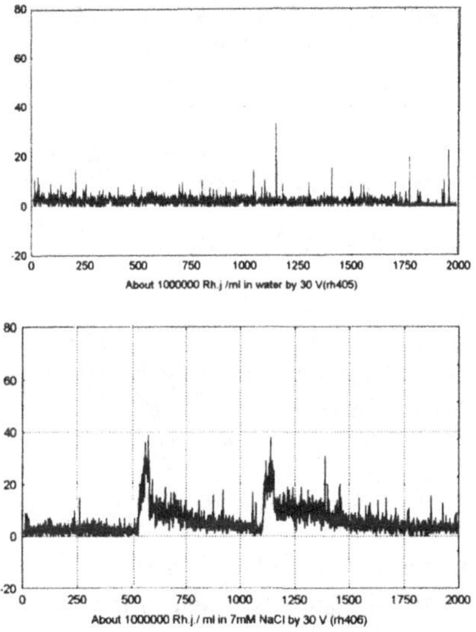

Figure 5. Electro-luminescence from water containing 10^6/ml bacteria but without ions (top) and with 7 mM NaCl. Stimulated by 30Vfor 5s.

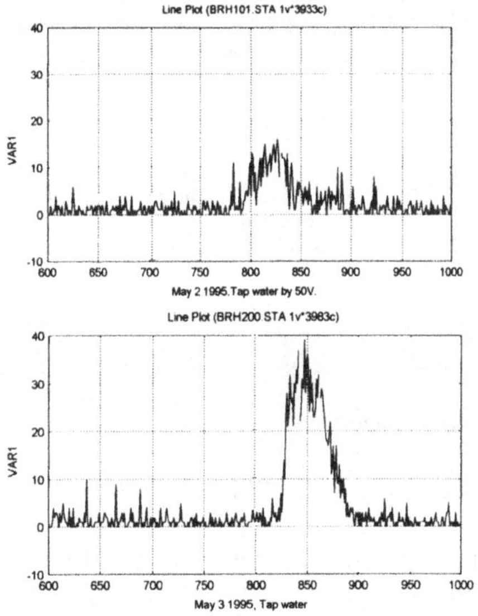

Figure 6. Electro-luminescence from tap water measured on different days. Stimulated by 50V for 5s.

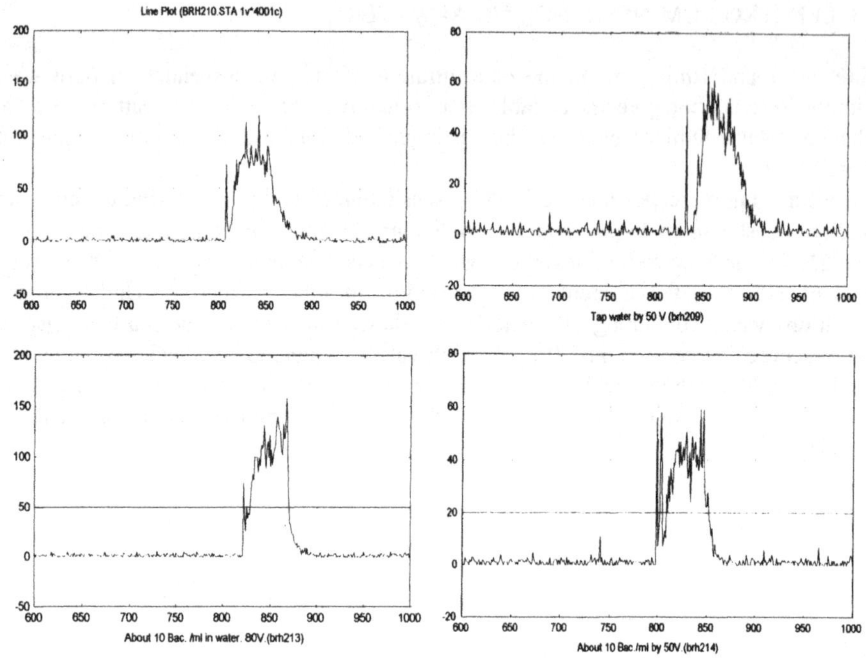

Figure 7. Electro-luminescence from tap water stimulated by 50V (right side) and by 80V (left side).
The top shows tap water, the bottom shows tap water with about 10/ml bacteria.

4. Discussion

4.1. POTENTIAL APPLICATION

Our experimental results show that the electro-luminescence method is a sensitive and reproducible method. The electro-luminescence from all materials tested in our studies showed sensitive changes in its mean photon counts, total photon counts during and after the electric stimulation according to the differences of the measured samples. Also the shapes of the photon emission patterns showed a relation to the changes of the samples. The examples presented here demonstrate that this method may become rather useful for monitoring bacterial contamination in food products like beer, drinks and milk, and in addition, for water antipollution.

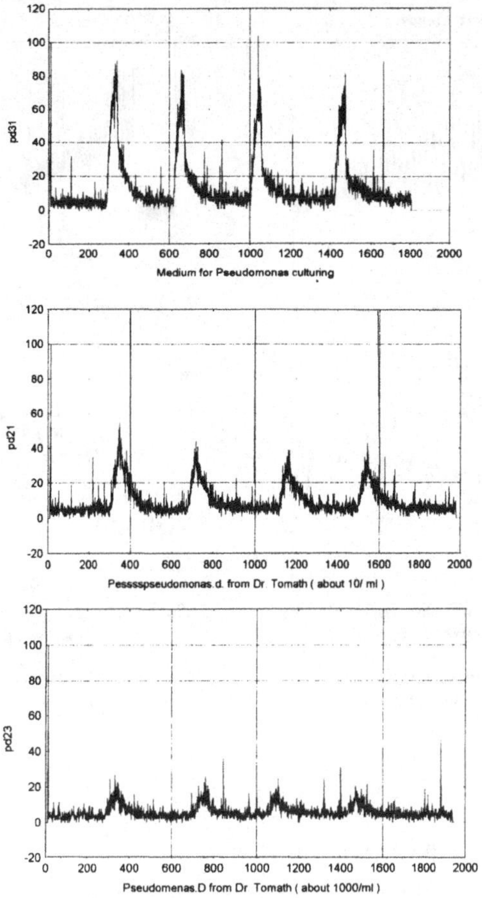

Figure 8. Electro-luminescence from cultured Psudomenas D. The top is the medium, the middle is the medium con-taining about 10/ml bacteria, the bottom is the medium containing about 10^3/ml bacteria

Electro-luminescence may also have some potential application in testing fruits quality. Fig. 9 and Fig.10 show the oscillations of the electro-luminescence from stems of Soybean and Bryophyllum. By the same electrical stimulation the difference of photon emission between the two plants can be distinguished. Even the same Bryophyllum plants show some differences in their photon emission pattern of electro-luminescence at different periods of a year (not presented here). Electro-luminescence from seedlings was reported to follow hyperbolic relaxation.

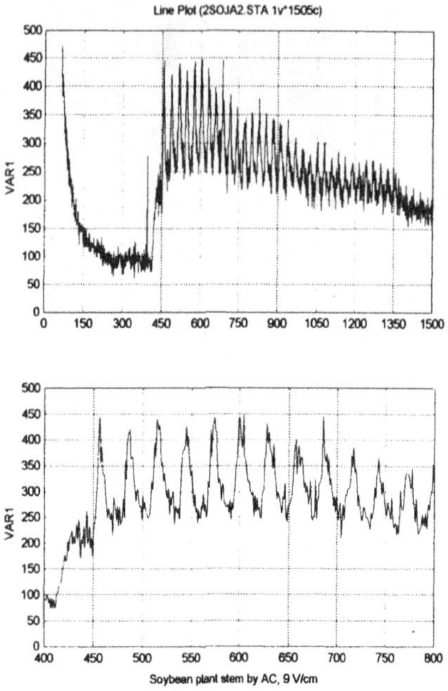

Figure 9. Electro-luminescence from soybean plant root stimulated by ac-current at 9V.

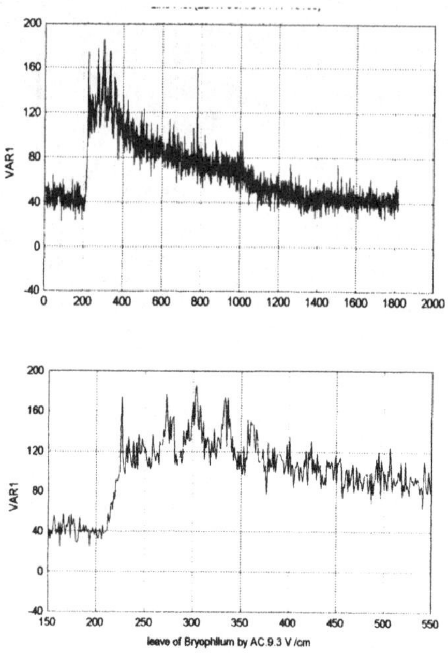

Figure 10. Electro-luminescence from Bryophllum plant stem stimulated by ac-current 9.3V.

4.2. COMBINATION OF PARAMETERS

From the photon emission patterns one can easily distinguish the difference between the controls and the solution containing bacteria. As shown in Fig.2, as the concentration of bacteria increases the photon counts decrease, but sometimes the photon counts increase as the concentration of bacteria increases depending on the combinations of the parameters shown in Fig.4.

There are several parameters which can be chosen in this method. For a given sample, the concentration of ions in the medium, the amplitude of electric voltage and the electric stimulation time should be optimized in order to get the best sensitivity. The experimental result shows that the most efficient combination of these parameters can easily be found. Depending upon the combination of these parameters, the electro-luminescence pattern of a sample can be related most sensitively to the concentrations of bacteria in the sample. Among these parameters, and with the exception of the concentration of ions, the most important one is the voltage. Fig.7 displays the electro-luminescence from the same samples but stimulated by different voltages. Fig.11 shows another similar result. From Fig.7 and Fig.11 we can see that as the voltage increases from 50 V to 80 V, the electro-luminescence of medium contaminating bacteria shows a reversed behaviour, it increases instead of decreasing.

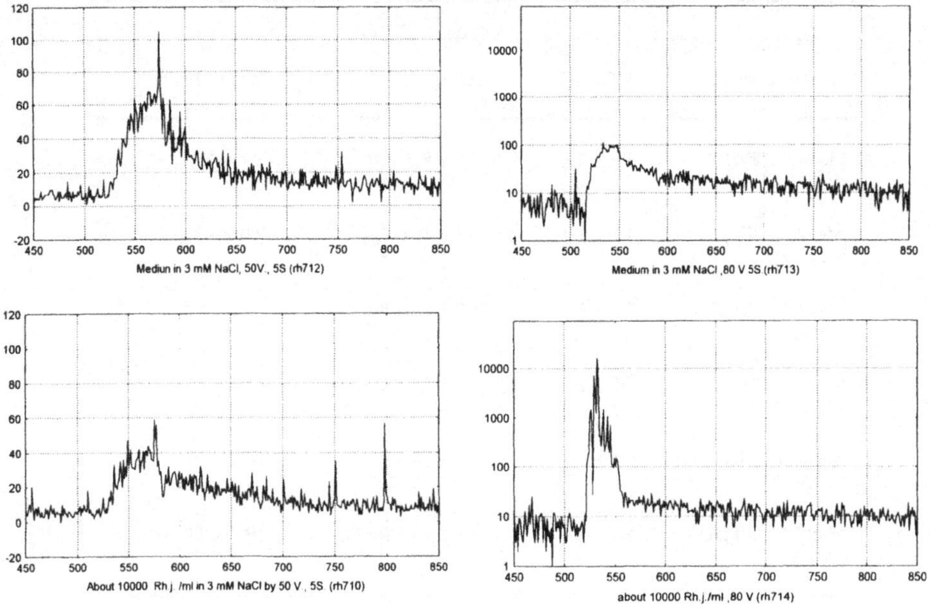

Figure 11. Electro-luminescence from Rh.j. suspended in water with 3 mM NaCl, stimulated by 50V (left) and by 80V (right), showing that when stimulated by 50V., as the concentration of bacteria increases the photon emission decreases but when stimulated by 80V. the emission increased.

For a given sample it is a good combination to use a low voltage. However if the intensity is too low, it is not possible to see a difference between the medium and the bacterial contamination (not presented here).

As shown in Fig.12 the electro-luminescence pattern can be described by the following three parts: before the electric stimulation, during the stimulation and within a short period after the stimulation. If we compare the mean counts of each part we find that in the third part they drop obviously down with increasing bacterial contamination. Table 2 shows the similar analysis from the measurements in Fig.2 and the same samples but stimulated now by 15V. From table 2 one can see that the photon counts during and after the stimulation (ES) are not linearly related to the intensity of the applied voltage. As the concentration of bacteria increases the difference of the mean counts/100ms can be identified for both the voltages, but the sensitivity is better if one uses 30V instead of 15V.

TABLE 2.

Sample (File name)	Before ES	Counts/ 100mS During ES	Within 10s or 5s after ES
Medium 15V,10s (RH 207)	5.7 ± 3.1	32.5 ± 8.7 (100%)	8.4 ± 4.4
About 10^2 Rh.j./ml (RH208)	4.9 ± 3.9	20.7 ± 6.8 (85%)	7.0 ± 3.7
About 10^5 Rh.j./ml (RH211)	5.5 ± 3.7	28.8 ± 8.6 (89%)	10.5 ± 4.3
Medium 30V, 5s (RH201)	5.5 ± 2.7	79.5 ± 19.1(100 %)	24.8 ± 8.8
About 10^2 Rh.j./ml (RH202)	5.1 ± 2.8	61.5 ± 16.6 (77%)	20.5 ± 10.9
About 10^3 Rh.j./ml (Rh203)	5.0 ± 3.7	51.4 ± 11.5 (65%)	19.7 ± 5.8

4.3. PRINCIPLE OF ELECTRO-LUMINESCENCE FROM AQUEOUS SOLUTION

Our experiments showed that the electro-luminescence from aqueous solution containing bacteria is a complex result of different processes including the following:

(1) Because electrolysis of organic material, there appear cations (A^+) on the anodes and anions (A^-) on the cathodes. After diffusing into the solution they move to the opposite electrodes. When they meet each other, discharges take place and emit light which can be described by:

$$A^- \rightarrow A^* + A \quad (1)$$

(2) Reactions take place on the electrodes. For instance in NaCl solution, on the cathode are the following reactions:

$$Na^+ + e^- \rightarrow Na \quad (2)$$
$$2H_2O + 2\,e^- \rightarrow H_2 + 2OH^-$$

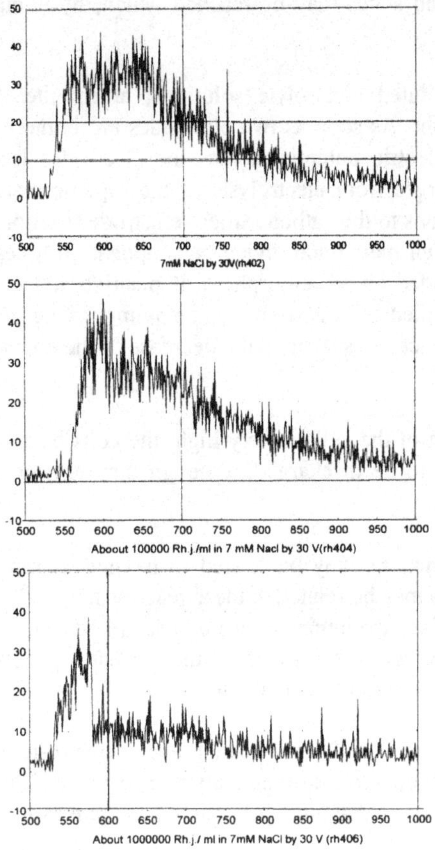

Figure 12. Electro-luminescence from water containing 7mM NaCl stimulated by 30V, showing that the emission pattern depends on the concentration of bacteria from 0/ml (top), 10^4/ml (middle) and 10^6/ml (bottom)

and on the anode:

$$2Cl^- \rightarrow Cl_2 + 2e^- \qquad (3)$$
$$2H_2O \rightarrow O_2 + 4H^+ + 4e^-$$

In these reactions excited states may be reached which, by going back to the ground states, may emit light.

Considering very diluted electrolyte solutions, at infinite dilution the ions are independent of each other. As soon as two electrodes are immersed in such a solution, the application of an electric potential across the electrodes leads the negative ions (including negative charges from electrolysis of the organic material) to move to the anode and the positive ions to the cathode where electrode reactions take place. Because of the structure of the cell membrane, living cells appear with negative charges on the surface and are surrounded by an atmosphere of positive and negative charges. If an electric potential is applied to the solution containing living cells, the free moving charged particles are reduced, resulting in the decrease of the above two reactions.

(3) If the concentration of the cells is very high, the cells have more chance to touch the electrodes and to be discharged, although they are too large to move under normal conditions.

(4) Some electrode reactions may be related to oxygen as reported in [12-13], and uncharged particles may be related to these reactions.
The final results of electro-luminescence of a sample depend on the combination of these processes. Therefore the shape of the emission pattern also gave us some information about the system being tested.

(5) Like P.elegans some cells show that the "electro-luminescence" is more complicated than biophoton emission after an electric stimulation and giving a hyperbolic relaxation behaviour [6]. Similar results were also reported with seedlings [11].

In order to understand its principle some more studies have to be carried out.

5. Acknowledgements

We are particularly grateful to the Familie-Ernst-Wendt-Stiftung for its interest and its financial support.

I am personally very grateful to Prof.Dr.F.A. Popp for inviting me to work in Kaiserslautern since 1993, so that I could perform these experiments, though the idea came already up by 1989. Thanks are also due to Dr.W.D.Yu for his help in electro-luminescence measurements of P.elegans and chicken embryonic cells. Many thanks to Miss T.Sun for the counting of bacterial concentration , to Mr.Yin Y. for his help in the measurements of Pseudomonas diecinuta and to all the staff members of Dr.Popp's Laboratory for many and various helps.

6. References

1. Van Der Zee H., Huis in't Veld, J.H.: Rapid and alternative screening methods for microbiological analysis, *J. AOAC Int* 80(4) (1997), 934-940.
2. Harris, C.M., and Kell, D.B.:, The estimation of microbial biomass, *Biosensors* 1(1) (1985), 17-84.
3. Koch, A.: *Growth Measurement in Manual of methods for general bacteriology*, Gerhardt, Murray, Costilow, Nester, Wool, Krieg and Phillips (eds.), American Society for Microbiology, 1981, pp. 179-206.
4. Wolff, L.F., Anderson, L., Sandberg, G.P., Reither, L., Binsfeld, C.A., Corinaldesi, G., Shelburne, C.E.: Bacteria concentration fluorescence immunoassay (BCFIA) for the detection of periodontopathogens in plaque. *J.Periodontol.* **63** (1992), 1093-1101.
5. Mieg, C., Mei, W.P., and Popp, F.A.: Technical notes to biophoton emission in F.A. Popp, K.H. Li and Q. Qu (eds.) *Recent Advances in Biophoton Research and Its Applications,* World Scientific Publishing Co. Ltd., Syngapore-London, 1992, pp. 197-205
6. Chang, J.J.,.Yu, W.D, and Popp, F.A.:): Studies on Photon Emission from P.elegans Induced by Electric Stimulation, *Acta Biophys. Sinica* **10** (1994), 641-644 (in Chinese with English abstract).
7. Popp, F.A., and Chang, J.J.: (199510 18) *Verfahren zum Nachweis Mikrobieller Infection* (European Patent Issued No:195 38 768.6).
8. Chang, J.J., and Popp, F.A.:(19960528*) Sensitive Method for Fast Detecting Microbial Contaminations* (Chinese Patent.Issued No.96105260.0).
9. Chang, J.J. , and Han, X.: Structural Studies on Rhizobium japonicium 1132-2 by Electron Microscopy, *Acta Microbiol. Sinica* **21** (1987), 318-332 (with English abstract)(27).
10. Tsein, H.C.: *Can. J. Microbiol.* **23** (1977),1274-1284.
11. Mathew, B.G., and Kumar, S.: *Experientia* 48 (3)(1992), 309-310.
12. Santhanam, K.S.V., and Haram, N.S.: Differential pulse volt biolominescence of quenched cells of Lampito-mauriti, *Bioelectrochem. Bioenerg.***28** (1992), 213-219.
13. Limaye, N.M., and Santhanam, K.S.V.: Oscillatory bioluminescence of Lampito maurito coelomocytes in pulsed dc electrolysis: oxygen controlled excited state production, *Bioelectrochem. Bioenerg.***24** (1990), 249-256.

BIOLOGICAL ORGANIZATION: A POSSIBLE MECHANISM BASED ON THE COHERENCE OF "BIOPHOTONS"

J.J. CHANG*# AND F.A. POPP*

**International Institute of Biophysics, IIB,*
Station Hombroich, Vockrather Straße, 41472 Neuss, Germany
 # Institute of Biophysics,
Chinese Academy of Sciences, Beijing 100101, China

1. Introduction

The Russian biologist Alexander Gurwitsch [1] was the first who enunciated the idea that photons within biological tissues („mitogenetic radiation") may trigger cell growth in a way that they become the governing factor of growth regulation and „Gestaltbildung" in living systems. Despite a variety of experimental confirmation of the existence of „ultraweak" photon emission from organisms (i.e. by the discoverer of holography D. Gabor [2]), the scientific community forgot Gurwitsch's ideas and his outstanding work [3]. The main reasons why „mitogenetic radiation" fell into oblivion may have been the discovery of growth hormones and the molecular approaches to understanding biology, the weakness (or dubiousness) of the photon emission from living systems, and the chaos of the second world war.

As early as 1954, with the improvement of the photomultiplier technique, Italian physicists were successful in showing evidence of „light emission by germinating plants" [4], and later on, mainly Russian biophysicists (see, for instance, the review [5]) investigated the „dark luminescence" or „ultraweak spontaneous luminescence" of more than 90 species.

The first concrete ideas on the origin of biophotons came up then, when, by following the discoverer Alexander Gurwitsch, mainly radical reactions were considered as representing the source of low-level luminescence [6, 7].

One of us (F.A.P) started in 1972 to find correlations between optical properties of molecules and their carcinogenic activity [8], connecting growth regulation and photon emission with the hypothesis that cancer induction is due to the loss of coherence of a photon field within the living tissues which should originate essentially from excited states of DNA. He denoted this hypothetical coherent photon field as „biophotons", in order to express the fact that it is a quantum phenomenon (only a few photons) that is characteristic of living biological systems [8-10]. The scientific group around F.A.P. was then successful in developing the techniques of „biophotonics" up to a stage where the human body can be systematically investigated [11], in presenting consistent models

J. J. Chang et al. (eds.), Biophotons, 217-227.

on the origin and the biological significance of biophoton emission [12, 13], and in finding fields of practical application [14]. During this period of more than 20 years a variety of important papers on this topic has been published by scientists over the whole world, covering the search for the source(s) of this spontaneous luminescence, the investigation of external influences as well as of the biological significance of this phenomenon (for review, see, for instance [12-15]). At present the experts in this field no longer doubt that biophoton emission does exist and that it is correlated to manifold, if not to all biological processes. More and more papers are devoted to the question of coherence of this radiation, and international efforts in scientific groups are growing to investigate the mechanism and significance of this universal phenomenon.

In this paper we would like to summarize and to discuss a sequence of already published experimental results which, at a first glance, seem to be completely unlinked. However, the application of the coherence hypothesis [10, 12, 16] connects the results in a surprisingly simple and understandable way. At the same time, it provides a theory of understanding yet unrevealed biological phenomena as well as of gaining a new insight into basic experimental problems in this field. In addition, we hope to stimulate and support further experimental work in this field.

2. A Definite Sequence of Experimental Results

Schamhart [17] and Scholz [18] have shown that with increasing density of normal cells in a suspension the reemitted photon intensity („delayed luminescence") increases for a low number of cells as expected, but with increasing numbers of cells the reflected intensity mounds in a saturation and drops down to even lower values. This cannot be explained in terms of linear physics, since with increasing optical density of the tissue according to the laws of linear optics the saturation may be understandable, but not the decrease after saturation [19]. Surprisingly, tumour tissue behaves completely differently. With increasing cell density the reflected intensity increases in not a very different way from the normal tissue. However, later on with increasing cell density, the increase of reflected light after definite white light excitation grows in a nonlinear way to higher and higher values. Figure 1 shows the results of Schamhart. These findings have been several times confirmed by other groups [20]. It is worthwhile to note that the relaxation after light illumination (where it is not decisive whether white or monochromatic light is used) follows an hyperbolic (1/t) - law, (where t is the time) rather than an exponential exp (-t/T)- law, (where T is the decay constant) [12, 21]. Scholz confirmed that the delayed luminescence relaxation of normal tissue with increasing cell density conforms more and more to an hyperbolic function, while that of tumor tissue displayed with increasing cell density in suspension increasing deviation from hyperbolic decay and increasing agreement to an exponential one.

A second series of experiments showing similar results has been performed by Galle. He investigated the spontaneous luminescence (biophoton emission) of Daphnia magna in dependence on the numbers of animals in the quartz cuvette of fixed volume [22]. The animals were female only of the same genetics and of about the same size and development stage. Instead of obtaining increasing biophoton emission with increasing

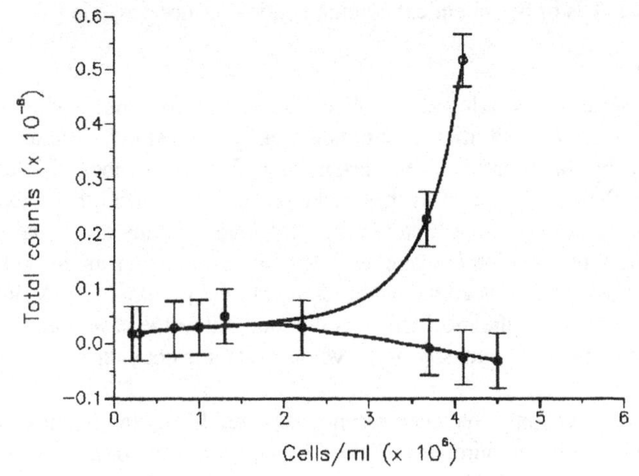

Figure 1. Normal cells display decreasing delayed luminescence with increasing cell density whereas tumor cells behave in the opposite way (see [17]).

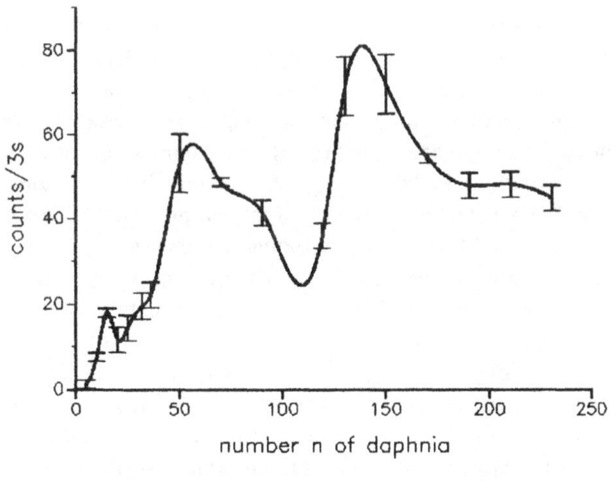

Figure 2. Daphnia display nonlinear change of biophoton emission with increasing density (see [22]).

numbers of animals, Galle observed in several experiments under strongly reproduced conditions maxima and minima of photon emission (see Fig.2). The minimum at about hundred animals corresponds to a „natural“ distance which is chosen by the animals if they are living in freedom. It has been shown that an interpretation of the results by „collision“-models or by „chemical communication“ is not possible [23].

A third series of experiments has been performed on Dinoflagellates and Thailand fireflies which show in addition to biophoton emission distinct bioluminescence (which is triggered by biophotons). The fascinating feature of their bioluminescence is synchronous flickering as soon as they make contact with each other. A careful analysis of the synchrony showed that it cannot be explained in terms of mutual excitation with light. A striking observation is that even if they are separated, synchronous bursts can be observed as soon as they become aware of an external perturbation. A further important factor in understanding the mechanism is the fact that in the case that a shutter between them is opened, the total intensity of the whole system is not simply the sum of both.

Similar observations have been recently reported [24]. Non-linear change of photon emission with increasing number of biological subjects have been confirmed with seeds [Private Communication]. Non-linear optical properties of cells have also been shown by B. Chwirot [25]. We have to conclude that biological systems display a variety of strange and up to now not registered optical effects which are correlated to their biological functions.

3. Synchronous Flickering of Dinoflagellates [26, 27]

In order to specify the nature of intercellular communication by light, we investigated again in detail the synchronous flickering of dinoflagellates. The results may help us to understand the mechanism of photonic communication. We constructed a light-double chamber in which two samples can be connected and disconnected by a shutter between them. Consecutive opening and closing of the shutter enables non-substantial communication between the samples or blocks it, respectively. At the same time, the two photomultipliers which are connected to each of the sample register their photon current. In the case that there is no difference in the biophoton emission between connected and disconnected stages of the samples, there is no indication of communication between them and on the contrary there should be some basis of mutual awareness in the case that there are significant differences between the stages of the connected and disconnected samples. Some years ago we found synchronous flickering between the samples in the case of visual contact between them. However, light scattering as a possible contribution to this effect could not completely be excluded. This was one of the reasons for constructing a coincidence circuit which works in the following way. One of the signal channels of the samples is used as a reference channel. As soon as in a time interval of T and T+Δt where after a photon from this reference channel has been emitted and the second channel of the other sample emits at least one photon, this event is registered as a coincidence. Otherwise, there is no registration of a coincident photon. This means that light scattering as a reason for synchronous flickering is excluded as soon as T is higher than the time which is needed by the reference light to scatter to the signal channel. Figure 3 shows the principle of this coincidence counting circuit. The idea of the measurement is then based on the registration of the counts in both of the channels plus the coincident counts c which are compared with the random coincidence rate z. As soon as c is equal to z, we know that communication is unlikely because photon emission does not deviate from randomness. On the contrary, if there are deviations from randomness we have ample indication of some „communication“ which may take place directly between the samples or between the samples and a common external event. In particular, if there is a characteristic difference between the stages of visual connection and disconnection, one has to conclude that the samples actually have a basis for becoming aware of each other.

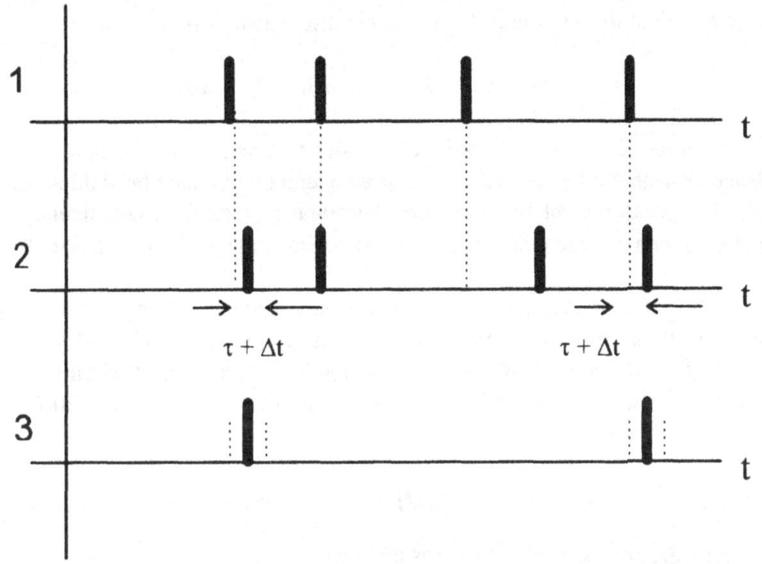

Figure 3. Coincidence Counting Equipment working in such a way that a coincidence is registered in channel 3 as soon as a photon is counted in channel 2 and at the same time at least one photon was counted in channel 1 within not longer than a time $\tau+\Delta t$ and not shorter than a time τ before this event in channel 2.

There is a further important point connected with this kind of measurement. Since the number of random coincidences is different for a coherent field and a chaotic one (where the coherence time is not much smaller than the preset measurement time interval ΔT), one can use this instrument for examining whether the samples emit coherent or chaotic light. One simply disconnects the samples in a way that one gets only random coincidences. Then one compares the registered coincidences with the calculated ones for both the cases of a coherent and a chaotic field. Agreement with one of these cases shows evidence of the coherent or chaotic nature of the radiation under investigation. Let us denote $Z(\Delta T)$ the registered number of coincidences in channel 3 and $n_2(\Delta T)$ the number of photons in channel 2. Then, the expected number of random coincidences $Z(\Delta T)$ is equal to

$$Z(\Delta T) = n_2(\Delta T)\, p(\Delta t,\, n_1 > 0), \tag{1}$$

where $p(\Delta t, n_1 > 0)$ is the probability of measuring at least one photon in the channel 1. This value is different for a fully coherent and a chaotic field of a coherence time which is not small compared to Δt. For a coherent field we have:

$$p(\Delta t,\, n_1 > 0) = 1 - \exp(-n_1), \tag{2}$$

while a chaotic field follows the following geometrical distribution:

$$p(\Delta t,\, n_1 > 0) = 1 - n_1/(1 + n_1) = 1/(1 + n_1) \tag{3}$$

In every time interval T and $T+\Delta T$ we are able then to compare the measured number of coincidences Z with the theoretical values of a coherent and a chaotic field and find out after sufficient measurement time whether the number of random coincidences agrees with a coherent or a chaotic field with a coherence time not small compared to Δt.

Figure 4 displays an example of two relaxing soy-beans in channel 1 and channel 2, where $p(\Delta t, n_1 > 0)$ has been calculated from the measured values $Z(\Delta T)$ and $n_2(\Delta T)$. During relaxation, the $p(\Delta t, n_1 > 0)$ follows accurately a Poissonian distribution whereas the geometrical one can truthfully be rejected. In that case the relaxation can be described by an hyperbolic-like relaxation

$$n_2(t) = \frac{A/(t + t_o)^2}{\sqrt{t}}, \tag{4}$$

where A and t_o are constant values and t is the time.
Figure 5 displays the agreement between theoretical and experimental results.

A further transformation is possible in order to fit the experimental results:

$$n_2(t) = \frac{B}{(t + t_o)^x} \tag{5}$$

4. Understanding in terms of Dicke's Theory

Nonlinear optical properties of matter are related to the coherence of photons. It is worthwhile to note that even chaotic light may display a coherence length that is comparable to the dimensions of cells in tissues. Actually, even an allowed optical transition, say a singlet transition of a typical life time of 10^{-9} s provides a coherence length of emitted light which is of the order of 10 cm, large compared to the size of a

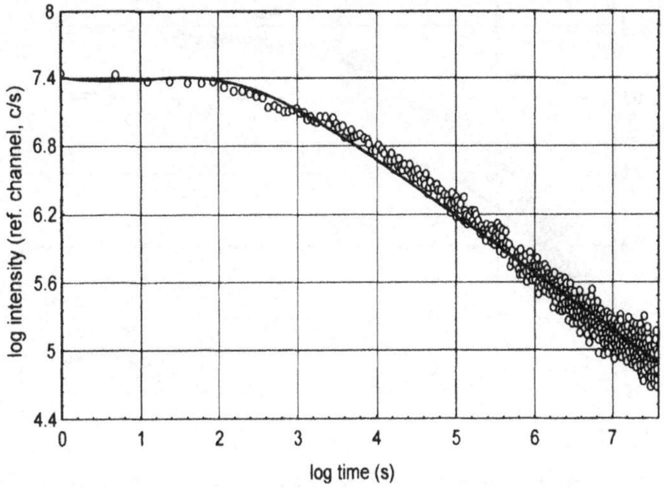

Figure 4a.

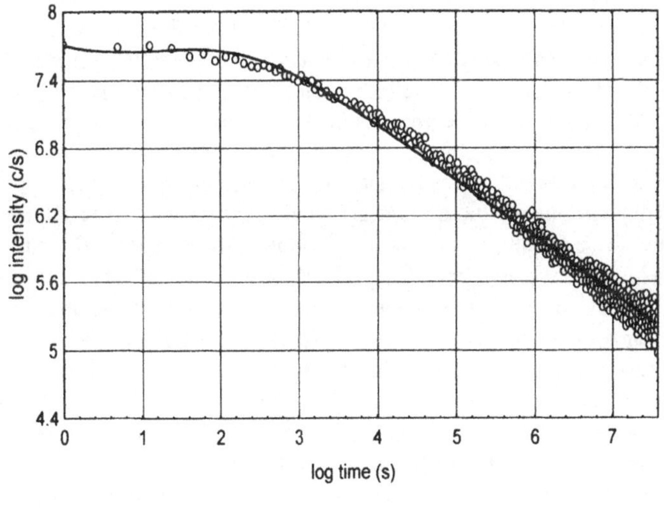

Figure 4b.

Delayed luminescence of soy-beans which are kept in two channels (Fig.4a. refers to the reference channel, Fig.4b. to the signal channel). The random coincidences during relaxation are measured. The relaxation function follows a quasi-hyperbolic decay (see eq.(4)).

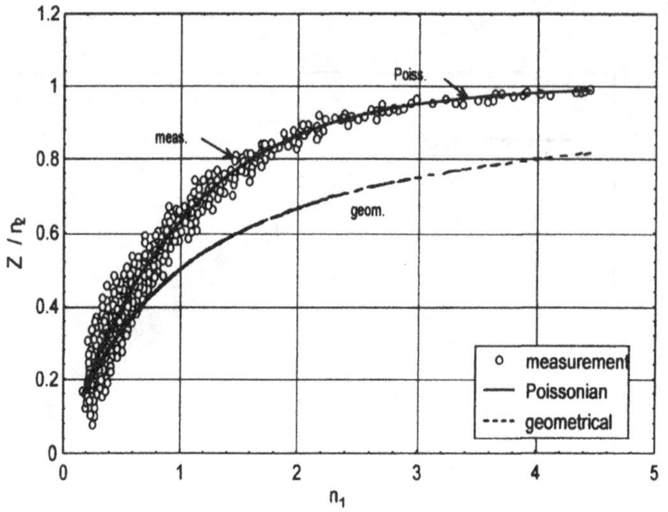

Figure 5. The random coincidences of the delayed luminescences of the soy-beans (fig.4.) follow over the whole time a Poissonian Photocount Statistic.

cell. Although this is not a real "coherent state", it shows that the phase relations of light does not get lost over the distance of a cell even if the light field is not coherent. Hence, it is impossible that a cell „sees" a photon. Rather, the absorption processes in a cell cannot be localized and assigned to definite molecules, at least not as long as the field is not coherent. Absorption of single photons in cells follows purely statistical laws where the probability of absorbing a photon is almost homogeneously smeared out over the whole cell. The situation changes dramatically as soon as an incoherent photon field turns into a coherent one, including non-classical squeezed light (see papers of Bajpai and Gu). Under these conditions perfect localization in the case that all the modes of the biophoton field are involved according to the (fv)=constant-rule [28] and perfect delocalization in the case of narrowing the field to isolated single modes, may take place. At the same time, the spatio-temporal structure of a cell may become partially or even completely ordered or disordered. In agreement with Prigogine's dissipative structures, the coherent photon field may fluctuate around phase transitions between „chaos" and „order" in dependence on the compensation of either chaotic or ordered influences. In this respect it should be mentioned that the Poissonian distribution of the photocount statistics of the biophoton field displays just this character. Actually, it may be regarded at the same time as a fully coherent as well as a fully chaotic field. The biological system does not allow us to observe whether it is fully coherent or rather chaotic, not before a „question" is asked. But if, for instance, an excitation by exposure to light illumination takes place, the „hyperbolic" relaxation dynamics reveals the coherent nature of the field [16].

In order to explain non-linear optical phenomena of biological systems, it is important to remind ourselves of a general but at the same time basic property of all biological tissues. They represent optically „dense“ matter which means that the intermolecular distances are small compared with the wavelength of light. Under these conditions the theory of R.H.Dicke [29] has to be applied in order to understand the interaction of light and matter. According to Dicke, spontaneous reemission of absorbed light is impossible as soon as the intermolecular distance is significantly smaller than the wavelength. Rather, the interaction of the pigment molecules and the photons split into two new „regimes“ of super- and subradiance. Superradiance corresponds to constructive interference of light waves cumulating up to coherent light flashes which are then emitted in relatively short time intervals. Subradiance is defined as the destructive interference of the light waves within the antenna-system of absorbing molecules. The result is „delayed luminescence“ of coherent light waves which relax according to hyperbolic functions. Just this situation is displayed in biological systems. Consequently, the explanation of (1) Schamhart's results on normal and tumor cells, (2) Galle's Daphnia experiments and (3) the synchronous flashes of Dinoflagellates follows the general theory of Dicke. Actually, with decreasing distances normal cells display an increasing probability of destructive interference within the extra-cellular space. This explains the decreasing delayed luminescence intensity with increasing density of normal cells in cell suspensions as well as the increasing agreement with an hyperbolic decay function. On the other hand, it allows an immediate understanding of the opposite behaviour of tumor cells which according to Dicke's theory display decreasing degrees of coherence with decreasing intercellular distances as well decreasing agreement to hyperbolic relaxation dynamics. However, Dicke's theory provides at the same time a basic understanding of Galle's results on Daphnia. Depending on the wave pattern of emitted biophotons, a complicated interference structure between the animals builds up. It is worthwhile to note that the minimum biophoton emission corresponds to the average distance which the animals take under natural conditions. This is in line with the assumption that the interference pattern is at the same time the communicative platform between the animals on which they regulate the swarm formation, or, more generally, the regulation principle for the formation of populations and the „Gestaltbildung“. It is understandable that the interference pattern of daphnia explains maxima and minima photon emission in dependence on mutual distances, while there is no possibility of explaining them in terms of „substantial“ messengers between the animals. There is no difficulty in understanding on this basis as well the coincident flashing of Dinoflagellates. As long as populations are disconnected they cannot establish a common coherent field. As a result, there is no mutual awareness. However, as soon as they are connected, an electromagnetic field of a complicated interference structure may build up, displaying the tendency to lower the average field intensity. The synchronous flashing is then simply a consequence of the common awareness of any distortion of the connecting field rather than a mutual „stimulation“.

5. Acknowledgments

We are particularly grateful to the Familie-Ernst-Wendt-Stiftung for its interest and its financial support.

6. References

1. Gurwitsch, A.G. and Gurwitsch, L.D.: *Die mitogenetische Strahlung, ihre physikalisch-chemischen Grundlagen und ihre Anwendung in Biologie und Medizin*, Gustav Fischer Verlag, Jena, 1959.
2. Reiter, T. and Gabor, D.: Zellteilung und Strahlung, *Sonderheft der Wissenschaftlichen Veröffentlichungen aus dem Siemens-Konzern*, Julius Springer Verlag, Berlin, 1928.
3. Kohn, A.: *False Prophets*, Basil Blackwell, Oxford and Cambridge, MA, revised paperback edition, 1988, pp. 23-26.
4. Colli, L., Facchini, U. et al : Further measurements on the bioluminescence of seedlings, *Experientia* **11** (1955), 150.
5. Ruth, B.: Experimental investigations on ultraweak photon emission, in F.A. Popp, G. Becker, H.L. König and W. Peschka (eds.), *Electromagnetic Bio-Information*, Proceedings of the symposium, Marburg, 5. September 1977, Urban & Schwarzenberg, München, Wien, Baltimore, 1979, pp. 107-122.
6. Zhurawlew, A.I. (ed.): Ultraweak luminescence in biology (in Russich), *Transactions of the Moscow Society of Naturalists* **39** (1972), Verlag Nauka, Moskau.
7. Seliger, H.H.: Applications of bioluminescence and chemiluminescence, in M.J. Cormier, D.M. Hercules and J. Lea (eds.), *Chemiluminescence and Bioluminescence*, Plenum Press, New York, 1973, pp. 461-478.
8. Popp, F.A.: A very significant correlation between carcinogenic activity of polycyclic hydrocarbons and certain properties of their transition states in the range of the lowest triplet states of the DNA, *Arch. Geschwulstforsch.* **47** (1977), Heft 2, 97-105.
9. Popp, F.A.: *Biophotonen: Ein neuer Weg zur Lösung des Krebsproblems,* Schriftenreihe Krebsgeschehen Band 6, 2.verbesserte und erweiterte Auflage, Verlag für Medizin Dr.Ewald Fischer, Heidelberg, 1984.
10. Popp, F.A.: Photon Storage in Biological Systems, in F.A. Popp, G. Becker, H.L. König and W. Peschka (eds.), *Electromagnetic Bio-Information*, Proceedings of the symposium, Marburg, 5. September 1977, Urban & Schwarzenberg, München, Wien, Baltimore, 1979, pp. 123-149.
11. Cohen, S. and Popp F.A.: Biophoton emission of the human body, *J. Photochem. Photobiol. B: Biol.* **40** (1997), 187-189.
12. Popp, F.A., Ruth, B., Bahr, W., Böhm, J., Grass, P., Grolig, G., Rattemeyer, M., Schmidt, H.G., and Wulle, P.: Emission of visible and ultraviolet radiation by active biological systems, *Collective Phenomena* **3** (1981), 187-214.
13. Popp, F.A., Gu, Q., and Li, K.H.: Biophoton Emission: Experimental Background and Theoretical Approaches, *Modern Physics Letters B* **8** (1994), Nos. 21 & 22, 1269-1296.
14. Popp, F.A., Gu, Q., and Li, K.H.: *Recent advances in biophoton research and its applications*, World Scientific Publishing, Singapur, 1992.
15. Popp, F.A., Gurwitsch, A.A., Inaba, H., Slawinski, J., Cilento, G., van Wijk, R., Chwirot, W.B., and Nagl, W.: Biophoton emission (Multi-author review), *Experientia* **44** (1988), 543-600.
16. Popp, F.A., and Li, K.H.: Hyperbolic relaxation as a sufficient condition of a fully coherent ergodic field, *International Journal of Theoretical Physics* **32** (1993), No. 9, 1573-1583.
17. Schamhart, D.H.J., and van Wijk, R. Photon emission and the degree of differentiation, in B. Jezowska-Trzebiatowska et al. (eds.), *Photon emission from biological systems*, World Scientific Publishers, Singapur, 1987, pp. 137-152.
18. Scholz, W., Staszkiewicz, U., Popp, F.A., and Nagl, W.: Light stimulated ultraweak photon reemission of human amnion cells and Wish cells, *Cells Biophysics* **13** (1988), 55-63.
19. Nagl, W., and Popp, F.A.: Opposite Long-range Interactions Between Normal and Malignant Cells, in T.W. Barret and H.A. Pohl (eds.), *Energy Transfer Dynamics*, Springer-Verlag, Berlin-Heidelberg, 1987, pp. 248-256.
20. Niggli, H.J.: The Cell Nucleus of Cultured Melanoma Cells as a Source of Ultraweak Photon Emission, *Naturwissenschaften* **83** (1996), 41-44.
21. Chwirot, W.B., Dygdala, R.S., and Chwirot, S.: Quasi-monochromatic-light-induced photon emission from microsporocytes of larch showing oscillation decay behaviour predicted by an electromagnetic model of differentiation, *Cytobios* **47** (1986), 137-146.
22. Galle, M., Neurohr, R., Altmann, G., Popp, F.A., and Nagl, W.: Biophoton emission from *Daphnia magna*: A possible factor in the self-regulation of swarming, *Experientia* **47** (1991), 457-460.

23. Galle, M.: Population density-dependence of biophoton emission from daphnia, in F.A. Popp, K.H. Li and Q. Gu (eds.), *Recent advances in biophoton research and its applications*, World Scientific Publishing, Singapur, 1992, pp. 345-355.
24. Brezovits, K.: *Aufbau einer Single-Photon-Counting-Meßanordnung zum Nachweis von Photonenkorrelationen mittels Koinzidenz*, Diplomarbeit, Atominstitut der österreichischen Universitäten, Wien, 1997.
25. Chwirot, W.B.: Ultraweak photon emission and anther meiotic cycle in larix europea (experimental investigation of Nagl and Popp's electromagnetic model of differentiation), *Experientia multi-author review on biophoton emission* **44** (1988), 594-599.
26. Popp, F.A., Chang, J.J., Gu, Q., and Ho, M.W.: Nonsubstantial Biocommunication in Terms of Dicke's Theory, in M.W. Ho, F.A. Popp and U. Warnke (eds.) Bioelectrodynamics and Biocommunication, World Scientific Publishing, Singapur, 1994, pp. 293-317.
27. Chang, J.J., Popp, F.A., and Yu, W.: Research on cell communication of *P. elegans* by means of photon emission, *Chinese Science Bulletin* **40** (1995), No. 1, 76-79.
28. Popp, F.A.: Some Essential Questions of Biophoton Research and Probable Answers, in F.A. Popp, K.H. Li and Q. Gu (eds.), *Recent advances in biophoton research and its applications*, World Scientific Publishing, Singapur, 1992, pp. 1-46.
29. Dicke, R.H.: Coherence in spontaneous radiation processes, *Physical Review* **93** (1954), 99-110.

DO WE ALWAYS NEED TO KNOW MOLECULAR ORIGIN OF LIGHT EMITTED BY LIVING SYSTEMS?

B.W. CHWIROT

Laboratory of Molecular Biology of Tumours, Institute of Biology and Protection of Environment, Nicholas Copernicus University, ul. Gagarina 9, PL 87-100 Toruñ, Poland

For several decades biological systems have been analysed by dividing them into smaller subunits: in the past it was organs, tissues, cells, cellular structures like for instance organella and finally macromolecular components and the so-called molecules. Why the so-called? Because very often the biologist's understanding of molecules is quite distant from their physical reality. As pointed out by Albrecht - Buehler [1] the biochemist looks at proteins as sets of symbolic subunits and functional groups, the molecular biologist thinks more of genes, coding sequences etc. while the cell biologist may be concerned with cellular distributions of some molecular species or their changes in a response to differentiation or development of cells.

Probably everybody accepts that the biological system is something different from just a complicated molecular system. At the same time there are no commonly accepted answers to many simple and quite obvious questions. One can ask for instance: When, during the studies of still smaller subunits does one switch from studies of biological systems to just chemical systems, or in other words from biology to chemistry?
or more generally:
What makes one to consider any given system a biological system and not just a more or less organised mixture of molecules?
An even more fundamental question may be: Is there a difference between the biological system and the living system?

Browsing through any set of recent papers from fields such as molecular biology, biophysics or biochemistry shows clearly that at least from a practical point of view the biological systems and the living systems should be thought of as different classes.
On the other hand, however, an ultimate goal of any biological research is to study the life and the living organisms and to discover the real meaning of life. The problems of defining the life and the living systems have been concerning people for ages and there is a little chance that anybody will provide the answers in a near future. For a sake of clarity we would like however to submit as a working definition that the living system is the biological system of any level of complication characterised by a dynamic temporal and spatial organisation of its components supported by a continuos intake and dissipation of energy allowing for its homeostasis, capable of replication and/or proliferation and evolution.

J. J. Chang et al. (eds.), Biophotons, 229-237.

Such a definition is by no means a strict one since it contains for instance an undefined biological system. However, it stresses the importance of homeostasis and of the integration of all physical and chemical processes into a functional whole. In that sense we follow here many earlier authors and the direct stimulation was for us the works of Albrecht-Buehler and Dyson [1,2].

Homeostasis provides means for sustaining the stationary state of the living system despite the perturbing actions of its environment and fluctuations due to the thermal noise. The perturbing actions of the environment may be due not only to different chemical and physical interactions but also to interactions between different living systems or different elements of a given system. For instance a bacterial infection will lead to a perturbation of the hemeostatic system of the host and can be looked upon as a competition between the two homeostatic systems for establishing the optimum conditions for a functioning of the two organisms. A similar conflict takes place when one of the cells of an organism undergoes a malignant transformation and attempts to break the regulatory actions of other cells and tissues and to establish conditions favouring its further development and proliferation at the expense of the homeostatsis of the parental organism.

The unexplained fact is that whether on a molecular or ecological or economical level the complex homeostatic systems are more prevalent and more effective that the simple ones. The homeostasis is usually achieved by means of a complicated network of interdependencies which is too complex for us to understand. Higher organism represent probably the highest achievement in terms of the systems of the homeostatic mechanisms. If one assumes, following Eigen, that they have to replicate their genetic information with the relative error of less than 10^{-8} (see for instance [2]) and takes into account their tolerance for variations in the environmental parameters one has to accept these mutually exclusive phenomena can be achieved only at the existence of a highly complex network of interlocking homeostatic mechanisms.

Is it really possible to unravel the homeostatic networks and to describe and to understand them in terms of molecular interactions? Most probably not, similarly as it still impossible to solve the Schroedinger equation for anything but the hydrogen atom. Of course such a statement does not mean neglecting the sense of molecular approach to studies of the living systems. The achievements of the molecular biology have been so far sufficiently impressive to make many people consider molecular analyses the only justified way of research on the living systems, especially at a cellular level. At the same time, however, most of the results obtained using the molecular approach have been either obtained in vitro or from in vivo studies involving mostly typically phenomenological approach i.e. looking for the results of such or another perturbation of the molecular machinery of cells, tissues or even organisms. However, can such studies lead to the understanding of the living system as a functional whole. It is clear that attempts to describe at the molecular level of any system composed of more than hundreds macromolecules have been so far unsuccessful and one obvious reason for this is that we just still are not able to analyse even a scale and a range of responses of such complicated systems to our perturbing actions. On the other hand it is the results of such studies that serve us as a base for developing concepts and theories regarding the living systems. At the same time we cannot produce a set of criteria defining something like a "properly functioning" living cell. Most probably, simply because of the techniques used, most of the results have been obtained from studies of living systems studied in conditions so different from their "normal" state that the data have often a little relevance to their natural counterparts.

The above discussion as well as examples from other fields of research show clearly that in biology of living systems there is a place not only for the molecular analyses but also for a biology of the whole living systems, similarly to a parallel and successful existence of a quantum physics and classical thermodynamics or micro- and macroeconomics. Only a parallel and mutually stimulating development of the two approaches may ensure a qualitative progress on our way to understanding the nature of life. Even now, many of studies claiming to be of a molecular nature deals with almost macroscopic objects like membranes or cellular organnella. moreover, quite often in the molecular studies, especially the biochemical ones we tend to forget that the more we decompose the system the we destroy its meaning [1]. This argument can be best illustrated by a simple fact that we are still far from being able to reconstruct a living cell from its components not to mention a construction of even a very simple and short living artificial cell. Moreover, for all the biochemical systems studied in vitro the efficiencies of the reactions are much lower than for the same systems operating in their natural environment and that several parameters used to characterise such systems in vitro like for instance pH or concentration are of no meaning for "molecular machines" of the cells.

It seems that generally there is no opposition to a concept of studying the living system such as the cell as a whole. The problem seems to be rather how to do it or more specifically: what experimentally accessible parameters can yield information on functioning of the homeostatic system or of a degree of its perturbation. Another important question is what parameters could be used for a determination of a transition from one to another state of homeostasis - an example of two homeostatic state of the same initial system might be a differentiated cell and a similar cell after a premalignant or a malignant transformation. Similarly, any pathological process can be considered a state of a perturbed homeostasis and to be able to measure the degree of such a perturbation would be very important both from a point of view of practical applications and of the fundamental studies.

Hence, what we need is macroscopic parameters that could be ideally measured in a non-invasive manner for the living systems such as a cell or a tissue without decomposing them into their building elements. Such parameters have been already found for really macroscopic living systems such as for instance human body; most commonly used is just a body temperature considered a measure of its homeostatic state, composition of the breathed air, blood, urine etc. Measurements of such parameters do not perturb the system, the loss of heat or of blood taken for a measurement have no influence on the homeostatic state of the body. Regarding the cells or tissues the chemical approach is of a limited use. The sensitivity of chemical methods is generally low and to get results one needs rather large amounts of a material. These methods cannot be of course totally excluded and may be useful in studies involving large biological systems or large populations releasing for instance to the environment substances useful for a determination of a physiological state, stage of differentiation etc. A continuous progress in the field of biochemistry, especially in increasing a sensitivity of immunoassays and optical detection of reaction products pave a way to many new applications and studies using the chemical parameters.

It seems, however, a very promising and not fully exploited yet research potential for studies of the whole and intact living systems is offered by optical methods based on measurements of the native ultraweak luminescence and of a fluorescence of native fluorophores of the cells and tissues, called the autofluorescence. The studies on possible applications of the ultraweak luminescence have been carried out for many years and the last decade brought a significant progress in developing new methods on a diagnostic potential of the autofluorescence.

The recent history of the research on the ultraweak luminescence began in 1950' with the report of Colli et al. [3] who used a photomultiplier to demonstrate in an objective way that etiolated seedlings of several plants did indeed spontaneously emit light. Since then this field of research has developed very dynamically (for reviews the reader is referred for instance [4-6]). Nevertheless, despite many new interesting results and observations the studies on the ultraweak luminescence still stimulate a continuos discussion on the biological importance the phenomenon as such and on the meaning of many of the data. The problem is well illustrated by the fact that after forty years there is still no commonly accepted formal definition of the phenomenon of the ultraweak luminescence. Abeles [5] suggested as a working definition that this luminescence is oxidatively driven production of light with an emission spectrum which ranges from 200 nm to 700 nm, intensities of 10 to 1000 photons s-1 cm-2 and a quantum efficiency of 10-3 to 10-14 photons per activated molecule. This description shows well some of the difficulties encountered by anybody who attempts to study the ultraweak luminescence. The light intensities measured in such studies are extremely low, especially if the objects of interest are small. Since a relative error of measurements carried out using a single photon counting technique are inversely proportional to a number of counts the integration times required for collecting statistically significant data are usually very long. If, additionally, one wanted to measure also the emission spectrum than the time required would increase many times. On the other hand, biological systems subject to such studies are not stable and change during the measurements either due to their natural developmental processes or due to their adaptation to the conditions within the measuring chamber.

The objections against the studies of the ultraweak luminescence most commonly regard the low specificity of the results obtained for the whole organisms or cellular systems, the lack of knowledge of the specific molecular emitters of the measured emissions and/or the role and the location of the emitters in a metabolism and a structure of the cells. Moreover, the results of several studies oriented on a determination of a nature and of a localisation of the sources of the ultraweak luminescence indicate that there are many possible emitters and besides some special cases it is impossible to assign the observed emissions to specific molecules or reaction systems. An exception here is the phagocyte luminescence already used in several clinical applications [7]. There is a little doubt that all those questions are of a fundamental importance for our understanding of the phenomenon of the ultraweak luminescence. Significantly, the continuos research on those fundamental questions brought recently qualitatively new results regarding some aspects of the cellular biochemistry. The experiments of Cilento et al. [8,9] demonstrated that a chemical generation of the excited states of some biomolecules may be important for driving metabolic processes and resulted in a new concept of "photochemistry without light". An important conclusion of that author was that not only the long living excited states can be efficiently generated in a molecular environment but also that just such processes may induce in biological systems reactions normally possible only in a presence of light. Another problem still waiting for the explanation is the origin of the ultraweak emissions in UV region of spectrum. The existence of such emissions have been reported by several groups for different organisms [10-13] and their origin cannot be explained by any of the chemical mechanisms proposed up to now.

At the same time it is now commonly accepted that both the intensity and the spectrum of the ultraweak luminescence depend strongly on the physiological state of the living systems, on their state of development and on the actions of the external factors, especially those that can be considered the stress factors. In fact, it is this extreme sensitivity of the ultraweak luminescence to all the above mentioned factors that often causes problems with a reproducibility of experiments and with the interpretation of the results. The body of available data demonstrating the strong dependence of the ultraweak luminescence on the state of the living systems under study is too large to discuss it here and the reader is referred to original publications and to the review papers on the subject (for instance [14-18]).

Another valuable source of information on the state of the living systems can be studies of the fluorescence of their endogenous fluorophores (from now on referred to as the autofluorescence). A very important feature of the autofluorescence as a source of the diagnostic information is that both the excitation and the detection are practically non-invasive. Intensities of the exciting light can be kept well below the threshold for damaging the cells and tissues and the optical detection requires neither a contact nor a sampling of any material as well as introduction of any chemical agents into the systems under study. Moreover, the fluorescence studies gives more freedom in controlling the experimental conditions than those based on the measurements of the ultraweak luminescence. The measurements of the autofluorescence require less time and by changing the excitation and detection bands one can obtain more information than from the measurements of the intensities of the total ultraweak emission only. Also the kinetics of the decays of the fluorescence and of the so called delayed fluorescence can yield very interesting data on the physiological state of the biological systems (see for instance [19-21]).

The basic idea behind all the methods employing the autofluorescence is that the absorption and emission spectra of any system depend on its physical and chemical properties. Endogenous fluorophores are molecules which are either structural components of the cells and tissues or participate in their metabolic processes. All of them interact with the local environment and the spectra of absorption and fluorescence of a living system depend both on a nature of the fluorophores and on optical properties of its all components as well as on pH, conformational changes of proteins, composition and concentration of body liquids etc. Therefore, any change of a metabolic function or a structure is reflected in a pattern of the spectra. Blood, for instance, may significantly modify the emission spectra by absorbing the exciting light and by changing in this way effective depth of its penetration into the body tissue and by reabsorbing the light emitted by the excited fluorophores. This dependence of the spectra on so many factors of different nature makes optical detection of pathologies difficult since the characteristic changes caused by the process of interest must be identified against a background of non - specific changes caused by a natural variability of biological systems.

Because of the extreme complexity of the living systems, very little is known about the origin of the observed spectra of autofluorescence and about the casual relationships between the changes in the spectral patterns and perturbations of the physiological state or occurrence of pathological processes. For these reasons, the potential of autofluorescence as a source of diagnostic information has been neglected for many years and more attention has been given to possible applications in diagnostics and therapy of exogenous fluorophores artificially introduced into a body (for instance photodiagnostics and dynamic phototherapy using hematoporphyrin derivatives HpD [22-23]). The first studies on the possibilities of using laser - induced fluorescence for detecting malignant tissues in humans were first carried out in 1987 [24,25]. Since then, research on developing diagnostic methods based on measurements of autofluorescence has been undertaken by several groups in different countries of the world. The research work is oriented on diagnosing tissues accessible to optical observation either in a direct way (skin, cervix) or using endoscopes and special fibre probes (gastrointestinal tract, bronchial tissue, urinary bladder and blood vessels). The main effort is being made in the field of early detection of malignant and premalignant lesions and in the following discussion we will concentrate on this subject.

The most spectacular success of autofluorescence diagnostics has been its application for detection of neoplastic lesions within the bronchial tree. The first feasibility studies performed in vitro have demonstrated clear differences in spectra of normal and malignant lung tissues [26,27]. Despite those exciting early results the work in this field has, during the next few years, been dominated by studies on fluorescence diagnostics based on using HpD. However, this fluorescent marker of malignant tissues is known to cause a potentially serious and prolonged skin photosensitivity lasting for several weeks [28]. An unexpected breakthrough was a demonstration by the group of Palcic and Lam [29], that it was possible to detect even very small areas of lung cancer and dysplasia without HpD, basing solely on analysis of the autofluorescence of the bronchial tissues. In this approach typical bronchoscopes are used both for the illumination and for the observation of the tissues. Further studies oriented on optimisation of diagnostic parameters of the method by selecting the best conditions for the excitation and the measurements of the autofluorescence [30,31] resulted in a new diagnostic modality. The LIFE system (Lung Imaging Fluorescence Endoscope) uses the 442 nm line of a He-Cd laser for the excitation. The diagnostics is based on a digital analysis of the images recorded simultaneously in two spectral bands. The system is now commercially available and has been tested in several clinics in the USA and other countries. Tests have shown that while using white light bronchoscopy only 22% of the in situ carcinomas were detected while with using fluorescence imaging the sensitivity improved to 78%. For dysplasia only 5% of the lesions were found with a classical white light examination while 45% were detected with the fluorescence imaging. The specificity of white light bronchoscopy was 94% and for the fluorescence endoscopy it was 82% [32].

The success of the autofluorescence ability to detect lung cancer is an extremely good illustration of the potential of the fluorescence approach. However, as far as other cancers are concerned, work is still at a stage of preliminary studies and of optimisation of the procedures for diagnosing malignancies both in situ and in vitro. To our best knowledge, the only other example of a completely elaborated diagnostic procedure is the autofluorescence method of detecting neoplastic lesions in tissues of the cervix, uterus and ovaries [33]. The method involves excitation of the autofluorescence with ultraviolet light (300 nm) and the diagnostics are based on a ratio of the intensities measured at 340 nm and 440 nm. A patent for the method was issued in 1990.

Most of the recent work in the field of the fluorescence diagnostics is devoted to detecting malignancies of the gastrointestinal tract, especially of colonic tissues. The incidence of gastrointestinal cancers is second only to lung cancer. These tissues are easily accessible with endoscopy and thus for a spectroscopic investigation. More recent studies [see for instance 34-36] were oriented mainly on malignancies of colonic tissues. Similar studies have been undertaken for colonic and gastric malignancies by our group in Toruñ in 1993. Our program consists in using an approach based on digital imaging of the spectrally resolved fluorescence instead of analysing the whole spectra of the autofluorescence. Initial stages of the research were carried out in vitro using specimens excised during surgical operations of patients treated for carcinoma. The preliminary results are very promising [37,38].

Although many fundamental questions regarding both the origin of the autofluorescence and of its spectra remain unanswered the research in this field enters at this time a stage for clinical tests. Such tests have already been started or are about to begin for several methods of diagnosing colonic and bladder malignancies elaborated by academic research groups and high-tech companies [39].

Of all human tissues the skin is the most easily accessible for fluorescence examinations. Two groups have reported studies oriented on the detection of malignancies arising in this tissue. Lohmann et al. [40,41] attempted to use skin autofluorescence to detect melanomas by differentiating between melanomas, dysplastic nevi and nevi. Their studies, involving a large sample of patients, demonstrated that in principle it was possible to discriminate normal nevi from melanomas and dysplastic nevi. However, it seems that their approach, based on absolute determination of the intensities in the spectra of the autofluorescence and involving a point by point scanning of the suspected areas of the skin, is too susceptible to methodological errors to find a way to clinical applications.

In Toruñ, during the past two years, our group has been working on elaborating an optical method for detecting melanomas via digital analysis of the images of the skin autofluorescence recorded in specially selected bands of the spectrum. Up to now, more then 700 patients have been examined using this new procedure. Both a sensitivity and a specificity of our method compare very well with similar parameters of the classical diagnostic method. We have submitted application for the issuance of a patent.

References:

1. Albrecht - Buehler G. (1990) In defense of" nonmolecular" cell biology. *Int. Rev. Cyt.* **120**, 191.
2. Dyson F. (1989) *Infinite in all directions*. Pelican Books, Harmondsworth.

3. Colli L., Facchini U., Guidotti G., Dugnani-Lonatti R., Orsenigo M. and Sommariva O. (1995) Further measurements on the bioluminescence of the seedlings. *Experientia* **2**, 479.
4. Cadenas E. (1984) Biological chemiluminescence. *Photochem. Photobiol.* **40**, 823.
5. Abeles F.B. (1986) Plant chemiluminescence. *Ann. Rev. Plant. Physiol.* **37**, 49.
6. Slawiñski J. (1988) Luminescence research and its relation to ultraweak cell radiation. *Experientia* **44**, 559.
7. Lilius E.-M. and Marnila P. (1992) Photon emission of phagocytes in relation to stress and disease.*Experientia* **48**, 1082.
8. Cilento G. (1988) Photobiochemistry without light, *Experientia* **44**, 572.
9. Cilento G. and Adam W. (1988) Photochemistry and Photobiology without light, *Photochem. Photobiol* **48**, 361.
10. Konev S.V., Lyskova T.I. and Nisenbaum G.D. (1966) Very weak bioluminescence of cells in the ultraviolet region of the spectrum and its biological role, *Biophysics* **11**, 410.
11. Quickenden T.I. and Que Hee S.S. (1976) The spectral distribution of the luminescence emitted during growth of the yeast Saccharomyces cerevisiae and its relationship to mitogenic radiation, *Photochem. Photobiol.* **23**, 201.
12. Tilbury R.N. and Quickenden T.I. (1988) Spectral and time dependence studies of the ultraweak bioluminescence emitted by the bacterium Escherichia coli, *Photochem. Photobiol.* **47**, 145.
13. Chwirot W.B. and Dygda³a R.S. (1991) Ultraweak photon emission in UV region during microsporogenesis in Larix europaea Mill., *Cytobios* **65**, 25.
14. Chwirot W.B., Dygda³a R.S. and Chwirot S. (1985) Ultraweak photon emission during microsporogenesis in Larix europaea, *Cytobios* **44**, 95.
15. Oonuki M. and Ryuzaki M. (1992) Photon emission during cleavage of frog eggs, *Photochem. Photobiol.* **56**, 567.
16. Popp F.-A. (Ed.) (1988) Multi-author Review: Biophoton Emission, *Experientia* **44**, 543.
17. Van Wijk R. and van Aken J.M. (1992) Photon emission in tumour biology, *Experientia* **48**, 128.
18. Popp F.-A., Li K.H. and Gu Q. (Eds.) (1992) *Recent advances in biophoton research and its application, World Scientific*, Singapore-New Jersey-London-Hong Kong.
19. Veselova T.V., Veselovsky V.A., Rubin A.B. and Bochvarov P.Z. (1985) Delayed luminescence of air-dry soybean seeds as a measure of their viability, *Physiol. Plant.* **65**, 493.
20. Musumeci F., Triglia A., Grasso F., Scordino A. and Sitko D. (1994) Relation between delayed luminescence and functional state in soya seeds, *Il Nuovo Cimento* **16D**, 65.
21. Chwirot W.B. and Popp F.-A. (1995) White-Light-Induced Luminescence from Normal and Temperature Sensitive Saccharomyces cerevisiae, in Beloussov L.V. and Popp F.-A. (Eds.) *Biophotonics. Non-equilibrium and Coherent Systems in Biology, Biophysics and Biotechnology.* Bioinform Services Co., Moscow.
22. Dougherty T.J. (1991) Tumor Detection and Treatment: Hematoporphyrin Derivative and Photofrin II, in Goldman L. (Ed.) *Laser Non-Surgical Medicine. New Challenges for an Old Application*, Technomic Publishing Co., Inc., Lancaster - Basel, 1991
23. Spinelli P., Dal Fante M., Marchesini R. (Eds.) (1992) *Photodynamic Therapy and Biomedical Lasers*, Excerpta Medica, Amsterdam - London - New York - Tokyo, 1992
24. Yuanlong Y., Yanming Y., Fuming L., Yufen L. and Paozhong M. (1987) Characteristic Autofluorescence for Cancer Diagnosis and Its Origin, *Lasers Surg. Med.* **7**, 528.
25. Alfano R.R., Tang G.C., Pradhan A., Lam W., Choy D.S.J.and Opher E.(1987) Fluorescence Spectra from Cancerous and Normal Human Breast and Lung Tissues, *IEEE J. Quantum Electron.* **QE-23**, 1806.
26. Tang G.C., Pradhan A., Sha W.L., Chen J., Liu C.H., Wahl S.J. and Alfano R.R. (1987) Pulsed and cw Laser Fluorescence Spectra from Cancerous, Normal and Chemically Treated Normal Human Breast and Lung Tissues, *Appl. Opt.* **28**, 2337.
27. Razum N., Balchum O.J., Profio A.E. and Carstens F. (1987) Skin photosensitivity duration and intensity following intravenous hematoporphyrin derivatives HpD and DHE. *Photochem. Photobiol.* **46**, 925.
28. Lam S., Hung J. and Palcic B. (1990) Detection of Lung Cancer by Ratio Fluorymetry with and without Photofrin II, *SPIE* Volume **1201**, Optical Fibers in Medicine 561.
29. Hung J., Lam S., LeRiche J.C. and Palcic B. (1991) Autofluorescence of Normal and Malignant Bronchial Tissue, *Lasers Surg. Med.* **11**, 99.
30. Palcic B., Lam S., Hung J. and MacAulay C. (1991) Detection and Localization of Early Lung Cancer by Imaging Techniques, *Chest* **99**, 742.

31. Lam S., MacAulay C. and Palcic B. (1993) Detection and Localization of Early Lung Cancer by Imaging Techniques, *Chest* **103**, 12S-14S
32. S. Lam's presentation from the New Developments in Bronchoscopy, 61st American College of Chest Physicians Annual Assembly, New Orleans, USA (1994) (personal communication)
33. Alfano R.R., Pradhan A., Tang G.C., Das B.B. and Yoo K.M. (1991) Optical Spectroscopy May Offer Novel Diagnostic Approaches for the Medical Profession, p. 55 in L. Goldman (Ed.) *Laser Non-Surgical Medicine. New Challenges for an Old Application*, Technomic Publishing Co., Inc., Lancaster - Basel.
34. Kapadia C.R., Cutruzzola F.W., O'Brien K.M., Stetz M.L., Enriquez R. and Deckelbaum L.I. (1990) Laser - Induced Fluorescence Spectroscopy of Human Colonic Mucosa. Detection of Adenomatous Transformation, *Gastroenterology* **99**, 150.
35. Schomacker K.T., Frisoli J.K., Compton C.C., Flotte T.J., Richter J.M., Nishioka N.S., Deutsch TF. (1992) Ultraviolet Laser - Induced Fluorescence of Colonic Tissue: Basic Biology and Diagnostic Potential, *Lasers Surg. Med.* **12**, 63.
36. Bottiroli G., Croce A.C., Locatelli D., Marchesini R., Pignoli E., Tomatis S., Cuzzoni C., Di Palma S., Dal Fante M., Spinelli P. (1995) Natural Fluorescence of normal and Neoplastic Human Colon: a Comprehensive ''ex vivo'' Study, *Lasers. Surg. Med.* **16**, 48.
37. Chwirot B.W., Chwirot S., Jêdrzejczyk W., Jackowski M. (Eds.) (1995) *Tissue Spectroscopy. Optical Methods for Early Diagnostics and Localisation of Human Cancer*, Pomerania, Toruñ.
38. Chwirot B.W., Chwirot S., Jêdrzejczyk W., Jackowski M., Raczyñska A.M., Winczakiewicz J. and Dobber J. (1997) Ultraviolet Laser - Induced Fluorescence of Human Stomach Tissues. Detection of Cancer Tissues by Imaging Techniques, accepted for publication in *Lasers Surg. Med.*
39. Gwynne P. (1994) A New Window on Biomedicine with Spectroscopy, *Biophotonics* **1**, 52.
40. Lohmann W. and Paul E. (1988) In situ Detection of Melanomas by Fluorescence Measurements, *Naturwissenschaften* **75**, 201.
41. Lohmann W., Nilles M. and Boedeker R.H. (1991) In situ Differentiation Between Nevi and Malignant Melanomas by Fluorescence Measurements, *Naturwissenschaften* **78**, 456.

THE PHYSICAL BACKGROUND AND THE INFORMATIONAL CHARACTER OF BIOPHOTON EMISSION

F.A. POPP* AND J.J. CHANG*#

*International Institute of Biophysics, IIB,
Station Hombroich, Vockrather Straße, 41472 Neuss, Germany
#Institute of Biophysics
Chinese Academy of Sciences, Beijing 100101, China

1. Introduction

At the present time it is generally impossible to show evidence of the „informational character" of any signals, since agreement about what „information" is does not exist. According to Shannon's accepted theory, one of the necessary conditions of „information" is a reduction of the lack of knowledge about an expected outcome. In that case we would have to say in every case what outcome, what knowledge, what lack and what reduction we speak about.

We have to mention at this point that Shannon's information theory works only because it refers to the rather special condition of an objective observer whose knowledge is exactly known. This knowledge concerns the calculable probabilities of a limited set of random events. Consequently , the actual realization of one of the preset possible events reduces the uncertainty by the countable amount of actually **not** realized and by the limited number of outcomes. The essential point is that this uncertainty can be exactly quantified, which provides the scientific requirement of objectivity, reproducibility and quantification in the case of Shannon's information definition.

But what about „ultraweak" photon emission from biological systems [1], about which we know at present not very much more than of its existence and some common characteristics for all living systems. We know that this phenomenon concerns a weak, but permanent photon current of a few up to some hundred photons per second and per square-centimeter of the emitting surface. The term „biophotons" (which we introduced for this phenomenon in 1976 [2]) does not mean that the photons have extra-physical or metaphysical character. Like „bioluminescence" which specifies luminescence of biological systems, „biophotons" refer to biological systems. The phenomenon is characterized by the emission of single photons which points more to a biological quantum phenomenon rather than to ordinary luminescence. The spectral distribution is continous and flat, covering a range from at least 200 to 800 nm. In a quasistationary state the living systems emit the photons without stronger fluctuations. Under this condition the variance takes the lowest value which is always found around the mean value. It is not completely clear at present whether the variance can take even lower

J. J. Chang et al. (eds.), Biophotons, 239-250.

values than the mean value or not. In general, the photocount statitics (PCS) approaches under quasi-stationary conditions a Poissonian distribution. A second remarkable physical property of photon emission from living systems is the characteristic relaxation function after exposing the subject under study to external light illumination. The intensity then drops down **not,** as one might expect, according to an exponential function exp(-t/T), where t is the time and T decay constant, but always in excellent proportionality to an hyperbolic relaxation function (1/t), irrespective of whether the excitation is monochromatic or white. Typical relaxation times are of the order of seconds to minutes. It has been shown that Poissonian photocount statistics (PCS) together with an hyperbolic relaxation are sufficient conditions for a fully coherent field, indicating some optimization of the informational role of biophotons [3].

Nevertheless, despite the proof of coherence of biophoton emission, there is still a lot of disputation on questions of the source(s), the mechanism, the actual degree of coherence and the biological significance of biophoton emission. While some scientists still follow the „imperfection theory", which claims that the biophoton emission originates from stochastic, but rather rare optical transitions of metabolic events, some others (including most scientists from our research groups) see more reason to state that biophotons originate essentially from exciplex systems of the DNA displaying a coherent basis of intracellular and intercellular communication [1]. In this way biophotons would provide the most significant channel of biological regulation. Hence, the main sources of disagreement concern the „informational" character of biophotons. While an „informational " role does not exist within the framework of the imperfection theory, for the „coherence" theory biological information would be one of the most fundamental characteristics of this phenomenon. However, what are the arguments, and how can one prove the validity of the imperfection or the coherence theory of biophotons?

In order to bring some light to the problem and its possible solution, let us follow the strategy of discussing some basic biological problems which are still unsolved with respect to the hypothesis that biophotons may provide a solution. It is clear that this can never be a proof of the speculations. However, this discussion has to be the first step toward a proof. It will lead to some surprising insights into the role of „information" in biological systems and the connection to biophotons and to the experimental evidence.

2. The Basis of Chemical Reactivity

It is well known that chemical compounds do not react in the ground state. Actually, at zeropoint temperature where matter rests in the ground state no reaction takes place at all. Under thermal equilibrium reactivity increases according to the Arrhenius-factor exp(-Ea/kT), where Ea is the activation energy for exciting the electronic level of an intermediate product called transition state-complex, and kT is the mean thermal energy of the surrounding photon field at temperature T. The simple reason for this dependence of chemical reactivity is the increase of the number of thermal photons of energy Ea with increasing temperature according to the Arrhenius factor with the consequence that the chemical reaction can take place only by borrowing suitable photons of energy Ea from the thermal field in order to excite the necessary transition state complex of the reaction. This intermediate product decays then into the energetically more favourable final stable products. This basic mechanism is involved in any kind of chemical reaction, including enzymatic biochemical reactions. As a consequence one has to note that without photons metabolic events in cells would not take place at all. The advantage of the inclusion of enzymes in biochemical reactivity lies in the fact that the necessary activation energy e_a of the enzyme-substrate-transition state complex is lower than the Ea of the substrate transition state alone. This means that the possible number of photons is much higher for activating the enzyme-substrate complex than for the substrate alone since the Arrhenius-factor $\exp(-e_a/kT)$ is higher than the corresponding $\exp(-Ea/kT)$ of the non-enzymatic reaction. Thus, the biochemical reaction is not only controlled by the availability of specific enzymes, but even more by the presence of suitable photons within the surrounding photon field. We have to conclude that the basic regulator of chemical reactivity is a photon, and all its physical properties influence the reactivity. Moreover, only photons can give us the answer to the crucial question how all the chemical reactions in a cell can be triggered at the right place at the right time. Actually, the reaction rate in a human cell is of the order of 100 000 reactions per second. One may object that a few „biophotons" are not enough to always provide the necessary activation energy for all these reactions. However, a rough estimation shows that even a few photons, and nothing else realistically, can provide the necessary activation energy to all the biochemical reactions which take place in a cell. Let us take, for instance, a reaction time of 10^{-9} seconds and assume that a photon is borrowed from the surrounding photon field to activate this reaction. After this short time interval of 10^{-9} seconds the photon is again available for the next reaction. Among others Cilento [4] showed that this procedure of „photochemistry without light" takes place for many, if not all biochemical reactions. Under thermal equilibrium conditions this borrowing process is governed by the randomness of the black-body radiation. However, far from equilibrium, the borrowing procedure becomes sensitively dependent not only on the number of photons and the spectral distribution, but also on the momenta, the polarization directions and coherence properties of a photon field. Under random non-equilibrium conditions the chemical reaction rate increases simply according to the number of photons of energy Ea, that is according to the Arrhenius factor $\exp(-Ea/kT)$ where T has to be substituted by the spectral excitation temperature $\theta(Ea)$. θ represents the temperature which the system under examination would have in the case that it emitted the spectral photon intensity according to black-body radiation. In this way the measured spectral intensity of photons can be expressed in terms of a definite temperature θ. For the case that θ does not change with the frequency or the activation energy, the system is identical to a thermal equilibrium system and, on the other hand, the frequency dependence of θ characterizes the deviation from thermal equilibrium. For coherent or squeezed states the situation gets more complicated since interference patterns of the photon field will provide a spatio-temporal structure of the possible supply with activation energy. However, let us estimate first the change of chemical reaction rate under the assumption of the most simple model of a chaotic biophoton field of just the intensity that can be measured. The experimental results of biophoton intensity show evidence that the number of available photons in the field is in spite of its low intensity many orders of magnitude higher than under thermal equilibrium conditions at physiological temperatures. Actually, the number of biophotons is at least

10 orders of magnitude higher in the red part of the spectrum and increases linearly with the frequency to about 20 orders of magnitude in the blue and violet part. The excitation temperature increases linearly with the frequency, taking values between 400 and 1000 K (Fig.1). Thus, black body radiation at physiological temperatures has 10 to 20 orders of magnitude lower intensities in the spectral range under consideration than biophoton emission. Consequently, the signal/noise ratio and the availability for chemical reactions increase proportionately to the tremendous factors 10^{10} to 10^{20} in this range of activating the electronic states of molecules. The biophoton field makes 10^{10} to 10^{20} photons more available for governing the biochemical reactivity than a thermal field of physiological temperature T would do. Let us stress, therefore, that it is unrealistic to believe that biophotons do not have a dramatic influence on biochemical reactivity.

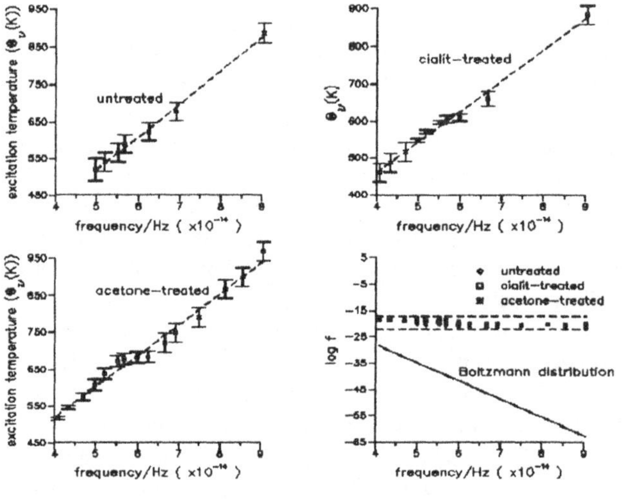

Figure 1. Spectral Biophoton Emission of Cucumber Seedlings in terms of the excitation temperature $\theta(\nu) = h\nu/k\left[\ln\left(2\nu^2/I_\nu \cdot c^2\right)\right]^{-1}$ where I_ν is the spectral biophoton intensity of frequency ν.
It turns out that $\theta(\nu)$ increases linearly with ν, even if the seedlings are poisoned with Cialit or acetone (upper right figure and lower left figure).
This means that $f(\nu) = \exp\left(-h\nu/k\theta(\nu)\right)$ is constant.

The only realistic conclusion of this experimental fact is that only biophotons provide the answer to the unsolved question of where the right activation energy at the right time at the right position in a cell comes from in order to enable a living system to coordinate its manifold reactions. It is now easy to estimate the information which is necessary to regulate the biochemical reactivity in a cell. According to Shannon's definition we calculate this information in terms of the necessary distinguishable states which the photon field has to take in order to transfer the right activation energy at the right time to the right position. The information per reaction is then equal to the negative value of the logarithm of the ratio of the reaction volume v and the volume V of a cell (or an organization center of cells) times the ratio of the necessary momentum of the energy transferring photon and the available momenta of the present photon field times other probabilities which may be necessary for triggering a definite reaction. A rough estimation is based on a reaction center of the size of $v = (10^{-7}\text{cm})^3$ which gives for a cell volume V of $(10^{-3}\text{cm})^3$ the probability $v/V = 10^{-12}$. A corresponding order of magnitude is obtained for the probability of providing the right momentum $\Delta p/p = \Delta\nu/\nu = 10^5\ \text{Hz}/10^{15}\ \text{Hz} = 10^{-10}$, such that the order of magnitude, R, of the information which is necessary to trigger all the reactions in a cell is of the order of:

$$R = r\,(ld\,(V/v) + ld\,(\Delta p/p) + \ldots) \cong 10^{10} * 100 \cong 10^{12}\ \text{bit/per cell and per day}, \qquad (1)$$

where r ($=10^{10}$) is the number of reactions per day.

In order to examine the question of whether biophotons can transfer this information, one has to analyse this photon field in terms of its coherence. This is not easy, since there are no tools available for investigating the photon field within the cell, and even as far as one can measure biophotons outside, conclusions about the field inside are not always justified. Nevertheless, in order to make the best of the situation, it is necessary to establish models, and draw conclusions that can lead to experimental investigation. For examination of the informational content it is useful to introduce the concept of the resonator value of a model cavity which represents the information of the biophoton field. The Q-value describes the number of reflections which a light beam of frequency ν may take before it is absorbed and dissipated to heat. Since the light beam can often be modulated by switch-on/switch-off processes as it is reflected, the Q-value describes at the same time the number of bits which can be stored on this light wave and transferred to the outside. The Q-value is at the same time a measure of the coherence of the light beam since the coherence time τ increases with the Q value according to:

$$\tau = Q/\nu \qquad (2)$$

where $Q = \nu \cdot U/i$ where U is the stored energy and i the power loss of the cavity. A cavity of the tremendous information of 10^{12} bit could be, for instance, the DNA, since the base pairs (and non-linear connections of the base pairs) may well provide this information. A Q-value of 10^{12} corresponds to a cavity of a coherence time of 10^{-3} s in the visible range, which is not an unrealistic model. One can say therefore that if the DNA (and possibly also other biomolecules) works as the resonator of biophotons in the cell it is very likely that all the information which is necessary to trigger the chemical reactions can be transferred.

3. Growth Regulation

It is well known that about 10^7 cells/s are lost in a human body by sudden cell death. This surprisingly high cell loss rate is compensated for rather exactly by the cell division rate, in order to avoid serious disease like abnormal swelling or shrinking of tissues, including cancer. One may speculate that the loss of a cell is always registered by the neighbour cell(s) which divides in order to replace the dead cell. However, this cannot explain the fact that considerable parts of some organs, such as the liver, can be replaced in a way that exactly the original size and shape of the whole organ is reconstituted. Let us therefore ask what the most optimized way of substituting dying cells would be. Take an average loss rate of 10^7 cells/s. The signal for informing other cells about the death of one cell has to transfer the message of a cell death in a time interval that is as short as it can be to prevent many more cells from dying. This means that the resolution time in consecutive cell death events is 10^{-7} s. Take the case that only the nearest neighbours are informed. This means that the signal has to travel a distance of at least 10^{-3} cm during the time of 10^{-7} s. The velocity has to be at least of the order of 10^{-3} cm/10^{-7} s = 10^4 cm/s which is much faster than the diffusion velocity of messenger molecules. However, it would fit the velocity of sound. Moreover, take the case that the message is an holistic one, informing the whole body in order to allow an optimized reaction including the response regarding the shape and size of the organs. Then the message has to travel over distances of the order of a meter. This means that the velocity has to be of the order of 1m/10^{-7} s = 10^7 m/s which can only be light. In other words, if growth regulation of biological systems is based on information originating from the death of cells, it is not possible to explain the regulation by messenger molecules. Rather, either sound waves, or more likely electromagnetic waves are suited for transfering the necessary messages. Again we come to the conclusion that biophotons may work for regulation functions of the biological system, i.e. the cell division rate which compensates cell loss. We do not confine ourselves here to the optical spectrum. The frequency distribution may even reach to the Megahertz range or even extremely low frequencies. A decision on this issue does not have to be made for this discussion. What is important to see is that electromagnetic interactions have to take the role of regulators of a biological system in order to explain many, if not all regulatory functions.

Consequently, we expect some correlation between growth and biophoton emission. Actually, Figure 2 shows on the left side the biophoton emission of germinating seeds over the whole period of germination, and on the right side, the increase in mass during this time. There is obviously some correlation between the photon emission and the increasing mass of cells. However, it seems that the biophoton emission is more finely structured than the increase in mass, indicating that the photons are the regulators of growth instead of the growth rate being the reason for increasing biophoton emission. Actually, the increase in mass does not follow an exponential law but a differential equation of the form

$$\dot{N} = CN(1 - BN) \quad (3)$$

with the solution:

$$N = 1/(B+(1-B)\exp(-CN)) \quad (4)$$

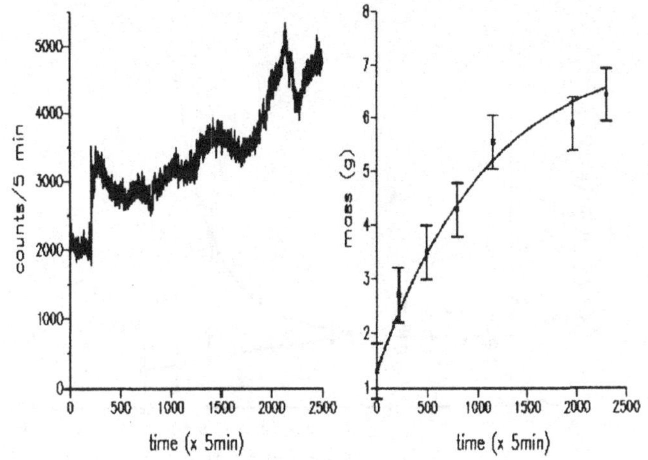

Figure 2. Biophoton Emission of Growing Seedlings during germination (left) compared with the growing mass (right).

where $\dot{N}$ is the growth rate, N the cell number and B and C constant parameters.

This model is based on the idea that not only nearest neighbour interactions play a role in growth regulation but at the same time holistic influences originating from the information potential among all the cells. Since C is a frequency of cell communication, it is interesting to look at the magnitude of this frequency. In accordance with the cell loss rate of 10^7 cells/s, C can be actually adapted to the growth curves of men for values between 10^{-6} to 10^{-8} s^{-1}. The value B describes the reciprocal cell number of an adult. It is evident that a decrease of B - including a continous decrease with growing cell number- may be a sufficient condition for cancer growth.

Actually, Figure 3 shows the difference of delayed luminescence emission of biophotons with increasing number of cells for normal cells (lower curve) and tumour cells (upper curve). As expected, the biophoton emission has just the opposite characteristic for normal cells than for tumour cells. Whereas normal cells show decreasing emission with an increasing number of cells, the photon emission of tumour cells increases, displaying thus a qualitative, not only a quantitive difference. It is in accordance with the idea of a coherent communication not only between neighbour cells but among all the members of a cell population. As soon as the integration of a new cell into the population by cell division does not result in an increasing coherence of the system, the body grows cancer tissue. Consequently, the model of a coherent biophoton field, providing the basic communication of the cells in an organism, helps us to understand cancer growth in terms of rather fundamental properties of a coherent field.

It is worthwhile to note that the coherent basis of biophoton emission is neither a necessary nor a sufficient condition of „information"-exchange of cells. Cells can also

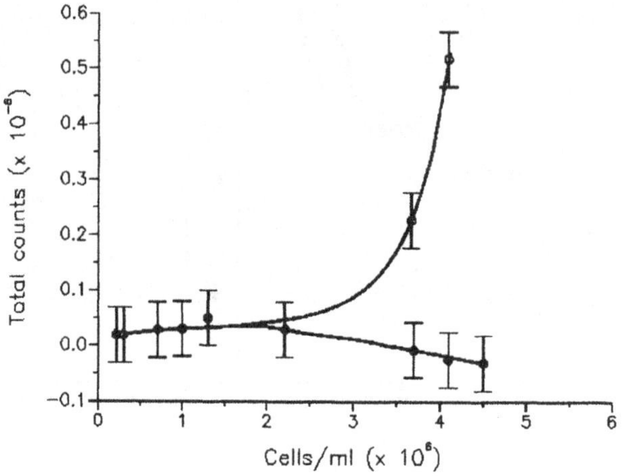

Figure 3. Delayed luminescence of cancer cells (upper curve) and normal cells (lower curve) independence on cell density.

communicate by means of incoherent light, for instance switching on and off an incoherent light source in a definite time sequence.On the other hand, even if a coherent biophoton field exists among cells, they are not forced to modulate in such a way as to transmit information. However, if a coherent biophoton field exists, the efficiency of communication is much higher than for a non-coherent field. If a coherent field exists, it would be unlikely that it is not used for communication. These arguments are the reason why the proof of the coherence of biophotons is of fundamental importance

In some papers we showed that the biophoton field is coherent [3]. The proof is based on the well-established experimental evidence that the photocount statistics of a stationary biophoton field follow a Poissonian distribution. In other words: the probability of measuring 0,1,2,....,n,... photons in a present time interval follows at any instant a Poissonian distribution. However, not only a coherent field but also a chaotic field displays a Poissonian distribution as soon as the preset time interval within which the photon counts are always registered is large compared with the coherence time of the chaotic field. Consequently, after showing evidence that the photocount statistics follow a Poissonian distribution, we do not know whether the subject of investigation is a chaotic field with a rather small coherence time or a fully coherent field. In order to solve this problem, we have to investigate the relaxation dynamics of biophotons after exposure of the biological system to an external light illumunation. An ergodic chaotic field relaxes according to an exponential function for single modes, whereas a fully coherent field (where every emitter is phase-coupled to every other rescattering center of the system) relaxes according to a 1/t-law, where t is the time after excitation of the field. From the Poissonian statistics of the photocount distribution we know that the field is ergodic. Since it is well established that all biological systems after illumination with monochromatic or white light display a 1/t-relaxation of „delayed luminescence", it is evident that the biophoton field can only be coherent. Coherence also includes minimum-uncertainty states like squeezed states [3].

4. Organization and Communication

Let us add an example where one gets some deeper insight into the mechanism of communication between cells and organisms [5]. Figure 4 shows the biophoton emission of Daphnia magna where the number of female inbred animals of about the same size has been increased step by step. Instead of a continous increase of biophoton emission with increasing number of (almost) identical animals, the biophoton emission displays minima and maxima.

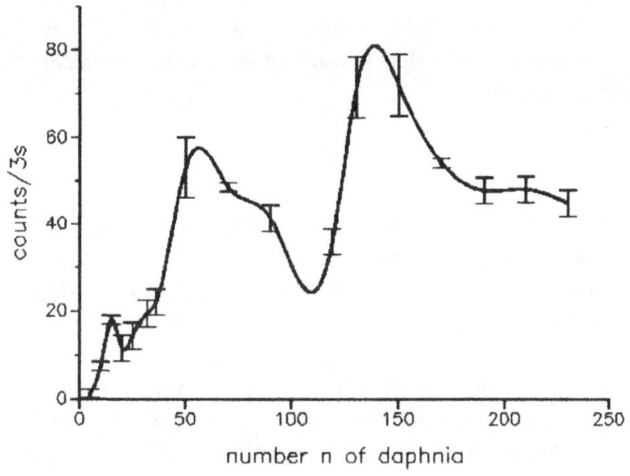

Figure 4. Biophoton emission of daphnia in dependence on the number of daphnia.

The most distinct minimum has to be assigned to a mean distance between the animals which is taken by them in free nature. For a careful analysis of this phenomenon let us propose the following mechanism behind this effect which cannot be understood in terms of chemical messengers. According to the theory of Dicke, antennae systems show constructive and destructive interference of the absorbed light as soon as the distance between the absorbers gets small compared with the wavelengths. Actually, in a biological system, the distance between neighbouring DNA base-pairs is much smaller than the wavelength of light. Consequently, Dicke's theory has to be applied to the exciplex system of DNA which is the most probable candidate for biophoton emission [1]. As a result, the Daphnia (or every other biological system) exhibit mutual interference patterns of their biophotons, based on destructive interference of their specific wave patterns. This model explains its dependence on the mutual distance of the animals as well as the interference pattern of Figure 4. At the same time, this mechanism provides the most effective basis for communication. Take the case that different biological systems emit different, but species-specific, patterns of electromagnetic waves which they may compensate mutually by destructive interference. This provides the most sensitive field of communication. Actually, the organisms are able to identify each other by the control of the capacity of destructive interference of their wave patterns. At the same time, every fluctuation around the ideal destructive interference, corresponding to the vacuum state, has to be interpreted as „information" of the highest possible signal/noise-ratio.

This capacity for using the phase information in order to organize the metabolic events as well as the „Gestalt"-formation of biological systems seems to be elementary for living systems.

An interesting result which has been obtained recently in our laboratory is an experiment of Süßmuth and Vogel (see paper Vogel and Süßmuth, this volume). They obtained an absorption of the light emission of the surrounding medium by growing bacteria according to an exponential law exp $(-t/\tau)$, where τ agreed with the cycle time of the bacteria. The analysis shows that this effect cannot be explained by the ordinary Beer-Lambert's law following:

$$dI = \rho_M A dx - \tau_M I dx, \tag{5}$$

where ρ_M describes the photon density of the medium, A the space area, x the depth, τ_M the absorbance of the medium and I the intensity.

A metabolic effect can be excluded simply because of the rather low cross section of this reaction which is of the order of 1 to 10 reactions/bacterium and per day. A possible interpretation is provided by destructive interference of the single bacteria which are able to use the phase information of emitted light of the emitting molecules of the nutrient. They may cancel the light amplitudes within the coherence volume around these molecules in order to find a distribution over the medium in such a way that they optimize the efficiency of nutrition. The mechanism may be based on a phase-conjugation-like mechanism (Popp, F.A. et.al., in preparation). Actually, if one provides the following dependence:

$$dI = (\rho_M - \alpha\rho_B) A dx - \tau_M I dx, \tag{6}$$

where $\alpha\rho_B$ describes the destructive interference cross-sectional absorbance of bacteria of density ρ_B, one is in line as well with the capacity of destructive interference in the coherence volume of the nutrition molecules as with the most efficient distribution of the bacteria over the medium. Then one gets the right and the observed dependency of light emission and density of growing bacteria within the medium. It is clear that α gets dependent on the distribution and density of bacteria themselves as soon as the interference pattern gets disturbed.

Similar and even more convincing and exciting are observations which have been published in addition to Galle by Schamhart [6], Scholz [7] and Chwirot [8]. They display the non-linear optical properties of biological systems in terms of biological effects such as growth regulation, swarm-formation and Gestaltbildung.

An other convincing result of this kind is the synchronous flickering of fireflies or of dinoflagellates [9, 10].

5. Consciousness

From the point of view of the coherence theory of biophoton emission the most essential basis of life is the „active" capacity of the living system for constructive and destructive interference of biophotons. The question remains what mechanism enables a living system to develop this property which is certainly not common in the non-living world and which is generally unknown. We think that DNA allows us to give an answer to this crucial question. There are already a lot of experimental results which point to DNA as the main source of biophotons [11]. Take the case that the exciplex system of the DNA is able to provide limited vacuum states in such a way that the electric field and its gradient always vanishes between two base pairs of the DNA, this medium then works like matter of refractive index 0 between a double layer with extremely high polarizability. This means that an incoming wave is reflected in such a way that it cancels the incident wave field while the transmitting wave doubles its amplitude. This mechanism which corresponds in quantum optics to a definite state of squeezed light derives its „biological significance" from the possible capacity of the system to tune its antenna system in a way that it adjusts its vacuum state exactly to the nodal points of the wave propagation. The phenomenon is similar to the known phase conjugation effect of polarizable matter. There are then two decisive properties of the living state which enable biological evolution: first, the rather tiny vacuum states of the exciplex system which provide the capacity for reacting to the phase of electromagnetic waves of wavelengths much longer than the distance of neighboured base pairs and second, the flexibility which is necessary to „tune" into the nodal planes of the electromagnetic field. The evolutionary principle is then the extension of coherent states by tuning into the actually existing coherent states, in order to expand the coherent electromagnetic „bio"-field into a more and more extended basis of higher and higher degrees of coherence. It is obvious that this process is linked to the development of consciousness. Let us speculate that what we call „consciousness" is the process of „tuning into" this coherent field which exists within us as well as within at least a part of the environment [12]. The boundary between the external world and the „I" may then be just fixed by the the border between destructive and constructive interference, both of which have to happen with respect to the energy conservation law. The Daphnia experiments and other „strange" observations of biophoton emission invite this proposed thesis.

In these terms, consciousness is a permanently proceeding process which transforms actual information into potential one, in order to transfer in turn, potential into actual. This optimizes the survival capacity by gaining certainty and by activating creativity.

6. Acknowledgments

We are particularly grateful to the Familie-Ernst-Wendt-Stiftung for its interest and its financial support.

7. References

1. Popp, F.A., Li, K.H., and Gu,Q.: *Recent Advances in Biophoton Research and its Applications*, World Scientific, Singapore-London, 1992.
2. Popp, F.A.: *Biophotonen*, Verlag für Medizin Dr. Ewald Fischer, Heidelberg, 1976.
3. Popp, F.A. and Li, K.H.: Hyperbolic Relaxation as a Sufficient Condition of a Fully Coherent Ergodic Field, *International Journal of Theoretical Physics* **32** (1993), 1573-1583.
4. Cilento, G.: Photochemistry without light, *Experientia* **44** (1988), 572-585.
5. Galle, M., Neurohr, R., Altmann, G., Popp, F.A., Nagl, W.: Biophoton emission from Daphnia magna: A possible factor in the self-regulation of swarming, *Experientia* **47** (1991), 457-460.
6. Schamhart, D.H., van Wijk, R.: Photon emission and the degree of differentiation, in B. Jezowska-Trzebiatowski et.al. (eds.), *Photonemission from biological systems*, World Scientific, Singapore-London, 1987, pp. 137-152.
7. Scholz, W., Staszkiewicz, U., Popp, F.A., and Nagl, W.: Light stimulated ultraweak photon Scholz, W., reemission of human amnion cells and Wish cells, *Cell Biophyics* **13** (1988), 55-63.
8. Chwirot, B. and Dygdala, R.S.: Ultraweak photon emission and anther mitotic cycle in larix europea, *Experientia* **44** (1988), 594-599.
9. Popp, F.A., Chang, J.J., Gu, Q., and Ho, M.W.: Nonsubstantial Biocommunication in Terms of Dicke's theory, in M.W.Ho, F.A.Popp and U.Warnke (eds.), *Bioelectrodynamics and Biocommunication*, World Scientific, Singapore-London, 1994, pp. 293-317.
10. Chang, J.J., Popp, F.A., and Yu, W.: Research on cell communication of P.elegans by means of photon emission, *Chinese Science Bulletin* **40** (1995), 76-79.
11. Rattemeyer, M, Popp, F.A. and Nagl, W.: Evidence of photon emission from DNA in living systems, *Naturwissenschaften* **68** (1981), 572-573.
12. Popp, F.A.: Evolution as the Expansion of Coherent States, in B.Rubik (ed.), *The Interrelationship Between Mind and Matter*, Marketing Graphics, Southampton, PA, 1992, pp. 249-281.

PHOTON EMISSION OF CEREAL SEEDS, "BIOPHOTONS", AS A MEASURE OF GERMINATIVE ABILITY AND VIGOUR

B.F. ZEIGER

Biophoton Laboratory, International Institute of Biophysics,
Technopark II, 67661 Kaiserslautern, Germany.

Abstract

Experimental investigations of the spontaneous emission of cereal seeds point to a superfluid vacuum structure of the radiation field coupled to biological matter. The superfluid vacuum provides an unified understanding of germinative ability (potentiality of growth) and vigour (performance potential) of seeds as expressed in the process of germination. The findings may be used as basis for a fast, reliable and non-invasive method to determine germinative ability and vigour of seeds but also may be applied to improve the storage capacity of seeds.

Key words: Biophotons, germinability, superfluid vacuum

1. Introduction: Biophotons, Vitality and Biological Rest

The ultraweak photon emission (biophotons) in the visible part of the spectrum is a universal property of biological systems (POPP, GU, and LI, 1994). It is linked to basic biological actions such as changes in the DNA conformation (RATTEMEYER, POPP and NAGL, 1981), different stages of cell development such as sprouting of soya beans (GU and POPP, 1994) and cooperative effects between or within organisms (GALLE, 1993, COHEN and POPP, 1997). Biophoton measurement has been used to monitor the state of human health, environmental pollution, food quality and the germinative ability of seeds. The foundation for the latter application is studied in this paper because of its practical importance for plant cultivation, agriculture, and certain branches of the food industry.

In general, biophoton emission seems to be a measure of the organisational order related to vitality (GU and POPP, 1994). The term "vitality" is used here to denote all properties based on the dynamical relationship between the wholeness of a biological system and its individual parts. This self-interaction at the root of vitality manifests in activity and growth of a biological system. In case of seeds "vitality" is the generic term for "vigour", which relates to activity, and "germinative ability", which relates to growth.

J. J. Chang et al. (eds.), Biophotons, 251-297.

Seed vitality expresses itself in the process of germination. Germination can be described in many different ways (JANN and AMEN, 1977):

"*Morphologically*, germination is the transformation of an embryo into a seedling. *Physiologically*, germination is the resumption of the metabolism and growth which were earlier depressed or suspended... *Biochemically*, germination is the sequential differentiation of oxidative and synthetic pathways."

From a *physical* point of view seed germination is the sequential emergence of higher levels of organisation from the state of biological rest. Each layer of higher organisation is contained in the least excited, hypometabolic state not in an actualized but in a potential form (state of "latent life").

The emergence of new qualities and modes of behaviour from an undifferentiated state of wholeness can be described within the framework of quantum mechanics, where new qualities emerge due to restrictions imposed on the holistic symmetry represented by the group of all unitary transformations of the state space. The quantum mechanical term for emergent evolution is "sequential spontaneous symmetry breaking" (PRIMAS and GANS, 1979).

The purest respresentative of the quantum mechanical reality of any system is the "ground state", which is the state of least excitation of all the ("Eigen"-) states defined by the HAMILTON-operator (representing the total organizing power of the system). The ground state of a quantum mechanical system with infinitely many degrees of freedom is analogous to the "vacuum state" of a field system (AITCHISON, 1985). Both the electromagnetic field and the radiation field coupled with molecular matter are systems having an infinite number of degrees of freedom. The term "vacuum state" implies that in this state of the field there exists no real excitations or particles. In this respect the vacuum state matches with the classical notion of "empty space".

However, the "quantum vacuum" (MILONNI, 1994) has as an additional feature, infinitely many "virtual" excitations persisting even at the temperature absolute zero. This "zero-point energy" can be described in the language of fluctuations familiar from thermodynamics. The "physical vacuum" is then full of virtual fluctuations which are responsible for basic dynamical effects, e.g.the spontaneous light emission of molecular matter.

The state of deep rest of plant seeds displays all the essential properties of the quantum mechanical ground state. It is noted at this point, that such a kind of ground state is always defined relative to the respective material structure or, in case of seeds, relative to the species, variety, seed lot or individual seed considered.

Furthermore, the vacuum properties are also a common feature of all macroscopic coherent ground state phenomena such as superradiance, superfluidity, or superconductivity (PRIMAS, 1968; COCHRAN, 1971). These quantum effects operating on the macroscopic scale are related to the Third Law of Thermodynamics which ultimately is a quantum law (SEWELL, 1986).

The macroscopic quantum effects known from physics may serve as models for the different "macroscopic" forms of biological rest, e.g.(1) the hypometabolic state of seeds or (2) the hypometabolic state of human physiology which corresponds to the state of least excitation of consciousness (WALLACE et al, 1971).

The bridge between microscopic (e.g.molecular) and macroscopic (e.g. biological) quantum phenomena isprovided by the principles of coherence (LI and POPP, 1996; LI, 1996; Popp, 1996).

TABLE 1. Qualities of states of least excitation

	forms of biological rest:	
quantum vacuum	(1) plant seeds	(2) pure consciousness (human)
ground state	deep rest	silence
pure potentiality	latent life	pure wakefulness
infinite life time	longevity	timelessness
non-change	stabilty	ever the same
fluctuations	vigour	creativity
autonomous	self-contained	self-sufficient
self-interaction	homeostasis	self-referral
coherence	high density	perfect order

The correspondence between the quantum vacuum and biological rest provides an interdisciplinary approach (HESKE, JORDAN, and MEYER-ABICH, 1954; POSTLE,1976) to the nature of seeds. Both the experimental and theoretical investigations of biological rest are guided by the quantum theory of the vacuum, which locates dynamism and order in the state of rest (SOUNDERS and BROWN, 1991). The types of order existing on the level of the quantum vacuum are described as "structure of the vacuum"(RAFFELSKI and MÜLLER,1985). The observed nature of the biophoton emission from cereal seeds points to a "superfluid" vacuum structure of the radiation field coupled with biological matter, which is shown in the following to provide an unified understanding of germinability and vigour which is of great practical significance.

2. Experimental results: Spontaneous Photon Emission of Cereal Seeds

Since dry seeds are in a state of deep rest they are characterized by a very low metabolism rate and a slow rate of respiration (MAYER & POLJAKOFF-MAYER, 1975). It is difficult to monitor the metabolism and respiration of dry seeds directly. Attempts to find a relation between the photon emission of seeds and their vitality have been successful in the case of induced photon emission ("delayed luminescence") after a short stimulation with white or monochromatic light (VESELOVA et al, 1985). However all the attempts to correlate the very low sponttaneous emission of seeds with their vitality failed, becuse of the limitations of the photon detection devices. By using a new temperature regulated photon counting system we have now found quantitative relations between the vitality of seeds and their photon emission.

2.1. METHOD: TEMPERATURE REGULATED PHOTON COUNTING

A stable and precisely defined temperature during the photon measurement of seeds is very desirable, because the spontaneous emission of seeds depends strongly on the temperature of the sample. Also the degree of hydration of seeds (which depends on the temperature maintained over a long period of time) influences the photon emission. Controlled and uniform drying before measurement is required to get comparable results. There is a rule that by a 1% increase in the moisture content of seeds the induced photon emission decreases about half of its value (VESELOVA et al, 1985). Furthermore, when seeds are exposed to light (e.g. daylight) they absorb a certain amout of energy which they release in the dark as delayed luminescence. To get comparable values of spontaneous emission from differet seed samples one has to uniformely keep them for a certain time (about 1 hour for cereal grains) in the dark within the measuring chamber.

In order to determine the spontaneous photon emission of seeds at temperatures different than room temperature, a measuring chamber has been developed which allows us to heat up a metal cuvette with the sample to temperatures between 0°C and 60°C. Within this temperature range the viability of cereal seeds is not affected.

To increase the sensitivity of photon detection the usual quarz window towards the detector has been replaced by a metal grating, tight enough to keep the grains in the cuvette.The volume of the cuvette is 10 ml. After filling the cuvette with the seeds a temperature sensor is placed in the sample. The temperature of the sample can be preset as a constant value in time and/or can be changed as a function of time. The rate at which the temperature can be changed is about 1.5 °C per minute for grains. The temperature accuracy is 0.5°C, and the temperature deviation over time 0.005°C.

Details of the photomultiplier and electron counter as well as some technical aspects of the measurement procedure have been described repeatedly in the literature (e.g. POPP et al. 1994).

2.2. VITALITY OF SEEDS: VIGOUR AND GERMINATIVE ABILITY

The vitality of seeds reveals itself in the germination process. Seed quality testing focuses on two main aspects of vitality: germinative ability and vigour. Instead of "germinative ability" often the term "viability" is used in the literature. Therefore "vitality" is choosen here as generic term for both "germinative ability" and "vigour". Roughly, germinative ability describes the "potentiality for growth" while vigour relates to the "performance potential" which determines the seedlings emergence in the field.

The failure of the germination counts in the biological germination test to predict field performance can be attributed to the lack of the specific conditions which the seed faces in the field during germination. Since the ability of seeds to respond to environmental influences must never be left out of consideration, so in addition to the germination test vigour tests are needed to determine the actual emergence under field conditions. However, field tests are time consuming and impractical because of the land and labour involved (PANDEY, 1992). Therefore methods of seed testing are demanded where the seed itself can tells its own story in a simple, fast, accurate, and reliable way involving both vigour and germinative ability but without undergoing germination. To fulfill all these requirements seed quality testing has to rely solely on the holistic nature of seeds. Biophoton measurements seem to be a way to realize all these requirements in a natural way and may therefore be a candidate for holistic seed quality testing.

2.2.1. *Germinative Ability*

Germinative ability is defined as the ability of seeds to produce a normal seedling under optimal growth conditions which are standardized and listed in the International Rules for Seed Testing (ISTA, 1985). The emergence of a seedling is due to an irreversible increase in the size of embryonic cells. The most important condition for elongational growth is the availability of water. The supply of oxygen and a specific temperature range are further requirements. The temperature takes an optimum value if both germinative ability (growth) and rate of germination (performance) are maximal. These temperatures are listed by ISTA (1985).

Environmental signals (e.g. light) are instrumental in overcoming certain barriers (e.g. inactive enzymes) which may prevent in case of "dormancy" viable seeds from germinating (AMEN, 1968).

As a typical example durum wheat was studied very carefully with the method described above. The germinative ability of the durum wheat investigated was determined by counting the germinated seeds or seedlings after 4 days of germination at 25°C. The resulting percentage of germinated seeds is referred to as the biological germinative ability. The value of the biological germinative ability (g) of the durum wheat samples is listed in table 2 from which the following rule is derived.

Rule 1 High germinative ability (g) corresponds to a high intensity of spontaneous photon emission (n):

$$n \sim g \tag{1}$$

The experimental conditions under which this rule is valid will be given below.

Since biologists are very reluctant in using the term "rule" (or "law") it is noted at this point, that this term has been choosen here and in the following sections because of the agreement with the theoretical considerations. At present, the data base of the rules is rather small, so further research is necessary to substatiate them experimentally and find out regions of validity, exceptions and influences of changed conditions .

2.2.2. *Vigour of Seeds*

To determine the vigour of seeds several methods have been developed. The simplest assessment is by counting germination at an early and a final stage. Because highly deteriorated seeds germinate at a slower rate than do less deteriorated ones the rate of germination is a measure of vigour. Other important classes of vigour tests conducted in the laboratory are stress tests and accelerated aging approaches. All tests of seed vigour determine the level of performance including conditions which are not optimal. Qualities related to vigour are rapidity of emergence, uniformity of growth, stand establishment and final yield.

Different cereal species may differ in their vigour a property called "genetic vigour". By comparing the ripening time of differnt cereals (RENZENBRINK, 1981) with their time of sprouting (ISTA, 1995) one finds as a rather crude rule that long vegetation time of different ceareal species corresponds to fast sprouting (Figure 1).

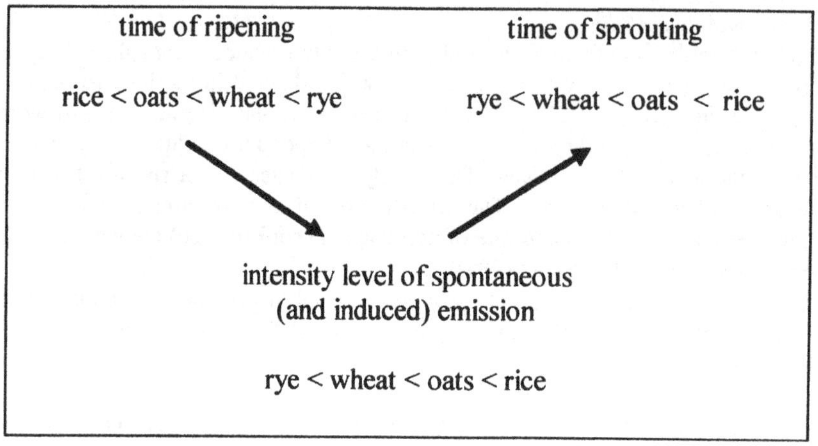

Figure 1. Ordering of cereals according to ripening time, sprouting time and photon emission.

The relations described by Figure 1 invite the following interpretation:

(1) Longer maturation time gives rise to deeper rest as measured by the intensity of photon emission.
(2) The deeper the rest the shorter is the time seeds need for sprouting; fast sprouting is indicative of high vigour.

As pointed by KRUSE (1997), this interpretation is exact within a cereal species (among varieties) but rather crude between species.
Because "genetic vigour" is the basis for "physiological vigour" (POLLACK and ROOS, 1972) different climatic or weather conditions have a related effect on seed vigour:

(3) Wheat from southern Europe (like Italy) has a much lower level of spontaneous emission than wheat from northern Europe (Finland), This observation is in agreement with HUTTER (1994), who observed a lower induced emission from barley grown in northern Europe indicating a reciprocal relationship between spontaneous and induced emission.
(4) Wheat exposed to more sunshine during ripening has a lower spontaneous (and a higher induced) emission.

The rule for seed vigour following from this observation is:

Rule 2 High vigour (D) correspond to low photon emission (n):

$$n \sim \frac{1}{D} \qquad (2)$$

2.3. TEMPERATURE DEPENDENCE OF THE PHOTON EMISSION FROM WHEAT

In order to understand the nature of biological rest by way of a non-destructive investigation, the response of seven wheat samples (Triticum durum) towards a moderate heating was studied.

TABLE 2. Specification of seven durum wheat samples

Sample	Variety	Region of growth	Harvest 1995	% Germs
1	Agridur	Foggia/Italy	July 4	84%
2	Grazia	Foggia/Italy	July 4	82%
3	Neodur	Foggia/Italy	July 4	92%
4	Vitron	Foggia/Italy	July 4	92%
5	Neodur	Voghera/Italy	July 15	56%
6	Creso	Ancona/Italy	July 7	64%
7	Brindur	Voghera/Italy	July 15	32%

The wheat grains were provided by the Research Center of Nestlé, Ltd., Switzerland. Before measurement the seeds were cleaned of stones, damaged seeds and awn. The measurements were performed at 25°C, 30°C, 35°C, 40°C, 45°C, and 50°C by keeping the samples for about 60 minutes at the respective temperatures. This time interval was chosen on the basis of a pilot test indicating no significant change in emission. The stability of the photon emission points at a low drying effect (dehydration) during this time interval. All measurements were done during the same period at night.

2.3.1. *Temperature Dependence of Emission*

The average value of the biophoton intensity from the seven durum wheat samples measured at 25°C, 30°C, 35°C, 40°C, 45°C, and 50°C are displayed in Table 3. After the measurement the germinative ability of the samples was determined in the manner described above. The value of the biological germinative ability (g) is again included in Table 3.

Two rules can be derived from the data in Table 3.

A first rule (rule 3) displayed in figure 2 describes the temperature dependence of the photon emission for seeds of a fixed vitality.

Rule 3 (on the temperature dependence of the photon emission from seeds):

> The photon emission (n) from a given sample of wheat grains under the (general conditions of this experiment) is an exponential function of the temperature (T).

$$n = B\, exp(C\, T) \quad or \quad ln\, n = C\, T + ln\, B \qquad (3ab)$$

B is a parameter which depends on the number of seeds measured (10g) and the time interval of measurement (60min), both conditions are kept the same in all measurements. The slope (C) of the straight line defined by this rule has the dimension of a reciprocal temperature. The average value of C for the seven durum wheat samples is 0.108 ±.004 $(°C)^{-1}$.

TABLE 3. Spontaneous photon emission of seven different durum wheat samples ordered by decreasing biological germinative ability g. (The emission of the empty cuvette has been substracted)

sample	photon counts at different temperatures (counts/sec)						g
	25°C	30°C	35°C	40°C	45°C	50°C	% Germs
4	14	24	42	73	126	215	92
3	12	22	40	70	122	210	92
1	9	16	29	50	89	156	84
2	9	14	25	46	82	143	82
6	8	13	23	40	67	110	64
5	4	8	16	31	54	91	56
7	6	11	19	32	54	87	32

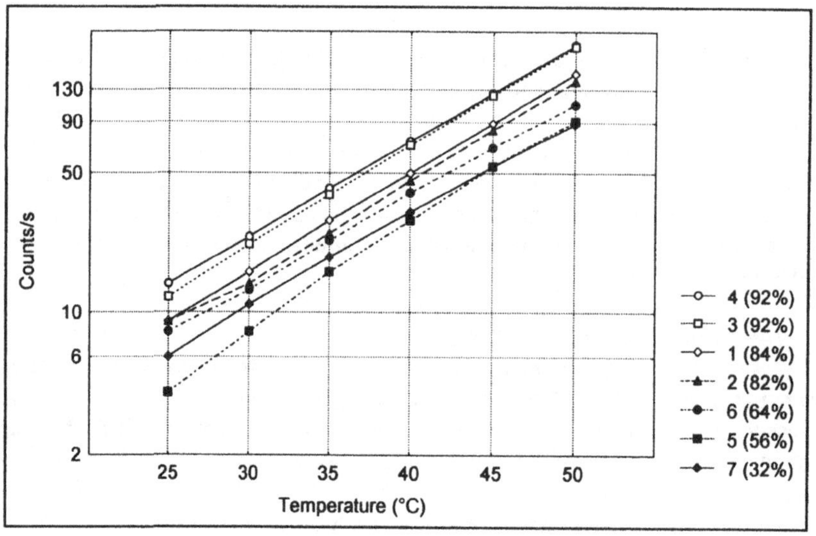

Figure 2. Temperature dependence of the spontaneous photon emission from seven wheat samples (logarithmic scale).

From Table 3 a second rule (rule 4) can be derived describing the relationship between photon emission and biological germinative ability (g):

Rule 4 (on the relationship between biological germinative ability and spontaneous photon emission):

> The spontaneous emission (n) of viable seeds at a given temperature leads to an ordering of seed samples $n_1 < n_2 < ...$ which corresponds to the respective order according to their biological germinative ability (g) in percent: $g_1 < g_2 < ...$

$$n_1 < n_2 < ... \longleftrightarrow g_1 < g_2 < ... \tag{4}$$

Rule 4 is only valid at fixed temperatures within a range where the germinative power of the seeds is not affected. Certainly, temperatures between 25°C and 50°C are not damaging the germinability of wheat (TEUBNER,1983). Although the spontaneous emission of seeds is still very low in this temperature range it is significantly higher than the background noise.

Another aspect of rule 4 is its probable independence from the varieties of wheat compared. This may indicate that climatic conditions are more important in their effect on spontaneous emission than slight genetic variations. But the database is too small to really substatiate such a conclusion.

Because photon emission is an extensive quantity which depends on the number of seeds, while biological germinative ability is an intensive quantity, which is independent of the number of seeds, the application of rule 4 requires that the number of seeds in the samples be fixed.

2.3.2. *Germinability and Spontaneous Emission*

Rule 4 may be incorporated into the temperature dependence of the emission as described by rule 3, because the pre-exponential factor B has to be proportional to the biological germinative ability (g) for rule 4 to hold:

$$B = b\,g \tag{5}$$

Thus our observations support the hypothesis that B contains a factor g which is indicative for germination capacity.

The proportional constant b depends in a non-linear way on the number of seeds (see section 2.4 for a discussion of the non-linear effects due to wave interference). For the seven durum wheat sample by measuring 10g of seeds an average value of $b = 4 \pm 1$ counts/sec was found, indicating a similar level of performance of the measured seven samples.

A similar level of performance is further a prerequiste for rule 4 to be valid.

By inserting equation (5) into equation (1) we arrive at a general relationship between photon emission(n) and biological germinative ability (g):

"Emission law" $$n = g\,N(T) \tag{6}$$

with $0 < g < 1$ and $N(T) = b \exp(C\,T)$

The relationship between germinative ability and spontaneous photon emission allows a non-invasive determination of the relative germinative ability of seeds.

2.3.3. *Water Content of Seeds and Photon Emission*

In order to find out how the history of the seeds, especially their water content, affects the parameter C, an experiment with seed samples of different water content (10.48 to 12.82%) has been performed. In the experiment the stepwise heating from 25 to 50°C was repeated a second time, with the second sequence immediately following the first one. This experiment used 10 wheat samples (triticum aestivum) from Finland. Even in samples with similar water content the parameter C of the photon emission changed between the two sequences pointing to the necessity for uniform pre-drying to arrive at consistent results. In the first heating sequence, the parameter C had an average value of C = .1003, while in the second sequence the average value of C decreased to C =.0896, corresponding to a decline of 10%.

Measurements with the induced photon emission of wheat seeds confirm that only after a short uniform drying (e.g. for 16 h at about 50°C) the correct vitality can be determined. Otherwise the vitality may be masked by non-uniform water content. This has become an important methodical principle in the application of photon emission measurements to determine seed vitality.

The relationship between photon emission and germinability may also be masked by differences in vigour. While for the durum wheat samples from Italy the difference in variety was not important, the wheat from Finland followed rule 4 only for samples of the same variety, because of the big difference in vigour of the varieties.

2.4. EFFECT OF HYDRATION AND SUBSEQUENT DEHYDRATION OF SEEDS

Another way of studying the state of rest is by awakening it, with a short uptake of water and then returning it to rest by drying the seeds (see Figure 3 for details).

2.4.1. *Awakening of Seeds*

Measurement of the spontaneous emission of vital seeds that imbibed water for about 1/2 hour shows a marked increase in the emission through hydration (Table 4).
It is important to note that contrary to spontaneous emission induced emission markedly decreases during the phase of imbibition.

Table 4 shows that both warming up and hydration of seeds increases photon emission. Even though the detailed mechanisms behind these two processes are different the overall effect is the same, as will become clear through the theoretical interpretation.

2.4.2. *Returning to Rest*

After imbibition seeds of the wheat variety Contra are warmed up to 50°C and kept at this temperature in the measuring chamber for 3½ hours. The change in the photon emission during this drying process is shown in figure 4. For comparison figure 4 also shows the effect of the same drying procedure on seeds which had been killed by autoclaving.

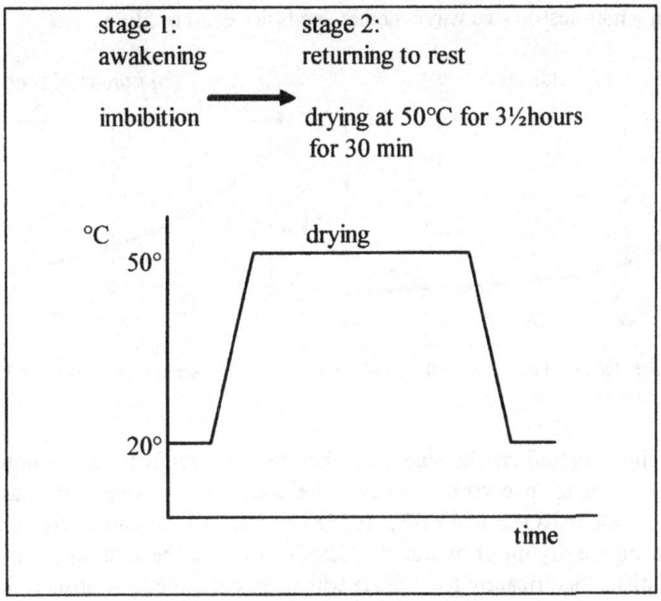

Figure 3. Stages of measurement on seed hydration followed by dehydration: awakening and returning to rest.

TABLE 4. Effect of a 30 min water uptake of wheat seeds (triticum aestivum, variety Contra) on spontaneous photon emission

	vital seeds		
	dry	hydrated	
20°C	16	80	counts/sec
50°C	109	400	counts/sec

The effect of the drying process on the photon emission of hydrated seeds shows that the state of order of vital seeds differs significantly from the state of order of dead seeds.

Figure 4a demonstrates that the spontaneous emission of vital seeds decreases hyperbolically, while that of dead seeds decreases exponentially in time during drying. The "negation" of the state of rest through imbibition reveals the fundamental nature of seeds: whether they behave coherently (vital seeds) or not (dead seeds). Therefore the photon emission pattern accompanying the drying of hydrated seeds may be used to determine in a non-destructive way whether seeds are dead or alive.

(a) vital seed (b) non-vital seed

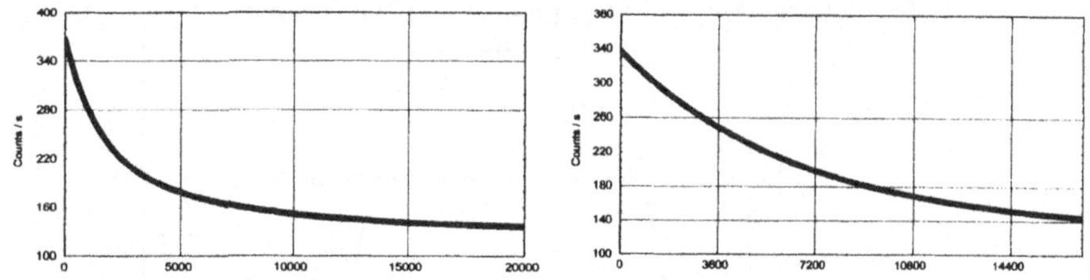

Figure 4. Drying of hydrated vital (a) and non-vital (b) wheat seeds kept at 50°C for 3½ hours.

This finding contradicts the statement that the first, rapid phase of imbibition is a purely physical process occurring in exactly the same way in living and dead seeds (e.g. .MAYER & POLJAKOFF-MAYER, 1975). Rather we conclude from the photon emission during the drying of hydrated seeds that even in the first stages of imbibition living seeds differ significantly from those which are dead. Several observations support this conclusion:

Seasonal rhythmicity in the imbibition rate has been observed with bean seeds. SPRUYT et al (1987) showed that the control mechanism for the water uptake exactly follows the lunar cycle. Rhythmic behaviour has also been observed by NEUROHR (1992). According to NEUROHR the circa-annual rhythm in the water uptake of beans is in agreement with the yearly rhythm of the induced emission observed with cress seeds indicating a dependence of the germination control system on external rhythms. This conclusion is supported through the theoretical analysis (see sections 3.3.3. and 3.4.1.).

The hyperbolic decrease of the induced photon emission during drying of vital seeds points to the vacuum state of a coherent field as the control system of the reversible initial phase of seed hydration. This invites a deeper look into the vacuum structure of seeds (see section 3).

2.5. EFFECT OF GAMMA IRRADIATION ON THE SPONTANEOUS PHOTON EMISSION OF CEREAL SEEDS

In 1990/91 several experiments were performed in our laboratory to study the effect of gamma (and beta -) irradiation on the spontaneous emission of cereal grains. Even though the main purpose of these investigations was the deteriorating short and long-term effects of those irradiations, the results also have significance for understanding the vitality of seeds.

One knows today that "cereal grains bombarded with large doses of gamma rays to induce chromosome breakage still may germinate and develop into seedlings ("gamma plantlets")...Germination and early seedling establishment can therefore occur without cell division" (BEWLEY and BLACK, 1984).

It is already well understood and widely accepted that the first stage of germination which culminates in radical emergence (the basis for the biological definition of the germinative ability) is not controlled by DNA but rather by a control mechanism which developed on the mother plant during maturation.

"But what is the nature of this control? What specific events are essential for radical elongation to commence, and what internal signals are responsible for setting this process in motion?" are the questions asked by BEWLEY and BLACK (1984) and they confess that "definite answers cannnot be given to these fundamental questions".

The photon emission during the drying of hydrated seeds already was pointing to the vacuum state as the main source of order in seeds and - as will be clear in section 3.3- the effect of gamma irradiation on seed emission further confirms this interpretation.

The short-term effect of irradiation treatments of different cereal species is displayed in Table 5 which presents the mean values of the measurements performed on oats, rye, barley, rice and wheat grains.

TABLE 5. Spontaneous emission (counts/sec) of different cereals after gamma irradiation of different intensities (Gray)

Cereal Species	0 Gray	1 Gray	10 Gray	250 Gray	(IK Gray)
Rice	52	106	47	--	(168.8)
Oats	24	40	--	17	
Barley	14	26	11	10	
Wheat	10	13	4	--	
Rye	3	4	2	3	
Emission	low	high	low	low	(high)

This response pattern is confirmed by extensive experiments using rice grains and irradiation intensities up to 20 KGray. In the latter case it was found that higher intensities than 250 Grey give rise to a dramatic increase of biophoton emission. (The emission values for the irradiation of rice with 1KGray is included in Table 5).

The pattern of the induced emission after irradiating the same cereals further confirms the above findings with spontaneous emission by showing a complementary pattern: maxima (minima) in the spontaneous emission correspond to minima(maxima) in the induced emission. However, the sequence of the cereal species remains the same after gamma irradiation both for spontaneous and induced emission.

The characteristic dependence of the spontaneous emission of seeds on the irradiation intensity can be interpreted as follow:

First there is an increase in emission because the energy input is felt as a small perturbation only. From a certain critical value of energy changes in the internal structure takes place culminating in a new stable state. A further increase in energy is then again felt as a small perturbation resulting in increased emission.

This response pattern which follows the pattern described by the ARNDT-SCHULZ-law can by understood by the classical model of a forced, damped oscillator (Popp and Löliger, 1989). Seen quantum mechanically the dispersion-like dependence of the photon emission on gamma irradiation intensities can be taken as an indication of the superfluid nature of the biophotons (as will be elaborated in section 3.3).

2.6. INDICATIONS OF COLLECTIVE COHERENCE IN SEED LOTS: STORAGE CAPACITY AND LONGEVITY

The experiments discussed so far are indicating that seeds are in a coherent state of rest:

(a) the reaction of seeds towards a moderate heating(emission law) can be related to the ability of seeds to maintain their coherent state under thermal excitations (as explained in section 3.1).
(b) the reaction of seeds to hydration (awakening) and subsequent dehydration (returning to rest) as expressed in the relaxation pattern of the photon emission indicates in the case of hyberbolic relaxation a state of coherence.
(c) the reaction of seeds to irradiation with gamma rays follows a pattern typical for systems dominated by quantum coherence (as explained in section 3.3).

In fact, all the experimental results with ultra-weak photon emission accumulated so far fit well with the assumption that the measured biophotons by coming from a coherent source are coherent as well (POPP, GU, and LI, 1994).

An important consequence of this assumption is the effect of wave interference. Constructive or destructive interference may be used to define and test the state of coherence of a "population" of molecules, cells or organisms. The respective paradigmatic experiments were done by GALLE (1993) who discovered an interference-like dependence of the photon intensity on the number of individuals in a daphnia population. The biophoton intensity displays minima and maxima dependent on the density of the animals in the population. When the population showed patterns indicating "superorganismic collective states" it correlated with minima of photon emissions which may be explained by destructive interference.

The three cereal experiments described above (section 2.2. -2.4.) relate to the same amount of grain (about 10g). One might assume that the photon emission of a seed lot is a linear function of the number of seeds, but in fact seed lots show a non-linearity due to collective effects.

By following HARRINGTON, the term "lot" is taken from the seed trade where it is used to identify a group of seeds to which an identifying number is given so that it is possible to retrace the history of a particular lot of seeds through the records of the companies that have handled the seeds. A lot may consist of many tons of seeds or be only a few grams. A lot may consist of genetically uniform seeds or may be a blend of several harvests of a variable cultivar. The lot may contain only uninjured fully mature seeds, or seeds of variable maturity, variously injured and containing inert matter and weed seeds from improper cleaning (HARRINGTON, 1972).

Now several experiments are summarized which indicate a non-linear dependence of photon emission on the number of seeds in a lot:

(1) The induced photon emission from wheat, oats, rice and spelt samples were investigated comparing in each case two varieties: "good" quality grains and "poor"

quality grains. The induced light emission after stimulation with white light plotted as a function of the mass of the sample showed a clear difference between poor quality grain and good quality grain: While the poor quality grain show a clear linear dependence of the induced emission on sample mass the good quality grain gives an emission pattern largely independent of the mass (LÖLINGER, 1989).

(2) The comparison of the induced light emission of two lots of cress seeds (Lapidum sativum) with an age difference of one year further confirmed the previous finding. From each of the two seed lots 2.5g , 5g , 7.5g and 10g where measured. The results show a clear difference in the mass dependence of induced emission for the old and new cress seeds (POPP, 1990; NEUROHR,1992).

(3) A further study compared the mass dependence of emission of a normal charge of maize with a charge containing 30% artificially de-vitalized seeds. It was observed that for the fully viable charge both spontaneous emission and induced emission per mass remained nearly constant (with a tendency to decrease) with increasing mass. The maize charge with 30% autoclaved seeds (giving a 30% decrease in germinability in the biological germination test) had less non-linearity in the mass dependence of induced emission as compared with the fully viable maize lot (POPP, 1990).

Figure 5 summarizes the outcome of the three experiments in a schematic way (adopted from LÖLINGER, 1989).

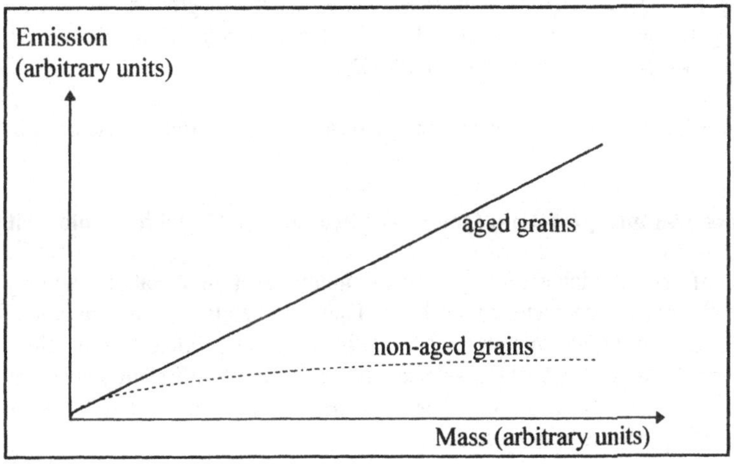

Figure 5. Mass dependence of the induced emission from aged and non-aged grains in arbitrary units (LÖLINGER, 1989).

All these experiments show that the photon emission of a seed lot reflects the loss in vitality of even a small percentage of seeds in the lot.

This supports the traditional practice of removing poor-quality seeds from the lot prior to storage. Once a lot of fully mature seeds has been cleaned and only sound seeds of high density remain in the lot, then through proper storage conditions vitality can be maintained. But at present in the literature there are conflicting statements on the longevity potential of seeds. So the longevity of seeds requires further research.

The principles of collective coherence (SEWELL, 1986) notably the N^2-effect applied to seed lots may shed new light on the question of seed longevity and storage capacity:
As soon as a large number (N) of vital seeds forming a lot are coupled in a way that the intensity of their radiation is proportional to N^2, then the square root of N seeds determines the coherence of the whole lot ("square root of N"-Law).
But the same law describes also the coherence reducing influence of a few de-vitalized seeds on the whole lot as the above mentioned experiments indicate. Thus both creation and destruction of collective coherence follows the same law.

Already collective effects in seed lots are described in the literature on the preservation and storage of grains and seeds (MÜNZING, 1993; MÜNZING, 1997). The principles of collective coherence may add now a completely new dimension to the problem of seed storage and longevity.

The coherence of a seed lot has t w o basic effects:

(a) it maintains the integrity of the individual seeds and thereby promotes longevity,
(b) the coherence creates a protective armor against deteriorating environmental influences (MEISSNER effect).

According to BOGOLIUBOV a macroscopically occupied ground state of N weakly interacting Bosons which is created and annihilated by virtue of the "the square root of N"-law, has a superfluid character (SOLOMON, 1971).

This further substantiates the assumption of a superfluid ground state of seeds.

3. Theoretical Interpretation: Quantum Vacuum as a Model for Biological Rest

The state of rest of viable seeds is the minimum level of metabolic activity which maintains the holistic character of seed life. The seeds` state of rest is assumed here to be correlated with the vacuum state of the radiation field coupled to biological matter. The task is to find out whether the vacuum state can have a sufficiently rich structure to account for the vitality of seeds. The analysis of the vacuum state as a model for biological rest proceeds in three steps:

1. The vacuum zero-point energy as meeting point of quantum mechanical and thermodynamical behaviour of seeds.
2. The fourfold nature of the gap between quantum mechanical and thermodynamical behaviour.
3. The superfluid vacuum integrating material structure and the radiation field.

By virtue of these three approaches the major aspects of biological rest are related to different functions of the vacuum state.

3.1. QUANTUM VACUUM AS THE JUNCTION POINT OF THERMODYNAMIC AND QUANTUM MECHANICAL BEHAVIOUR

The quantum mechanical ground state is always the lowest state in the energy level "ladder" to which an assembly of particles can descend. Such a ladder structure is not only typical for atoms and molecules, where it has been discovered, but also can be assumed to be applicable to living organisms (SITIKO and GIZHKO, 1991). This assumption is confirmed by the experimental discovery of characteristic eigenfrequencies of the human body in response to frequency changes of external electromagnetic fields in the mm band range (SITIKO et al, 1988).

In principle the allowed eigenstates of a system are obtained by solving the time-independent SCHRÖDINGER-equation. The transitions of particles or quasi-particles (collective excitations) from the ground state to excited states and vice versa creates a certain distribution of the particles among the available energy levels. A natural measure of the degree to which particles have left the ground state is the ratio between the total number of particles N to the number of particles which occupy the ground state (n):

$$Z = \frac{N}{n} \tag{7}$$

This ratio is called in statistical physics "partition function" and measures the extent to which particles occupy excited states and therefore are more easily subject to thermal excitations. The partition function is a pure number, which ranges from unity when all the particles are in the ground state to an infintitely large number when increasingly fewer particles remain in the ground state:

$Z = 1$: all the particles are in the ground state (N=n)
$Z > 1$: particles have left the ground state (N>n)

The fundamental role of the partition function is due to its relation to the "path integral" over all possible configurations of a system (FEYNMAN and HIBBS, 1965). The path integral provides an unified view of Quantum Mechanics, Field Theory and Statistial Mechanics. In quantum field theory the path integral is connected to the vacuum persistance functional which measures the stability of the vacuum state towards perturbations (GREINER and REINHARDT, 1993).
The reciprocal of the partition function

$$g = \frac{1}{Z} = \frac{n}{N} \tag{8}$$

then measures the degree or tendency with which particles remain in the ground state.

When the ground state of the system has the necessary kind of coherence allowing for germination, g can be identified with the germinative ability.
At this point this is an assumption which will be further substantiated in the following sections.
By virtue of the above assumption:

g=1 represents the highest germinative ability because it relates to a coherent ground state situation.
g=0 represents a situation far away from the ground state.

Thus, in terms of the partition function high germinative ability is a ground state phenomenon and related to spontaneous symmetry breaking. When through aging part of the seed structure has become separated from the coherence of the ground state thermodynamic aspects of behaviour become increasingly important and this reduces the germinability.

If n are the photons related to the ground state then the difference N - n can be interpreted as the photons related to excited states. Because of N - n = (1 - g) N the germinative ability can also be expressed through the ratio of these two kinds of photons:

$$g = \frac{1}{1 + \frac{N - n}{n}} \quad \text{or} \quad \frac{N - n}{n} = \frac{1 - g}{g} \qquad (9)$$

From the thermodynamic free energy (F)defined by

$$F = - k\,T \ln Z = k\,T \ln g \qquad (10)$$

and related to the internal energy E and the entropy S by

$$F = E - TS \qquad (11)$$

and with the assumption that the photons n which are measured are related to the internal energy E by exp(E/kT) it follows that

$$n = g \exp (S/k) \qquad (12)$$

This is just the empirically found temperature dependence of the photon emission from seeds, if in addition we set:

$$S/k = C\,T \qquad (13)$$

The conclusion is:

the temperature dependence of the photon emission simply describes the tendency of the system to remain in the ground state if it gets subjected to thermal excitations in the physiological range.

An entropy proportional to T means that seeds in their reaction to heat behave as if the temperatures (up to 50°C) are low ("near" absolute zero) and therefore quantum behaviour is dominating (GASSER and RICHARDS, 1974). This also is in accordance with the observation that seeds can be kept for days at temperatures near absolute zero without losing their viability (BÜNNING, 1953).

When entropy is proportional to T, then the heat capacity for reversible processes is also proportional to T:

$$\text{heat capacity} = \frac{\partial E}{\partial T} = T\frac{\partial S}{\partial T} = C\,k\,T \tag{14}$$

This again means that only the few particles which can reach empty energy levels in the range of kT near the ground state are excited by warming up seeds. The situation is analogous to the electron contribution to the heat capacity of metals (see KITTEL, 1971).

The emission law in the interpretation just given further says that the germinative ability is completely determined by the ratio of the photons in the ground state to all the photons existing at a certain temperature $T < 50°C$. The spontaneously emitted photons are according to this picture the photons "leaking out" from a state we label the "ground state".

This "leaking out from the ground state" warrants some remarks, because two kinds of mechanisms can be considered here:

(1) Spontaneous emission in the usual sense is only possible from states which differ from the zero energy state. The ground state of biophotons as described above must therefore be separated from the "true" zero-energy ground state by an energy gap. This energy gap has to be large enough to account for the observed photon emission in the visible range of the spectrum (section 3.3).

(2) Leaking out of photons through macroscopic quantum tunneling is another possible mechanism. Macroscopic quantum tunneling is a necessary but not a sufficient condition for macroscopic quantum coherence (LEGGETT, 1986).

In both cases different "ground states" are distinguished which are separated by an energy barrier.

The shifting between the ground states of different energy is done by a processs called renormalisation, which is illustrated by the following argument showing the equivalence between the quantum mechanical ground state and zero temperature state: In terms of the energy eigenstates of the Hamilton operator H the partition function is

$$Z = trace\ exp\left(-\frac{H}{kT}\right) = \sum_i exp\left(-\frac{e_i}{kT}\right) \tag{15}$$

e_i are the energy eigenstates with the label i counting the states starting from the ground state energy for which i=0. If the ground state energy differs from zero by a certain amount d, we can eliminate this energy by setting $E_i = e_i - d$ so that $E_0 = 0$. It follows

$$Z = exp\left(-\frac{d}{kT}\right)\sum_i exp\left(-\frac{E_i}{kT}\right) \quad (16)$$

By inserting (16) into (11) the free energy reads:

$$F = d - kT\,ln\sum_i exp\left(-\frac{E_i}{kT}\right) \quad (17)$$

The ground state energy d separated from the partition function by way of renormalisation has the meaning of a "zero-point energy" (GODNEW,1963).
For $T \rightarrow 0$ follows

$$F(T=0) = E(T=0) = d \quad (18)$$

This result invites two interpretations of the zero point energy: It is the energy of the quantum mechanical ground state F(T=0) and it is the energy of the system at absolute zero temperature E(T=0).

In this sense the ground state of a system is the meeting or junction point of quantum mechanical and thermodynamic behaviour:

At T=0 the system occupies only the ground state.

This is the basis of the Third Law of Thermodynamics (WILKS, 1961) the universal principle of creating order by reducing random excitations, which allows for quantum effects operating even on a macroscopic scale.

As already recognized by SCHRÖDINGER (cited by COCHRAN, 1971) and further elaborated by POPP (1984) the Third Law of Thermodynamics plays a fundamentel role in understanding biological systems because of their ability to maintain and even increase order in spite of the disorderly (entropic) influences from the environment.

3.2. FOUR-FOLD NATURE OF THE GAP BETWEEN THE QUANTUM AND THERMAL BEHAVIOUR OF THE RADIATION FIELD

The role of the state of least excitation in the response of seeds to a moderate heating can be described in a more detailed way by taking recourse to the mechanics taking place in the gap between the ground state of the radiation field and its thermal excitations. The gap mechanism presented here was first proposed by POPP in order to understand the temperature dependence of photon emission from seedlings (SLAWINSKI and POPP, 1987). The mechanism is based on the distinction between the ground state of the radiation field (supposed to represent a BOSE-condensate) and the thermal excitations of the BOSE-condensate. This may also be described as a balance between "coherent" and "incoherent" photons which are denoted in this section by n_c (= n) and n_i (= N-n) respectively. The gap between the quantum and thermodynamic reality then has four dynamical components (see also Figure 6):

(1) The ground state of the radiation field described by a temperature, T_c, which characterizes the phase transition to the BOSE-condensate of photons. When the BOSE-condensation is connected with a population inversion as in a Laser, then T_c is a negative temperature (SEWELL,1986).
(2) The point of balance between BOSE-condensation and thermal dissipation. The thermal excitations are characterized by the absolute temperature (T).
(3) The chemical potential $\mu(n_c)$ of the coherent photons to remain in the coherent state.
(4) The chemical potential $\mu(n_i)$ of the incoherent photons to emerge from the coherent state.

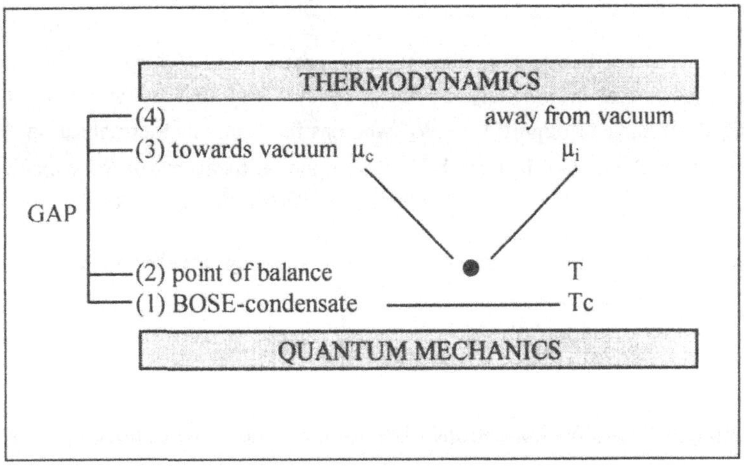

Figure 6. The four-fold nature of the gap between quantum mechanical and thermodynamic behaviour.

3.2.1. *Information-Entropy of the BOSE-Condensate*

"Information-entropy" is a term introduced by HAKEN(1977) to denote the use of statistical entropy as a measure of the "theoretical possible information". The statistical entropy S is defined by $S = k \ln W$ with W the non-normalized probability of all the "realizations" or "possible events". Theoretical possible information is not information in the usual sense, for example when biologists speak about the information content of the DNA. Such real information relates to a certain "meaning" be it semantic or pragmatic. Theoretical possible information on the other side is "potentiality of information" and the statistical entropy is then a measure for "all possibilities " inherent in the BOSE-condensate.

As has been shown, for instance, by FEYNMAN (1975), the probability of BOSE-condensation increases with $n_c!$, where n_c is the number of coherent photons. The probability of condensing a photon coherently into a definite microstate of energy hν is therefore proportional to $n_c(h\nu)!$, where nc is the number of coherent photons of energy hν. On the other hand the probability of thermal dissipation follows the Boltzmann factor exp(-hν/kT). Since, as an average, one photon of energy hν decays thermally into n_i = hν/kT "thermal" quanta of energy kT, the term hν/kT of the Boltzmann factor can be substituted just by this average number n_i of incoherent photons into which a coherent photon of energy hν decays thermally.

For the known laws of probability theory of balanced multiplicative processes, the total probability takes the form

$$W(h\nu) = W_i W_c \tag{19}$$

where W_i describes the incoherent part of the process and W_c the coherent one. The boundary conditions of (14) are obvious: In the case that no coherent photon is thermalized, we have to require $W = W_c$, whereas for complete thermalisation $W = W_i$. Therefore W_i and W_c can, in turn, be looked upon as measures of the capacity of the system for coherent photon storage and thermal dissipation, respectively.

Apart from terms which are not essential for understanding, we obtain then

$$W(h\nu) = exp\left(-\frac{h\nu}{kT}\right) n_c! = exp(-n_i) n_c! \tag{20}$$

The "unbounded" information-entropy of the BOSE-condensate of photons is finally

$$S = k\,(ln\, n_c! \; - n_i) \tag{21}$$

3.2.2. *Balance between BOSE-Condensation and Thermal Dissipation*

The mechanism for the transition from coherent(c) to incoherent(i) photons can be formulated in terms of entropy (S) in the following way:

$$dS_{ic} = -\frac{\mu_i}{T} dn_i - \frac{\mu_c}{T_c} dn_c \tag{22}$$

where μ_i and μ_c are the chemical potentials of the incoherent and coherent photons, respectively. That a chemical potential can be assigned to thermal (incoherent) photons follows from the provision that the thermal photons under consideration originate from the coherent source.

The essential assumption now is that the transition from coherent to incoherent photons balances the entropy of the system in a stationary state such that

$$d\,S_{ic} = 0 \tag{23}$$

The effect of (23) is a homeostatic regulation producing no entropy.
From (22) and (23) then follows

$$n_i = -\left(\frac{\mu_c}{\mu_i} T_c\right) \cdot T\, n_c \tag{24}$$

where we fixed the number $\mu_i = 0$ at $T = 0$, in agreement with the Third Law of Thermodynamics. It is noted that

$$\left(\frac{\mu_c}{\mu_i T_c}\right) < 0 \tag{25}$$

because either T_c is negative (LASER-Model) or μ_c is negative (spontaneous BOSE-condensation).

Note also that for the limiting case $\mu_c = 0$ equation (22) requires $\mu_i = 0$ which describes the thermal equilibrium.

The stationary requirement (23) of course also holds for the information-entropy of the BOSE-condensate:

$$d\,S \quad = k\, d\, ln\, W = 0 \quad (for\ varying\ n_c) \tag{26}$$

This condition means that only the maximal number of realisations is relevant.

After insertion of equation (24) into (21) and calculating the change in information-entropy with changing n_c one obtains:

$$\frac{d\, ln\, n_c\,!}{d\, n_c} = C\, T \quad \text{with} \quad \frac{\mu_c}{\mu_i T_c} \tag{27}$$

As a result we note that the constant C is completely determined by the fourfold nature of the gap between the quantum mechanical and thermodynamic behaviour of the radiation field.
By applying STIRLING's formula in the approximation $l\,n\ n! = n\ ln - n$ valid for $n > 100$, (27) becomes:

$$ln\, n\ = C\, T\ \ or\ \ n = exp\, C\, T\ \ or\ \ n/exp(C\, T) = 1 \tag{28a,b,c}$$

Because of the structural agreement of the ratio (28c) with the empirical emission law (6) as well as with its partition function interpretation (12) the ratio may be identified with the germinative ability g, being one in this case: $g=1$.
The equations (28a.b,c) then say that a large number of coherent photons indicate maximal germinative ability.

If the number of coherent photons decreases, other approximations of STIRLING's formula have to be applied which result in a value for g smaller than one : $g < 1$.

According to this analysis based on the nature of the gap between quantum mechanical and thermodynamic aspects of the radiation field, the decrease in the germinative ability is related to a decrease in the number of coherent photons.

Furthermore, the factor exp(C T) in the emission law is solely a consequence of the fourfold gap mechanism describing the relationship between coherent and thermal raditation.

3.3. SUPERFLUID VACUUM STRUCTURE OF SEEDS

In a complete description biophotons cannot be separated from the underlying material structure even though with the formation of molecular matter the total electromagnetic field has split into radiation field and an electrostatic field with the latter being responsible for the formation of atoms and molecules (COULOMB gauge). The radiation field which is forming a sea of virtual and real photons is still accompanying ("dressing") every molecular state (PFEIFER, 1981). "Even a so-called free molecule is interacting with the rest of the world by the unavoidable coupling with the electromagnetic radiation field. Due to the peculiar nature of this coupling, the effect of this coupling can be of qualitative importance." (PRIMAS, 1981) Because of the highly complex energetic structure of biological macromolecules the interaction with the radiation field has far reaching consequences as revealed by research on the ultra-weak photon emission of biological matter (for a recent review see POPP, GU, and LI, 1994).

The basic experimental results on the ultra-weak photon emission from biological matter accumulated so far have led to the assumption that biphotons are in a "trapped" or "condensated" state by way of exciplex formation (POPP, GU, and LI, 1994). Because the exciplex bound photons are coupled with the vibrations of the underlying molecular structure (phonons) in addition elementary excitations ("polaritons") exist (LI, 1992).

The observed "dispersion-pattern" of the photon emission due to gamma irradiation of seeds (see section 2.5) and the indications of collective coherence in seed lots (see section 2.6) favour the hypothesis:

> Biophotons of seeds are elementary excitations of a BOSE-condensate of photons being in a "superfluid" state.

The superfluid state is usually described in a "graphic" way by two inter-penetrating "fluids" (LEIGHTON, 1959, ENZ, 1974): the "superfluid" condensate having no entropy and the "ordinary fluid" or "first fluid" of elemenatary excitations which carries

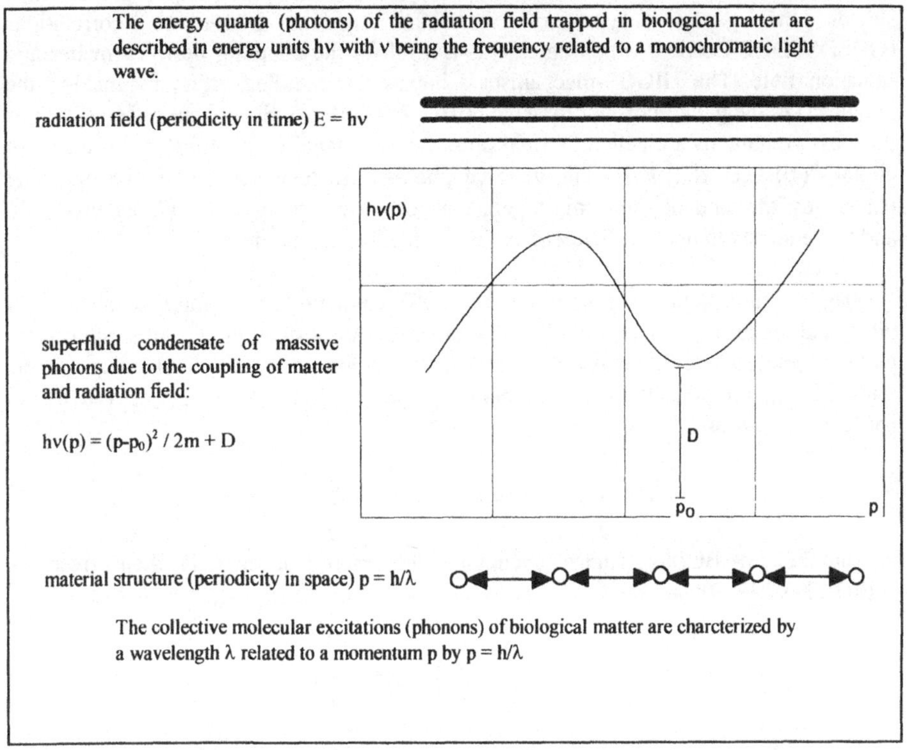

Figure 7. Assumption of a superfluid radiation field by virtue of a dispersive coupling to its own molecular structure ("self-interaction" in the COULOMB-gauge) in schematic representation.

entropy. The ordinary fluid thereby is considered as a wave propagation of temperature and entropy excitations quite different from the ordinary diffusion propagation by heat (ENZ, 1974).

To theoretically substantiate the hypothesis of a superfluid radiation field, four different aspects have to be clarified: first (3.3.1.) the general conditions under which a BOSE-condensate of photons may exist in seeds and then (3.3.2.) the elementary excitations present in the photon-condensate have to be described, which turn out to be candidates for the observed biophotons. Thirdly (3.3.3.) the conditions of a superfluid state of photons have to be specified which integrate the photonic condensate with its excitations into one coherent state. To be able to verify the superfluid hypothesis of seeds, finally (3.3.4), a relationship has to be postulated between the basic parameters of the superfluid excitations (biophotons) and the observable aspects of seed vitality.

3.3.1. *BOSE-Condensation of Photons in Biological Matter*

Inside matter, photons in general have a non-zero mass(PREPARATA, 1995, ENZ, 1997a). Without taking recourse to any molecular mechanism, the non-vanishing, effective mass of photons can be understood by means of a HIGGS-mechanism which shields the gapless, symmetry restoring, bosonic long-range correlations (GOLDSTONE-Bosons) which emerge as a result of the coupling between matter and radiation field. The HIGGS-mechanism achieves its shielding effect by making the photons (i.e. the gauge bosons) massive (JIBU, et al, 1996). Seen from different angles the massive photons are called "virtual photons" or "tunneling photons" or "evanescent photons"(JIBU,et al. 1996). The massive photons are then subject to two opposing tendencies: the tendency to come together in the lowest energy state ("condensation") and the tendency to occupy different excited states ("evaporation").

The starting point for describing the BOSE-condensation of massive photons in biological matter is the well established statistical law for distributing indistinguishable particles obeying the BOSE-EINSTEIN-statistics among a given set of energy levels (labled by j). Provided the total number of particles N and the total energy E of the particles are constants

$$N = \sum_j n_j = const. \quad and \quad E = \sum_j e_j = const. \tag{28}$$

the number of BOSE-particles occupying the energy level j is then given as (LEIGHTON,1959):

$$n_j = \frac{g_j}{exp\left(\mu N / kT\right) exp\left(e_j / kT\right) - 1} \tag{29}$$

g_j is the "degeneracy", i.e. the number of possible distinctions, of a given energy level. For the sake of simplicity the g_j are assumed here to be one for all j: $g_j = 1$. But in view of the actual germination process the degeneracy is very important, because it allows for spontaneous symmetry breaking which is equivalent to a removal of degeneracy.

Different molecular mechanisms have been proposed to explain BOSE-condensation of photons. According to POPP et al (1994) the photons may be trapped in biological matter by exciplex formation. The photons bound as exciplexes are in a "dressed" state, so that a certain finite mass can be assigned to them. Furthermore it can be assumed that the total number of photons N is fixed by the number of exciplexes.

As required by the Third Law of Thermodynamics all excitations are in the ground state at T=0. The number of particles in the ground state $N_o = N(T{=}0)$ is then described by:

$$N_o = \frac{1}{exp\left(\mu N\right) - 1} \quad or \quad \mu N = ln\left(1 + \frac{1}{N_o}\right) \tag{30}$$

For all particles to be in the ground state ($N_o \rightarrow \infty$) as it is the case for BOSE-condensation the chemical potential μ has to approach zero $\mu \rightarrow 0$.

But because N_o has a singularity for $\mu \rightarrow 0$ it is necessary to treat the lowest energy level separately from the others:

$$N(T) = N_o + \sum_{j>0} \frac{1}{exp\left(\mu N + e_j\right) - 1} \qquad (31)$$

A critical temperature T_C can then be defined (WILKS, 1961)

$$T_c = \frac{h^2}{2\pi m k}\left(\frac{N}{2.612 V}\right)^{3/2} \qquad (32)$$

which describes a situation where the particle waves do not overlap anymore and where there is equality between thermal and kinetic energy (ENZ, 1997a).
The photons of the non-condensated, "gaseous" state

$$N_{gas} = N - N_o \qquad (33)$$

are then given by

$$N_{gas} = N\left(\frac{T}{T_c}\right)^{3/2} \qquad (34)$$

and condensated photons N_O are described by

$$N_o = N - N_{gas} = N\left[1 - \left(\frac{T}{T_c}\right)^{3/2}\right] \qquad (35)$$

According to (35) all particles are in the ground state (BOSE-condensate), when (T/T_C) approaches zero. Certainly this is true for $T \rightarrow 0$. But this condition is not of interest here, because the photon condensate in biological matter should exist also in the physiological temperature range.

However, T/T_C also approaches zero when T_C is very large. According to (32) this situation is realized in the following cases:
(a) reduction of the volume V,
(b) increase of the particle density N/V,
(c) very small mass m.

These conditions are present in seeds because of their small volume and high density and because the photons are trapped as exciplexes. So BOSE-condensation may exist in biological matter by virtue of the very high critical temperature T_C. A high critical temperature for photon-condensation is in agreement with the thermodynamic model developed by POPP(1984), who suggested a condensation temperature T_C as high as the sun temperature to explain biological photon condensation. Also JIBU et al (1996) estimated a very high critical temperature for the BOSE-condensation of massive photons by studying the interaction between radiation field and the electric dipole field of intra- and extracellular water.

The salient point here is, that condensated photons may exist in biological matter at a finite non-zero temperature provided the photons have a finite mass (ENZ, 1997a).

In the BOSE-condensate the individual photons(oscillators) are in a coherent state but further specification is needed with regard to the nature of the excitations (section 3.3.2.) and regarding the real ground state of the whole condensate (3.3.3.).

3.3.2. *Excitations in the BOSE-Condensate: Biophotons*
The existence of massive photons due to the coupling between molecular matter and radiation field does not only specify a finite non-zero temperature below which the BOSE-condensate exists but also allows us to formulate general properties of the elementary excitations of the photon condensate which relate to the equality of thermal and kinetic energy at the onset of evaporation (characterized by the critical temperature T_c).

The equality of thermal and kinetic energy leads to the assumption that in the system of tapped photons there may exist at physiological temperatures elementary excitations whose nature can be described by an energy-momentum relationship of the form:

$$E_o(p) = \frac{(p - p_o)^2}{2m} = h\nu(p) - D \qquad (36)$$

m is the effective mass of the trapped photons and D is the energy gap separating the excitations from the condensated state or from a collective coherent ground state as specified in the next section. At this stage D is a function of T_c. The minimal value of momentum p_o is related to the energy gap D.

There is experimental evidence through gamma-irradiation of cereal seeds for such a dispersion behaviour (see section 2.5).

The existence of an energy gap D in the excitation spectrum means that direct transitions between the excitations and the condensate are forbidden. According to LANDAU and LIFSHITZ (1969) such a dispersion behaviour is typical for superfluid Helium; and ENZ (1974) has given arguments that such a behaviour has general significance in being characteristic for a large class of superfluid systems.

Contrary to the fixed number of photons in the condensate, the total number of photonic excitations cannot be assumed to be constant, because they may be created or destroyed at any time.

According to LANDAU and LIFSHITZ (1969) the excitations of the condensate can therefore be approximated by non-interacting and non-conserved bosons. As a consequence the well known distribution law for the boson occupation number n(p) holds:

$$n(p) = \frac{1}{exp(h\nu/kT) - T} \qquad (37)$$

This approximation for n(p) is valid for the long-wavelength thermal excitations of the condensate and can be further simplified if the photonic quasi-particles existing at physiological temperatures are described by (36). Due to the existence of a large energy gap D to the ground state, $h\nu(p) >> kT$ the excitations approximately follow the BOLTZMANN distribution law:

$$n(p) = exp\,(-h\nu(p)/kT) \tag{38}$$

With (36) the distribution law (38) for the quasi-particles reads

$$n(p) = exp(-D/kT)\, exp\,(-E_o(p)/kT) \tag{39}$$

Of course for $p=p_o$ follows $E_o=0$ and the number of excitations at the minimum ($p=p_o$) of the dispersion relation (36) is

$$n = n(p=p_o) = exp(-D/kT) \tag{40}$$

This means: the energy gap D defines the stationary number of photonic excitations of lowest energy, which have two opposing properties:

(1) The excitation ground state photons share the coherence of the BOSE-condensate,
(2) But due to their separation from the true ground state of the condensate they may carry entropy away from or through the system to maintain the collective coherence.

Because of these two properties the photonic excitations can be identified with the observed biophotons.

Both the BOSE-condensate and its excitations exist by virtue of the condition of vanishing chemical potential $\mu=0$. But the meaning of $\mu=0$ is different in both cases: In case of the BOSE-condensate it defines a coherent but singular ground state occupied by all the photons. In case of the photonic excitations $\mu=0$ holds because of the total particle number being undefined as is the case for a photon gas.

The total number of excitations in the vicinity of p_o then is:

$$N = Z\, exp(-D/kT) \tag{42}$$

with $$Z = \sum_p exp\left(E_o(p)/kT\right)$$

or

$$n = g\,N \quad with\ g = 1/Z \tag{43}$$

$Z=1$ or $g = 1$ means all excitations are in the ground state at p_o for which $E=0$. Shifting from D to $E_o=0$ is equivalent to a renormalisation of the ground state energy due to a change in the reference point.

The partition function Z can be approximately calculated by replacing the sum by an integral over p ranging from $-\infty$ to $+\infty$ and using the condition that $p_o^2 >> 2\ m\ kT$ (BRENING,1975).
With the quantum rule of momentum $p = h/\ \lambda$ (LANDÄ, 1965) It follows for the reciprocal of Z :

$$g = \frac{x_m \lambda_o{}^2}{V} \quad with\ x_m = \frac{2\pi^2 h}{\sqrt{2\pi\, mkT}} \tag{44}$$

λ_o being the wave length of the quasi-particles described by (36), m is their "inertial" mass, V the coherenc volume and T the external temperature. Since by definition $0 < g < 1$ the allowed range of the parameters is restricted.
g as reciprocal of the excitation partition function measures the stability of the ground state (of the excitations) towards thermal fluctuations. g equal to one (g=1) then means that the coherence of the excitation ground state n=N is dominating the behaviour of all the particles.

Coherent ground states respond to restrictions by creating new modes of behaviour. This fundamental principle relating coherence and formation of structures is explained in section 3.4.

Therefore g includes in its structure all the parameters related to the basic types of restrictions: collective modes (λ_o, volume or cavity (V), random activity (T), and inertia or damping(m).
Because the product $x_m\lambda_o{}^2$ has the dimension of volume, a "ground state volume" V_o can be defined as:

$$V_o = \frac{2\pi^2 h \lambda_o^2}{\sqrt{2\pi\, mkT}} \tag{45}$$

Equations (44) then has the simple form

$$g = V_o/V \tag{46}$$

The same situation which is described by equation (43) in terms of photons, equation (46) describes in terms of volume.
When all excitations are in the ground state then $V_o = V$ and the coherence extents over the whole volume. But when g decreases also the coherence volume V_o decreases.
The stability of the ground state of photonic excitations can therefore be evaluated alternatively by the photon ratio n/N or the volume ratio V_o/V.

3.3.3. *Collective Coherent Ground State of Biophotons*

Now the collective character of the interaction between matter and the radiation field is included. Thereby a collective coherent ground state of the radiation field is defined which at the same time reveals the integrative role of the energy gap D between the collective coherent and the excitation ground state of photons condensed in matter.

In general a BOSE-condensate need not show the kind of macroscopic coherence that characterizes systems as superfluids or superconductors (PREPARATA, 1995). It is only by virtue of an additional collective effect that a non-zero "order parameter" or "wave function" of the ground state exists allowing macroscopic phase coherence.

To achieve macroscopic phase coherence the coupling between matter and electromagnetic field has to be such that the matter system performs collective oscillations in phase with a particular condensated (coherent) state of the electromagnetic field.

The principle requirement for a collective coherent ground state is the amplification of the electromagnetic coupling by the square root of the number of atomic radiators (N) in the coherence volume:

$$\text{collective coupling} \qquad G_c \sim \sqrt{N} \tag{48}$$

This principle of collective coherence has been called in section 2.6. the "square root of N"-law and can be traced back to the work of BOGOLIUBO (SOLOMON, 1974). In particular, the amplification of spontaneous photon emission by a N^2 coherence factor due to the interaction of N molecules with a common radiation field has been discovered by DICKE (1954) who coined the term "superradiance" for coherence in spontaneous radiation processes.

When applying the term "superradiant" to the collective coherent ground state of photons tapped in matter, one should be aware that these photons are not observable directly (ENZ, 1997). Only the excitations with energy gap ,i.e. the biophotons, are observable.

PREPARATA (1995) has studied the principles of collective coherence for various other forms of the electromagnetic interaction in atomic matter (and even extended the principle to nuclear matter).

The energy of the collective coherent ground state of a BOSE-condesate, E_C, is in general lower than the ground state energy, E_O, of the weakly (perturbatively) coupled excitations (PREPARATA, 1995; ENZ, 1997): $E_C < E_O$. The excitation ground state has been described in section 3.3.1. using a critical temperature T_C. The collective coherent ground state therefore is the real groundstate of a BOSE-condensate in the case of strong electromagnetic coupling.

In striking difference to a Laser, population inversion is unfavorable to collective coherence. Rather, collective coherence occurs preferably at low excitations (ENZ, 1997).

When space-dependence is neglected both the collective ground state and the excitation ground state of a BOSE-condensate (see section 3.3.2.) are coherent states of the harmonic oscillators associated with space independent modes (PREPARATA, 1995).

The order without phase coherence underlying the notion of a BOSE-condensate is preserved in the perturbative ground state because of the free nature of the photonic excitations.

The transition from the perturbative to the collective coherent ground state takes place only when the system is structured in small domains which are allowed to emit electromagnetic radiation and matter fluctuations independently (PREPARATA, 1995). This confirms the importance of forming groups or compartments for creating coherent many-body system.
In this way the principle of collective coherence (45) may become useful in improving the storage capacity of a seed lot.

Also instrumental in creating a collective coherent ground state is the principle of stepwise transition to the collective ground state , which is a generalisation of OSWALD'step rule.

The superfluid state is characterized by the coexistence of two kinds of ground states: the perturbative coherent ground state of the "free" photonic excitations and the non-perturbative, collective coherent ground state of the strongly coupled photons.

TABLE 6. Qualities of the two ground states characteristic of a superfluid condensate of photons in biological matter

	collective ground state (c)	excitation ground state (o)
type of order:	long distance collective coherence	space-independent coherent modes
phase coherence:	yes	no
coupling type:	strong	weak
photons:	trapped photons (non observable)	photons leaking out (observable)
character of the ground state:	non-perturbative	perturbative
fluctuations:	coherent(no gap)	incoherent(with gap)
models:	DICKE theory	oscillator model, mean field model

The structure which integrates the collective ground state, which is the real ground state, with the excitation ground state is the energy gap between the two ground states. Because the energy of the real ground state (E_C) is always lower than the energy of the excitation ground state (E_O) the difference defines a gap d_O (existing at absolute zero temperature)

$$E_o - E_c = n\, d_o = D_o \qquad (for\ T=0) \tag{46}$$

with n being the number of particles excited from the "superradient" to the excitation ground state.

3.3.3.1. *Quantum Gap and Genetic Vigour.* The gap d_o defined by (46) can be interpreted as the minimum energy needed to create from the collective ground state a (quasi-) particle fluctuating incoherently around the excitation ground state. In distinction to the gap between the quantum mechanical and thermodynamical behaviour (see section 3.2. and Figure 8) the gap defined by (46) is a purely "quantum mechanical gap" or "unmanifest gap".

In accord with (36) the energy of the perturbative ground state (at zero temperature) can be normalized to zero (E_o=0) and the value of d_o then is given by E_c, which has to be negative. Explicit expressions for E_c have been derived in the two level approximation both by PREPARATA (1995) and ENZ (1997) showing a dependence on the following basic parameters of collective coherence:

(i) the stationary oscillatory frequencies ν which the radiation field assumes in the collective coherent ground state,
(ii) the degree of excitation (z) of the matter system which depends on these frequencies,
(iii) the coupling constant (G_c) of matter and radiation field,
(iv) a mass term (m) related to the penetration depth due to spatial damping.

ENZ (1997) in confirming results derived by PREPARATA gives the following simple expression for the minimum total energy of the collective coherent ground state and hence for the quantum gap d_o:

$$d_o = E_{c,min} = 1 - \cos z - \frac{G_c^2 \sin^2 z}{4(1+m)} \tag{47}$$

where G_c characterizes the coupling between matter and radiation field, subject to the following condition

$$G_c^2 > \frac{4(1+m)}{1+\cos z} \tag{48}$$

This condition says that population inversion which occures for $z > \pi/2$ (LASER condition), is unfavorable for collective coherence, which only requires low excitation: $0 < z < \pi/2$.

The "quantum gap", d_o, described by the energy of the collective coherent ground state is then an interplay of three fundamantal aspects of collective coherence: the degree of atomic excitation (z), the $\sqrt{N}$ coupling (G_c), and the photon damping factor (m).

The "quantum gap" can be interpreted as "genetic vigour", which is discussed from a different angle in section 3.4.2.

A further aspect of the quantum gap is to protect the collective coherent ground state against thermal fluctuations.

3.3.3.2 *Thermodynamic Gap and Physiological Vigour.* The effect of changes in temperature and density on the energy gap integrating superradiant ground state and excitation ground state can be understood by using the mass density as introduced by LANDAU in case of superfluid Helium (LANDAU and LIFSHITZ, 1969; PREPARATA, 1995):

The total mass density of trapped photons ρ(N) is the sum of the densities for the superradiant ground state ρ(s) and for the photonic excitations ρ(n), therefore

$$1 = \frac{\rho(n)}{\rho(N)} + \frac{\rho(s)}{\rho(N)} \tag{49}$$

When ρ(n)/ ρ(N) increases then ρ(s)/ ρ(N) decreases and vice versa.

This leads to a renormalisation of the "microscopic" parameters as G_C and z by the factor ρ(s)/ ρ(N) and thus to a density dependent gap energy D(ρ):

$$D(\rho) = d_o \frac{\rho(s)}{\rho(N)} = d_o - d_o \frac{\rho(n)}{\rho(N)} \tag{50}$$

D(ρ) describes the population shift between the collective coherent ground state (s) and the excitation ground state (n).
Equation (50) can be interpreted as a mathematical formulation of the biological observation that genetic vigour (d_o) is the basis of physiological vigour (D(ρ)).
ρ(n)/ ρ(N) approaching 1 means all the photons have left the collective coherent state and the value of the gap D(ρ) approaches zero. When this happens the excitation ground state becomes unstable and the photons can be easily excited further. This was the situation analysed in section 3.2.

In case of seeds, a change in the ratio ρ(n)/ ρ(N) is either due to a change in temperature and/or a change in water content. So the gap depends on both the temperature and the density separately:

$$D(\rho,T) = D(\rho,0) - d_o \frac{\rho(n)}{\rho(N)} \tag{51}$$

Equation (51) interpretes the experimental observation of Table 4 (section 2.4.1.) by implying, that both with increasing temperature and increasing hydration the gap effectively decreases because in both cases the ratio ρ(n)/ ρ(N) increases.

The gap D(ρ, 0) is defined as

$$D(\rho,0) = D(\rho,Tc) + d_o \tag{52}$$

T_C is the critical temperature introduced in section 3.3.1. and $D(\rho,T_C)$ is the gap related to the photonic excitations described in section 3.3.2 .
Equation (52) then says that in the collective coherent situation the quantum gap energy (d_o) has to be added to the excitation gap $D(\rho,T_C)$.

The methodical importance of this discussion is the confirmation of the necessity to strictly control both water and temperature, when the biophoton emission is used to compare seeds of different vitality.

Because the superradient ground state is characterized by spatial coherence the gap is space dependent as well.

The total energy gap D , which depends on the three macroscopic parameters: temperature (T), the photon mass density (ρ) and space coordinates, may be related to the physiological vigour of seeds.

3.3.4. *Biophotons and Vitality of Seeds*

As PRIMAS (1968) has noticed, macroscopic quantum effects despite their unique properties can only be detected experimentally if one specifically searches for them.

The assumption of a superfluid structure for the radiation field of seeds can be given the following biological interpretation which opens the hypothesis to a direct verification:

(a) The superfluid part of the radiation field provides an unobservable, silent background field correlating all seed parts and also the seeds with the whole environment.
(b) The excitations (biophotons) which are separated from the superfluid ground state by an energy gap are the carriers of information-entropy.
(c)The energy gap integrates the stabilty of the collective coherent ground state with the flexibility inherent in the excitations.

A large energy gap and a corresponding low photon emission characterize a situation of rest or performance potential. Because this is just the meaning of the term "vigour" in seed physiology the following correspondence can be postulated:

Postulate 1 energy gap ⟷ seed vigour

The observed photon emission (biophotons) is basically related to the energy gap separating excitation ground state and collective coherent ground state. Therefore it is a direct measure of vigour to which it is inversely related (rule 2). The distinction between genetic vigour and physiological vigour also naturally fits into this picture because of the gap having two distinct aspects: quntum mechanical (quantum gap) and thermodynamic aspects (thermal gap).

The correspondence between germinability and the photonic partition function aready has been explained in a heuristic way in section 3.3.1:

Postulate 2 stability of coherence ⟷ germinative ability

The germinative ability is proportional to the observed biophoton emission when the total number of photonic excitations can be experimentally controlled. This is the case when samples of similar seeds under exactly the same conditions are compared.

The superfluid model further specifies postulate 2:

(1) According to (46) the germinative ability can be determined by the ratio of ground state volume (V_0) to total volume (V). When the total volume V is identified with the volume of the embryo then (46) provides a biophysical basis for the tetrazolium test. In the tetrazolium test the living parts of the embryo are distinguished from the dead ones via the diffent reactions towards triphenyl tetrazolium chlorid. The living parts turn red by hydrogenation while the dead parts remain colorless (HAMPTON and McKRONY, 1995). In this manner two "volumes" are compared.

(2) By definition (45), the ground state volume $V_o = V_o(\lambda_o, \lambda_m)$ is related to two basic factors, λ_o and λ_m: λ_o is the wavelength of the photonic quasi-particle waves which depends on the mass density, which in turn changes by virtue of matter transported through the boundaries (membranes) of the coherence volume. λ_m can be interpreted as "longitudinal extensibility" of the coherence volume which increases when the damping mass decreases. λ_o and λ_m may provide an explanation of two parameters introduced by LOCKHART(1965) to specify the permeability and extensibility of boundaries allowing for elongational growth of plant cells and organs.

(3) Since the volume V is related reciprocally to a density ρ it follows from (44) that g ~ ρ. This means that germinative ability (g) is always proportional to seed density ρ , a result which provides a further confirmation of the vacuum model, because more and more evidence indicates that the seeds of greatest longevity in a given lot are those with the greatest density. As a rule, only seeds of high density should be kept for long storage (HARRINTON,1972).

In the superfluid model the biophotons emitted from seeds are coherent excitations which carry entropy through or away from the system to maintain the state of collective coherence.

Both vigour and germinability are reflected in the biophoton emission because photon emission and vitality are only different manifestations of a state of macroscopic collective coherence. This will be elaborated upon in the following section 3.4.

Further practical consequences of the assumption of a superfluid vacuum structure are summarized in the concluding section 4.

3.4. COHERENCE AND BIOLOGICAL STRUCTURE

In view of the practical applications of the biophoton emission of seeds the three basic steps of analysing biological rest from the perspective of the quantum vacuum are summarized (sse also Figure 8):

1. The quantum vacuum as junction point between quantum and thermal behaviour is responsible for the stability of the response of seeds to thermal excitations (emission law).

2. The fourfold nature of the gap between quantum mechanical and thermodynamic behaviour reveals the details of the response pattern of seeds to thermal excitations (nature of the constant C in the emission law).
3. The superfluid vacuum structure of seeds allows insights into the holistic nature of vitality by relating germinative ability and vigour of seeds to the energy gap between the BOSE-condensate and its excitations: gap stability relates to the germinative ability and the gap value to vigour.

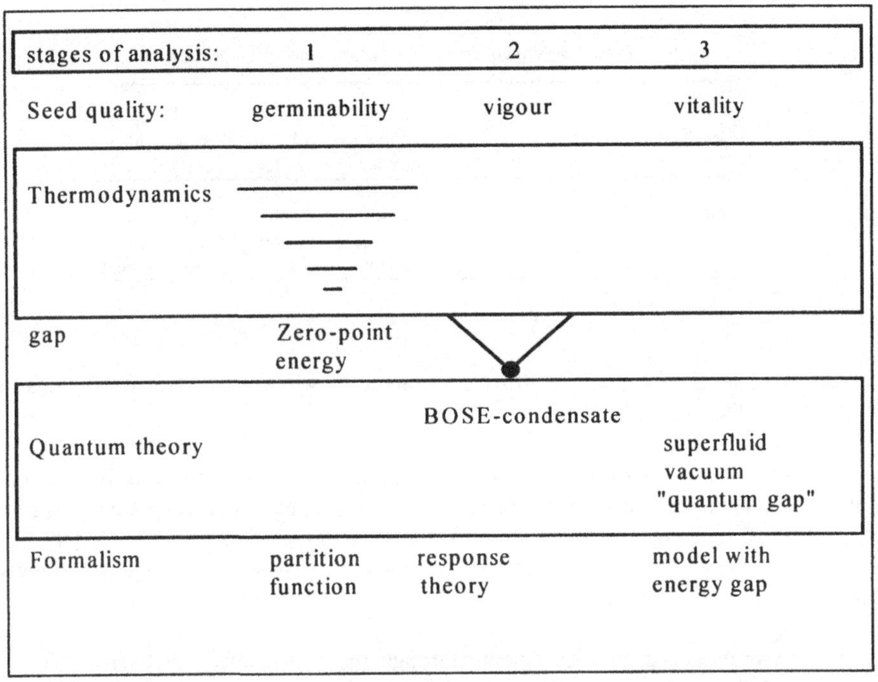

Figure 8. The three stages of understanding biological rest in terms of the quantum vacuum.

Our understanding of biological rest cannot be separated from the following general biological phenomena which are the basic challenges of biophysics:

"Biological systems have always been described by a variety of hierarchicallevels...(but) the hierarchical structure of a system is never an intrinsic property of a system, it rather arises from a particular way we interact with the system".(PRIMAS 1981).

"In biological systems there is no clear delimitation between microscopic and macroscopic elements. Both phenomena harmonically exist together in order to perform biological functions" (LI, 1996).

"Phenomena such as phase transitions, dissipative structures, and even biological growth. must be, in some sense collective, in that they involve the cooperation of an enormous number of particles"(SEWELL,1986).

3.4.1. *Three Levels of Coherence*

The key for understanding the fundamental biological phenomena i.e. "hierarchy", "function" and "growth" is the notion of "coherence" as has been independently recognized by "generalized quantum mechanics with superselection rules" (e.g.PRIMAS, 1981), the "coherent state approach"(e.g. POPP et al, 1994) and the "quantum theory of infinite systems" (e.g. SEWELL, 1986). To shed further light on the fundamental importance of the vacuum model for understanding biological rest, the three approaches are classified in terms of coherence (Table 7 summarizes the salient points)

TABLE 7. Three levels of coherence in relation to biological phenomena

	I	II	III
levels of coherence :	quantum coherence	coherent states	superfluid states
biological phenomena:	wholeness, emergence, hierarchy	function, response, pattern	collective coherence, development

On every level of coherence structures emerge by virtue of certain restrictions or boundary conditions (LI, 1994).

I. *Quantum Coherence (Wholeness)*. Coherence in its most elementary form is an inbuilt feature of quantum mechanics realized as coherent superposition of pure states. In case of unrestricted superposition the "silent world of quantum mechanics"(PRIGOGINE and STENGERS,1980) knows no observable phenomena because everything is infinitely correlated with everything else (EINSTEIN-PODOLSKI-ROSEN-Correlation).

> Observable properties emerge through the restrictions which an observer imposes on the unrestricted coherent superposition of quantum states (PRIMAS, 1981).

The classical notion of space-time which defines the way the observer interacts with the world, leads to "superselection rules" which restrict coherent superposition to states existing at the same instant of time and to elementary objects(particles) having the same mass, spin and charge. Based on the notion of space and time are the parallelism of quantum description and wave description (LANDE, 1965), the existence of PLANCK's constant (h) and HEISENBERG uncertainty relations including all their consequences (e.g.the vacuum fluctuations). The HEISENBERG uncertainty relations describe quantum coherence by means of incompatible properties. Coherent superposition and incompatibility of certain observables are equivalent aspects of quantum coherence. In the algebraic description the HEISENBERG uncertainty relations correspond to commutator relations between observables (e.g. of total energy (H), position (x) and momentum (p)). Thereby a commutator algebra is defined specifying quantum coherence.

For harmonic oscillators represented by the total energy operator $H = p^2 + w^2q^2 = h\omega (P^2+Q^2)$ the uncertainty product of position and momentum has a non-vanishing minimal value in the ground state. This minimum value defines the zero-point energy of the oscillator. Because field systems can be described by infinitely many oscillators, they have an infinite zero-point energy in the ground state (ITZIKSON and ZUBER, 1980).

> By modifying the boundary conditions of the infinite zero-point energy of a field system through the mere presence of bodies (even macroscopic ones or uncharged particles) certain forces acting among them are created (CASIMIR effect).

Vacuum fluctuations are a manifestation of quantum coherence and have definite structural effects by virtue of the various kinds of forces they induce between bodies (CASIMIR forces, VAN DER WAALS forces). These forces are responsible for the states of aggregation and structural stability of biological matter.

II. *Coherent States (Oscillator Model).* Consistent with the general HEISENBERG uncertainty relations "coherent states" are defined by the minimum value which the uncertainty product takes in the ground state. Coherent states are also called "quasi-classical states" because of the absence of certain quantum correlations (MERZBACHER, 1970). The requirement of a minimal uncertainty product of the ground state allows the introduction of a coherence length and a coherence time for the respective system and thereby defines a "coherence space-time" independent from PLANCK's constant and applicable in macroscopic dimensions (LI, 1994):

> Structures emerge as interference pattern in the coherent space-time region due to certain constraints (POPP et al, 1994)

The natural models for coherent states are harmonic oscillators with the appropriate boundary conditions. In algebraic notation the minimal uncertainty condition of harmonic oscillators means $\omega Q + iP |vac\rangle = 0$. The operator $\omega Q+iP$ has the meaning of a lowering operator (annihilation operator) describing transitions in the direction of the ground state $|vac\rangle$. Any other coherent state can then be expressed by a unitary transformation of the ground state $|c\rangle = \exp(-|c|^2/2) \exp(c\, a') |vac\rangle$. In particular any coherent states of the radiation field can be generated by unitary transformations of the coherent vacuum state but also by the various eigenstates $|n\rangle$ of the number operator $N = a'a$ which are not coherent states (KLAUDER and SKADERSTAM, 1985).

The degree of coherence of a light beam is determined by the coherence of its source. In a source of coherent light (Laser) the atoms and the radiation field vacuum mutually influence each other (DICKE's Laser-model). It is the resonant reponse of atoms to the radiation field which leads to coherent spontaneous radiation. But the precise details of the radiation in terms of frequeny and spatial distribution are strongly influenced by the macroscopic environment or cavity conditions (HAROCHE and KLEPPNER, 1989).

> Different cavities by modifying the vacuum and thereby the atom-vacuum interaction, change the spontaneous emission (cavity quantum electrodynamics).

From the perspective of cavity quantum electrodynamics the interfaces within seeds (e.g. the interface between matter and vacuum) may play an important role in determining the specific nature of the emitted radiation.

III. *Superfluid States (Collective Coherence).* In view of the coherent collective cooperation in systems of Boson-particles the notion of coherence can be further specified by incorporating the "square root of N"-law into the formalism of the state generating algebra. The "square root of N"-law describes the creation (and destruction) of coherence when the particles of the assembly are coupled in such a way that their coherence creating effect is proportional to N^2. Experimental evidence for this effect in seed lots has been described in section 2.4.

The inclusion of the "square root of N"-law in the frame work of coherent states is simply achieved by treating the spectrum generating lowering and rising operators (a and a') as equal to the square root of the particle number operator of the ground state $N_o = < a_o', a_o >$ (SOLOMON, 1971):

$$a_o' \sim a_o \sim \sqrt{N_o} \qquad (53)$$

This assumption which is called BOGOLIUBOV-approximation describes both the annihilation and creation operators on an equal footing. The BOGOLIUBOV-approximation is surprisingly accurate even when the particle number is very small (BASSICHIS and FOLDY, 1964).

The superfluid vacuum state is a macroscopic occupied coherent state. Everything that can possibly happen is contained in the structure of the vacuum by virtue of the symmetry of the Hamiltonian supplemented by groups which are not symmetry groups of the Hamiltonian but whose algebras are spectrum generating LIE-algebras (SOLOMON, 1971). The related LIE groups enables one to write down the energy eigenvalues and evaluate the eigenstates, as well as suggesting what form additional correction terms of the Hamiltonian should take.

Everything which actually happens in the superfluid state is by way of unitary transformations, which are merely a rotation in the space of the algebra.
"Unitary" means that the transformations maintain the total information. For the superfluid vacuum any change is only an excitation of the non-changing purely quantum mechanical reality (the superfluid BOSE-condensate). This excitation may carry entropy (disorder) through the system but produces no entropy. In this way the system remains unaffected by environmental disorder and does not create disorder itself. Because the superfluid vacuum integrates and maintains change on the level of the non-changing condensate it promotes longevity. Therein lies the fundametal importance of this approach for seed storage (see section 2.5.).

3.4.2. *Space-Time Coherence*

There is yet another aspect of the superfluid vacuum which is of very fundamental importance for understanding the nature of biological rest: the connection between biological functions and basic cosmological factors. ZIOUTAS (1996) interprets the biophoton emission as a sign of the cosmological "dark"(non-luminous) matter reaction with biological matter. In biorhythms following seasonal, circadian and other astrophysical frequencies organisms seem to measure in some manner and to some extent astronomical times. A possible mechanism suggested by ZIOUTAS is the enhancement of the permanent, oscillating flux of dark matter by biological matter due to the N^2-factor of collective coherence.

In astrophysics dark matter is made responsible for the observed inhomogeneity in the distribution of the large clusters of luminous matter in the visible universe and can therefore be related to the unmanifest background space-time geometry.

"Empty space" is thus not only a physical reality being described as "quantum vacuum" but must also have a topological structure allowing for the relationships between objects in terms of distances, angles or velocities. This geometrical aspect of the vacuum is classically described by EINSTEIN's theory of gravitation (also called theory of general relativity). The empirically well established foundation of EINSTEIN's theory is the equivalence of gravitational and inertial mass.

However, this equivalence causes great problems when one wants to answer the question (HALLIWELL, 1991): what happend before the very instant of creation? A question similar in spirit to the question of a biologist: what is the basis of the germination process?

A possible answer to the cosmological question has been given by SINHA (1976) by proposing a superfluid vacuum. The invisible, dark matter corresponds then to the superfluid component of the vacuum, while the visible matter to the elementary excitations of the superfluid, with the energy described by (36):

$$h\nu(p) = \frac{(p - p_0)^2}{2m} + d_o$$

The gap energy d_0 stands for the gravitational mass $M = d_0/c^2$ of the particles described by (36) and the mass parameter m for the inertial mass (FALK, 1968). In general (36) describes a situation where the gravitational and inertial mass of the superfluid excitations differ from each other (FALK, 1968). Thus, for the superfluid vacuum the principle of equivalence between inertial and gravitational mass is not valid. The superfluid vacuum relates to a reality of "space-time coherence" or "infinite correlation" for which WHEELER (1962) has coined the term "pre-geometry".

The state of biological rest of seeds seems to be related to the field of pre-geometry rather than to the space-time structure of the classical observer: The germinative ability expressing thereby the "inertial" mass m of collective excitations while vigour the "gravitational" mass M: Germinative ability, the power of maifestation,is high when the "inertia" is low, as predicted by (44). Vigour on the other hand is proportional to "gravitation" since it characterizes the degree of rest, which has to be large for high levels of performance.

Furthermore, the switch from the vacuum controlled first awakening of a seed to the subsequent DNA controlled growth taking place at a certain stage in the development of a seedling receives a natural interpretation in terms of the superfluid vacuum:

At a certain stage in the germination process the DNA becomes the means of transforming the structure of the vacuum into biological actions. BISCHOF (1995) uses the term "antenna geometry" to describe the potential of the DNA structure to assume the role of a "physiology of the vacuum" at a certain stage during germination.

This is parallel to that stage in cosmological evolution where the material structure of the universe starts to express the structure of the vacuum.

This kind of thinking already has inspired the application of RIEMANN-geometry to describe the information content of the DNA (FINDLEY, et al 1985). The basic construct of FINDLEY's genetic cosmology is a 65-dimensional differentiable manifold in a coordinate structure such that the manifold points represent the number of each codon typ in the DNA molecule, plus the evolutionary time of the DNA. The evolutionary motions in the informational space-time manifold are assumed to be following the principle of least action (geodesics). The corresponding field equations are governed by a field whose exact physical meaning still has to be identified. But the relationship between DNA structure and the regulating function of its radiation field point at the biophotons as a canditate for this field:

In the DNA the digital, genetic information seems to be coupled to a wealth of analog information also imprinted in its structure, which may - via the radiation field- be responsible for the control and expression of genetic information as well as the synchronisation of the various biological rhythms (POPP, 1984). Very little structural changes may have dramatic effects in the control function of the radiation field (POPP, 1984, LAIBEL and POPP, 1995).

Both from the theoretical and the practical perspective the emphasis of further research should be on the superfluid vacuum structure of seeds. The integration of structure and radiation field in the superfluid state may be the key to a deeper understanding of seed behaviour which can open the door (again) to methods of plant cultivation which are not damaging the connection to the cosmic rhythms and laws of nature in which life on earth is embedded (FAGAN, 1995).

4. Conclusion: Biophoton Emission and Seed Quality

The practical application of biophoton emission to determine the vitality of seeds is based on the correspondence between biological rest, vitality, and biophotons, so that any one of the three can be used as measure of the others:

$$\text{biological rest(seeds)} \quad \longleftrightarrow \quad \text{vitality} \quad \longleftrightarrow \quad \text{biophotons.}$$

Simple rules based on this correspondence allow a non-destructive and rapid testing of seed quality by means of biophoton emission (Table 8).

Vitality and biophotons are structurally related representations of the same holistic reality. This can be understood by virtue of the vacuum state model of biological rest. The vacuum model of biological rest also includes an approach to seed quality in terms of macroscopic collective coherence which may become of practical importance in the field of seed storage and longevity (Figure 9).

TABLE 8. Biophoton emission as measure of seed vitality

seeds vitality	spontaneous emission (n) indicates	rule
vigour(D) (genetic and physiological) Dmin < D < Dmax	level of rest	$n = \exp(-D/kT)$ $D >> kT$
germinability(g) $0 < g < 1$	thermal stability	$g = n / b \exp(CT)$ "ceteris paribus"

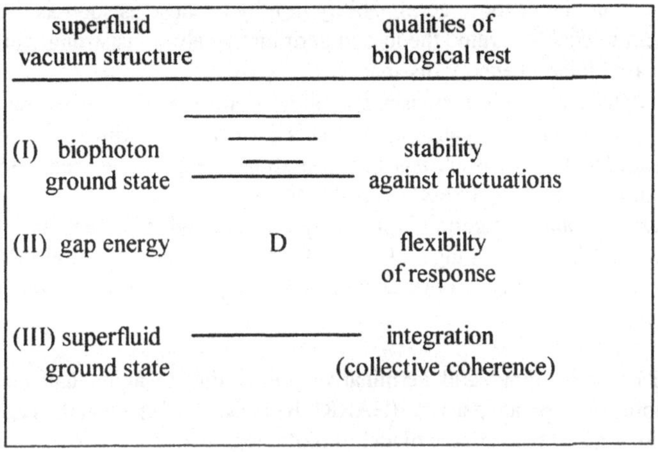

Figure 9. Basic qualities of biological rest related to the three-in-one structure of the superfluid vacuum.

The central concept for understanding seed vitality is the existence of an energy gap separating the biophotons from the zero-energy state of the trapped photons. The existence of a large energy gap is the pre-requisite that quantum coherence is not blurred out by thermal disorder. A large energy gap corresponds to a high degree of quantum coherence, to a low level of random activity and hence to a low level of biophoton emission. In this sense the overall level of biophoton emission is a measure of the performance potential or vigour of seeds:

High quantum coherence = large energy gap

The energy gap also determines the reaction towards thermal fluctuations. This reaction evaluates the stability of coherence underlying biological growth (germinative ability). High coherence or high germinability means that thermal excitations are easily carried through the system and therefore lead to a high biophoton emission as a compensating response:

High quantum coherence = less dissipation

This picture of vitality allows a simple interpretation of basic observations which are of great practical importance in the field of seed storage:

(1) Great vigour and high germinative ability is always related to a large energy gap, because only when the energy gap is large compared to kT is the behaviour dominated by quantum coherence.
(2) Through aging vigour is always affected first, because the reduction of quantum coherence is related to a decrease in the energy gap. This is confirmed by the observation that "a loss in vigour, however it is measured, precedes the loss in germinability (HARRINGTON,1972).
(3) Seeds low in vigour loose germinability rapidly because as soon as the vigour has reached a critical low value the loss in germinative ability is strongly enhanced due to increased internal energy dissipation.
(4) Water uptake of seeds increases the photon emission due to a lowering of the energy gap which may cause aging. This is in accord with the experience that during seed storage moisture has a deteriorating effect first on seed vigour and then on germinative ability (HARRINGTON,1972).
(5) Temperatures above a certain value may advance aging by activating excitations lying within the energy range kT around the energy gap. Because this effectively lowers the energy gap, vigour decreases accompanied by an increase in photon emission.

The decline of both vigour and germinative ability due to aging and environmental factors follows a typical pattern (HARRINGTON, 1972), which can be easily understood in terms of the vacuum model of biological rest.

In spite of all methods to slow down the aging process of seeds by controlling environmental factors such as moisture and temperature, all efforts of the past were faced with the "inevitability of deterioration" (HARRINGTON,1972). The superfluid model of biological rest provides now a new look at storage capacity and longevity of seeds by applying the principles of collective coherence.

A technology working at the superfluid level of collective coherence has the potential to stop aging and promote longevity to the maximum possible degree, and it seems worthwhile to study such a technology in greater depth.

5. Acknowledgements

This research was originally suggested and has been continuously supported by Prof. F.-A. Popp, International Institute of Biophysics, Kaiserslautern and Neuss, Germany.
Prof. Ch. Enz, University of Geneva, contributed insights into the role of massive photons in the superradiant phase transition.
M. Bischof, International Institute of Biophysics, Germany, provided important information on the quantum vacuum.
Dr. M. Kruse, University of Stuttgart-Hohenheim, Institute of Prof. Steiner, contributed valuable insights on seed physiology.
Dr. K. Münzing, Bundesanstalt für Getreideforschung, Detmold, Germany, confirmed the existence of collective effects in seed lots.
Financially, the support of Nestlé, Ltd., Switzerland, is gratefully acknowledged

6. References

Aitchison, I.J.R.(1985) Nothing's plenty - The vacuum in modern quantum field theory, *Contemporary physics* **26**, 333 - 391.
Amen, R.D.(1968) A model of seed dormancy, *The Botanical Review* **34**, 1 - 30.
Bewley, J.D., and Black, M (1984) *Seeds - Physiology of Development and Germination*, Plenum Press, New York.
Bischof, M.(1995) *Biophotonen*, Zweitausedeins, Frankfurt/Main.
Brenig, W.(1975) *Statistische Theorie der Wärme*, Springer,Berlin and Heidelberg.
Bünning, E.(1953) *Entwickungs- und Bewegungsphysiologie der Pflanze*, Springer, Berlin and Heidelberg.
Ching, T.M. (1973) Metabolism of germinating seeds, in T.T.Kozlowski(ed.), *Seed Biology*, vol. 2, Academic Press, New York and London.
Cohen, S., and Popp, F.-A.(1997) Biophoton emission of the human body, *J. Photochem. Photobiol B:Biol.*. **40**, 187-189.
Cohen, S., and Popp, F.-A.(1997) Low-level luminescence of the human skin, *Skin Research and Technology* **3**, 177-180.
Cochran, A.A. (1971) Relationship between quantum physics and biology, *Foundations of Physics* **1**, 235 - 250.
Dicke, R.H.(1954) Coherence in spontaneous radiation processes, *Physical Review* **93**, 99 -110.
Enz, Ch. P.(1974) Two-fluid hydrodynamic description of ordered systems, *Reviews of Modern Physics* **46**, 705 -753.
Enz, Ch. P. (1997) On PREPARATA's theory of a superradiant phase transition, *Helvetica Physica Acta* **70**, 141 - 153.
Enz, Ch. P. (1997a), private communication.
Fagan, J.(1995) *Genetic Engeneering*, MUM Press, Iowa.
Falk, G.(1968) *Theoretische Physik*, Band II, Springer, Berlin, and Heidelberg.
Feynman, R.P., and Hibbs, A.R.(1965) *Quantum Mechanics and Path Integrals*, McGraw-Hill, New York
Feynman, R.P., Leighton, B.B., and Sands, M.(1975) *The FEYNMAN Lectures on Physics*, vol.3, Addison Wesley, London.
Findley, A.M., McGlynn, S.P., and Findley, G.L.(1985) Application of differential geometry to molecular genetics. *Journal of Biological Physics* **13**, 87 - 94.
Galle, M. (1993) Untersuchungen zum dichte- und zeitabhängigen Verhalten der ultraschwachen Photonenemission von parthenogenetischen Weibchen des Wasserflohs *Daphnia magna*, Dissertation, Saarbrücken.
Gasser, R.P.H., and Richards, W.S. (1974) *Entropy and Energy Levels*, Clarendon Press, Oxford.
Godnew, L.N. (1963) *Berechnung thermodynamischer Funktionen aus Molekülдaten*, VEB, Berlin.
Greiner, W. and Reinhardt, J. (1993) *Feldquantisierung*, Harri Deutsch, Frankfurt/Main.

Gu, Q., and Popp, F.-A.(1994) Biophoton emission as a potential measure of organizational order, *Science in China(series B)*, **37** 1009 -1112.

Haken, H. (1977) *Synergetics*, Springer, Berlin, Heidelberg, New York.

Halliwell, J.J. (1991) Quantum cosmology and the creation of the universe. *Scientific American*, December, 28 - 35.

Hampton, J.G. and D.M. TeKrony (1995) *Handbook of vigour test methods, ISTA*, Zürich.

Haroche, S. and Kleppner, D. (1989) Cavity quantum electrodynamics. *Physics Today* **25**, 24 -30.

Harrington, J.A.(1972) Seed storage and longevity, in T.T. Kozlowski(ed.), *Seed Biology*, vol.3, Academic Press, New York and London.

Heske, F., Jordan, P., and Meyer-Abich, A. (1954) *Organik*, Haller, Berlin.

Hutter, K.-J. (1994) Einsatz der Biophotonenanalyse für die Braugersten-Qualitätskontrolle, *Brauwelt* **134**, 1404 -1409.

ISTA-International Seed Testing Association(1995) *Internationale Vorschriften für die Prüfung von Saatgut*, Zürich.

Itzykson, C. and Zuber, J.-B. (1980) *Quantum Field Theory*, McGraw-Hill, New York.

Jann, R.C. and Amen, R.D. (1977) What is germination? in A.A. Khan(ed.),*The Physiology and Biochemistry of Seed Dormancy and Germination*, North Holland Publishing Company, Amsterdam, New York, and Oxford.

Jibu,M., Pribram, K.H., and Yasue, K.(1996) From conscious experience to memory storage and retrieval: The role of quantum brain dynamics and Boson condensation of evanescent photons, *International Journal of Modern Physics*, 1735 - 1754.

Kittel, Ch. (1971) *Introduction to Solid State Physics*, Wiley, New York and London.

Klauder, J.R. and Skagerstam, B.-S. (1985) *Coherent States*, World Scientific, Singapore.

Kruse, M. (1997) private communication.

Landau, L.D. and Lifshitz,E.M.(1969) *Course of Theoretical Physics* Vol.5- Statistical Physics, Pergamon Press, London.

Landä A.(1965) *New Foundations of Quantum Mechanics*, University Press, Cambridge.

Leible, M., and Popp, F.A. (1995) Unterschiedliche Biophotonenemission im Fall gentechnischer Veränderungen in K.-H. Engel, G.A. Schreiber and K.W. Bogl(eds): *Entwicklung von Methoden zum Nachweis mit Hilfe gentechnischer Verfahren hergestellter Lebensmittel.*

Leggett, A.J. (1086) The superposition principle in macroscopic systems, in R.Penrose, and C.J. Isham (Eds.) *Quantum Concepts in Space and Time*, Clarendon Press, Oxford.

Leighton, R.B. (1959) *Principles of modern physics*, McGraw-Hill, New York.

Li, K.-H. (1992) Coherent radiation from DNA molecules in F.-A.Popp, K.-H.Li, and Q.Gu (eds.) *Recent advances in biophoton research and its applications*, World Scientific, Singapore.

Li, K.-H. (1994) Uncertainty principle, coherence and structure in R.K.Mishra, D.Maaá, and E.Zwierlein (eds) *On Self-Organization*, Springer, 1994.

Li, K.-H.(1996) Coherence-a bridge between micro- and macro-systems in Ch.Zhang, F.-A.Popp and M.Bischof (eds), *Current Development of Biophysics*, Hangzhou University Press.

Li, K.-H.(1996) and Popp, F.-A.(1996) Coherence, de Broglie's and the EPR paradox in Ch.Zhang, F.-A.Popp and M.Bischof (eds), *Current Development of Biophysics*, Hangzhou University Press.

Lockhart, J.A. (1965) An analysis of irreversible plant cell elongation, *Journal of Theoretical Biology* **8**, 264 - 275.

Lölinger, J. (1989) Bio-luminescence, progress report 6 (17.4.1989) Nestlé Ltd., Lausanne.

Mayer, A.M., and Poljakoff-Mayer,A. (1975) *The Germination of Seeds*. Pergamon Press, Oxford.

Merzbacher, E. (1970) *Quantum Mechanics*, Wiley,New York and London.

Milonni, P.W. (1994) *The Quantum Vacuum*, Academic Press, Boston.

Münzing, K. (1993) Aufbereitung und Lagerung, in W. Seibel and W.Steller(eds), *Spelz- und Schälgetreide*, Behr's, Hamburg.

Münzing, K. (1997) private communication.

Neurohr, R. (1992) Untersuchungen zur spontanen und induzierten Lichtemission von Kresse und Kressesamen, Dissertation, Saarbrücken.

Pandey, D.K. (1992) Conductivity testing of seeds, in: H.-F.Linskens and J.F.Jackson: *Seed Analysis*, Springer, Berlin, Heidelberg, New York.

Pfeifer, P. (1981) A non-linear SCHRÖINGER equation yielding the shape of a molecule`by spontaneous symmetry breaking, in K.E.Gustafson and W.P.Reinhardt (eds), *Quantum Mechanics in Mathematics, Chemistry, and Physics*, Plenum Press, New York.

Pollock, B.M., and Roos, E.E. (1972) Seed and seedlings vigour, in T.T.Kozlowsski (ed.), *Seed Biology*, vol 1, pp 313 -387, Academic Press, New York and London.

Popp, F.-A. (1984) *Die Biologie des Lichts (Biology of light)*, Paul Parey, Berlin and Hamburg.

Popp, F.-A., and Lölinger, J. (1989) *Bericht über Lumineszenz-Messungen an Lebensmitteln zur Erfassung von Qualitätsmerkmalen*, (research report),Kaiserlautern.

Popp, F.-A. (1990) Versuche zur Lichtemission qualitativ verschiedener Samenkörner, research report, Kaiserslautern.

Popp, F.-A. and Li, K.-H. (1993) Hyperbolic relaxation as a sufficient condition of a fully coherent ergodic field, *International Journal of Theoretical Physics* **32**, 1573 -1583.

Popp, F.-A., Gu, Q., and Li, K.-H. (1994) Biophoton emission - experimental background and theoretical approaches, *Modern Physics Letter B*, **8**, 1269 - 1296.

Popp, F.-A. (1996) Evolution as the expansion of coherent states, in Ch.Zhang, F.-A.Popp and M.Bischof (eds), *Current Development of Biophysics*, Hangzhou University Press.

Postle, D. (1976) *The Fabrics of the Universe*, Macmillan, London.

Preparata, G. (1995) *QED Coherence in Matter*, World Scientific, Singapore.

Prigogine, I,. and Stengers,I.(1980) *Dialog mit der Natur*, Piper, München and Zürich.

Primas, H.(1968) Zur Theorie großer Moleküle, *Helvetica Chimica Acta* **53**, 1037 -1051.

Primas, H., and Gans, W. (1979) Quantenmechanik, Biologie und Theorienreduktion, in B.Kanitschneider (ed.), *Materie-Leben-Geist*,Duncker & Humbolt, Berlin.

Primas, H. (1981) *Chemistry, Quantum Mechanics and Reductionism*, Springer, Berlin, Heidelberg, and New York.

Rafelski, J. and Moller, B. (1985) *Die Struktur des Vacuum*, Deutsch, Frankfurt/Main.

Rattemeyer, M., Popp, F.-A., and Nagl, W. (1981) Evidence of photon emission from DNA in living systems, *Naturwissenschaften* **68**, 572 - 573.

Renzenbrink, U. (1981) *Die sieben Getreide*, Geering, Dornach.

Saunders, S., and Brown, H.R. (1991) *The Philosophy of Vacuum*, Clarendon Press, Oxford.

Sewell, G.L. (1986) *Quantum Theory of Collective Phenomena*, Clarendon Press, Oxford.

Sinha, K.P., Sivaram, C., and Sudarshan, E.C.G. (1976) Aether as superfluid state of particle-antiparticle pairs. *Foundations of Physics* **6**, 65 - 69.

Sinha, K.F., Sivaram, C., and Sudarshan (1976) The superfluid vacuum, time-varying cosmological constant, and nonsingular cosmological models. *Foundations of Physics* **6**, 717 - 725.

Sitiko, S.P., Andreev, E..A., and Dobronravova, I.S., (1988) The whole as a result of self-organization , *Journal of Biological Physics* **16**, 71 -74.

Sitko, S.P., and Gizhko, V.V. (1991) Towards a quantum physics of the living state, *Journal of Biological Physics* **18**, 1 - 10.

Slawinski, J., and Popp, F.-A. (1987) Temperature hysteresis of low level luminescence from plants and its thermodynamical analysis, *Journal of Plant Physiology* **130**, 111 - 123.

Solomon, A.I. (1971) Group theory of superfluidity, *Journal of Mathematical Physics* **12**, 390 - 394.

Spruyt, E., Verbelen, J.-P., and De Greef, J.A. (1987) Expression of circaseptan and circannual rhythmicity in the imbibition of dry stored bean seeds, *Plant Physiology* **84**, 707 - 710.

Teubner, R.(1983) Zur Qualitätsbestimmung von Nutzpflanzen, insbesondere Medizinalpflanzen, mit Hilfe der ultraschwachen Photonenemission, Dissertation, Göttingen.

Veselova, T.V.,Veselovskii, V.A., Krasnovskii, A.A., and Lichschteld, J.K.(1985) Delayed luminescence of seeds. *Biophysics* **30**, 775 - 777.

Wallace, R.K., Benson, H., and Wilson, A.F. (1971) A wakeful hypometabolic physiological state, *American Journal of Physiology* **221**, 795 -799.

Wheeler, J.A. (1962) *Geometrodynamics*, Academic Press, New York.

Wilke, J. (1961) *The Third Law of Thermodynamics*, Oxford University Press, Oxford.

Zioutas, K. (1996) Bioluminescence a signature for dark matter reactions, in Ch.Zhang, F.-A.Popp and M.Bischof (eds), *Current Development of Biophysics*, Hangzhou University Press, Hangzhou.

BIOPHOTONS AND NONCLASSICAL LIGHT

Q. GU

International Institute of Biophysics, Technopark II,
Sauerwiesen 6, 67661 Kaiserslautern, Germany, and
Physics Department, Northwest University, Xi'an, 710069, China

Abstract. A model has been presented for the description of biophoton activity, which is based on the quantum theory of the interaction of a photon field with a phonon reservoir. The model is used to analyse the dynamics of the nonclassical effects induced by the Schrödinger-cat-state, especially involving aspects of the entropies of light field. A new quantity, so-called photon statistical entropy, is predicted to be able to characterise coherence properties of biophotons, which can be obtained by means of measurement of spontaneous emission. Experiments on a leaf are then performed and compared with a non-living radiator; the biophotons are found to have significantly higher coherence than ordinary light. As an application, biophoton emission of soybean samples with and without genetic treatment is investigated in terms of the photon statistical entropy concept.

Key words. Nonclassical light, entropies of light field, coherence of biophotons.

1. Introduction

It is well-known that the electric field strength for a nearly monochromatic plane wave of frequency ω may be written as $E(t) = E_0[a\exp(-i\omega t) + a^+\exp(i\omega t)]$, where E_0 is a constant including the spatial field distribution. In the quantum theory of radiation the complex amplitudes a and a^+ are quantum mechanical operators; a is called an annihilation operator and a^+ a creation operator. They are nonhermitian and obey the boson commutation relation $[a, a^+] = 1$. In terms of a and a^+, we may define two hermitian operators x and p as [1, 2]

$$x = a^+ + a \quad \text{and} \quad p = i(a^+ - a). \tag{1}$$

J. J. Chang et al. (eds.), Biophotons, 299-321.

Essentially, x and p are the dimensionless coordinate and momentum operators, respectively. In terms of x and p the electric field strength may be written as $E(t) = E_0(x\cos\omega t + p\sin\omega t)$. Thus x and p may be identified as the real amplitudes of the two quadrature components of the field. Furthermore, we deduce the commutation relation for x and p to be $[x, p]= 2i$. The relation for the uncertainties in x and p is then given by $\Delta x\Delta p \geq 1$, where $(\Delta u)^2 = \langle u^2\rangle - \langle u\rangle^2$ ($u = x$ or p).

A coherent state is known to be a minimum-uncertainty state. In fact, the fluctuations in the two quadratures of a coherent state are equal and minimise the uncertainty product allowed by Heisenberg's uncertainty principle, i. e. $\Delta x = \Delta p = 1$. In this sense the coherent states are the quantum mechanical states closest to a classical description of the field. The quantum fluctuations in a coherent state are equal to the zero-point fluctuations. These zero-point fluctuations represent the standard quantum limit to the reduction of noise in a signal. Even an ideal laser operating in a pure coherent state would still possess quantum noise due to zero-point fluctuations.

Other class of minimum-uncertainty states are possible which have less fluctuation in one of quadratures than a coherent state. The fluctuation in the other quadrature must then be greater than that of a coherent state in order to satisfy the requirement of the uncertainty principle. Such states are called *squeezed states*, which may be defined by $\Delta x < 1 < \Delta p$ (or $\Delta p < 1 < \Delta x$). It is obvious that the squeezed states may have an intriguing potential in ultra-low-noise optical communication systems. In the present optical communication systems which use coherent beams of laser light propagating in optical fibres, the ultimate limit to the noise is given by the quantum noise or zero-point fluctuations. If, instead, beams of squeezed light were used to transmit information in the quadrature that had reduced fluctuation, the quantum noise level could be reduced below the zero-point fluctuation. In principle, the quantum fluctuation in a squeezed state may approach zero as a limit; this means that squeezed light may have an infinite signal-to-noise ratio.

The statistical properties of light fields may be analysed by techniques similar to classical probability theory, which is based on the Glauber-Sudarshan P representation:

$$\rho = \int P(\alpha)|\alpha\rangle\langle\alpha|\, d^2\alpha \tag{2}$$

Essentially, the Glauber-Sudarshan P representation is to represent the density operator ρ of a radiation field as a statistical mixture of density operators $|\alpha\rangle\langle\alpha|$ for pure coherent states $|\alpha\rangle$, where $P(\alpha)$ serves as a distribution function. The representation has found wide-spread applications in quantum optics because the taking of quantum mechanical averages resemble classical averaging procedures. In fact, provided that $P(\alpha)$ exists as a positive nonsingular function, the expectation value of any operator f may be written by use of (2) as

$$\langle f\rangle = \int P(\alpha) f(\alpha)\, d^2\alpha\,, \tag{3}$$

where $f(\alpha) = \langle \alpha | f | \alpha \rangle$, which resembles the classical form for the average value of the observable $f(\alpha)$. The distribution function for chaotic fields is represented by $P(\alpha) = (1/\pi N)\exp(-|\alpha|^2/N)$, with N as the mean photon number, while coherent states have a distribution function in the form $P(\alpha) = \delta(\alpha - \alpha_0)$. Squeezed states, on the other hand, have no nonsingular distribution function. The statistical properties of such fields cannot be calculated by using the classical form (3). Squeezed states are, therefore, an example of a nonclassical light field. To be precise, a nonclassical light field is defined as one that has no nonsingular Glauber-Sudarshan P representation.

Another example of a nonclassical light field is displayed by fields which have sub-Poissonian photon statistics, which means a *narrow* photon distribution compared to Poissonian distribution with an equal mean photon number. In terms of the normalised second-order correlation function $g^{(2)}(0)$, a coherent light field with Poissonian statistics has $g^{(2)}(0) = 1$. A chaotic light which has increased intensity fluctuations has $g^{(2)}(0) = 2$. Since $g^{(2)}(0)$ represents the probability of two photons arriving simultaneously, $g^{(2)}(0) > 1$ is referred to as photon bunching. A light field with sub-Poissonian photon statistics will have $g^{(2)}(0) < 1$, an effect known as photon antibunching. For fields which display photon antibunching, the distribution functions in terms of Glauber-Sudarshan P representation are highly singular. In this sense we say that such fields are nonclassical. Squeezed states and antibunching as two typical examples of nonclassical effects of light fields have been observed experimentally in some quantum optical systems [3].

A question may now be asked: Are there any nonclassical effects in biological systems [4] ? This question is motivated by the following fact. Biological systems are essentially open systems, which exchange matter, energy and information with their environment; such systems are thus far away from thermal equilibrium. An organism is highly organised such that all of its components may be correlated with each other and behave cooperatively in metabolic process to maintain the organism as a whole in a high degree of orderliness. Accordingly, there must be something among the components, which is responsible for transmitting bio-information and which works with an extremely high efficiency in order to regularise the component's activities and to guarantee long-range correlation in the organism. What could play such a role? Instead of materials, it should be some kind of "field" acting within living matter [5].

Biophoton emission has been known as a universal phenomena in the biological kingdom [6-10]. It concerns a ultraweak emission of radiation from whole organisms and their tissues, cells, or molecules, with an intensity of the order of $10 - 1000$ photons/$(\mathrm{s} \cdot \mathrm{cm}^2)$ in the spectral range from 200 to 800 nm, at least. Biophoton emission is considered to originate from a delocalized electromagnetic field within a living matter, which may be responsible for transmitting bio-information. Thus unlike of a chaotic field, the biophoton field is likely to operate in a state with a high degree of coherence, even in a nonclassical state with extremely high signal-to-noise ratio [11,12], in order to ensure an extremely high efficiency of informational transfer in life activity. In fact, the coherent nature of biophoton field has been indicated in some biological systems, i. e., the interference effects from a swarm of *Daphnia magna* [13], the sub-Poissonian photon counting statistics for a great collection of *Gonyaulax polyera* [6], and the higher-order nonclassical effect in mungbean seedlings [10].

In this paper a model has been presented for the description of biophoton activity, which is based on the quantum theory of the interaction of a photon field with a phonon reservoir. The model is used to analyse the time development of nonclassical effects induced by the Schrödinger-cat-state. It also suggests a new technique of demonstrating statistical properties of biophotons *via* measurement of spontaneous emission. In connection with the model, the experiments on biophoton emission from a leaf have been performed and compared with a non-living radiator. The results show that biophotons have a significantly higher degree of coherence compared to ordinary light. Finally, an example of the applications of the new technique is mentioned.

2. Model: Interaction of the Biophoton Field with a Phonon Reservoir

According to the coherence theory of biophoton emission, the most essential role is played by the DNA. In particular, standing vibrational waves in the DNA lattice may provide the necessary conditions for biophoton emission [5]. Starting from this fact, we will establish a model, in terms of quantum optics, for the description of biophoton activity. It involves, in principle, the interaction of photons with phonons as a consequence of quantization for vibrational waves. We represent standing vibrational waves in the DNA lattice as a reservoir, which is regarded by a large collection of phonons occupying a wide variety of modes. On the other hand, we consider for simplicity a *single* mode of the biophoton field. The interaction of the biophoton field with the DNA phonon reservoir is diagrammed in Fig. 1.

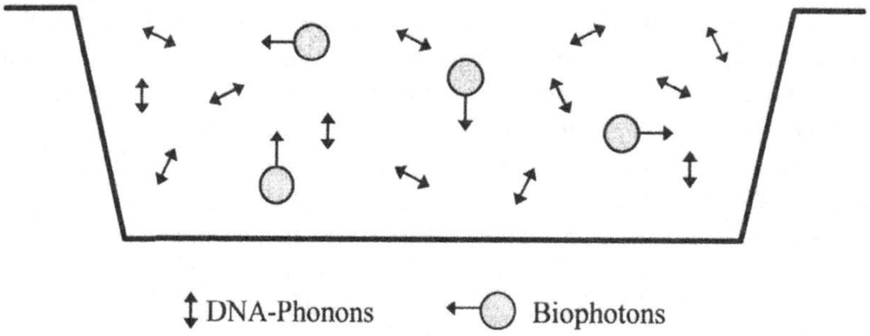

Figure 1. Diagram of interaction of biophotons with DNA-phonons. Biophotons originate from a delocalized and quantized electromagnetic field within a living matter. DNA-phonons, as a multi-mode excitation of standing vibrational waves in the DNA lattice, may provide the necessary conditions for biophoton emission.

We assume that interaction between the field and reservoir may be described in the rotating-wave approximation, so that the Hamiltonian of the entire system can be written in the form

$$H = \hbar\omega\left(a^+a + \frac{1}{2}\right) + \hbar\sum_j \omega_j\left(b_j^+ b_j + \frac{1}{2}\right) + \hbar\sum_j\left(\kappa_j b_j a^+ + \kappa_j^* b_j^+ a\right) \tag{4}$$

where the first part describes the single-mode field, b_j and b_j^+ are respectively annihilation and creation operators for the phonons in jth mode of frequency ω_j, satisfying the boson commutation relations $[b_j, b_k^+] = \delta_{jk}$, $\hbar\omega/2$ and $\hbar\omega_j/2$ are the corresponding zero-point energies, and κ_j is the coupling constant of the interaction. We assume that the reservoir spectrum is so flat that the mean number of reservoir oscillators (phonons) in the mode j, given by $\langle b_j^+ b_j\rangle$, is independent of j. We then neglect the thermal radiation of the system. Under these approximations, the master equation for the density operator ρ of the field is written in the interaction picture as

$$\frac{d\rho}{dt} = \frac{\gamma}{2}\left(2a\rho a^+ - a^+a\rho - \rho a^+a\right) \tag{5}$$

where γ is the damping constant [14]. The master equation (5) describes generally a non-thermal decay of a single-mode radiation field in a vast phonon reservoir, and it has, for any time t, the formal solution:

$$\rho(t) = \exp\left(\hat{L}t\right)\exp\left\{\frac{\hat{J}}{\gamma}\left[1 - \exp(-\gamma t)\right]\right\}\rho(0) \tag{6}$$

where $\hat{J}$ and $\hat{L}$ are superoperators, defined by their actions on the density operator through

$$\hat{J}\rho = \gamma a\rho a^+ \text{ and } \hat{L}\rho = -\frac{\gamma}{2}\left(a^+a\rho + \rho a^+a\right) \tag{7}$$

For a given initial density operator $\rho(0)$, we may write an analytical expression for the time-dependent density operator by use of (6) and (7). Then we can calculate, for instance, the photon number distribution of the field, which is defined by

$$P(n) = \langle n|\rho|n\rangle, \tag{8}$$

being the diagonal elements of the density operator in photon number states $|n\rangle$. We can also calculate the expectation value $\langle f\rangle$ of any operator $f(a,a^+)$ as a function of a and a^+, according to $\langle f\rangle = \mathrm{Tr}[\rho f(a,a^+)]$. In fact, we have the mean photon number:

$$N = \mathrm{Tr}\left(\rho a^+ a\right), \tag{9}$$

the normalised photon number variance:

$$V = \left[\langle a^+ a a^+ a\rangle - \langle a^+ a\rangle^2\right] / \langle a^+ a\rangle \tag{10}$$

and the fluctuations of the quadratures of the field, which are given by [15]

$$\begin{pmatrix} (\Delta x)^2 \\ (\Delta p)^2 \end{pmatrix} = 1 + 2\langle a^+ a\rangle \pm 2\,\mathrm{Re}\langle a^2\rangle - 4\begin{pmatrix} \mathrm{Re}^2\langle a\rangle \\ \mathrm{Im}^2\langle a\rangle \end{pmatrix} \tag{11}$$

As it is well known, the density operator of a quantum system can be used to determine all the quantum statistical properties of the system.

3. Dynamics of the Schrödinger-Cat-State

Let us consider an initial state, which is interesting in biological systems, and analyse its dynamics determined by Eq. (6). Universally, there is such a state that appears as a superposition of two components which are mutually opposite but correlated with each other. These components can be found to occur extensively in biological systems, for instance, "active" and "listless" components, "healthy" and "ill" components, "positive" and "negative" components, and "Yin" and "Yang" components, etc. Such state as a philosophic concept is earliest introduced by Schrödinger, so-called Schrödinger-cat-state, which is a quantum superposition of two macroscopically distinguishable states and may be diagrammed as

Schrödinger-cat-state can now be formulated in different ways, one of them gives [12]

$$|\mathrm{C}\rangle = \frac{1}{\sqrt{A}}\left[\,|\alpha \exp(i\phi_1)\rangle + |\alpha \exp(i\phi_2)\rangle\right], \tag{12}$$

which is a superposition of two coherent states with the same amplitude α but with the deferent phases ϕ_1 and ϕ_2, where $A = 2(1 + q\cos\varepsilon)$ is the normalisation constant with $q = \exp[-2\alpha^2 \sin^2(\phi/2)]$, $\varepsilon = -\alpha^2 \sin\phi$, and $\phi = \phi_1 - \phi_2$. The state (12) has been investigated generally from the quantum optical point of view. It is found to demonstrate the various novel features, such as the constructive and destructive interferences in mean photon number, sub-Poissonian photon statistics, an oscillation of photon distribution (in particular, $g^{(2)}(0) = 1$ may yields a strong oscillation of photon distribution, instead of a Poisson distribution), the squeezing in quadratures of the field, the variety of aspects of a vacuum state (it may be a special case of coherent states, number states or chaotic states, corresponding to $g^{(2)}(0) = 1, 0 \text{ or } 2$, respectively), and the "most quantum state" with an infinite product of the uncertainties in the two quadratures (in contrast with a coherent state which has the minimum uncertainty product and is thus the closest one to a classical field). All of these natures originate from quantum interference between the two components in the state (12).

We consider further the time development of the state (12) governed by (5). For this purpose, we first write the density operator (6) with the initial state (12) as

$$\rho(t) = \sum_{k,j=1}^{2} P_{kj}(t) |\Psi_j(t)\rangle\langle\Psi_k(t)| \tag{13}$$

where

$$|\Psi_j(t)\rangle = |\alpha \exp[(-\gamma t/2) + i\phi_j]\rangle, \tag{14}$$

$$P_{kj}(t) = Z_{kj}^{\;1-\exp(-\gamma t)} / A, \tag{15}$$

$$Z_{kj} = \langle \alpha \exp(i\phi_k) | \alpha \exp(i\phi_j) \rangle \tag{16}$$

Based on (13), we will reveal the quantum statistical properties of the field from the different points of view as follows.

3.1. GENERAL PROPERTIES

By using Eqs. (8-11, 13-16), we deduce analytical expressions for the following quantities: the photon number distribution $P(n,t)$, the mean photon number $N(t)$, the normalised photon number variance $V(t)$, and $F(t)$ which is the common minimum value of the fluctuations $(\Delta x)^2$ and $(\Delta p)^2$ [12]. These quantities are given, respectively, by

$$P(n,t)=\exp\left[-\alpha^2\exp(-\gamma t)\right]\frac{\alpha^{2n}\exp(-n\gamma t)}{n!}\frac{1+q^{1-\exp(-\gamma t)}\cos\{\varepsilon[1-\exp(-\gamma t)]-n\phi\}}{1+q\cos\varepsilon}, \quad (17)$$

$$N(t)=\alpha^2\exp(-\gamma t)\frac{1+q\cos(\varepsilon-\phi)}{1+q\cos\varepsilon}, \quad (18)$$

$$V(t)=1-2\lambda q\exp(-\gamma t)\frac{\cos(\varepsilon-\phi)+q\cos^2(\phi/2)}{[1+q\cos\varepsilon][1+q\cos(\varepsilon-\phi)]}, \quad (19)$$

$$F(t)=1-2\lambda q\exp(-\gamma t)\frac{q+\cos\varepsilon}{(1+q\cos\varepsilon)^2} \quad (20)$$

where $\lambda=2\alpha^2\sin^2(\phi/2)$. It is clear that $V(t)<1$ and $F(t)<1$ correspond to sub-Poissonian photon statistics and squeezing effects, respectively.

Let us first discuss two limiting cases of the results (17-20) as follows. (1) As $t=0$, they reduce to the previously obtained results [12]. (2) For $t\to\infty$, the field approaches a vacuum state with $\rho(\infty)=|0\rangle\langle 0|$, characterised by $P(n,\infty)=\delta_{n0}$, $N(\infty)=0$, and $V(\infty)=F(\infty)=1$, which shows the features of a coherent state. We next discuss a particular case $\phi_1=\phi_2$ that leads Eq. (12) to a coherent state. The quantities shown in Eqs. (17-20) are now give by $P(n,t)=\exp[-N(t)][N(t)]^n/n!$, $N(t)=\alpha^2\exp(-\gamma t)$ and $V(t)=F(t)=1$. These show that the field remains in time-development a coherent state with a Poissonian photon distribution and an exponential decay of the mean photon number. This is a natural consequence of the employed Hamiltonian (4) which preserves coherence for all times [16].

The dynamics of the Schrödinger-cat-state (12) is shown generally in Fig. 2. One can see that the time behaviour of the mean photon number $N(t)$ is almost the same for both fields. However, they exhibit a certain difference in the photon number distribution $P(n,t)$, namely the small peak at large n values, as shown for $t=0.2$. It is this difference that results in the two fields having completely different photon statistical properties at any time: one is bunching field with super-Poissionian statistics while another is an antibuching field with sub-Poissionian statistics. One also can see that for the present case the antibuching effect appears together with the squeezing effect, which is to be regarded as a completely nonclassical effect. Since both fields are dissipated by the reservoir, their intensities and fluctuations reduce commonly, finally approaching the vacuum state.

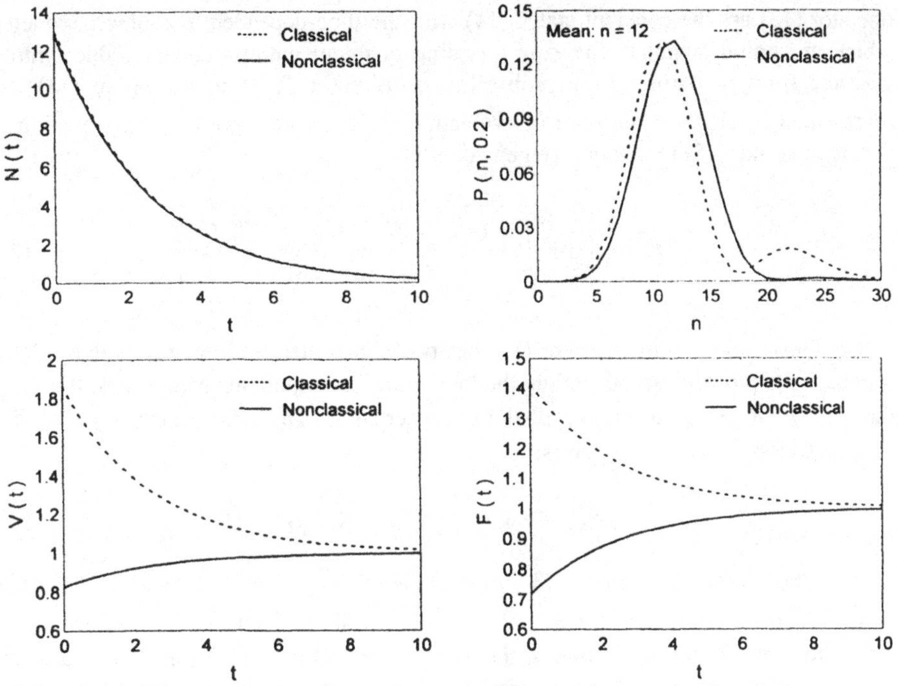

Figure 2. Dynamic aspects of the Schrödinger-cat-state characterised by (17-20) with $\gamma = 0.4$. The initial states are taken as: (1) $\alpha = 4, \phi = 0.168$ and (2) $\alpha = 4, \phi = 0.142$, which have the equal mean photon number $N(0) = 13$; the former is a classical field with $V(0) = 1.86$ and $F(0) = 1.42$, while the latter is a nonclassical field with $V(0) = 0.83$ and $F(0) = 0.72$. Both fields present their own nature with the passage of time and evolve towards the vacuum state.

3.2. QUANTUM ENTROPY

The density operator (13) describes a mixed state for $t > 0$, since $\mathrm{Tr}\rho^2(t) \neq 1$. This suggests to analyse further the properties of the field in terms of quantum entropy. The importance of this point lies in the fact that the quantum statistical properties of a radiation field with a mixture of coherent and non-coherent signals may be best described by a quantum entropy [17], which is defined generally in terms of density operator ρ as

$$S = -\mathrm{Tr}\rho \ln \rho, \tag{21}$$

subject to the normalised condition $\mathrm{Tr}\rho = 1$. Since the component states in the density operator (13) are the coherent states (14) with the time-dependent complex parameter, which are non-orthogonal, the corresponding quantum entropy cannot reduce into a classical form in terms of the probability distribution $P_{k_j}(t)$ appearing in (13) and represented by (15). For purpose of evaluating the trace we solve the eigenequation of density operator (13) to obtain its eigenvalues [18]:

$$R^{\pm}(t) = \frac{1}{2}\left\{ 1 \pm \sqrt{1 - \frac{1 + q^2 - q^{2\exp(-\gamma t)} - q^{2[1-\exp(-\gamma t)]}}{(1 + q\cos\varepsilon)^2}} \right\} \tag{22}$$

Obviously, the density operator (13) has two eigenstates, and the eigenvalues $R^{\pm}(t)$ represent the probabilities of finding the field in the corresponding eigenstates. It is clear that $R^{+}(t) + R^{-}(t) = 1$ displays probability conservation. The quantum entropy (21) then reduces to the following two terms:

$$S(t) = -R^{+}(t)\ln R^{+}(t) - R^{-}(t)\ln R^{-}(t) \tag{23}$$

As the limiting cases of (23), we have for $t = 0$ and $t \to \infty$ that $R^{+}(0) = R^{+}(\infty) = 1$ and $R^{-}(0) = R^{-}(\infty) = 0$, whence $S(0) = S(\infty) = 0$, i. e. we have pure states. In fact the field is known in the Schrödinger-cat-state (12) for $t = 0$ and in the vacuum state $|0\rangle$ for $t \to \infty$. Provided the existence of a time t satisfying $1 + q^2 - q^{2\exp(-\gamma t)} - q^{2[1-\exp(-\gamma t)]} = (1 + q\cos\varepsilon)^2$, at which we have that $R^{\pm}(t) = 1/2$, the quantum entropy of the field reaches ln2, which is the greatest possible value of entropy of a two-state system.

The details of the time-dependence of the quantum entropy $S(t)$ given in (23) are shown in Fig. 3(a). For $t > 0$, the two fields are both in the mixed states [$S(t) > 0$], one of which corresponds to a nonclassical field with the sub-Poissonian photon statistics and squeezing effects (see Fig. 2). We thus arrive at the significant conclusion that a nonclassical light may operate in a mixed state (must not be in a pure state). It suggests a wide range of possibilities to generate nonclassical states in actually optical systems. On the other hand, this makes it understandable that biophoton field operates in a nonclassical state, because it is probably difficult to image biophoton field as a pure state of nonclassical light, like the ideal two-photon coherent state. In addition, one can see that for any particular time $t > 0$, the value of $S(t)$ for the nonclassical field is lower than that of the classical field. This shows that the nonclassical field has a lower uncertainty to be found in the eigenstates of the density operator. We thus arrive at another significant conclusion that quantum entropy may serve as a measure of noise and order in the field: the lower value of quantum entropy the lower noise and the higher order.

For the present case in which the density operator has only two eigenstates, the probability difference $R(t) \equiv R^{+}(t) - R^{-}(t)$ characterises the certainty of finding the

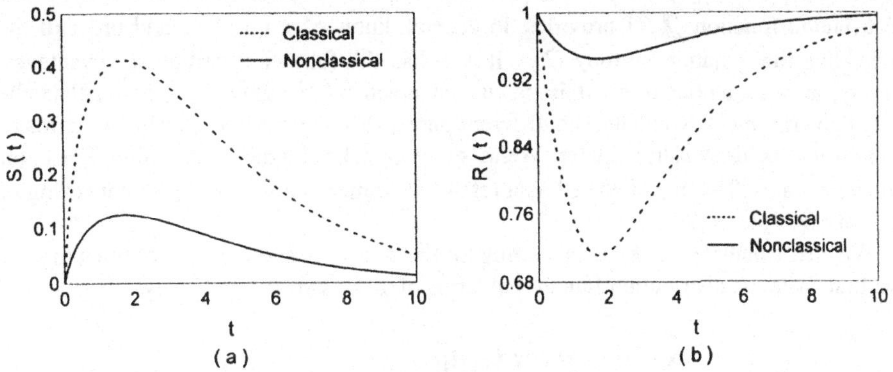

Figure 3. Dynamic aspects of the Schrödinger-cat-state characterised by: (a) the quantum entropy $S(t)$ from (23) and (b) $R(t)$ as the difference between the probabilities $R^{\pm}(t)$ represented in (22). The initial conditions are the same as in Fig. 2.

field in the eigenstates. The time development of $R(t)$ is shown in Fig. 3(b), where the value for the nonclassical field is higher than that of the classical field for any given time. This demonstrates the expected fact that the nonclassical field has a higher certainty to be found in the eigenstates of the density operator.

3.3. WEHRL'S ENTROPY

The quantum entropy (21) has the problem that it gives the same value for all pure states, whether or not they are classical or nonclassical. Wehrl introduced a new definition of the entropy for a radiation field in terms of coherent states by [19]

$$W = -\frac{1}{\pi}\int Q(\beta)\ln Q(\beta)\, d^2\beta\,, \tag{24}$$

where $Q(\beta) = \langle\beta|\rho|\beta\rangle$ are the diagonal elements of the density operator in coherent states $|\beta\rangle$, subject to the constraint $\int Q(\beta)\, d^2\beta = \pi$. We recognise that $Q(\beta)$ is exactly the distribution function corresponding to the antinormal ordering of operators, and is called the Q representation. The distribution function $Q(\beta)$ involves the x,y-representation through (1) and therefore contains information on the amplitude of the field, in particular, on the fluctuations in quadratures of the field. The expression (24) as a definition of entropy is reasonable because: (1) the Q representation is always non-negative, unlike other representation such as P and W representations, and (2) $Q(\beta)$ is a bounded function lying within unit. Thus $Q(\beta)$ serves as a standard distribution function since $0 \leq Q(\beta) \leq 1$. Wehrl's entropy as a measure of the uncertainty of the distribution function $Q(\beta)$ provides, in general, knowledge on noise and order of the field, like the quantum entropy (21). It also has the following particular advantages. Firstly, as a c-number form, it is calculated much more easily, in general, than the quantum entropy is. Secondly, the different pure states may be distinguished in terms of it, as it will be demonstrated later. Wehrl's entropy takes its minimum value $W = 1$ for coherent states. This may be used as a reference to measure the degree of squeezing of the squeezed states [20].

We next calculate the Wehrl's entropy for the density operator (13). For this purpose, we first write the distribution function in terms of $\beta = r\exp(i\theta)$ as follows:

$$Q(\beta,t) = \frac{\exp\{-[r^2 + \alpha^2 \exp(-\gamma t)]\}}{2(1+q\cos\varepsilon)}\left[Q_1(\beta,t) + Q_2(\beta,t) + Q_{12}(\beta,t)\right] \tag{25}$$

where

$$Q_j(\beta,t) = \exp\left[2\alpha r \exp(-\gamma t/2)\cos(\theta - \phi_j)\right] \ (j = 1 \text{ or } 2), \tag{26}$$

$$\begin{aligned} Q_{12}(\beta,t) = {} & 2q^{1-\exp(-\gamma t)} \exp\left[2\alpha r \exp\left(-\frac{\gamma t}{2}\right)\cos\left(\theta - \frac{\phi_1+\phi_2}{2}\right)\cos\left(\frac{\phi_1-\phi_2}{2}\right)\right] \\ & \bullet \cos\left\{\varepsilon[1-\exp(-\gamma t)] - 2\alpha r \exp\left(-\frac{\gamma t}{2}\right)\cos\left(\theta - \frac{\phi_1+\phi_2}{2}\right)\sin\left(\frac{\phi_1-\phi_2}{2}\right)\right\}. \end{aligned} \tag{27}$$

Obviously, Q_1 and Q_2 are associated, respectively, to the two coherent states defined in (14), and Q_{12} originates from the quantum interference between them. By using (25-27), the Wehrl's entropy (24) can be calculated numerically and the results are shown in Fig. 4(a), where the two different initial pure states (classical and nonclassical) have been distinguished. As $t \to \infty$, the fields reach the vacuum state $|0\rangle$, which, as a special case of coherent states $|\beta\rangle$ corresponding to $\beta = 0$, has $W(\infty) = 1$. The more interesting fact is that at any given time $t \geq 0$ the value of the Wehrl's entropy for a nonclassical field is lower than that of a classical field.

In order to understand further the nature of the Q representation, we next discuss the dependence of distribution function $Q(\beta,t)$ shown in (25) on the complex parameter β. For simplicity, we consider only its radius component:

$$Q(r,t) = \frac{r}{\pi}\int_0^{2\pi} Q(\beta,t)\, d\theta \tag{28}$$

which is normalised: $\int_0^{\infty} Q(r,t)\, dr = 1$. The numerical results are shown in Fig. 4(b). One can see that for equal mean values of r the nonclassical field has a narrower distribution

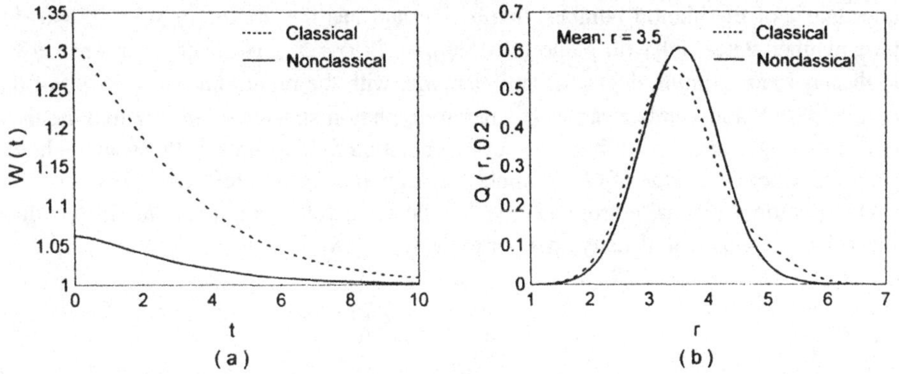

Figure 4. Dynamic aspects of the Schrödinger-cat-state characterised by: (a) the Wehrl's entropy $W(t)$ from (24) and (b) the radius distribution function $Q(r,t)$ from (28) for $t = 0.2$, where Eqs. (25-27) have been used for the calculations. The initial conditions are the same as in Fig. 2.

compared to the classical one. This is the root of the existence of the squeezing effects in the nonclassical field. As mentioned above, the Q representation really contains the information on the fluctuations in quadratures of the field.

3.4. PHOTON STATISTICAL ENTROPY

The difficulty with the quantum entropy (21) and Wehrl's entropy (24) lies in the fact that they remain unmeasurable at present. Actually, neither eigenvalues of a density operator or its Q representation have yet been found to be measurable by the present quantum optical technique, although some possible methods have been suggested [18]. Recently, some researchers [18, 21] have proposed

$$G = -\sum_{n=0}^{\infty} P(n) \ln P(n) \tag{29}$$

as an alternative measurement of entropy for the radiation field on account of the measurability of photon number distribution $P(n) = \langle n|\rho|n\rangle$. Essentially, such an entropy is introduced by techniques similar to Wehrl's entropy. The distribution function $Q(\beta)$ employed in Wehrl's entropy and the probability distribution $P(n)$ in the entropy (29) are the expectation values of the density operator in coherent states $|\beta\rangle$ and in number states $|n\rangle$, respectively. $Q(\beta)$ differs from $P(n)$ in that the former involves the amplitude of the field (a, a^{+}-representation), while the latter involve the number of photons ($a^{+}a$ – representation). Therefore the entropy (29) is essentially a "photon statistical entropy". Obviously, photon statistical entropy works as a measure of uncertainty of the photon number distribution and has the minimum value $G = 0$ for pure number states $|m\rangle$ with photon distribution $P(n) = \delta_{nm}$. Number states are known to display the maximum degree of antibunching, with the minimum value $V = 0$ of the normalised photon number variance. Therefore, photon statistical entropy may be used to measure the degree of antibunching of the radiation fields with sub-Poissonian photon statistics, where the value of G for number states will serve as a reference [18].

The photon statistical entropy (29) yields the same value as a quantum entropy does for radiation fields with density operator in the form [18]

$$\rho = \sum_{n=0}^{\infty} P(n)|n\rangle\langle n|, \tag{30}$$

which is a statistical mixture of pure number states $|n\rangle$. This means that the photon statistical entropy (29) is exactly applicable to particular fields, such as phase-average, mixed optical fields [18, 21]. On the other hand the photon statistical entropy (29) results, for a pure state with definite phase, in the same value as quantum entropy does for the corresponding phase-averaged state. For example, for a coherent state of the density operator $\rho = |\alpha\rangle\langle\alpha|$, it gives

$$G = N(1 - \ln N) + \exp(-N)\sum_{n=0}^{\infty}\frac{N^n \ln n!}{n!} \tag{31}$$

with N as the mean photon number. This just is the result of the quantum entropy [18] for the phase-averaged coherent state [14] characterised by the density operator (30) with $P(n)$ as a Poissonian distribution.

The coherent state represents a coherent signal with a definite phase while the phase-averaged coherent state represents a coherent signal with no knowledge of the phase. At optical frequencies the latter situation represents the coherent field, for instance, a laser above threshold, to a good approximation [14]. This means that the photon statistical entropy (31) is applicable for a coherent *optical* field in a considerable degree.

In fact, photon statistical entropy may contain phase information provided that the photon distribution of the field is phase-dependent. An example is displayed by the photon distribution (17), which contains the phase difference ϕ. For this distribution, Eq. (29) can be calculated numerically and the results are shown in Fig. 5(a). One can see that, like the Wehrl' entropy, the two different pure states at $t = 0$ have been distinguished. As $t \to \infty$, the fields approach the vacuum state $|0\rangle$, which, as a special case of number states $|m\rangle$ corresponding to $m = 0$, displays $G(\infty) = 0$. In addition, the photon statistical entropy in terms of the photon distribution (17) is shown as a function of the mean photon number $N(t)$ given by (18) in Fig. 5(b), where $G(N)$ displays a monotonic increase with N. This indicates that in the present case, the mean photon number is a measure of the photon statistical entropy and may reflect the uncertainty of

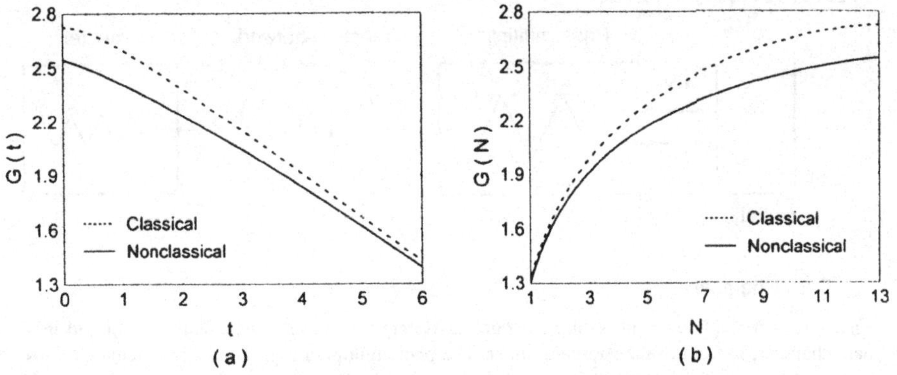

Figure 5. Dynamic aspects of the Schrödinger-cat-state characterised by the photon statistical entropy (30) as a function of: (a) time t and (b) the mean photon number (18), where the photon distribution (17) used for the calculations. The initial conditions are the same as in Fig. 2.

the photon distribution. The relation between G and N can be measured experimentally for an actual radiation field, as it will be demonstrated later.

At the conclusion of this section, the following points should be emphasised for the two fields as characterised in Figs. 2-5: (1) They have the same intensity behaviour but the completely different quantum statistical properties: one is classical and the other nonclassical. (2) For any given time, the quantum entropy, the Wehrl's entropy and the photon statistical entropy all display lower values for a nonclassical field than a classical one, indicating that these entropies are mutually equivalent in the present case. (3) These entropies may serve as a measure of noise and order in the field - in other words, as a measure of its degree of coherence.

4. Experiments: Biophoton Statistical Properties

The above-mentioned theoretical investigations will be connected here with experimental observations on biophoton emission. In particular, we will experimentally demonstrate biophoton statistical properties by comparison with a non-living radiator.

4.1. INSTRUMENT

Photon emission from a sample is detected by means of the single-photon-counting-system, and its principle is shown in Fig. 6. The instrument operates mainly with the photomultiplier. Its cathode has a diameter of 4.4 cm and is sensitive within the spectral range of 200 to 800 nm. The entrance for photons entering the photomultiplier has diameter of 6 cm.

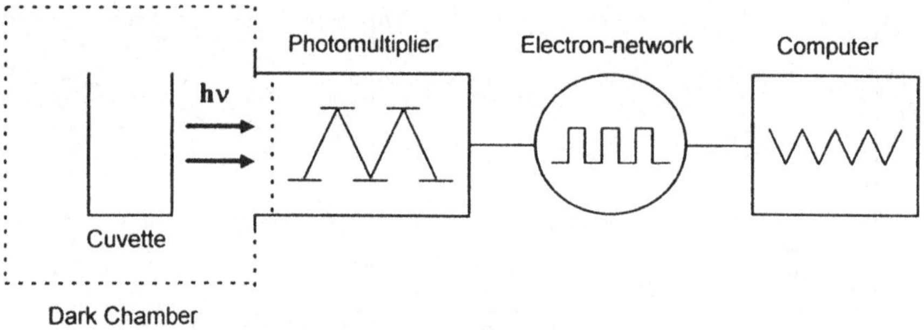

Figure 6. Principle of a single-photon-counting-system. The cuvette with sample is located in a dark chamber, in front of the photomultiplier. The photomultiplier registers the impinging photons and transforms them into electric signals, which are treated then by the electronic network (amplifier, discriminator and counter). A computer is used to record and to display the resulting data.

4.2. EXPERIMENTAL DESIGN

Without any kind of stimulation, the spontaneous emission of radiation from the samples is measured with a time interval of counts, so-called dwell time and denoted by Δt. The measurement consists of a sufficiently large number M of observed values of counts; in this process the photon emission is required to be quite stationary. Based on these M, values a normalised distribution $P(n)$, the probability of finding number n of counts, is obtained as a function of Δt. Then the average value of any quantity $f(n)$ as a function of n can be calculated by use of relation

$$\langle f \rangle = \sum_{n=0}^{n_m} f(n)P(n) \tag{32}$$

where n_m is the maximum of the counts in the M values. In particular, the mean value of counts and the photon statistical entropy are obtained as $f(n) = n$ and $f(n) = -\ln P(n)$, respectively. Furthermore, the normalised photon number variance V can be represented by $V = [\langle n^2 \rangle - \langle n \rangle^2]/\langle n \rangle$.

4.3. SAMPLES AND MEASUREMENT CONDITIONS

The measured samples are a piece of fresh leaf as a living sample and the non-living radiator. The latter is a standard source of light (called "Lichtzeiger" in German) with a very stable intensity. The leaf was cut from a gum tree (rubber plant). and had a total area of about 35 $cm^2$and a thickness of about 0.5 mm. It was fixed exactly in front of the photomultiplier, with the measured area of 28 cm^2 (being the area of the entrance for photons entering the photomultiplier). The samples were measured at the room-

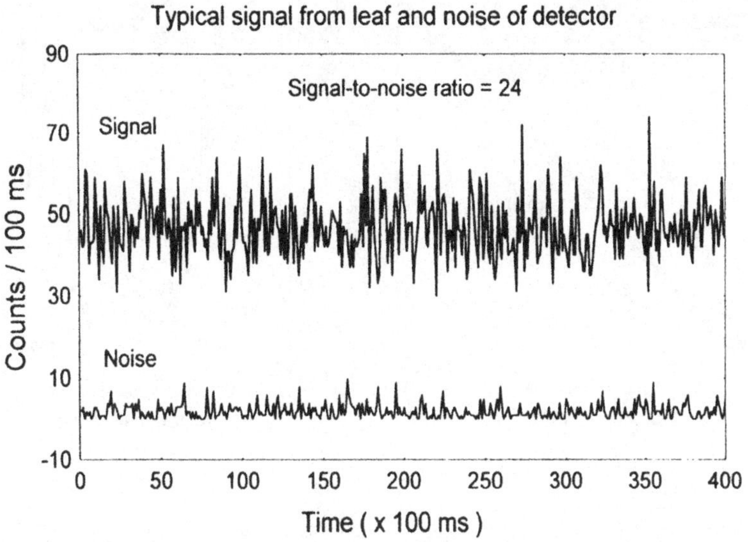

Figure 7. A typical signal detected from the leaf in stationary emission is compared with the noise of the detector, where the total number of observed values of counts is $M = 400$ and the dwell time is taken as $\Delta t = 100ms$.

temperature. The number of observed values of counts is taken as $M = 400$ for every measurement.

4.4. RESULTS

A typical signal is shown in comparison with the noise of the detector in Fig. 7. The results concerning the stationary emission are shown in Fig. 8, which involves changes of dwell time Δt. Before these measurements, we spent 1.5 hours in measurements on the same leaf, with the fixed dwell time $\Delta t = 100ms$. During this mean time the photon emission was found to be in decay.

However, in every measurement for getting 400 observed values, which took $40s$, the photon emission was quite stationary, like that shown in Fig. 7. The results concerning the decay process are shown in Fig. 9, where the data corresponding to stationary emission also are included for comparison.

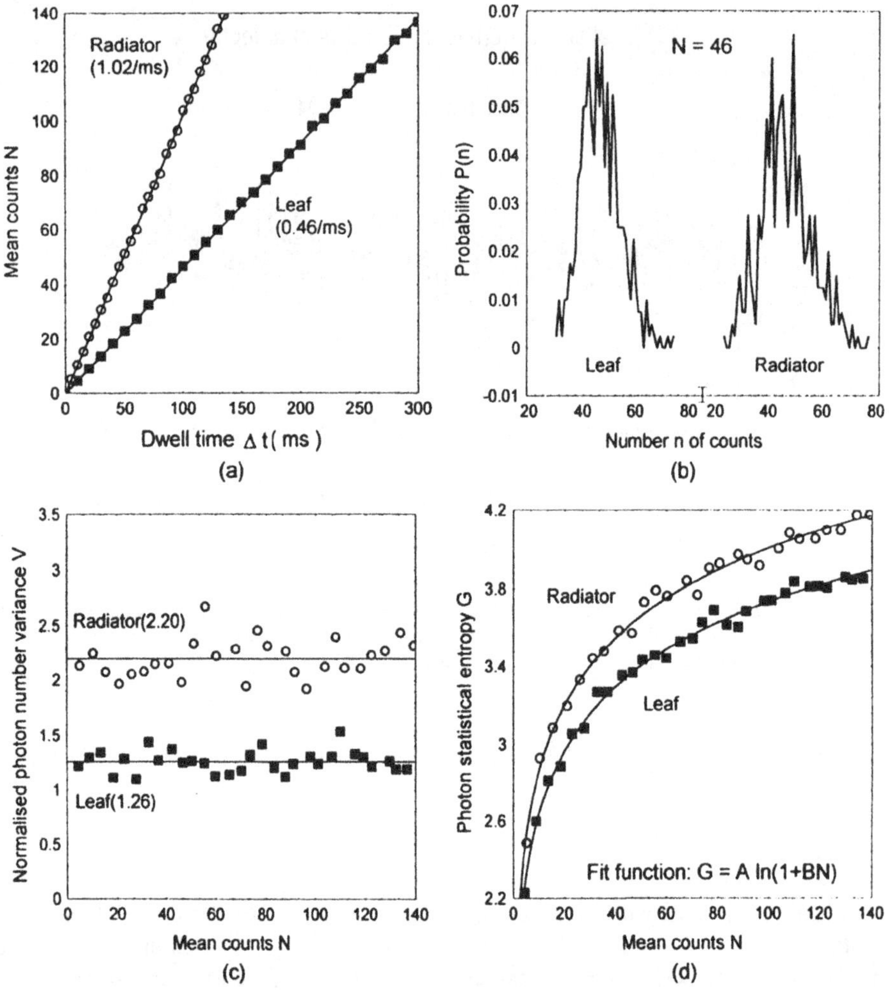

Figure 8. The comparison in photon statistical properties between the leaf and the radiator. (a) Changes of the mean counts N with dwell time Δt, where the two linear relations demonstrate that the samples operate in stationary photon emission for over the total measurements. The two numbers in brackets are the emission rates of the samples. (b) Typical probability distributions of number n of counts, where $\Delta t = 100ms$ for the leaf and $\Delta t = 45ms$ for the radiator result in an equal number of mean counts, $N = 46$. The former is based on the data shown in Fig. 7. (c) With change of N, the normalised photon number variance V displays a fluctuation around 1.26 for the leaf and around 2.20 for the radiator. (d) Changes of the photon statistical entropy G with N, which are found to be in agreement with those shown in Fig. 5(b). The observed values are fitted by the entropy as a function of photon intensity, which is one of consequences of the semiclassical theory of biophoton emission [22], where $A = 0.48, B = 25.5$ for the leaf, and $A = 0.50, B = 31.3$ for the radiator.

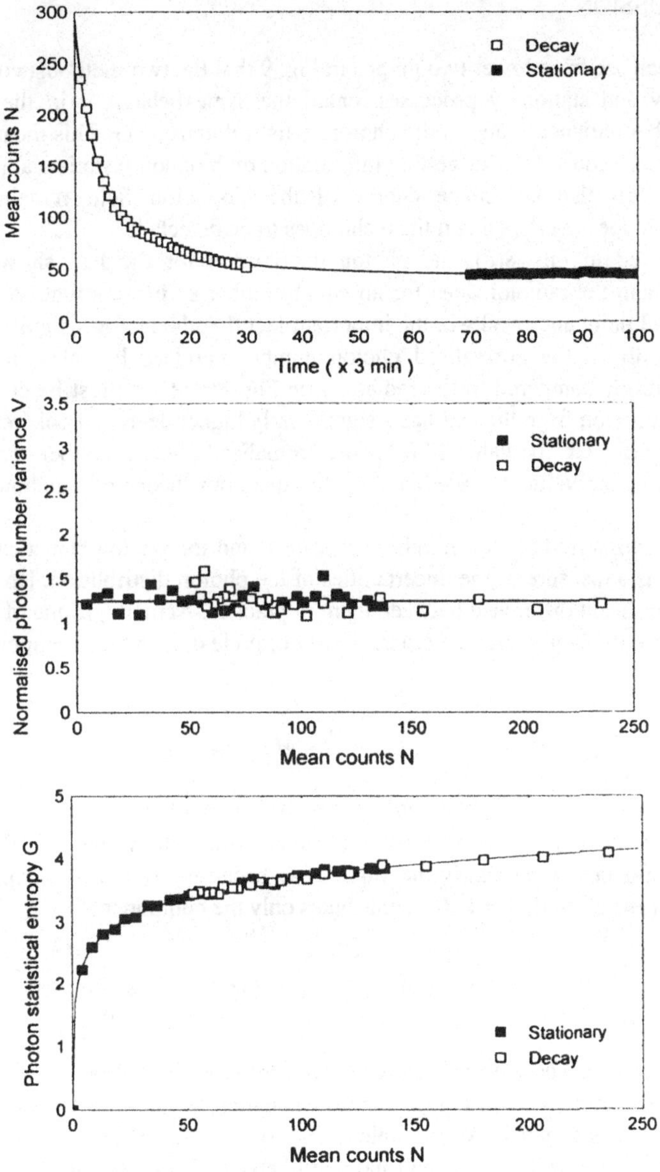

Figure 9. The statistical properties of biophotons from the leaf. Both decay and stationary processes of biophoton emission are included, each consisting of 30 values of N. The decay involves the fixed dwell time $\Delta t = 100ms$, while the stationary corresponds to the data in Fig. 8. Both decay and stationary processes are found to follow the same behaviour in V and in G.

4.5. DISCUSSION

Firstly one can see from lower two graphs in Fig. 9 that the two methods, corresponding to the decay and stationary processes, entail the same behaviour in the normalised photon number variance V and in the photon statistical entropy G. This means that these two methods are equivalent for getting information on biophoton statistical properties. It reflects the fact that coherence degree of the biophoton field remains relatively stationary and does not depend on the techniques to be detected.

As displayed in Fig. 8(b), the photon distribution for the leaf shows a smaller uncertainty than the radiator does for an equal number of mean counts N. This is the most essential point and results in the important fact that the leaf has significantly lower values not only in the normalised photon number variance but also in the photon statistical entropy compared to the radiator [see Fig. 8(c,d)]. It must be concluded that the photon emission from the leaf has a significantly higher degree of coherence than the radiator. In particular, the value 1.26 for the normalised photon number variance of the leaf is close to the value 1.2 predicted by the quantum theory of biophoton emission [11].

Both the normalised photon number variance V and the photon statistical entropy G can be used as a measure of the uncertainties of the photon distribution. However, there may be a significant difference between them in practice. Actually, if one of M observed points is unusually larger than the others, it may provide quite a high component

$$nP(n) = \frac{n_m}{M} \tag{33}$$

to the mean value $\langle n \rangle$. This point influences the mean value $\langle n^2 \rangle$ more significantly. Consequently, it may play a considerable role to determine the value of V. But this point can never influence significantly the value of G because any unusual point appearing once, with probability $P(n) = 1/M$, contributes only the component

$$-P(n)\ln P(n) = \ln M/M \xrightarrow{\text{large } M} \frac{1}{M} \tag{34}$$

to the total entropy. Therefore, G is insensitive for these unusual points compared to V. This is one of reasons why in Fig. 9 G shows a smaller deviation between the observed and predicted values than V. An example is displayed in Fig. 10, which shows the lager difference in the influence of the unusual point between V and G. On the other hand, it has been proved theoretically that the component entropy $[-P(n)\ln P(n)]$ as a function of n takes its maximum value at the most probable photon number n_p if the mean value satisfies $N \geq 2$ [18]. This means that in $n_m + 1$ components of the photon statistical entropy, one that contributes most to the total entropy is $[-P(n_p)\ln P(n_p)]$ for an actual light field, which corresponds to the peak probability $P(n_p)$.

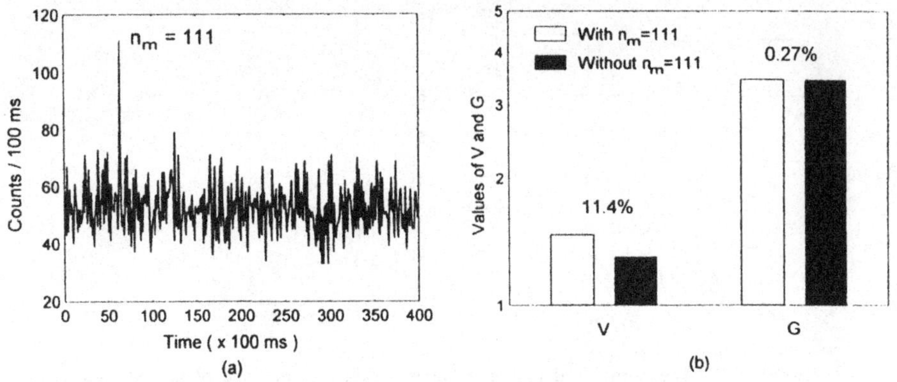

Figure 10. (a) Original data of spontaneous emission of biophotons from the leaf, which correspond to the last point of the decay process shown in the top of Fig. 9, where the biggest value $n_m = 111$ is unusual. (b) Changes in the normalised photon number variance V and in the photon statistical entropy G after the biggest value is removed, where the two percents represent their relative decrease. The sensitivity of the G to the unusual point is found to be more than 40 times lower than the V.

4.6. APPLICATIONS

Generally, a probability distribution $P(n)$ of the observable n yields a value of entropy in the form (29), which depends apparently on the distribution $P(n)$ rather than on the average value of n. Such an entropy, unlike the corresponding standard deviation determined by the mean values $\langle n \rangle$ and $\langle n^2 \rangle$, does almost not take error in measurement of observable n into account, therefore, can display a more reliable and more accurate measure of uncertainty in observable n than the standard deviation [18, 23]. Especially, Eq. (29) as a *measurable* photon statistical entropy can be used to characterise the quantum statistical properties of the light field: the lower value of it, the lower noise of the field and its higher order. Thus photon statistical entropy as a new quantity may be expected to have a wide application in biophoton analysis. An example is displayed in Fig. 11, where the photon statistical entropy shows its advantage over the normalised photon number variance.

5. Conclusion

We use the model of the interaction of a single-mode radiation field with a vast phonon reservoir to describe biophoton activity, which is based on the potential correlation of biophotons with the DNA-phonons. As an example, we discuss the dynamics of the Schrödinger-cat-state as a superposition of two coherent states with the same amplitude but with different phases. One can see that the quantum statistical properties of the

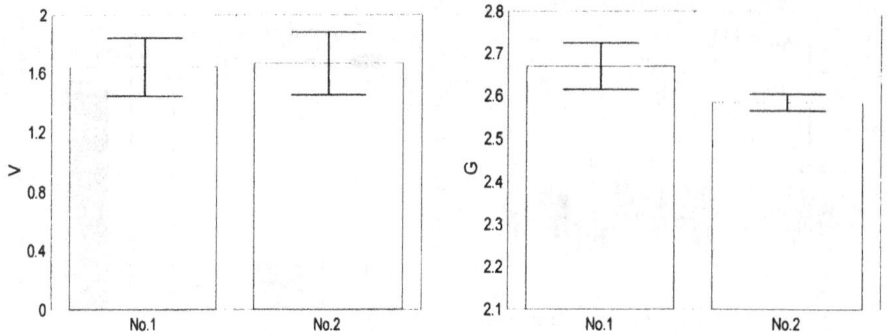

Figure 11. The normalised photon number variance V and the photon statistical entropy G, which are from the measurements of spontaneous emission of biophotons of two samples, provided by GREENPEACE in Hamburg, Germany, No.1: the traditional soybean and No.2: genetically treated from No.1 (Roundup-Ready-Soybean, also genmanipuliert). The temperature of measured samples is 26°C, their volume 17 cm^3, the dwell time $\Delta t = 100\,\text{ms}$, and the number of observed points $M = 1800$. The original data of 4 repetitions are used for the evaluations. The significant difference between the two samples are found in the photon statistical entropy G.

fields can be characterised not only by the general quantities, such as the mean photon number, the normalised photon number variance and the fluctuations of quadratures of the field, but also by the entropies, including the quantum entropy, the Wehrl's entropy and the photon statistical entropy. These entropies may serve as a measure of noise and order in the field [24] - in other words, as a measure of its coherence.

Based on the theoretical background, we experimentally investigated biophoton statistical properties by means of the measurement of spontaneous emission of radiation from the piece of fresh leaf in comparison with a natural radiator. The results show: (1) The photon distribution for the leaf has a smaller uncertainty than the radiator for an equal number of mean counts. (2) With the changes of the mean value of counts, the normalised photon number variances remain around 1.26 for the leaf and around 2.20 for the radiator. (3) For any given mean value of counts the photon statistical entropy for the leaf displays a significantly lower value compared to that of the radiator. All of these demonstrate clearly that biophotons have a significantly higher degree of coherence than ordinary light. As an application, the photon statistical entropy is used to distinguish the two soybean samples with and without genetic treatment.

The biophoton field as a delocalized electromagnetic field within a living matter is considered responsible for transmitting bio-information. It is likely to operate in a state with a high degree of coherence, even in a nonclassical state with an extremely high signal-to-noise ratio, in order to satisfy the requirement of extremely high efficiency of informational transfer in life activity. However, our experiments have not yet indicated any nonclassical effects of detected signals. One of the possible reasons is that the detected signals are a mixture of biophotons with the chemiluminescence induced by chlorophyll in the leaf; and the latter is known to have less coherence. The further experiments are in preparation.

6. References

1. Gu, Q.: Squeezing in a two-photon Jaynes-Cummings model, *Acta Phys. Sin.* **37** (1988), 752-759
2. Gu, Q.: Photon antibunching in light from a single-mode laser, *Chin. Phys. Lasers*(USA) **15**(1988), 615-619
3. Meystre, M. and Walls, D. F.(eds.): *Nonclassical Effects in Quantum Optics*, American Institute of Physics, New York, 1991.
4. Gu, Q.: Ultraweak photonemission from living systems, *Science* (China) **41** (1989), 35-40.
5. Popp, F. A.: Coherent photon storage of biological systems, in F. A. Popp, U. Warnke, H. L. König and W. Peschka (eds.), *Electromagnetic Bio-Information*, Urban & Schwarzenberg, München-Wien-Baltimore, 1989, pp 144-167.
6. Popp, F. A.: Some essential questions of biophoton research and probable answers, in F. A. Popp, K. H. Li and Q. Gu (eds.), *Recent Advances in Biophoton Research and its Applications*, World Scientific, Singapore, New Jersey, London, Hong Kong, 1992, pp 1-46.
7. Gu, Q.: The ultraweak photon emission from biological systems, *Quantum Electronics* (China) **5** (1988), 97-108.
8. Gu, Q.: Analogy between life and the laser, *J. Infrared & Millimetre* (USA) **8** (1989), 83-89.
9. Gu, Q. and Popp, F. A.: Nonlinear response of biophoton emission to external perturbations, *Experientia* **48** (1992), 1069-1082.
10. Popp, F. A., Gu, Q. and Li, K. H.: Biophoton emission: Experimental background and theoretical approaches, *Modern Phys. Lett. B* **8** (1994), 1269-1296.
11. Gu, Q.: Quantum theory of biophoton emission, in F. A. Popp, K. H. Li and Q. Gu (eds.), *Recent Advances in Biophoton Research and its Applications*, World Scientific, Singapore, New Jersey, London, Hong Kong, 1992, pp 59-114.
12. Gu, Q.: Quantum interference between coherent states, in L. V. Beloussov and F. A. Popp (eds.), *Biophotonics*, Bioinform Services Co., Moscow, 1995, pp 115-135.
13. Galle, M.: Population density-dependence of biophoton emission from *Daphnia*, in F. A. Popp, K. H. Li and Q. Gu (eds.), *Recent Advances in Biophoton Research and its Applications*, World Scientific, Singapore, New Jersey, London, Hong Kong, 1992, pp 345-355.
14. Louisell, W. H.: *Quantum Statistical Properties of Radiation*, Wiley, New York, 1973.
15. Gu, Q.: Squeezing in an one-atom maser, *Chin. Phys. Lasers* (USA) **17**(1988), 347-351.
16. Glauber, R. J.: Classical behavior of systems of quantum oscillators, *Phys. Lett.* **21** (1966), 650-652
17. Barnett, S. M. and Phoenix, S. J. D.: Entropy as a measure of quantum optical correlation, *Phys. Rev. A* **40** (1989), 2404-2409.
18. Gu, Q.: Entropy of radiation fields in the mixed states, in preparation.
19. Wehrl, A.: On the relation between classical and quantum-mechanical entropy, *Rep. Math. Phys.* **16** (1979), 353-358.
20. Lee, C. T.: Wehrl's entropy as a measure of squeezing, *Opt. Comm.* **66** (1988), 52-54.
21. Perinova, V., Krepelka, J., Perina, J., Luks, A. and Szlachetka, P.: Entropy of optical fields, *Optica Acta* **33** (1986), 15-32.
22. Gu, Q.: Semiclassical theory of biophoton emission, in preparation.
23. Deutsch, D.: Uncertainty in quantum measurements, *Phys. Rev. Lett.* **50** (1983), 631-636.
24. Gu, Q. and Popp, F. A.: Biophoton emission as a potential measure of organisational order, in E. G. Jung and M. F. Holick (eds.), *Biologic Effects of Light 1993*, Walter de Gruyter, Berlin, New York, 1994, pp. 425-443, and *Science in China B* **37** (1994), 1099-1112.

COHERENT NATURE OF BIOPHOTONS: EXPERIMENTAL EVIDENCE AND PHENOMENOLOGICAL MODEL

R.P. BAJPAI

Institute of Self Organising Systems and Biophysics,
North Eastern Hill University Shillong, 793022, India
and International Institute of Biophysics
Vockrather Straße, D-41472, Neuss, Germany

Abstract

Light induced delayed luminescence signals emitted by living systems have a characteristic non-exponential decay behavior. The photon flux changes by 2 or 3 orders of magnitude during a decay. The decay has a long tail in which photons are emitted with an almost constant flux. Biophoton emission is the common name for radiation emitted in both regions. The biophoton signal is more intense in photosynthetic systems. We have measured the conditional probability of zero photon detection in a time Δ in the range (10 μs-10ms) in the biophoton signals emitted by leaves. We measured the probability in only 0.1s for the values of Δ less than 100μs. We have also determined the dependence of the probability on signal strength in decaying signals emitted by different types of leaves at various temperatures in the range (15-50)^{0}C. The dependence is different for coherent and thermal radiation, and is more pronounced for signals with a strength in the range (0.1-5) counts per Δ. We found the same dependence in all leaves. The observed dependence was in the form of a smooth curve ; the curve was in agreement only with that predicted for coherent radiation and it establishes the coherent nature of biophotons. The lower limit of the coherence time was given by the time required to measure the probability; it was 100ms in the decay region and 10s in the tail. The efficacy of the method was tested by determining the nature of optical radiation emitted by a semi-conductor diode; this radiation was not coherent. The incoherent nature showed up when the strength of the signals became less than 2.5 counts per Δ. The diode radiation also showed photon bunching.The coherent nature of biophotons requires a quantum description of the biophoton field; which is presented in the form of a model. The model assumes that an electromagnetic field exists around a living system; the field is specific to the system and is in a squeezed state of photons. The state of the field and its evolution are given by a frequency stable damped oscillator. The model calculates the shape of the biophoton signal;which comes out to be the sum of a hyperbolic decay term and a constant term. The relative strength of the two terms is situation specific and varies over a wide range.

J. J. Chang et al. (eds.), Biophotons, 323-339.

1. Introduction

Biophoton emission [1] is the emission of optical radiation of ultra weak intensity by living systems. A biophoton signal has a characteristic shape that does not correspond to an exponential decay. The shape has two distinct parts-a decaying part and a constant part [2], the former being observable only for a small time interval. The photon flux changes by 2 or 3 orders of magnitude in the decaying part. The intensity of the biophoton signal is much stronger in photosynthetic systems[3] , wherein the decaying part is observable for a few minutes even with a not very sensitive detector [4]. As a result, the decaying part has been studied mostly in photosynthetic systems and is called delayed luminescence[5]. The shape of the signal could be reproduced by a sum of exponential decay terms, the number of which was not definite however. The decay constant and strength of the exponential terms were sensitive to many physiological and environmental factors and varied in an unpredictable manner[6]. They could not be related to the known properties. It has not been possible to identify the constituents of a system responsible for decay. All groups working in the area, have experienced this difficulty. Popp[1] has noticed a remarkable property of the decaying part; its shape can be represented by a single hyperbolic decay term.

The constant part constitutes the tail of the signal and is observable for many hours. The tail has always been mysterious. The mysterious character stems from the fact that it can not be obtained from a sum of any number of exponential decay terms. It has to represent a new phenomenon, which was called biophoton emission. The name implied a connection between photon emission and "life". The decaying part was later included in the phenomenon, once its non exponential decay character was recognized. The two parts are still identified by different names. The decaying part is called light induced biophoton emission, and is usually observed after the stimulation of the system by light. The constant part is called spontaneous biophoton emission, which has no necessary connection with light stimulation although the possibility of light stimulating the emission of constant part can not be completely ruled out. The stimulated signal may have a long lasting tail that will make the separation of the signal into two parts arbitrary.

Photon emission is visualized in a semi-classical picture as a transition of some sub-unit of a system, from an excited state to one of lower energy state. The sub-unit can be an atom, a molecule, a bio-molecule or any other complex substructure. The system has many identical sub-units. All of them excite and de-excite independently in the semi-classical picture. Let S be the number of excited sub-units at any time t, then dS the number of sub-units making a transition in a small time interval dt will be given by

$$dS = -\lambda S\, dt \qquad (1)$$

where λ is the transition probability of the sub-unit and is determined by the dynamics of the system. Wigner and Weisskopf have shown that λ can be taken to be time-independent [8], whence eq.(1) can give only exponentially decaying signals [9]. Signals observed in non-living systems usually decay in an exponential manner. They confirm the validity of the semi-classical picture, independent decay of sub-units and the Wigner and Weisskopf approximation. There is no place for a signal with non exponential decay character in the semi classical picture[10].

One can modify the semi-classical picture to obtain a non exponential decay. A hyperbolic decay is obtained, for instance, by making any of the following assumptions (i) many decay modes with different values for λ,(ii) a time dependent λ, and (iii) a very strong pairwise correlation among excited sub-units. All the assumptions are unsatisfactory for one or other reason in the case of biophoton emission. There is very little predictive power when many modes conspire to produce the hyperbolic decay behavior. The relative intensities of different modes will change in the course of the decay and the decay will lose its hyperbolic character if some modes are filtered out. Preliminary observations with band-pass filters on leaves do not support both these consequences. The relative ratios of blue, green and red components in the signal remain nearly unaltered during 200s in a decay, and different components of the signal continue to decay non exponentially. A hyperbolic decay also emerges if the transition probability has a time dependence of the form $(1+\lambda_0 t)^{-1}$; but it is singular and its justification is more troublesome. The presence of correlation among decaying sub-units will modify eq.(1). If the modification replaces S by S^2 in the right hand side of the equation, then it will lead to a hyperbolic decay. Such a modification will be needed if there exists a very strong pairwise correlation among sub-units. The modification has another consequence; the signal intensity will grow as the square of the number of sub-units in the system. Since the number of sub-units increases with size of the system, the signal intensity should also grow with the size in a system; which has never been observed. All three assumptions can not produce the constant part of the signal. We thus see that the semi-classical picture, even after modifications, can not explain biophoton emission.There is a need of a new approach.

A new approach is also required to understand the functioning of a living system. a living system has many spatially separated parts organized in a definite structure; which act in a cooperative manner while performing biological functions involving regulation, switching action, rhythmic activity, and pattern formation. The cooperative functioning requires the presence of some controlling mechanism and fast communication channels for instructions and information transfer. Chemical molecules carry instructions and information; they are indeed the recognized channels of substantial communication, but are too slow and their capacity to carry information is limited. They seem inappropriate to generate the level of cooperativity required for the functioning of a biological system. There could be other communication channels. These channels are not recognized and their nature is not understood, they are responsible for "non-substantial communication". The word non-substantial indicates that no material agent is involved in information transfer and the strength of the signal carrying information is weak. Biophotons fulfil both criteria and can act as channels for "non substantial communication". Information transfer by photons is fast, economical and efficient. A photon wave can carry information in its amplitude and phase, while a photon wave packet which is made up with many waves of different frequencies, can carry information in its spectral distribution as well. It is possible to transmit a large amount of information through phase and spectral distribution of a photon wave packet of ultra-weak strength. We suspect that the biophoton signal carries information. We can not ascertain it because of our inability to measure the phase and spectral distribution of a biophoton signal. We can only get some indirect evidence in support of our suspicion. The different photons in a signal containing information will be correlated and will exhibit non classical features[11]. Their nature will not be thermal or stochastic and they will require a quantum description for their state, dynamic evolution, and origin. Photons having non classical features are generically called coherent photons. Classical and semi-classical descriptions of coherent photons may lead to contradictions. The coherent photons originate from the decay of correlated sub-units of a system. The correlated sub-units do not de-excite independently, so the shape of the signal will not be exponential. One wonders if the coherence of the biophotons was responsible for the failure of earlier descriptions.

Li has also argued that biophotons have to be coherent in nature in order to provide a support base to holistic models [12].Experiments were performed in the past to detect the coherence of biophotons and some of the results were encouraging. It was observed that the strength of biophoton signal did not increase linearly with the increase in the cell density, but showed an oscillatory character [13]. The oscillations observed in any intensity is a typical interference phenomenon that requires coherence of the field. In another experiment, attenuation in the intensity of biophoton signal passing through a dispersive medium was measured. It was found to be at least an order of magnitude less than the attenuation observed using an incoherent light signal of similar intensity [14]. Coherent photons shows a smaller attenuation of intensity because scattering losses are reduced in a beam with a precisely defined direction of propagation. Photo count statistics of biophotons was measured in a few samples of cells. The observed distributions of photo counts were mostly Poissonian. There were a few rare examples of sub and super Poisson distributions [1]. The distribution of photons in a quantum coherent state is Poissonian; and it can be sub, super or simply Poissonian in a quantum squeezed state[15]. Photons in thermal equilibrium are called thermal photons and they have a Bose-Einstein distribution. Unfortunately, the strength of the signals in all these experiments was rather weak and comparable to the background noise. Consequently, measured quantities had large errors and were not discriminating enough. They only gave an indicative evidence of the coherent nature of biophotons.

Conclusive evidence of the coherence requires measurements of the correlation among emitted photons. The most important quantity is the conditional probability of subsequent photon emission (or equivalently of zero photon detection) in a time interval Δ. The measurement of this probability can discriminate between thermal and coherent photons if the signal strength [16] is around 1 count in the interval Δ. This probability also exhibits photon bunching effect; its value for thermal photons becomes twice the value it attains for coherent photons when signal strength goes to zero. Arecchi et al [17] measured this probability to differentiate between coherent laser radiation and quasi thermal radiation. We have also made similar measurements with biophotons emitted by many different leaves at various temperatures. The strength of our signals covered the entire range used by Arecchi et al. We used an improved technique, that reduced the background noise to a negligible level and eliminated the errors of individual measurements in decaying signals.

The measurement of a probability requires the outcome of nearly 1000 similar events. The outcome of an event in our case is specified by the number of photons detected in a definite time interval. The time interval was fixed by a mechanical shutter in earlier measurements and could not be reduced to less than 0.1s. It was too large a time for a single event and one could not observe many similar events in the decay part of a biophoton signal. Consequently, the probability in the decay region has not been measured so far. We circumvented the problem by using an electronic data acquisition system that reduced the time of a single event to 1μs in the measurement of photo count statistics and to 10μs in the measurement of conditional probability.We could take intense source that reduced the noise in discriminating signals, and could ignore the effect of decay in signals having a characteristic decay time of 10s or more.We were able to determine the conditional probability in 0.1s, and to repeat the measurements continuously. This enabled us to eliminate the errors of individual probability measurements and to determine the dependence of the probability on signal strength.

2. Materials and Methods

We used the experimental set up [18] fabricated by Prof. Popp in his laboratory for investigating the role of biophotons in signal communication between living systems [19]. The details of the experimental set up and measuring procedure are given in the Ref. [18].The measuring system consisted of two photo multiplier tubes (EMI 9558 QA) placed apart at distance of 12.5 cm and a coincidence device [18].All of them could register counts. The photo multiplier tubes detected photons emitted from two different sources in any preset time duration. They were two independent channels for measurements, and were coupled by a coincidence device (DASY Lab DAP-DLL). The coupling identified the channels as photon and reference channels. It was possible to make either of the two channels as photon or reference channel. The coincidence device during its active state, registered the counts detected in the reference channel. The device was activated by the detection of a count in the photon channel; each activation was for a time interval Δ only. The device became active and inactive many times and registered only a fraction of the counts detected in the reference channel.

Let N, R, and Z be the number of counts detected, respectively, in the photon channel, reference channel and coincidence device in a time duration T. The counts determine n(Δ) ,the average signal strength; and $P_0(\Delta)$,the conditional probability of zero photon detection in an interval Δ after the detection of a photon in the channel. The signal strength was given by

$$n(\Delta) = \frac{N}{T}\Delta \tag{2}$$

and the conditional probability by

$$P_0(\Delta) = \left(1 - \frac{Z}{R}\right) \tag{3}$$

The expression for P_0 (Δ) requires an explanation. One needs to note that the coincidence device becomes inactive if no photon is detected during a time interval Δ. As a result, the probability of not detecting a photon in the interval Δ was equal to the probability of finding the device in the inactive state. The latter probability was measured directly. It was given by the fraction of the counts of the reference channel which were not registered by the device. We were thus, able to measure the conditional probability without any detecting the outcome of individual events. Each event gave the detection behavior of the photon channel in the interval Δ just after the detection of a photon. It had two possible outcomes. Either a photon was detected or was not detected. The experimental set up measured the cumulative outcome of N events. If it is further assumed that photon emission is a Poisson process and number of counts detected was proportional to the number of photon emitted by the source, then P_0 (Δ) was also equal to the probability of zero photon emission in a time duration Δ. Both assumptions are implicitly assumed in single photon measurements. We shall also assume them. Accordingly, the probability of photon emission in the interval Δ was given by (1- P_0 (Δ)), and was equal to Z/R, the fraction of counts actually registered in the device. It is further pointed out that the use of a photo multiplier tube in the reference channel is not essential. One can replace it by a signal generator. We have used both of them in our measurements.

R, T, and Δ were adjustable parameters. R was adjusted by changing the strength of the source in the reference channel (or by altering the frequency of signal generator). A large value of R was desirable to reduce fluctuations in the measured ratio Z/R .We found R greater than 100 to be satisfactory. The values of T and Δ could be varied in steps of 10μs. There was a need to impose two restrictions in their values: $\Delta \ll T$ and $T \ll \tau$,where τ was the time characterizing the decay behavior of the signal. The first inequality ensured that a large number of events were included in any measurement of the probability. We chose $T > 1000\Delta$, so that nearly 1000 events were included in the measurement. The second inequality ensured that the strength of the signal did not change in the duration T, and that various events considered in the measurement of the probability were of a similar nature. We used T=100ms in the decay region because $\tau \approx$ 10s for leaves. We used T=10s in the tail region where signal was almost constant. We also imposed another requirement in the choice of parameters. The requirement was to include signals of strength in the range (0.1-5) in any measurement. The signal strength was given by the product of N and Δ / T. The value of N was determined by the source and it variedwith time in a decaying signal. We ,therefore, used different values of the ratio Δ /T for investigating different regions. We used the values of T in the range (100ms-10s), and of Δ in the range (10μs-10ms) in our measurements. The measurement of probability for each set of parameters was continuously repeated 2000 times.

Leaves of different sizes from different plants were used - both fresh ones as well as those plucked up to one week earlier. Each leaf was stimulated by an exposure to white mercury light for 10s. The stability of the delayed luminescence signal was checked in each case by re-exciting the same leaf[20]. The dependence of the probability of zero photon emission on signal strength was measured in the manner described earlier. The leaves. We also measured the probability of zero photon emission in radiation emitted by a commercially available 3 mm As-Gl green light emitting diode, henceforth called a micro lamp, which was a non biological source of optical radiation. The strength of signal in the micro lamp depended linearly on the firing potential. A decaying signal was obtained by supplying the firing potential through a discharging condenser. The strength and decay constant were adjusted to obtain a signal similar in these characteristics to a biophoton signal emitted by a leaf. The probability of zero photon emission in the radiation of the lamp and of a leaf were measured one after the other under identical measuring conditions. We did not succeed in obtaining a suitable decaying signal in a purely thermal source. The signal was very sensitive and unstable. We were only able to determine photo count statistics in a purely thermal source.

3. Results

The salient features of our measurements are summarized below:

1) The measurement of N and R gave the shape of the delayed luminescence signals from two different sources for 200s or more. The shape had the usual hyperbolic character. The values of Z/R did not depend upon Z and R, but depended upon N and Δ through a single variable $n(\Delta)$, which measured the probability of photon emission during the time Δ in the photon channel. The dependence of the probability on signal strength was the same in biophoton signals from different leaves. We plotted all observed values of Z/R as a function of $n(\Delta)$ from different leaves and for various values of Δ in a scatter plot. All values were found to lie around a single curve. The data points determined the shape of the curve. The shape resembled that predicted by the Glauber theory of coherent photon detection[21].The scatter plot of points in a limited region, obtained in a single set of continuous measurement, is given in Fig. 1, which also contains the predicted curves for thermal and coherent radiation. The parameters used in this measurement were Δ=1ms and T=100 ms; these parameters were not optimal and large fluctuations were expected. The figure indicates that most of the points are very close to the curve predicted for coherent radiation, and quite far from the curve predicted for thermal radiation. The agreement between observed points and the curve predicted for coherent radiation was much better in cases where $T \geq 1000\Delta$.

2) The probability of photon emission in a decaying signal varied by more than two orders of magnitude. Therefore, we divided the observed probabilities by the predicted probabilities for thermal and coherent radiation, thus obtaining two probability ratios from each observation. The probability ratio for coherent radiation was always around 1, whilst for thermal radiation was quite often different from 1 and varied with signal strength. The probability ratios at the two ends of our measuring conditions are depicted in Fig.2 and Fig.3. The parameters of these measurements were (Δ=10μs and T=100 ms) and (Δ=10ms and T=10 s). Fig.2 is a scatter plot ,while Fig.3 is a line plot of the data .

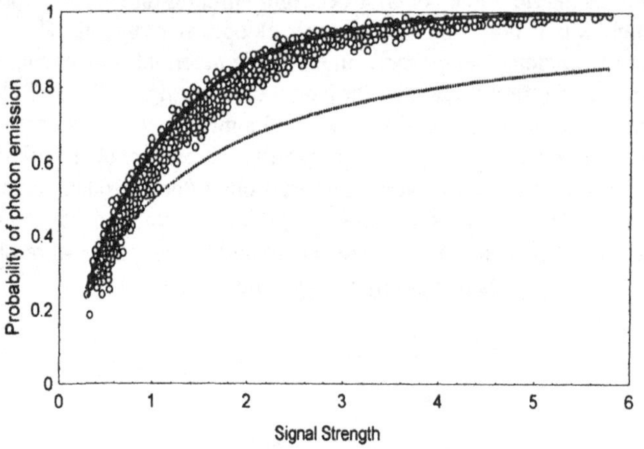

Figure. 1: **Dependence of (1- P_0) on signal strength:** The measured probability of photon emission in 1ms is plotted against the signal strength n(Δ). The dotted and continuous curves give, respectively, the predictions for thermal and coherent radiation . The measuring parameters were T=100 ms and Δ = 1ms .

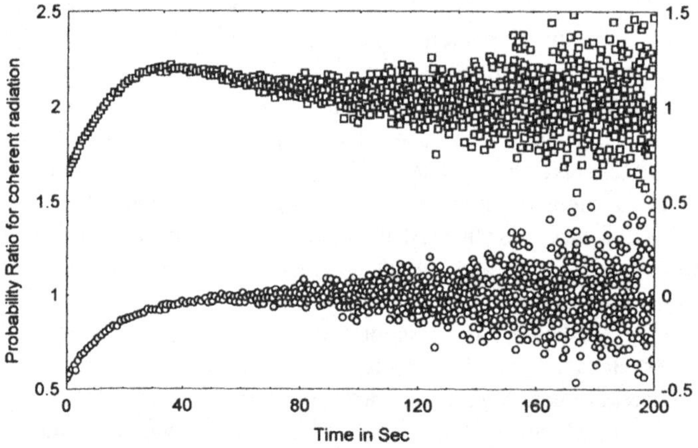

Figure 2: **Effect of non linearity in detection :** The probability ratios of photon emission in thermal and coherent radiation are plotted in a decaying signal. The right side Y axis corresponds to thermal radiation. The signal strength was more than 10^6counts/s for 60m. The measuring parameters were T=100 ms and Δ=10μs.

The choice of parameters of Fig.2 was such that the difference in the two probability ratios was perceptible only in the initial stage of the decay. These measurements can establish coherent nature of the radiation only during the first 120s, since at longer times the two probability ratios became indistinguishable. *The overlapping values of the two probability ratios are separated in the figure by displacing the scale of the probability ratio for thermal radiation.*

One finds the probability ratio for coherent radiation to be less than 1 for initial 50s in this leaf; but we do not consider it to be indicative of incoherence. The probability ratio was observed to be less than 1 in a few other leaves as well. This occurred only for a small duration at the initial stage of the decay, and we attribute it to the high intensity of the signal. The observed counting rate in all such cases was more than 10^6 counts/s. Both photo-multiplier and coincidence device become dysfunctional at high counting rates. The detection in the photo multiplier becomes non linear and the dead time of the coincidence device ($\approx 1\,\mu s$) becomes appreciable. The data points at counting rates higher than 10^6 counts/s were spurious in our view, and should be ignored. We have presented the spurious data points to highlight the apparent non coherent nature arising from the instrumental deficiencies. The probability ratio for coherent radiation was always around 1 in less intense signals.

The two probability ratios were very different in the long tail region for the above choice of parameters. We measured them for a long time and have depicted only a part of the data in Fig.3. The data points in this region correspond to measurements at one signal strength. The figure exhibits the results of repeated measurements. The probability ratio of the full data set was 1.00 ± 0.02 for coherent radiation, and 1.26 ± 0.07 for thermal radiation. The figure demonstrates the coherent nature of radiation only after 3m. The figure also shows an apparent incoherence in the first 3m. It is much longer than 50s shown in Fig.2, and arose because of the choice T=10s, which was too long a time to ignore the change in the signal strength in the first 3m. The instrument was measuring the probability from dissimilar events; the result was destined to be erroneous.

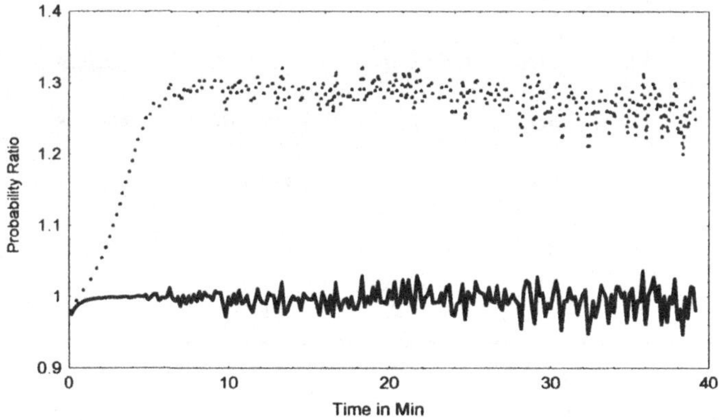

Figure. 3: **Coherent nature of radiation:** The ratio of observed to predicted probability (1- P_0) of photon emission in thermal(dotted line) and coherent radiation (continuous) are plotted as a function of time. The behavior is shown only for 2250s. The behavior was similar beyond this time. The measuring parameters were T=10s and Δ =10ms.

The data represented in Fig.(2) and (3) were obtained from the same leaf using different measuring parameters. The two figures establish the coherent nature of radiation emitted at the two ends of a biophoton signal. The data points obtained with other measuring parameters similarly bring out the difference in other regions and establish the coherent nature of radiation in the entire signal.

3) The results were similar in all leaves under different conditions. The temperature of a leaf was varied in the range (15-50)^{0}C. The time between measurement of probability and plucking of the leaf was in the range (1m -7days). The intensity of the signal did change but the dependence of the probability of zero photon emission on signal strength remained unaltered. The observed probability always suggested a coherent nature to radiation.

4) The mean and standard deviation of the data set gave the overall behavior of the two probability ratios. These are given in the Table 1 for different choices of parameters. Spurious data points with counting rates higher than 10^6 counts/s were ignored in the calculations. The probability ratio for coherent radiation remained essentially 1 at different values of signal strength. In contrast, the probability ratio for thermal radiation was different from unity and it varied with signal strength. The variations were more pronounced in cases that gave more data points with discriminating signal strengths.

TABLE 1. Probability ratios of zero photon emission P_0 in a leaf for coherent and thermal radiations and in a micro lamp for coherent radiation.

S.No.	T	Δ	Leaf(Coherent)	Leaf(Thermal)	Lamp(Coherent)
1	100ms	10μs	1.00± 0.11	1.05± 0.12	1.52± 0.26
2	100ms	20μs	1.00± 0.08	1.08± 0.10	1.41± 0.31
3	100ms	30μs	1.00± 0.06	1.11± 0.08	1.56± 0.34
4	100ms	50μs	0.99± 0.07	1.11± 0.10	1.27± 0.28
5	100ms	0.1ms	0.99± 0.02	1.24± 0.14	1.11± 0.21
6	1s	1ms	1.00± 0.07	1.14± 0.10	1.02± 0.17
7	10s	10ms	1.00± 0.02	1.26± 0.07	

5) The behavior of the probability ratio for coherent radiation was compared in signals emitted by a micro lamp and a leaf. The ratio in both samples was around 1 for signals of strength higher than 2.5 counts per Δ. The ratio began to increase in the micro lamp when signal strength was decreased below 2.5 counts per Δ, and approached the value 2 at signal strengths of around 0.05. There was no further increase in the ratio when the signal strength was decreased below 0.05. This is a typical photon bunching effect

shown by the radiation of the micro lamp. The biophoton radiation of similar strength emitted by the leaf did not exhibit such a behavior. The probability ratio for coherent radiation observed in the two samples with the parameters ($\Delta = 20\mu s$ and $T = 100ms$) is depicted in Fig. 4. The probability ratio remained around 1 in both samples only for about 10s. It subsequently increased in the micro lamp and became nearly 2 after 190s. The probability ratio in the micro lamp depended only on the signal strength and behaved in a predictable manner in other time intervals in the range ($10\mu s \leq \Delta \leq 0.1ms$). The overall behavior of this probability ratio observed during 200s under different measuring parameters is also given in Table 1, which indicates the non coherent nature of the radiation emitted by the micro lamp. The average value of the probability ratio depended upon measuring time interval and was different from 1 for $\Delta < 0.1ms$. We did not succeed in measuring the dependence of the probability of zero photon emission on signal strength in the radiation emitted by an incandescent lamp because of the sensitivity of the signal to applied voltage. The signal strength varied approximately as the eighth power of the applied voltage. It was impossible to produce a measurable and decaying signal from it. We did measure photo count statistics of radiation emitted by a hot metal tip and a leaf in intervals of a few micro seconds. The statistics of photons was Poissonian in case of the leaf and binomial in case of metal tip.

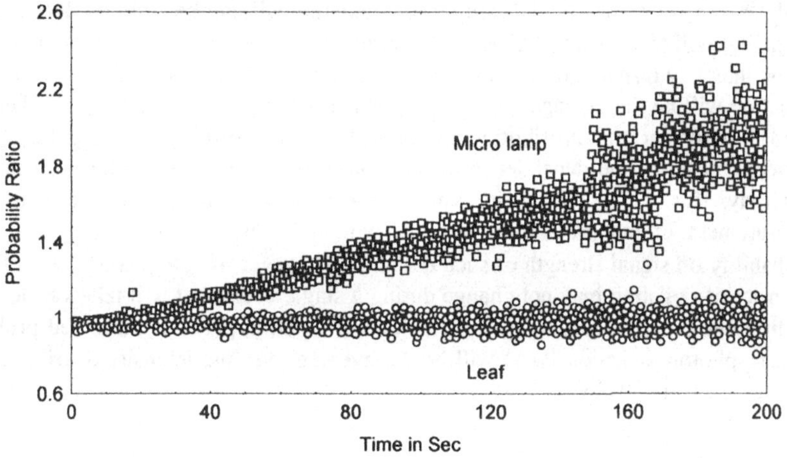

Figure. 4: **Radiation of two different natures:** The probability ratio of photon emission ($1\text{-}P_0$) for coherent radiation from a micro lamp and a leaf. Both sources emitted radiation of comparable strength. The measuring parameters were $T=100ms$ and $\Delta=20\mu s$.

4. Discussion

Glauber [21] developed the theory of single photon detection. Mandel applied it to calculate the conditional probability for not detecting a subsequent photon in a time interval by a semi-classical method [22]. The same expression for the probability has been obtained by following quantum mechanical methods [23]. The probability is given by

$$P_0(\Delta, I) = exp(-n(I)) \tag{4}$$

,where I is the intensity of the radiation and n(I) the signal strength measured as the number of photon detected in the interval Δ. The calculation assumes a single mode of radiation of constant intensity. The intensity of the radiation varies with time; the variation is different for different types of radiation. The intensity of coherent radiation does not change, at least during the coherence time τ_c. It varies subsequently and ultimately get thermalized. Coherent radiation become thermal ,say, after a time τ_{th} ,called thermalization time. Janossy [24] has shown that the complex amplitude of a thermal radiation field has a Gaussian distribution with zero mean value. The distribution function ρ(I) [23] of the intensity in thermal radiation is given by

$$\rho(I)\, dI = exp(-I / I_0)\ dI / I_0 \tag{5}$$

, where I_0 is the average intensity of the field. The variation in the intensity has to be taken into account in comparing with experiments, the predictions of the theory of single photon detection. The parameters τ_c and τ_{th} play a crucial role in the subsequent analysis. If $\tau_{th} < \Delta$, then the intensity of the radiation will not be constant during a single event, but will take various values in the duration Δ . One must take the thermal average of the intensity distribution in calculating any quantity of an event. The intensity of the radiation will be the average intensity I_0 , which will be same for all events; there is no need to consider the distribution of intensity among different events. Eq. (4) will correctly describe the dependence of probability on signal strength for such radiation. Similarly, if $\tau_c > T$, then the intensity of the radiation will not change in the entire measurement of the probability. Eq. (4) will again give the correct dependence of probability on signal strength of such radiation. However, if $\tau_c > \Delta$ and $\tau_{th} < T$, then the intensity of radiation may not change during a single event, but will take various values in different events according to the distribution function ρ(I).The measured probability of zero photon detection $P_0(\Delta)$ will be the average over the intensity distribution. The average is given [23] by

$$P_{0T} = \frac{1}{\left(1 + n(\Delta)\right)} \tag{6}$$

,where the subscript T is added to indicate that various quantities are thermally averaged over a time τ_{th}. Other choices of τ_c and τ_{th} do not lead to analytic expressions. The measured probability will probably lie between the two expressions. In case of a multi-mode radiation field, one has to take an appropriate average over the spectral distribution, which again can not be done analytically, though eq. (4) remains valid for a mixture of multi-mode coherent radiation fields.

The measured probability of zero photon detection in the biophoton signal of leaves shows an excellent agreement with eq. (4) in all measurements. The agreement implies that either $\tau_{th} < \Delta$, or $\tau_c > T$. The former alternative requires $\tau_{th} < 10\mu s$, the minimum value of Δ used in the measurements. If it were so, then the instantaneous intensity will change very rapidly, and more than one million thermalization cycles will span a single measurement of the probability with T=10s. A large number of thermalization cycles should reduce the fluctuation in repeated measurements of the probability. One expects the fluctuation to be around the inverse of the square root of 1 million i.e. 0.1%. only. The standard deviation of the data given in Table1 for this choice was 2.1%, which is an order of magnitude higher. If the intensity of an individual event was not thermally averaged, then one expects fluctuation of around 2% in a data set of 2000 events. A small value of τ_{th} is problematical for any classical equilibrium model of delayed luminescence. The problem stems from a need to generate two widely different time scales characterized by τ_{th} and τ. A single chemical reaction has one equilibrium constant and can generate only one time scale. Generation of two time scales will require two (or more) reactions along with a mechanism to coordinate them in the temperature range of $(15\text{-}50)^0$C. We do not have any other evidence for the existence of such a mechanism. We therefore, believe that the alternative $\tau_{th} < 10\mu s$ does not hold good and for biophotonic radiation we indeed have $\tau_c > T$. Since we used $T \leq 100$ms in the decay region and $T \leq 10$s in the tail region, the coherence time of the biophotonic emission should be more than 100ms in decay region and more than 10s in tail region.

5. Quantum Description in a Phenomenological Model

The coherent nature of biophotons is a non classical feature which requires a quantum description of the of the biophoton field. The full description should include origin, purpose, quantum state, and evolution of the field. Li[12] took the first step towards the quantum description and speculated that biophoton field might be in a squeezed state of photons. We have studied the consequences of the speculation and developed a partial description of the phenomenon in a model [26] that specifies that the state of the biophoton field and determines the shape of the biophoton signal. The model assumes that every living system is endowed with an electromagnetic field in a squeezed state of photons; the state and its evolution are given by a frequency stable damped oscillator. The observation of biophotons confirms the presence of the field while the shape of a biophoton signal shows the evolution of this field. A field has a coherent nature in both coherent and squeezed states, but only a field in a squeezed state can evolve in a manner that produces a decaying signal. Since the signal in light induced biophoton emission is decaying, one has to assume that the coherent biophoton field is in a squeezed state. A squeezed state and its evolution can be obtained from the equation of motion of a damped harmonic oscillator. The damping determines the nature of the squeezed state and the shape of the associated signals. One has to make assumptions about the form of damping. Our model assumes that the damping is time-dependent and of the type that permits solutions in the form $q(t)\exp(\pm i\omega t)$ for a mode frequency ω. It is an amplitude modulated wave, in which the amplitude $q(t)$ varies slowly compared to the mode frequency. Many different damping terms are still permissible; and they generate different functional forms of $q(t)$ and different shapes of the signal. We shall illustrate the working of our model by considering the damping suggested by Popp and Li [18] from semi classical considerations. The Hamiltonian of their model is given by

$$H = \frac{p^2}{2(1+\lambda_0 t)^2} + \frac{1}{2}(1+\lambda_0 t)^2 \omega^2 q^2 \tag{7}$$

, where λ_0 is a constant specific to the mode and system. The system can be quantised to obtain time dependent annihilation and creation operators b(t) and $b^+(t)$ of its particles called quasi photons. These operators differ from the corresponding photon operators a and a^+. The quasi photon operators at different times are related by a unitary transformation [27]

$$b(t) = \mu(t)b(0) + \nu(t)b^+(0) \tag{8}$$

,where, the parameters satisfy the relation $|\mu|^2-|\nu|^2=1$ and are given by

$$\mu(t) = \cos\omega t + \frac{\lambda_0}{2\omega}\frac{\lambda_0 t}{(1+\lambda_0 t)}\sin\omega t + i\frac{\lambda_0}{2\omega}\frac{\lambda_0 t}{(1+\lambda_0 t)}\cos\omega t - i\left\{1+\frac{\lambda_0^2}{2\omega^2(1+\lambda_0 t)}\right\}\sin\omega t \tag{9}$$

and

$$\nu(t) = \frac{\lambda_0}{2\omega}\left\{1+\frac{1}{(1+\lambda_0 t)}\right\}\sin\omega t + i\frac{\lambda_0}{2\omega}\frac{\lambda_0 t}{(1+\lambda_0 t)}\cos\omega t - i\frac{\lambda_0^2}{2\omega^2(1+\lambda_0 t)}\sin\omega t \tag{10}$$

The unitary transformation also gives the evolution of the coherent state of quasi-photons. The coherent sate of quasi-photons is an eigenstate of the operator b(t) and remains an eigenstate of the operator b(t) with the same eigenvalue at all times. The coherent sate of quasi-photons is also a squeezed state of photons. Our model assumes that biophoton field always remains a coherent state of quasi-photons. Such a field will be a squeezed states of photons with time dependent squeezing parameters. These parameters can be calculated for any initial state. Let us represent the initial state of the field at t=0, by the state vector $|\beta,\mu_0,\nu_0\rangle$, where β is the eigenvalue of the quasi-photon operator b(0) and μ_0 and ν_0 are squeezing parameters. The specification of the initial state relates photon and quasi-photon operators by an invertible linear relation

$$b(0) = \mu_0 a + \nu_0 a^+ \tag{11}$$

The state evolves and becomes another state $|\beta,\mu_t,\eta_t\rangle e^{i\phi(t)}$ at time t, where $\phi(t)$ is a time dependent phase factor. The squeezing parameters at a subsequent time t are given by

$$\mu_t = \mu(t)\mu_0 + \nu(t)\nu_0^* \tag{12}$$

and

$$\nu_t = \mu(t)\nu_0 + \nu(t)\mu_0^* \tag{13}$$

They relate b(t) with photon operators in a manner similar to eq.(11). All measurable quantities of the biophoton field can now be calculated. We shall illustrate the procedure by calculating the number of photon detected at any time. We first calculate the expectation value of photon number operator i.e.

$$\langle a^+ a \rangle = \langle \beta, \mu_t, \nu_t | a^+ a | \beta, \mu_t, \nu_t \rangle \ . \tag{14}$$

It is a simple algebraic calculation. The time dependence of the calculated value has two different time scales. The nature of the time dependence on these two scales is also different. One is non oscillatory and slow, and the other oscillatory and fast. The non oscillatory time dependence determines the relaxation or decay behavior of the system and has terms of the form $(1+\lambda_0 t)^{-m}$ with m = 0, 1, and 2. The oscillatory time dependence has frequency ω and is unobservable. It has to be averaged out by integrating the expectation value over a time interval $2\pi/\omega$. The variation of $(1+\lambda_0 t)$ during a time interval $2\pi/\omega$ can be ignored if $t \gg \frac{1}{\omega}$. The condition is satisfied for biophotons emitted as optical radiation with $\omega \approx 10^{15}\,s^{-1}$. The average expectation value of number operator gives the number n(t) of photons detected at any time t as

$$n(t) = B_0 + \frac{B_1}{(1+\lambda_0 t)} + \frac{B_2}{(1+\lambda_0 t)^2} \tag{15}$$

The coefficients B_0, B_1, and B_2 are determined within an over all normalization constant by explicit expressions of the parameters β, μ_0 , and ν_0. Since these parameters specify the state of the field, they will be situation specific. *Eq.(15) provides a new framework to analyze the data.* There is a need to perform experiments in different environmental and physiological conditions and to extract phenomenological information about the dependence of three coefficients on various factors. Some of the coefficients may turn out to be suitable response parameters of environmental and physiological conditions. We expect λ_0 to be less sensitive to environmental conditions since it is a system variable and should be independent of the state of the field; it may or may not depend upon the mode frequency. The aim of phenomenological analysis should be to extract its value.

The determination of the value of λ_0 from phenomenological analysis requires data at times satisfying $\lambda_0 t \approx 1$. If $\lambda_0 t \ll 1$, then the influence of λ_0 will not detectable, and one will observe only a constant photon flux. Similarly if $\lambda_0 t \gg 1$, then λ_0 will get absorbed in the coefficients B_1 and B_2. The values of λ_0 less than $10^6 s^{-1}$ can only be determined by using measuring time intervals of 1μs or more. In all such cases one can construct a dimensionless quantity λ_0/ω, which is less than 10^{-9} for such systems. We can use it as an expansion parameter to express n(t) in the following manner:

$$n(t) = \sum_{m=0}^{4} C_m \left(\frac{\lambda_0}{\omega}\right)^m . \tag{16}$$

Only first term contributes because of the smallness of the expansion parameter. It is given by

$$C_0 = |\nu_0|^2 + |\beta|^2 \left(1 + 2|\nu_0|^2\right). \tag{17}$$

This term gives a constant photon flux, and will be dominant if any of the two quantities $|\nu_0|$ and $|\beta|$ is greater than λ_0/ω. If both quantities are smaller than λ_0/ω then the terms of en(16) need to be rearranged, yielding in leading order

$$n(t) = |\nu_0|^2 + |\beta|^2 + \frac{\lambda_0}{\omega} Im(\mu_0 \nu_0) + \frac{\lambda_0^2}{4\omega^2}\left(1 + \frac{1}{(1+\lambda_0 t)^2}\right) + O\left(\frac{\lambda_0}{\omega}\right)^3 . \tag{18}$$

Eq. (18) gives the leading order contribution to the three coefficients. It may be noted that contribution to B_1 vanishes in the expansion. Thus we again get a relaxation behavior with only three parameters. Eq.(18) has two terms; one gives hyperbolic decay while other a constant photon flux. The relative strength of the two terms depends upon the choice of parameters and can vary over a wide range. Even only a hyperbolic decay is permissible!

We would like to stress that the form of Hamiltonian given in eq.(7) is not essential for the results contained in our model. Amplitude modulated solutions at the same mode frequency occur in many cases with different damping terms. We can construct a coherent state in each case and relate them with the squeezed state of photons. We can follow the procedure outlined above and obtain the shape of the signal, which is not very sensitive to the specific form of damping. The shape of the signal in many different damping terms is similar to the shape obtained in the above illustration. All of them can explain the coherent nature, relaxation behavior, specificity of the signals and inability to extract intensive and extensive variables from conventional analysis in the biophoton emission Future phenomenology will select the correct form of damping.

Finally, it must be confessed that the idea of the existence of electromagnetic field in a squeezed state around living systems came from rather wild speculations. We argued that living systems must have distinct advantages over non-living systems. An advantage acceptable from a physical point of view is the capacity to operate at the absolute limits set by physical laws. Physical laws have only two absolute limits, namely velocity of light and uncertainty relation. The first limit requires the use of zero mass particles. The photon is such a particle and living systems are well adept in using photons. The biochemistry is essentially photon chemistry. The second limit requires the use of minimum uncertainty states, which are coherent and squeezed states. The two limits may be combined. We therefore, speculated that living systems are capable of using photons in coherent or squeezed states. Squeezed states are more advantageous than coherent states for localization purpose and information transfer. Perhaps, some system in the remote past learned the art of producing a squeezed state of light, realized its importance in almost lossless transfer of information and used it. This could have been a turning point in the evolutionary struggle; the advantages of using squeezed light were too overwhelming. They resulted in the presence of light in a squeezed state around all living systems. Only future will tell if there is any grain of truth in our speculation.

References

1. Popp, F.A.: Some Essential Questions of Biophoton Research and Probable Answers , in *Recent Advances in Biophoton Research and its Application*, (Popp, F.A., Li, K.H., and Gu, Q. eds)World Scientific, Singapore, 1992, pp. 1-46.
2. Popp,F.A.,Gurwitsch,A.A.,Inaba,H.,Slawinski,J.,Cilento,G.,Wijk,R.van.,Chwirot,W.B.,and.Nagl,W.: Biophoton Emission (Multi author Review) *Experientia* **44** (1988), 543-600.
3. Strehler,B.L.,and Arnold, W. :Light production by green plants.*J.Gen. Physiol.* **34** (1951), 809-820.
4. Lavorel, J. : Luminescence, in *Bioenergetics of Photosynthesis*, (Govindjee, R. ed.) Academic Press, New York, 1975, pp.225-314.
5. Jursinic, P.A.: Delayed Fluorescence: current concepts and status ,in *Light Emission by Plants and Bacteria*, (Govindji, J.A., and Forg , D.A ., eds) ,Academic Press New York 1986, p.291.
6. Ruth,B.: Experimental Investigations on Ultraweak Photon Emission, in *Electromagnetic Bioinformation*, (Popp,F.A., Warnke,U., König,H.L., and Peschka,W. Eds.),Urban &Schwarzenberg, 1989, pp.128-143.
7. Beloussov, L.V.,and Popp, F.A.:*Biophotonics*, Bioinform Service Co.,Russia,1995.
8. Weisskopf, V., and Wigner, E.P.: *Z.Phys.* **63** (1930), 54 ; **65** (1931),18.
9. Fonda, L., Ghirardi, G.C., and Rimini, A.: *Rep. Progr. Phys.* **41** (1978), 99.
10. Popp, F.A., and Li, K.H: Hyperbolic Relaxation as a Sufficient Condition of a Fully Coherent Ergodic Field, *Int.J.Theo.Phys.***32** (1993), 1573-1583.
11. Loudon, R.: *The quantum Theory of Light*,. Clarendon Press, Oxford, 1975.
12. Li, K.H: Coherence in Physics and Biology in *Recent Advances in Biophoton Research and its Applications*, (Popp, F.A., Li,K.H, and Gu,Q. eds), World Scientific, Singapore 1992, pp.113-155.
13. Galle, M. ,Neurohr, R., Altmann, G., Popp, F.A ,and Nagl, W .:*Experientia* **47** (1991),457.
14. Popp, F. A., and Deny, J.: Biophotonen-Information und Chaostheorie in *Ganzheitsmedizin Zweiter Wiener Dialog* (H.Stacher ed) Facultas Universitätsverlag, Wien, 1991, pp.53-66.
15. Yuen, H.P.: *Phys. Rev.* **A13** (1976), 2226.
16. Perina, J.: *Coherence of Light* ,D.Reidel P.C., Dordrecht ,1985.
17. Arecchi, F.T., Gatti, E., and Sona, A.: *Phys. Lett.* **20** (1966) ,27.
18. Popp,F.A.,Gu,Q.,and Li,K.H.:Biophoton emission: Experimental Background and Theoretical Approaches, *Mod.Phys.Lett.B* **8** (1994), 1269-1296.
19. Ho, M.W.,Popp, F.A., and Warnke, U. : *Bioelectrodynamics and Biocommunications* ,World Scientific ,Singapore 1994 .
20. Bajpai, R.P., and Bajpai, P.K .:*J. Biolum. Chemilum.***7** (1992), 177.
21. Glauber,R.J.:Coherent and Incoherent states of the Radiation Field, *Phys.Rev.* **131** (1963), 2766 also. *Phys. Rev.* **130** (1963), 2521.
22. Mandel, L .:*Progress in Optics* **2**, 183 North-Holland Amsterdam 1963.
23. Walls, D.F., and Milburn , G.J.:*Quantum Optics* ,Springer-Verlag 1994.
24. Janossy, L.: *Nuovo Cimento* ,**12** (1959), 369; **6** (1957), 125.
25. Christen, H.R.:*Grundlagen der Organischen Chemie* Diesterweg, Frankfurt,1970.
26. Bajpai,R.P.,Kumar,S.,and Sivadasan,V.A.:Frequency-stable damped-oscillator model of Biophoton emission, *Frontier Perspectives* **6** (1997), 9.
27. Jannussis, A., and Bartzis, V. :*Nuovo Cimento* **102B** (1988), 33.

QUANTUM COHERENCE AND THE UNDERSTANDING OF LIFE

G. J. HYLAND
Department of Physics
University of Warwick
Coventry
CV4 7AL, U.K.
Email: G.J.Hyland@warwick.ac.uk

1. Introduction

An event 30 years ago this month (June) marked the dawn of a new era in Biology - an era which offers quite novel possibilities, not only for the understanding of the orderly functioning of metabolically active biosystems and their near-quantum sensitivities, but also, more recently, for an integration of consciousness into Physical Science. The event was the first international conference devoted to Theoretical Physics and Biology, and took place at Versailles under the auspices of *l'Institut de la Vie*. The conference was the brainchild of the institute's director, Maurice Marois, and the internationally renowned theoretical physicist, Herbert Fröhlich; it arose in response to a somewhat provocative assertion [1] from some quarters - and espoused by Wigner [2] - that living systems (and particularly their self-reproducibility) could not be understood in terms of the laws of contemporary physics, but required additional (so-called *biotonic*) laws[1], yet to be discovered.

In his opening address, Fröhlich [3] argued that whilst it is indeed *impossible* to understand the orderly functioning of active biosystems in terms of the (reductionistic) '*physics of yesterday*' (which had proved so successful in dealing with questions pertaining to structure in microbiology), this does *not* exhaust the possibilities offered by contemporary physics, *provided* the relevant *holistic* concepts can be found - a view independently advocated somewhat later by others [4]; establishing what these concepts might be, however, must be anticipated to be a highly non-systematic task (*c.f.* reference [5]). For active biosystems are dissipative, *open* systems, which are necessarily held far

---

1 In the light of contemporary developments, it is interesting to record that Wigner's own belief (at least at that time) in the existence of such laws stemmed from ..."*the overwhelming phenomenon of consciousness*", whose description ..."*clearly needs incorporation of concepts into our Laws of Nature which are foreign to the present laws of physics*" [2].

J. J. Chang et al. (eds.), Biophotons, 341-358.

from thermal equilibrium by the influx of metabolic energy on which their vitality depends; they are thus far beyond the regime of validity of any systematic approach based on linear response. Instead, progress can only be made by tentatively introducing a new concept, elaborating its consequences, and then testing them against experiment; a particularly close collaboration between theory and experiment is thus called for. The relevant *physical* concepts must, however, be anticipated to be of a rather subtle nature. For *discounting* the existence of 'biotonic' laws, the most remarkable functioning of active biosystems - such as their near quantum sensitivities, the control of cell division in tissue, or the enormous catalytic power of enzymes - forces one to conclude that these systems must have developed extraordinary ways of making use of the laws of *contemporary* physics; discovering what these ways might be, must, accordingly, be expected to require extraordinary ingenuity! Interpreting such remarkable functioning as manifestations of a certain 'order' of a non-obvious kind, Fröhlich was led in 1967 to consider the possible relevance of *coherence* - an intrinsically *holistic* (non-reductionistic) concept which was already known to underlie certain other subtle kinds of (non-spatial) order in various *macroscopic* quantum systems, such as superfluids and superconductors.

> ..."*I believe that this concept (i.e. coherence) may have a much wider range of application, in particular, in systems which are relatively stable but not near thermal equilibrium, and which show an organised collective behaviour which cannot be described in terms of an obvious (static) spatial order*" [3].

Despite their activity, biosystems are nevertheless relatively *stable*; indeed their structure is relatively *insensitive* to whether they are active or not. Accordingly, the distinction between active and non-active biosystems must be reflected in a relatively *few* degrees of freedom which *dominate the remainder* when the system is activated through metabolic energy supply. Since their orderly functioning is a *holistic* property, Fröhlich argued that these few dominating degrees of freedom must be in the nature of *collective modes* which get lifted far from the thermal equilibrium to which the remaining majority of modes continue to subscribe. It is these relatively few, strongly excited collective modes which must thus characterize the complex, global organization which is created and maintained by the (random) energy supply; the way they do so, he conjectured, is by supporting certain *coherent* excitations, of a kind not hitherto contemplated in biology, and whose very existence almost runs counter to intuition.

2. Dynamic and Static Coherent Excitations

In the quest for collective modes characteristic of biosystems - consideration of which, incidently, circumvents complications arising from the great structural complexity of such systems at the molecular level - Fröhlich turned to their *dielectric* and *elastic* properties. In particular, he noted that a ubiquitous feature of biomaterial at the cellular and sub-cellular levels is the existence electric dipolar units of various kinds -not least the cell membrane itself, across which is maintained an enormous electric field of magnitude $\sim 10^5$ *V/cm*; accordingly, it must be anticipated that any macromolecules (*e.g.* proteins) dissolved in the membrane would themselves become strongly polarized - *i.e.* acquire an induced electric dipole moment. In consequence of the considerable number of such dissolved proteins, it is possible for small sections of the membrane (between two proteins, for example) to vibrate separately from the rest - in particular, in such a way that local regions of the positive (outer) layer vibrate *against* the negatively charged (inner) layer, *perpendicularly* to the membrane, thereby behaving as local oscillating electric dipoles. The long-range Coulomb interaction between the totality of these dipoles, all oscillating at the same frequency, ν_o, but, in general, with different amplitudes (and phases) entails the existence of a relatively narrow band of *collective* longitudinal modes of electric oscillation which extend throughout the region[2] occupied by the local dipoles. These 'polarization waves' have a range of macroscopic wavelengths depending only *weakly* on frequency, which, in consequence of the long-range interaction between positively and negatively charged regions characterizing this excitation, never vanishes, *whatever* the wavelength. An estimate of ν_o can be obtained from the frequency of the longest wavelength sound wave which can be propagated perpendicular to the membrane; if d is the membrane thickness, then $\lambda_{max} = 2d$, whence $\nu = s/2d$, where s is the speed of sound. Taking $d \sim 10^{-6}$ cm and a typical value of $s \cong 10^5$ cm/s (corresponding to water) yields $\nu \sim 10^{11}$ Hz. It should be stressed that whilst the value of ν_o is essentially determined - *via* s - by *short*-range (elastic) forces within the membrane, it is *via* the *long*-range Coulomb interaction that these local dipole oscillators interact to form the band of collective polarization waves, whose width depends on the strength of this interaction. Sources of dipolar oscillations *other* than those associated with localised regions of cell membranes can readily be identified, such as, for example, proteins within the membrane which, as already mentioned, become strongly polarized. In addition, there are other giant molecules located within the cytoplasm which also have dipolar properties, in consequence of the frequently occurring H-bonds; a collective dipolar frequency in the region of $5 \times 10^{13} Hz$ has, for instance, been reported [6] for H-bonded amide structures, whilst collective frequencies as low as 10^9 Hz have been calculated for a number of giant breathing and rocking modes in double helix structures [7].

Treating the remainder of the system in which the dipolar units are embedded as an *elastic* heat bath held at constant temperature, Fröhlich [8] discovered - on the basis of a simple quantum mechanical model permitting both linear (single quanta) *and* non-linear (two quanta) interactions of short-range between this heat bath and the collective polarization modes - that energy randomly supplied above a certain rate, s_o, is *not* completely thermalized, but is, instead, partly channelled into the *lowest* frequency collective mode (ν_l), wherein it is stored in a purely mechanical fashion. This single mode thus becomes strongly (coherently) excited, far in excess of that appropriate to thermal equilibrium (involving single quanta exchange only) - *i.e.* for $s \gtrsim s_o$ the number of quanta with energy $h\nu_l$ increases far above the thermal equilibrium value, in a way somewhat reminiscent of Einstein condensation in an ideal Bose gas of *material* particles, where below a certain temperature, T_c, (and in thermal equilibrium) the number of *particles* in

---

2 This could, of course, include *many* cells if their local frequencies are all equal.

the lowest energy state increases dramatically[3]. In the bio-case, on the other hand, the macroscopically occupied ν_l mode represents a long-wavelength coherent oscillation of electric polarization which extends throughout the whole system of coupled dipolar oscillators; of particular importance is the collective polarization mode of zero wavevector ($\boldsymbol{k}=0$) - corresponding to a *giant dipole oscillation* of the whole system - since this mode can be excited by electromagnetic radiation of the same frequency (ν_l), provided the size of the oscillating biosystem is much *smaller* than the corresponding (vacuum) wavelength (c/ν_l).

Thus, counter to the intuitive expectation that random supply of external energy to a dissipative system at a finite temperature results in heating (and to a corresponding increase in entropy), the admitted non-linear couplings entail, instead, the establishment of a *highly ordered subsystem*, in which part of the incoming energy is stored in a purely mechanical way - a situation foreseen by Schrödinger [9] already in 1943 (and somewhat later by F. London [10]).... "*The living organism seems to be a macroscopic system which* ***in part*** *of its behaviour approaches to that purely mechanical (as contrasted with thermodynamical) conduct to which all systems tend as the temperature approaches the absolute zero and molecular disorder is removed*" [9].

It should, of course, be stressed that the establishment of such an entropyless subsystem (the $\boldsymbol{k} \cong 0$ coherent mode) does not entail any violation of the Second Law, since there is a corresponding *increase* in the entropy of the heat bath, so that the entropy of the *whole* system does indeed increase. A parallel with superfluidity (and superconductivity) is clearly apparent; indeed, just as in this case, wherein the ordered subsystem is the macroscopically occupied zero momentum condensate (described by an appropriate macroscopic wavefunction), so Fröhlich's strongly excited lowest frequency collective polarization mode is equivalent, quantum mechanically, to a (Bose) 'condensate' of polarization quanta of energy $h\nu_l$, between which there is long-range phase coherence.

It is essential to appreciate, however, the fundamental difference between these examples: superfluidity and superconductivity occur under conditions of thermodynamic equilibrium, whereas Fröhlich's coherent bio-excitation is an inherently *non*-equilibrium effect, the closest example to which, non-biologically, is, perhaps, the pumped laser. Failure to appreciate this difference has unfortunately resulted, in some quarters, in... "*a general scepticism about the possibility of quantum coherence effects having any relevance to such a 'hot' object as the human brain - or, indeed, any other biological system*" [11]; appeal to high temperature superconductivity in an attempt to mitigate this scepticism is only likely to compound the confusion still further[4], however.

---

3 See Appendix.

4 More worthwhile, perhaps, would be consideration of the *converse* - namely, the possibility of establishing room temperature superconductivity in non-biological matter as a *non*-equilibrium macroquantum phenomenon!

Now in consequence of the dependence of the dielectric self-energy (proportional to the square of the polarization) on the *shape* of the material, and the relative ease with which biological material can be *deformed*, a mechanism for *stabilizing* the coherent mode exists involving elastic deformation. Essentially, the associated macroscopic polarization field couples non-linearly to the (lower frequency) elastic modes of the heat bath, resulting in a *lowering* of the frequency of the coherent mode (*i.e.* in a softening) and an associated reduction in the value of the critical rate of energy supply [3,12]. Increasing further the rate of energy supply (*i.e.* the power), the coherent mode eventually goes 'soft', resulting in the establishment of a *static* displacement - *meta*-stabilized through higher order, non-linear couplings to the elastic field - in which the energy of the dynamic coherent excitation is effectively stored [13]. In contrast to the dynamic coherent excitation (for which only $\overline{P} \neq 0$), the static metastable state is characterized by a non-vanishing average polarization ($\overline{P} \neq 0$) and the system thus resembles[5] a ferroelectric. It should be realized that the existence of such metastable excited states with very high electric dipole moment is a quite general (*non*-linear response) property of any highly polarizable system (such as a giant molecule in the strong membrane field) provided it is sufficiently deformable elastically. Already by 1970 Fröhlich had suggested [14] that this static, highly polar state is realized by enzymes when in their activated state - the associated high internal electric field reducing their activation energy and thus ..."*resolving the essential mystery of their enormous catalytic power*" [15]. Additionally, excitation of the metastable state would give rise to the possibility of non-chemical storage and transport of energy, as is required in some biological processes.

3. A Long-range Selective Interaction

One might be tempted to conclude that the high field arising from the electric dipole moment of the metastable state gives rise to long-range forces; this might well be so in the case of molecules dissolved in the membrane. Brought into cell water, however, the high field is *screened* by dissolved ions within a short (Debye) distance, leaving only the possibility of *short*-range forces. This is no longer so, however, in the case of systems such as macromolecules which, provided they are separated by a distance, *R*, much greater than their individual spatial extensions, can interact through their giant dipole oscillations (*i.e.* polarization waves with $\boldsymbol{k} = 0$). In the simplest case of *two* such systems characterized by frequencies $\omega_{1,2}$, this interaction results in *two* frequencies $\omega_{\pm}$ of the *joint* system, the difference between ω_{+} and ω_{-} depending, amongst other things, on *R* and the longitudinal dielectric function, $\varepsilon_{\pm}$, of the intervening medium at frequencies $\omega_{\pm}$. At resonance ($\omega_1 = \omega_2$), two quite different possibilities exist of realizing an interaction $\propto R^{-3}$, which is, of course, of *much longer* range than the usual (London) R^{-6} dependence [16]:

---

5 A true ferroelectric is, of course, a stable (as opposed to metastable) system, below a certain temperature.

a) $\omega_{\pm}$ *are excited only to their thermal equilibrium values*: provided $\varepsilon_{+} \neq \varepsilon_{-}$, an R^{-3} interaction results, which is attractive (repulsive) for $\varepsilon_{-} < \varepsilon_{+} (\varepsilon_{-} > \varepsilon_{+})$ - *i.e.* the *sign* of the interaction is determined by the dielectric dispersion ($\partial\varepsilon_{\pm}/\partial\omega_{\pm}$) of the intervening medium; for $\varepsilon_{+} = \varepsilon_{-}$ the interaction weakens to an R^{-6} dependence.

b) *ω_{-} (ω_{+}) is strongly excited* ***well beyond*** *thermal by supercritical pumping* ($s \geq s_o$): attractive (repulsive) interactions $\propto 1/\ \varepsilon_{-}R^{3}\ (1/\varepsilon_{+}R^{3})$ here result[6] (provided $\varepsilon_{\pm} > 0$), whilst away from an resonance (*i.e.* if $\omega_1 \neq \omega_2$) the interaction again weakens to an R^{-6} dependence.

In case (b), the interaction can clearly be switched on or off according as whether or not the rate of metabolic energy supply to the two systems is sufficient to excite the coherent vibration at frequency ω_{-} , whilst the actual value of the required critical energy supply rate, in turn, depends - *via* the elastic stabilizing mechanism - on the deformability of the system, the degree of which must vary with its evolution; accordingly, it is possible that the associated force acts only during a certain *limited* period of the system's evolutionary history.

A potentially very important consequence of case (b) is that, through strong excitation of ω_{-}, a 'naturally' occurring repulsive interaction (case (a), with $\varepsilon_{-} > \varepsilon_{+}$) can be transformed into an *attractive* one; the possibility thus arises of 'facilitating' (local) chemical reactions - in the sense that through this selective attraction, particular systems (those with resonant frequencies) can be brought into sufficiently close proximity for them to be able to interact *chemically*, at a certain stage of their evolution. The actual distance over which the attraction operates, however, is controlled by the *frequency dispersion* of the intervening dielectric medium. In the case of water, for example, the frequencies of the coherent modes must be below 5 x $10^{13} Hz$ (where water absorbs strongly) if the range of the interaction is to exceed a typical cell diameter. This is indeed so, in the case of coherent excitations based on cell membranes ($\nu \sim 10^{11} Hz$), whence the possibility arises of selective interaction and communication *between* cells. In turn, these interactions offer quite novel possibilities for understanding certain aspects of biological control, such as the control of cell division in a fully grown organ, and hence its breakdown in the case of *cancer*[7] [17]. In the case of much higher frequencies, such as characterise certain proteins, the effective range of the interaction is *intra*-cellular, where it could control enzyme-substrate attraction, for example. Other possibilities include an understanding of the pairing of homologous chromosomes in meiosis [18], the attraction of nutrient molecules to

---

6 It should be noted, however, that since, within the context of the Fröhlich's particularly simple model for the establishment of (dynamic) coherent excitations, only the *lowest* frequency collective mode becomes strongly excited - only the *attractive* interaction (based on ω_{-}) is actually realizable here.

7 For possible experimental corroboration see second citation in Ref.17.

cells, *rouleaux* formation in blood [19], and, in the *repulsive* case, certain aspects of system immunity.

We thus see how Fröhlich's introduction of quite general holistic concepts of contemporary physics into open, dissipative biological systems leads - in conjunction with quite specific material properties (namely their remarkable *dielectric* behaviour and *elastic* deformability) - to the prediction, under appropriate conditions, of both dynamic and static *coherent excitations*, the existence of which had not hitherto been suspected, and which entail quite novel mechanisms of (non-chemical) biological control, communication and energy storage.

It is important to realise, however, that these coherent excitations were predicted using particular, highly simplified models, and it is perfectly feasible that *quite different* ones could lead to the *same* excitations - *i.e.* that a multicausal situation might prevail, such as is indeed sometimes the case in biology, where, in order to maintain some vital function if one apparent cause is removed, *another* takes over, thereby 'protecting' the functioning of the system. Thus, for example, whilst the *strong* excitation (*i.e. macroscopic* occupation) of a *single* mode, which characterizes Fröhlich's model, is a sufficient condition for coherence, it is *not* necessary - as has long been appreciated in other contexts pertaining to thermal equilibrium - *i.e.* coherent states need *not* be monochromatic [20], nor involve macroscopic occupation of individual modes [21] - features which might well be relevant to the coherency associated with biophoton emission (*vide infra*). The conditions under which such *generalized* coherence might be realized as a *non*-equilibrium feature of open biosystems is yet to be established, however.

Thus experimental confirmation of a certain coherent excitation cannot necessarily be taken as support for Fröhlich's *particular* model; clearly, closer collaboration between theory and experiment is called for than is perhaps usual in Physics, whose role is here to *suggest*, rather than to predict. Indeed, since biosystems have developed to fulfil certain specific functions, one is even permitted to ask questions which are strictly forbidden in Physics (except when dealing with machines or similar constructions), such as ..."*What is the* ***purpose*** *of these coherent excitations*?"

4. Experimental Evidence

Before considering the evidence in support of the conjectured coherent excitations in active biosystems, it should be noted that since (at the frequencies concerned) $h\nu_1$ is at least one order of magnitude small than kT (at room temperature), *very strong* excitation is required to overcome thermal noise. Thus, if vibrational states with frequencies of order $10^{12}Hz$ or above are sufficiently strongly excited by metabolic processes, then corresponding Raman lines should be found when the system is metabolically active, but not otherwise. An alternate approach would be to attempt to interfere with biological processes by exciting the systems from outside by coherent microwave irradiation of the appropriate frequency; both approaches have been followed.

Above $10^{12} Hz$, laser Raman effect has provided evidence [22] of metabolic excitation of a coherent mode in *Escherichia coli* bacteria - active cells being found to exhibit[8] high intensity spectra (Raman shift lines, ν_R) which are *absent* in non-metabolic conditions, associated with which are only *broad* bands. It should further be noted that these lines appear *only* at certain stages in the evolution of the system - precisely as is expected if excitation of the coherent mode has to fulfil a specific biological task. The non-thermal nature of the excitation is convincingly demonstrated by the observation [23] of an anti-Stokes/Stokes intensity ratio close to the maximum possible value of unity, rather than the thermal equilibrium value of about 0.5. It should be noted, however, that these experiments pose many difficulties connected, not only with the proximity of the coherent frequency to the water absorption band and the need to ensure extensive synchronization - particularly, at the high dilutions necessary to keep the cells in their active state - but also, with possible *adverse* effects of the laser field on the activity of the biosystem.

Other possible evidence of metabolically excited coherent excitations which should be mentioned includes the detection (spanning the *IR-UV* range) of *coherent*, ultraweak photon emission[9] (biophotons) from active biosystems [27], and the existence of a lower frequency coherent oscillating electric field (in the *kHz* $\rightarrow$ *MHz* range) near the surfaces of living, but not dead cells, as revealed by microdielectrophoresis [28] - both of which are maximal near mitosis.

---

8 It should be noted that given the proximity of certain water absorption bands, the very observation of the lines at all (involving an enhancement by a factor of order 10^5 above thermal) is, in itself, indicative of a *coherent* excitation involving many *synchronized* cells, all at the same stage of development, since the scattering intensity is then proportional to the *square* of the number of scattering units.

9 Assuming that the emission of coherent biophotons is connected in some way with the excitation *within* the system of Fröhlich's longitudinal coherent polarization mode, the recent consideration [24] of the emission of coherent *THz* radiation by coherent *LO* phonons in *GaAs* (excited by ultrashort (femtosecond) laser pulses) is of particular interest - especially since the requirement that the spatial extension of the laser-heated region be less than the wavelength of the radiation exactly parallels the condition for Fröhlich's strongly excited giant dipole ($\boldsymbol{k} = 0$) excitation to be weakly optically active -which condition is here realized *only* in consequence of the existence of inner surfaces and other spatial inhomogeneities.

It should be noted that the detection [25] of *coherent THz* vibrational modes modes excited by femtosecond optical pumping is not restricted to inorganic matter, but has been reported, over a broad range of temperatures, for the case of a protein embedded in its natural membrane [26].

If active biosystems do indeed 'make use' in some way of Fröhlich's dynamic coherent excitation, then a high sensitivity to irradiation is anticipated, particularly if the frequencies match. Over the past 20 years much experimental evidence has accumulated which is consistent with this. Particularly compelling is the effect on the growth and functioning of a variety of living systems of irradiation by coherent microwaves of intensity as low as pW/cm^2 (the associated electric field being here infinitessimal in comparison with the membrane field), which leave little doubt that the effects are *neither* thermal, *nor* due to the electric[10] field of the microwaves. It has been established [29], for example, that irradiation of *E. coli* bacteria by low power microwaves of *GHz* frequency actually *decreases* their rate of growth, whilst heating increases it! The frequency sensitivity of the effect is well illustrated by the observation that reproducible changes ($\approx 10\%$) in the average growth rate of yeast induced by irradiation at $42GHz$ are *obliterated* by detuning by 1 part in 10^4 [30]. That these highly resonant effects occur in a region where the optical constants show *little* frequency dependence is further strong indication that the effects are *non*-thermal; it should be remembered that heating is mainly due to absorption by cell water and is only very *weakly* dependent on frequency.

The reported effects can be readily understood [31], however, if the microwaves are considered to supply simply the *difference* in power between that already available internally from metabolism and the critical value necessary for the excitation of the coherent mode - *i.e.* that the microwave radiation acts as a '*trigger*' for events for which the system is already partly prepared. (Later we will encounter another possible triggering mechanism involving the externally induced collapse of a limit cycle.) The response of the system must thus depend in a step-like fashion on the intensity (power) of the microwave radiation, exhibiting zero response until the microwave power is sufficient to make up the deficit, and (within limits) *independent* of intensity thereafter - a feature again consistent with the *non*-thermal nature of the process, and one which would account for the lack of reproducibility sometimes encountered in attempts to corroborate irradiation experiments; for if the metabolic energy supply rate is *itself* adequate to excite the coherent mode, then the external radiation will be deprived of its triggering role. It must be anticipated, however, that it will take some time for the system to channel the supplied power into the mode that is ultimately excited coherently, although the transfer will be most effective when there is *resonance* between the microwave field and the coherent mode. All these features are corroborated experimentally - pre-effect irradiation times of the order of 1 hour being typically required at $10^{11}Hz$ [32]. Since there is presumably no evolutionary need for biosystems to be so frequency - sensitive to such weak external microwave radiation, it must be concluded that the systems themselves *do indeed make use, in some way, of vibrations in this frequency range.*

---

10 To be able to orient a typical polar macromolecular in solution against thermal noise, for example, requires external electric field intensities of the order of $10^6 W/cm^2$.

Evidence of a more indirect nature is to be found in the observation [19] of *rouleaux* formation by red blood cells, where, below a certain separation - of the order of μm (which is orders of magnitude greater than the range of chemical forces) - the cells rush towards each other to form coin-like stacks (*rouleaux*) at a much faster rate than is expected on the basis of Brownian motion. Upon removal of the membrane potential, however, the rate reduces to the Brownian value (as it also does when metabolism is inhibited), but increases again when the potential is restored. Furthermore, in the case of a *mixture* of mammalian cells, it is found that *like* cells tend to aggregate predominantly, in contrast to the prediction based on random Brownian motion. These effects fit perfectly with the properties of the selective long-range interaction between systems which support coherent vibrations of approximately equal frequency.

> "*It thus appears that the evidence for the existence of coherent excitations is very strong. So far, however, very little insight has been gained on their specific biological significance. Experiments of a different nature will have to be designed for this purpose. We are dealing with cooperative phenomena, and it would seem that multicellular systems should make use of coherent excitations to a larger extent than single cells. Investigations of differentiated tissues are thus highly desirable*" [33].

It is sometimes claimed that the credibility of the available experimental evidence is undermined by the lack of reproducibility sometimes encountered in the course of independent attempts at corroboration. A more positive attitude, however, would be to accept this situation more as a biological 'fact of life', than as an indictment. For it should be remembered that, in contrast to physics, the experimental conditions prevailing in the case of biology are quite different, both pragmatically and in principle. For quite apart from the applicability of conventional *physical* techniques being here possibly undermined by a lack of sample homogeneity, for example, or by difficulties in obtaining samples of suitable size and shape *without interfering* with their *in vivo* activity, a more fundamental difficulty can arise in the case of biosystems in consequence of their inherent *non-linearities*: they can exhibit *deterministic chaos* [34], whence even the slightest difference in sample preparation can result in vastly differing evolutionary behaviours.

5. Fröhlich's Brain-Wave Model

Of particular interest in connection with subsequent elaborations and applications of Fröhlich's ideas to the problem of *consciousness* - which is currently engaging much multidisciplinary attention - was his model [35] for the origin of brain-waves (*EEG*) which dates from 1974, and represents the pinnacle of his '*oeuvre biologique*' - incorporating, as it does, the 3 novel biological possibilities already introduced - namely: *i)* the dynamic coherent mode, entailing *ii)* the possibility of selective long-range interactions, and *iii)* the metastable (ferroelectric-like) highly polarized static excitation. The model was developed to circumvent difficulties attendant on assuming that the low frequency (~10*Hz*) *EEG* arises simply from oscillations of extended regions; for if this were the case, enormous amplitudes would necessarily be involved in order to overcome thermal noise.

It is known, however, that during certain periods many thousands of generators of *EEG* (~ 0.5 - 60*Hz*) act coherently over large regions [36], whilst external signals tend to correlate the phases[11] of the various Fourier components [37]. The extremely low frequencies of the *EEG*, in comparison with those associated with the coherent excitations (10^9 - 10^{13}*Hz*), suggests that the *EEG* must arise in some *secondary* way. Fröhlich ingeniously showed how this could be realized as a consequence of a subtle interplay between the long-range selective attractions and the (metastable) highly polar *static* coherent excitation, as applied to enzymes distributed over the Greater Membrane of the brain. Perhaps even more importantly, however, this same model offers a novel understanding of the ultrasensitivity of the brain to extremely weak external stimuli. In brief, the energy liberated from the chemical reaction between enzymes and substrates is considered to excite the enzymes to their highly polar metastable state, characterised by a *finite* life-time. The energy released upon (spontaneous) decay back to the non-polar ground-state is now assumed to excite a high frequency coherent mode in the substrate system, in consequence of which fresh substrates of the same type are attracted (*via* the associated frequency selective, long-range force) into the vicinity to replenish those which were destroyed in the course of earlier chemical reactions. In this (autocatalytic) way, a periodic (*Lotka-Volterra*) collective enzymatic process is established, provided the number of non-excited enzymes remains effectively constant - such as would be the case, for example, if their number density was sufficiently large. In consequence of the high polarization of the excited enzymes (which is absent in their ground-states), the activation/deactivation process is accompanied by a periodic polarization-depolarization, associated with which (provided the enzymes are reasonably ordered spatially) will be periodic *electric* oscillations. Their frequency is determined, not by the coherent mode but, by parameters of the (relatively slow) underlying *chemical* reactions - specifically (in the simplest approximation), those governing the spontaneous decay rate of the excited enzymes and the rate of attraction of new substrates; accordingly, this frequency can, in principle, be quite low.

Including an additional short-range interaction between the highly polar (activated) states of these spatially ordered enzymes - which tends to *favour* the establishment of the ferroelectric state - together the (opposing) *damping* effect of the electric current associated

---

11 Upon sensory stimulation, the frequency components of the *EEG* exhibit phase correlations which are absent in the unevoked (spontaneous) cortical spectrum. In particular, imposition of the *phase* characteristics of the evoked potential on the spontaneous signal is found to reproduce the evoked spectrum; this is quite contrary to the results expected if the *EEG* simply arose from the summation of neural firings, since this would entail *amplitude* (rather than phase) differences between evoked and unevoked spectra.

with the periodically changing polarization, results in a great enhancement in the richness of the underlying non-linear dynamics, permitting, in particular, the possibility of self-sustained oscillations of a particular amplitude and frequency (*limit cycles*), under certain conditions.

From consideration of the interaction of such a limit cycle with a weak, external oscillatory electromagnetic field, a rather subtle interpretation of the brain's remarkable sensitivity[12] to weak external stimuli is then obtained - namely, through an externally induced *collapse* of the limit cycle, whereby the stored energy is liberated - the amount of which could be sufficient to initiate certain physiological processes (such as nerve conduction), whose threshold far exceeds the energy available in the weak applied field alone; the role of the external field is thus simply to "trigger" the collapse. Details have been elaborated by Kaiser [41] in a series of papers; particularly interesting is the finding that *deterministic chaos* is possible when, in addition to the oscillatory field, a *static* background field is also present.

In connection with chaos, it should be stressed that the approximation under which Fröhlich's Brain-Wave model has been solved to date - namely, neglecting completely the dynamics of the unexcited enzymes, which thereby reduces the number of degrees of freedom to *two* - *precludes* the possibility of chaotic solutions in the case of an *un*driven system; given the empirical detection [42] of chaos in (unevoked) *EEG* signals, however, it is clear that this approximation must be lifted.

Although originally developed as a model for the generation of brain-waves, it is equally well applicable to other periodic enzymatic reactions, yielding associated limit cycle behaviour. In this connection, it is of interest to note that high frequency radiation is often found to have a much greater effect *after* modulation to lower frequency; this could be accounted for if the high frequency triggers the excitation of high frequency collective mode, whilst the lower (modulating) frequency acts to collapse the limit cycle, such that a much *larger* response is then obtained than would be realized from each *individually*.

6. Towards the Mind-Brain Interface

The well-known correlation between the *EEG* and varying degrees of consciousness effectively places Fröhlich's Brain-Wave model at the interface of mind and brain, permitting, in principle, identification of the model parameters which control the characteristics of the various *EEG* spectra, *and hence the different states of consciousness*; for some of these parameters can be influenced by temperature, and others chemically. In

---

12 Other examples of bio-ultrasensitivity include the human eye which is sensitive to a single light quantum [38], the human olfactory system which is sensitive to a single molecule of stimulant chemical [39], and the brains of certain fish which are sensitive to electric fields as low as $1 \mu V/cm$ [40].

addition to being characterised by different amplitudes and frequencies, ($\delta,\theta,\alpha,\beta,\gamma$), it is now known (as mentioned above) that *EEG* signals exhibit deterministic chaos, the dimension of which *increases* with increasing degree of conscious awareness [42]. The requirement that any realistic model for the origin of brain-waves must yield chaos of the observed dimension is a severe constraint; this is currently under investigation [43] in the case of Fröhlich's model. Searching for a more *qualitative* correlate of different degrees of consciousness, one might cite the dramatic increase in coherence between the two sides of the brain which is evident [44] in the *EEG* of subjects who have achieved the so-called 4th state of consciousness using advanced meditation techniques.

Early attempts [45] to find a physical basis of consciousness appealed in a rather non-specific way to Fröhlich's original high frequency coherent polarization mode, and, in particular, to its *macroscopic quantum mechanical* equivalent as a *Bose condensate*, constituted not of matter, but of the quanta associated with a collective excitation of some unspecified biological matter - a distinction which parallels, somewhat, mind as a 'condition' of brain. That consciousness, in turn, is a global condition of mind, then leads to the hypothesis that consciousness itself is connected in some way with the existence in the brain of a Bose condensate, in terms of which the 'binding problem' would receive a novel solution. Progress in this direction requires, however, identification of a *material* system, a collective 'condition' of which might constitute the (non-material) condensate, and - *a fortiori* - a material system which is *peculiar* to the brain. The majority of attention [46] is currently focused on *microtubules* in the cytoskeleton of neurons (wherein they are *quite distinct* from those in other cells). For microtubules essentially constitute, *via* their tubulin subunits, a rather regular array of coupled dipolar oscillators to which energy produced by *GTP* hydrolysis is delivered; they thus constitute an *intra*-cellular variant of Fröhlich's original system based on the dipolar oscillation of cell *membranes*. Much consideration [47] has been given to the possibilities offered by these structures, in particular with respect to sustaining coherence and to the establishment of optical neural holography [48]. Most novel, however, has been the conjecture that microtubules are the *microsites of consciousness* - a conjecture motivated by the empirical fact that anaesthetics, which *suppress* consciousness, are known to bind to microtubules [49] where, through their electric dipole moment, they could interfere with the conformational oscillations of the tubulin (dimer) proteins on which the coherent excitation is based. Exactly how such intracellular excitations are coherently integrated across the brain to produce macroscopic consciousness is much less clear, however, although some quite ingenious suggestions [11,46,50] have recently been made involving *gravitationally* induced collapse of the intratubule coherent excitations.

On the other hand, consistent with the existence of deterministic chaos in the *EEG*, there has been a move to entertain the possibility that consciousness might be understandable purely *classically* - as an emergent, self-organised state of a dissipative, highly complex microscopic neuronal system far from thermal equilibrium and subject to a non-linear dynamics [51]. The self-organised state is characterized by relatively few degrees of freedom - far fewer than in the *un*-organised state; the system is thus 'attracted' to a lower dimensional phase space whose dimension (the so-called embedding dimension) represents the number of active degrees of freedom whose *non*-linear dynamics underpin the self-organised state. The possibility of *deterministic chaos* in such a system endows it with great dynamic *versatility*, such as is necessary for the rapid dissemination of information throughout the system - *i.e.* chaos provides a rich, quasi-random background from which new, adaptive patterns may emerge...*Chaos allows us to change our mind*!

> "*The establishment of a chaotic regime would provide the system with a certain "impartiality", and with almost unbiased probability of reaching the various meaningful limit cycles. The great advantage of deterministic chaos as a reset mechanism, with respect to Brownian [that is, stochastic] noise, derives from the fact that it is possible to switch it on or off almost instantaneously through (a sequence of) bifurcations - sudden transitions from limit cycle activity to chaotic activity seen in many systems as the value of a controlling parameter is slightly altered*" [52].

Prior to the introduction of these (neo-classical) possibilities, it appeared that only *quantum* superpositions provided the required versatility, and so far there has been very little interaction between the two camps; an exception is Fröhlich's Brain-Wave model [35], where (neo-classical) deterministic chaos and macroscopic quantum excitations *coexist*: here the brain-waves, whose evoked spectra exhibit *chaos*, are an emergent property of the underlying periodic enzymatic reaction - a reaction which is sustained by a coherent excitation (*the macroscopic quantum state*) which is, itself, an emergent property of the collective modes of the material brain system!

7. Consciousness and the Quantum Mechanical Measurement Problem

These suggestions that macroscopic quantum effects are involved in certain aspects of brain function, and *consciousness* in particular, reminds one of the *converse* - namely the consideration which has been given from time to time to the possible role of a conscious observer in the context of the measurement problem in the quantum mechanics of inanimate microscopic systems, and of the related problem of establishing the demarcation between quantal and classical descriptions [53]. Although, as early as 1932, von Neumann [54] (and later, others) pointed out that quantum mechanics seemed to be "tailor-made" to rehabilitate consciousness within physical science (from where it had been banished since the development of classical mechanics), Heisenberg's (*post*-Copenhagen) 'propensity' interpretation [55] still maintained a completely objective stance, in which mind was kept out of physics. Not until 1962 did Wigner [56] point out that, in the final analysis, all we can know of the physical universe is the information which enters our conscious mind through our senses[13]; he thus suggested that the collapse of the wave-function occurs when

[13] It was, incidently, F. Fröhlich's chance remark [57] to Marois of a related statement by Wigner [2] - to the effect that according to standard quantum theory there is zero probability of self-reproducing states (*and hence life*) - that motivated Marois to approach Fröhlich; the result was the conference referred to at the beginning of this article.

information enters the conscious human mind[14]. If, as the foregoing considerations suggest, the brain is itself host to certain macroscopic quantum states, then this approach to wave-function collapse may require reconsideration.

More generally, however, *macroscopic quantum systems* occupy a position intermediate between the microquantal and macroscopic classical systems considered by conventional measurement theory, in that, by their very nature, they *already straddle* the micro-macro gulf which measurement theory attempts to bridge. For although macroscopically large quantum systems are still described by complex (macroscopic) wavefunctions, $\Psi(\mathbf{x},t)$, the usual probabilistic interpretation of the wavefunctions of *microscopic* quantum systems can here be dispensed with (in consequence of the macroscopically large number of particles which constitute the 'condensate' described by Ψ), and $|\Psi(\mathbf{x})|^2$ interpreted as the *actual* density of (condensed) particles at $\mathbf{x}$. Furthermore, unlike the case of a quantum system (either single or many-particle) described by a (*non-*macroscopic) Schrödinger wavefunction, where non-linearity enters only in the measuring process (collapse of the wavefunction), macro-quantum mechanics is *inherently* non-linear *already* at the level of the wave-equation - the non-linearity reflecting the self-interactions underlying the associated macroscopic coherence. It would thus be of some interest to extend consideration of the measuring process - which traditionally involves the interplay between a microscopic quantum system and a classical measuring device - to include the case of macroscopic quantum systems, even when such pertain to *non*-living systems in thermal equilibrium; indeed, the existence of macroscopic quantum systems seems to have been generally overlooked in many considerations (e.g. [53]) of the transition between the quantal and classical limits; an exception is the recent work of Li [59].

8. Epilogue

This contribution has attempted to show how the concept of *coherence*, which H. Fröhlich first introduced into Biology some 30 years ago leads - in conjunction with typical biomaterial properties - to the prediction of coherent excitations in living matter. The existence of these excitations, which had not hitherto been suspected, entail quite novel possibilities, not only of non-chemical control, communication, energy storage, and an understanding of the near-quantum sensitivity of many biosystems, but also - *via* their implications on certain aspects of brain function - for a *physically based* assault on the problem of consciousness. Furthermore, the predicted existence of macroscopic quantum states in the living brain could well have implications on the quantum mechanical measurement problem, which are yet to be explored. It is thus not inconceivable that out of

---

14 Many years later, this motivated Stapp [58] to develop a model of the mind-brain system based on the physical similarity between the brain and a macroscopic detection device of the type considered by Heisenberg [55]. In this model, certain Heisenberg 'events' which 'actualize' large scale patterns of neuronal activity in the brain are identified as the physical correlates of consciousness.

a deeper understanding of the physical nature of **consciousness** - wrought through the power of the concept of *biocoherence* - the new biology of today might indeed eventually yield valuable clues to the '*physics of tomorrow*'- a possibility entertained long ago by Delbrück. If realized, this would surely rank as Fröhlich's most far reaching legacy to the 21st century.

> *"The above considerations are meant to be highly speculative. They should demonstrate, however, that application of quantum mechanical concepts can lead to new points of view which might be used as guiding points in the search for undiscovered regularities."* [3].
>
> *- H. Fröhlich, 1967*

Appendix: Some qualitative remarks on equilbrium Einstein condensation in an ideal Bose gas of (material) atoms and its comparison with non-equilbrium coherent excitations in active biosystems.

A Bose gas is an assembly of particles, such as helium atoms (^4He), whose mutual interaction is considered to be very small - as is usual for an ideal gas. It is essential, however, that these atoms are treated according to the laws of quantum mechanics. This implies that for each particle, a de Broglie wave, whose wavelength is inversely proportional to the velocity of the particle, is relevant for the description of its motion. At high temperatures, when the velocities are relatively high, such a gas behaves exactly like a classical ideal gas - the de Broglie wavelength being then small compared with the mean separation of the atoms. As the temperature falls, the velocity decreases, increasing the mean wavelength which eventually becomes larger than the mean interparticle separation. When this stage has been reached, the atoms can no longer be considered as independent because, according to quantum mechanics, the position of a particle can only be defined with an uncertainty of the order of the Broglie wavelength. This lack of definition, however, is incompatible with the concept of a classical ideal gas. The gas of ^4He atoms resolves this dilemma by accumulating (condensing) atoms into a single energy state - that with the lowest (*zero*) momentum -until the number of remaining particles is small enough to ensure that their mean separation is *no longer* less than the de Broglie wavelength, whence they can behave in a normal way. This condensation starts at a critical temperature, T_C, and increases with decreasing temperature, until, as $T \rightarrow 0$, all the particles occupy the condensed state, which thus becomes *macroscopically occupied.* The system constituted by these condensed particles represents a highly ordered state which exhibits superfluid properties, which are undisturbed by the uncondensed (thermal) particles with which they coexist at $0 < T < T_C$ - *i.e.* when a bulk flux is introduced into the gas, the velocities of the *condensed* subsystem at different points remain *highly correlated*; the relevant order is thus, not spatial but, *motional.*

As T falls towards T_c, the chemical potential, μ - which here reflects the existence of a *fixed* total number of material particles (atoms) - increases from large negative values toward zero, becoming zero at T_c. In the biocase, on the other hand, the counterpart of the material atoms are the quanta of the polarization waves, the number of which is *not* fixed but which, in thermal equilibrium, depends on temperature; accordingly, in thermal equilibrium μ is here *zero.* The possibility of realizing a macroscopic occupation of a single mode still exists, however, but only as a *non*-equilibrium feature at a *fixed* temperature. For in consequence of a competition between the influx of metabolic energy (pumping) and the thermal exchanges with the surrounding heat bath, the number of quanta can become effectively fixed; accordingly, with increasing pumping (s), μ increases from zero and tends to $h\nu_l$ as $s \rightarrow s_0$ (This behaviour should be compared to that which obtains in the case of a fixed number of Bose atoms in thermal equilibrium where the value of μ increases to the *zero*, ground-state energy of the system, as $T \rightarrow T_c$ from above). At s_0 the number of quanta with energy $h\nu_l$ starts to increase above thermal, resulting in a *macroscopic occupation* of the mode ν_1, which means that this *single* mode becomes excited far in *excess* of that appropriate to thermal equilbrium.

9. References

1. Elsasser, W.M.: "*The Physical Foundations of Biology*", Pergamon, London, 1958.
2. Wigner, E.P.: "*Symmetries and Reflections*", Ch.15, Indiana Univ. Press, 1967.
3. Fröhlich, H.: in "*Theoretical Physics and Biology*", (Proceedings of the First International Conference on Theoretical Physics and Biology, Versailles, 1967), (Edited by M. Marois) North Holland, Amsterdam, 1969, pp.13-22.
4. Bertalanffy, von L., Beier, W., and Laue, R.: "*Biophysik des Fliessgleichgewichts*", Braunschweig, Vieweg, 1977.
5. Raskevsky, N.: *Bull. Math. Biology* **28** (1966), 283-308.
6. Careri, G.: in "*Theoretical Physics and Biology*", (Proceedings of the First International Conference on Theoretical Physics and Biology, Versailles, 1967), (Edited by M. Marois) North Holland, Amsterdam, 1969, p.55.
7. Prohofsky, E.W., and Eyster, J.M.: *Phys. Letts.* **50A** (1974), 329-330.
8. Fröhlich, H.: *Phys. Letts* **26A** (1968); 402-3, *Int. J. of Quantum Chemistry* **2** (1968), 641-649.
9. Schrödinger, E.: "*What is Life*?", CUP, 1944, pp.69-70.
10. London, F.: "*Superfluids*", Vol.I, John Wiley and Sons, NY, 1950 pp.8-9.
11. Penrose, R.: "*Shadows of the Mind*", OUP, 1994 p.351.
12. Bhaumik D. *et al*: *Phys. Letts* **62A** (1977),197-200.
13. Frohlich, H.: *Collective Phenomena* **1** (1973), 101-109.
14. Fröhlich, H.: *Nature* **228** (1970), p.1093.
15. Koshland, D.E., and Neet, K.E.: *Ann. Rev. Biochem.* **37** (1968), 359-380.
16. Fröhlich, H.: *Phys. Letts* **39A** (1972), 153-4.
17. Fröhlich, H.: *IEEE Trans. Microwave Theory and Techniques*, **26**(8), 613-617 (1978); M.S. Cooper, *Collective Phenomena* **3** (1981), 273-288.
18. Holland, B.W.: J. *Theor. Biology* **35** (1972), 395-397.
19. Rowlands, S.: in "*Biological Coherence and Response to External Stimuli*", (Edited by H. Fröhlich) Springer-Verlag, Berlin, 1988, pp.171-191.
20. Titulaer, U.M., and Glauber, R.J.: *Phys. Rev.* **145** (1966), 1041-1050.
21. Girardeau, M.D.: *J. Math. Phys.* **6** (1965), 1083-1098.

22. Webb, S.J., and Stoneham, M.E.: *Phys. Letts* **60A** (1977), 267-8.
23. Webb, S.J. *et al*: *Phys. Letts* **63A** (1977), 407-8.
24. Kutnetsov, A.V., and Stanton, C.J.: *Phys. Rev.* B **51** (1995), 7555-7565.
25. Kütt, W.A. *et al.*: *IEEE J. Quantum Electronics* **28** (1992), 2434-2444.
26. Vos, M.H. *et al.*: *Nature* **363** (1993), 320-325.
27. Popp, F.A.: in "*Electromagnetic Bio-Information*", (Edited by F.A. Popp) Urban and Schwarzenberger, München, 1979, pp.123-149.
28. Pohl, H.A.: in "*Coherent Excitations in Biological Systems*", (Edited by H. Fröhlich and F. Kremer) Springer-Verlag, Berlin, 1983, pp.199-140.
29. Berteaud, A.J.: *et al, Compt. Rend. Acad. Sci* **281D** (1975), 843-6.
30. Grundler, W. *et al:* in "*Biological Coherence and Response to External Stimuli*", (Edited by H. Fröhlich) Springer-Verlag, Berlin, 1988, pp.65-85.
31. Fröhlich, H.: *Phys. Letts.* **51A**, 21-22 (1975).
32. Smolyanskaya, A.Z., and Vilenskaya, R.L.: *Sov. Phys. Uspheki* **16** (1974), 571-2.
33. Fröhlich, H.: in "*Coherent Excitations in Biological Systems*", (Edited by H. Fröhlich and F. Kremer), Springer-Verlag, Berlin, 1983, pp.1-5.
34. Kaiser, F.: in "*Energy Transfer Dynamics*", (Edited by T.W. Barrett and H.A. Pohl) Springer-Verlag, Berlin, 1987, pp.224-236.
35. Fröhlich, H.: *Neuroscience Res. Prog. Bull* **15** (1977), 67-72.
36. Elul, R.: *Neuroscience Res. Prog. Bull* **12** (1974), 97-101.
37. McA Sayers B., *et al.*: *Nature* **247** (1974),481; **260** (1976), 461.
38. Rose, A.: *Image Technol* **12** (1970), 13.
39. De Vries, H., and Stuiven, M.: in "*Sensory Communication*" (Edited by W.A. Rosenblith) M.I.T. Press, Cambridge, Mass, 1961.
40. Bullock, T.H.: *Neuroscience Res. Prog. Bull.* **15** (1977), 17-22.
41. Kaiser, F.: *Nanobiology* **1**, 149-161 (1992) - contains references to earlier work dating from 1977.
42. Gallez, D., and Babloyantz, A.: *Biol. Cybern.* **64** (1991), 381-391.
43. Hyland, G.J., and Lewis, C.P.: - to be published.
44. Levine, P.H. *et al.*: in "*Scientific Research on the Transcendental Meditation Programme*", Collected Papers Vol.I, (Edited by D.W. Orme-Johnson & J.T. Farrow) MERU Press, Germany, 1976, pp.187-207.
45. Marshall, I.N.: *New Ideas in Psychology* **7** (1989), 73-83.
46. Hameroff, S.R.: *J. Consciousness Studies* **1** (1994), 9-18.
47. Jibu, M. *et al.*: *Biosystems* **32** (1994), 195-209.
48. Hameroff, S.R.: "*Ultimate Computing: Bimolecular Consciousness and Nano Technology*" - North Holland, Amsterdam 1987; K.H. Pribram, "*Brain and Perception*", Lawrence Erlbaum, N. Jersey, 1991.
49. Hameroff, S.R., and Watt, R.C.: *Anesth. Analg.* **62** (1983), 936-940 - see also N.P. Franks and W.R. Lieb, *Nature* **300** (1982), 487-93.
50. Nanopoulos, D.V.: invited talk at "*Physics without Frontiers*", Trieste, July 1995, and CERN-TH/95-128.
51. Pritchard, W.S., and Duke, D.W.: *Brain and Cognition* **27** (1995), 353-397.
52. Serra, R., and Zanarini, G.: "*Complex Systems and Cognitive Processes*", Springer-Verlag, New York, 1990, p.183,.
53. Ghirardi, G.C. *et al.*: *Phys. Rev. D.* **34** (1986), 470-91.
54. Neumann, J. von.: "*Mathematische Grundlagen der Quantenmechanik*", Ch. VI-1, 1932.
55. Heisenberg, W.: Daedalus **87** (1958), 99, "*Physics and Philosophy*", Harper Row, N.Y., 1958.
56. Wigner, E.P.: "*Symmetries and Reflections*", Ch.13, Indiana Univ. Press, 1967.
57. Fröhlich, F., and Hyland, G.J.: in "*Scale in Consciousness Experience*", (Edited by J. King and K.H. Pribram), Lawrence Erlbaum N.Jersey, 1995, pp.405-438.
58. Stapp, H.: "*Mind Matter and Quantum Mechanics*", Springer-Verlag, Berlin, 1993.
59. Li, K.H.: in "*Current Development of Biophysics*", (Edited by C. Zhang, F.A. Popp and M. Bischof) Hangzhou University Press, 1996, pp.287-305.

THE CONCEPTS OF COHERENCE AND 'BINDING PROBLEM' AS APPLIED TO LIFE AND CONSCIOUSNESS REALMS

Critical Consideration with a Positive Alternative

M. LIPKIND

Unit of Molecular Virology, Kimron Veterinary Institute
Beit Dagan, P.O. Box 12, 50250 Israel

1. Introduction

The title of the present paper may seem an ambitious amalgamation of two universal concepts - Life and Consciousness called here 'realms' - connected with two, also general, conceptual notions - coherence and 'binding problem' - which are not immediately clear unless appropriately defined. However, both the 'realms' - Life and Consciousness - also have no unequivocal meaning and notoriously have no strict definition. Consequently, the aim of the present paper is to find out whether the notions of coherence and 'binding problem' which have recently become of use in theoretical studies on Life and Consciousness are fruitful for analyzing relationship - hierachic if at all - between the Life and Consciousness realms. Therefore, the consideration of these notions will be followed by critical reflections and the positive solution based on the Gurwitschian field principle (Gurwitsch, 1944, 1954, 1991; Lipkind, 1987a, 1987b, 1992, 1996) will be considered. The hope is that such analysis will be finally succeeded in finding non-tautological definitions of both the Life and Consciousness.

2. Coherence

The notion of coherence, correctly, coherent states, originates from the physics glossary where it is either strictly defined (classical formal optics), or employed in formulation of certain principles (quantum mechanics). In the latter case, the concept of coherence is defined in accordance with different approaches expressed in the form of either principles and theorems like those by Heisenberg, Schroedinger, Feynman, Pauli, and Bell, or paradoxes like those by De Broglie and Einstein-Podolsky-Rosen (Li, 1992a, Li, 1995, Popp & Li, 1993). Confrontation of the above principles has resulted in some inferences which appeared to be very tempting for their employment for the analysis of Life and Consciousness (Li & Popp, 1992, 1996), for example, the 'squeezed state' determining the coherent state as that of minimum uncertainty (Li, 1996a) and the non-locality principle (Penrose, 1994; Li, 1996a, Li & Popp, 1996).

J. J. Chang et al. (eds.), Biophotons, 359-373.

The application of the concept of coherence to living systems has been dictated by the urge to understand unique specificity of the living state by means of alternative approaches - apart from the dominating dogmas of molecular genetics (Popp,1992a). However, the peculiarity of the situation is that the notion of coherence employed in biological systems has no formal definition anyhow differing from the formal definitions used in quantum physics (Popp, 1992b, Popp & Li, 1993; Zhang, 1994; Li, 1995; Zhang & Popp, 1996). However, the concept of 'biological coherence' pretends to 'explain' the intrinsic specificity of the 'living state' that exceeds the limits of the immediate 'physical' definition. The outcome is that the notion of the 'biological coherence' is instilled by rather emotional content without sufficient axiomatic basis. Therefore, because of such discrepancy, sometimes argumentation in such limits leads to negation of any unique specificity of living systems which is evident from the following citation: "We try to show that from the view point of physics, which aims to invoke general principles of the motion of matter in the space-time frame of our real world, the biological phenomena are **not too complicated** (boldface is mine) and can be explained as features of open systems far away from thermal equilibrium at macroscopic as well as at microscopic levels" (Li, 1996a, p. 42). Apart from a legitimate general assertion about the reduction in principle of living phenomena to physical fundamentals, there is a strange claim about non-complicatedness of biological phenomena. This is in contrast to persuasive evidence (widely accepted even by the most extreme materialist reductionist views) that biological phenomena even in the most primitive species are of supreme complexity. Therefore, the theorists of the coherence concept assert that namely the remarkable uniqueness of the Life phenomena still remains the topic for research: "The surprisingly essential feature of a biological process seems to be that, in spite of the vast numbers of molecules involved, it is ordered in time and space" (Li, 1996a, p. 42). This is a formal (and rather trivial) declaration about the intrinsic feature of the biological processes expressed on microlevel while no specific postulation is provided.

A particular example concerns application of the coherence concept for the analysis of biophoton emission which is a modern term (Popp, 1992a, 1992b, 1995) for the phenomenon of mitogenetic radiation discovered by A. Gurwitsch in 1923 (A. Gurwitsch & L. Gurwitsch, 1948). Because of the low intensity and broad spectral distribution of the biophoton emission, the biological significance of the coherence (Popp, 1995) in the authors' context is understood not merely as the ability for electromagnetic waves to interfere but as the binding of the emitted photons to their source (Popp, 1995). Therefore, the analysis of the photon emission parameters may gain insight into this source, namely, the processes "bearing" that emission. In the case of biophotons, their emission source is hoped to be that mysterious 'source of livingness'. According to the modern biophoton theory, DNA is that source (Li, 1992b, Popp, 1992a), and the biophoton emission originates from the coherent electromagnetic field within this source and functions in intra- and intercellular biocommunication (Popp et al., 1994). A more general theory of 'evolution as the expansion of coherent states' has been suggested (Popp, 1992c, 1996b).

In reflexion to the concept of coherence employed for the description of living processes, the above-mentioned discrepancy between the formal physical definition of coherence and much more general scope of Life phenomenology which is included into the same notion, should be emphasized. In practical language (Zhang, 1994; Zhang &

Popp, 1996; Li, 1996a), the coherence in living systems is reduced to the ideas of molecular orderliness of the intracellular molecular substrate providing coordination of biochemical reactions *in vivo* as opposed to chaotic molecular collisions proceeding *in vitro*. This means coordination of most if not all the processes occurring on the molecular level *in vivo*, which determines also 'coherent' realization of the processes observed on the cellular and morphological levels of the living systems. However, the ideas of orderliness and coordination have appeared as a result of generalization of rich store of knowledge reflecting the Life phenomenology expressed on molecular level. It seems that the coherence means the ideas of orderlinnes and coordination **themselves**, so that the notion of coherence serves as just a convenient designation (a kind of a term) of those ideas but not as a principle giving their 'explanation'. Therefore, any attempt to use the notion of coherence for theoretical substantiation (or, moreover, 'explanation') of the notions of orderliness and coordination will be inevitably tautological.

The coherence theory concerning the biophoton emission is an example: on one hand, the coherent biophoton emission is claimed to be dependent on the "cell organization", responsible for "nonsubstantial" intra- and intercellular biocommunication (Popp et al., 1994), on the other hand, the development (**formation**) of such 'biological' coherence is beyond the consideration. Hence, there is an 'explanatory gap' between the physically defined coherence and its 'biological' counterpart. By other words, the 'cell organization' coordinating processes on microlevel as described by the concept of the 'biological coherence' only wishfully depends on (derives from?) the real physical coherence.

The conclusion is that the concept of coherence uttered as a general principle pretending to describe the biological specificity and uniqueness but based on formal quantum mechanics definitions has no immediate value in this respect unless it is provided with additional postulation concerning biological specificity. The positive example, in this respect, is the idea of nonequilibrial molecular constellations whose existence was prognosticated from the suggested abstract field principle (the theory of biological field by Gurwitsch, 1944, 1954, 1992) becoming fruitful for the development of working hypotheses by means of which experiments on mitogenetic radiation were performed and supported the idea.

3. Binding Problem

The binding problem in neurobiology and cognitive science appears from a well established evidence on segregated nature of the brain which means that individual features of any perceived object are processed separately while the object appears unitary, so that the whole dynamically changing world is perceived as integral and coherent. The perceived unity cannot be derived from the objects of the physical world themselves since what enters the brain from outside is the stream of continually fluctuating sensory inputs. Therefore, the synthesis of all these disjuncted, dispersed and separately processed elements of information about the continually changing picture of the external world must be realized by binding together neurological states of the brain.

Historically, the binding problem was connected first with psychology being dependent on general philosophical views on spatio-temporal contiguity of mental representation of the external world (Hume, 1735/1958, 1777/1975). However, the modern version of the binding problem is expressed on the neurological level being based on the well documented evidence on the disjunctive way of processing of visible percepts. Namely, the visual data are processed separately within about fifty functionally segregated specialized cortical areas (Hubel & Livingstone, 1987; Livingstone & Hubel, 1987), each one being responsible for a specific feature, like movement, colour, texture, size, curvature, some topological properties like height/width ratio, stereoscopic depth, orientation of lines and edges, and so on (Ramachandran, 1990; Ramachandran & Anstis, 1986; Treisman, 1986; Zeki, 1992). At the same time, multi-modal association areas in the cortex in which single perceptual features could be unified into a final percept have not been found.

Correspondingly, the binding problem sounds differently in psychology and neurobiology. In psychology, it is formulated in a wider sense as integration and interpretation of the information coming from the previously disjuncted sensory inputs into consciously perceived representations (Hardcastle, 1994). Hence, the question is how disparate perceptual features can yield unified interpretations of the perceived objects.

The neurobiological version of the binding problem is naturally expected to simplify the picture since it is reduced to the question of how the various outputs from different processing modules are linked together to provide unified percepts. However, this question although being 'free' from all the cognitive attributes and mental epiphenomena, still exposes insuperable obstacles for the classic neuronal theory.

Usually, the main difficulty is connected with the binding problem *per se*, i.e. the question is how the disjuncted features of any perceived object are linked together. The common conclusion is that there is NO SOMETHING which is called either CENTRAL CORTICAL 'INFORMATION EXCHANGE' (Hardcastle, 1994), or 'MASTER MAP' (Treisman, 1986), or MULTI-MODAL ASSOCIATION AREAS (Damasio, 1989) in the cortex. Nevertheless, still there is a certain (rather intuitive) hope for future finding of 'grandmother' neurons and convergence zones (Hardcastle, 1994). However, I want to attract attention to another aspect of the visual binding problem which is a stumbling-block for its solution in the limits of the neuronal theory.

If roughly to consider only visual perception, a scheme is as follows. An object under being seen is projected upon retina and its visible image irritates the retinal receptors. Then there are two antidromically proceeding phenomena. The first one is connected with immutable histological fact that about 300 retinal rods (the 1st neuron of the visual network way) are structurally (histologically) connected with one ganglionar cell (the 3d neuron). Consequently, these 300 adjacent rods form a microarea upon retina which during the vision process may contain heterogeneous picture as projected from a (micro)part of the visible object (even if the latter is not moving). Thus, the axon of the ganglionar cell must conduct forward an impulse carrying such complex (integrated) information. This already is incompatible with the classic neuronal theory according to which the neuron firing means only the CONDUCTION of a signal which is triggered according to the all-or-nothing principle. However, at the same time, there is another - antidromic - process which consists of the above-described splitting of the perceived object's image into dozens of quite different 'components' of the vision (features of the perceived object) like shape, movement, colour etc. which are processed separately in distinct cortical areas. Totally incomprehensible is the possibility itself of coexistence and simultaneous realization of two antidromic processes: confluence (merging, junction, mayby synthesis) of a number (hundreds!) of axonal impulses BEFORE they come to the brain centres, on one hand, and in parallel occurring disjunction (splitting, breaking, maybe analysis) of the object as a whole - not into **parts** (portions, pieces) - but into so drastically different causally disconnected (somewhat categorial) features, like colour, form, movement etc, on the other hand. This insuperable obstacle for the neuronal theory which was analyzed first by Gurwitsch (1954) is considered now in the framework of fundamental computational mechanism of the brain. According to computational language, the problem is formulated as claiming that the objects or their different aspects have to be represented (to co-exist) within the same physical hardware that resulting in the '**superposition catastrophe**' (Von der Malsburg, 1987). Therefore, if to define the main postulate of the neuronal theory in the way that each mental representation ('symbol') of the external objects is represented by the corresponding subsets of coactive neurons within the same brain structure, then if more than one of such 'symbols' become active at a moment (which must occur within the real functioning pattern), they become superimposed by coactivation (structural 'overlapping' of the activated subsets). In such a case, any information carried by the 'overlapping' subsets must be lost, that being the mortal verdict for the whole foundations of the classic neuronal theory.

Thus, the binding problem, being not a purely theoretical construction but arisen from the very midst of the neurobiological and psychological reality has become a challenge for the general grounds of theoretical biology. In accordance with this, the alternative description based on the theory of biological field by A. Gurwitsch is suggested.

4. The Alternative Approach: The Theory of Biological Field by A. Gurwitsch

The Gurwitsch's theory of biological field and its application for analysis of living manifestations on all the levels of biological organization - molecular, cellular and morphological - can be described here only shortly (Gurwitsch, 1944, 1954, 1991, for English-published reviews see Lipkind, 1992, 1996) . The Gurwitschian field is irreducible to any known physical fields and only in this respect may be called vitalistic that inevitably inducing dislike and prejudice. The pity is that, according to the commonplace (better to say, vulgar) ideas of the vitalistic principle, the latter is a kind of an abstract metaphysical omnipotent 'vital spirit', 'organizing' ('animating') material components of a living system (organism), thus determining its structural appearance and coherent functioning. Such evidently tautological character of the vitalistic concept has no explanatory character and scientific value. The mode of action of the Gurwitschian field on the morphological level is defined as subjection of equipotential elements (cells) to integral morphogenetic field causing the cells' **spatial** orientation or/and movement. The Gurwitsch's theory has a form of a mathematical model with strictly defined postulates. The latter concern vectorial repulsive character of the field, its anisotropy, field sources, nature of elementary field "flash", formation of integral microfields and macrofields, dynamics of field tension and its association with metabolic rate. The Gurwitsch's field theory was employed for the adequate description of different living manifestations at each level of the biological organization (molecular, cellular and morphological), namely, morphogenesis, differentiation, histogenesis, mitosis, metabolic processes, neuromascular physiology and brain cortex functioning. Application of the Gurwitsch's field theory to psychic sphere was based on the analysis of the brain cortex cytoarchitectonics as a geometric continuum as opposed to the conventional neuronal theory which is based on the discrete neuronal firing and highly complicated network of neuronal interconnections. In the light of the Gurwitsch's theory, disturbances of highly reactive dynamic integral Gurwitschian field of the cortex is what specifically correlates with (is expressed by) the psychic phenomenology. Correspondingly, the theory "describes" such consciousness manifestations as perception, feelings, memory and recollection, and "explains" embryogenesis of consciousness and the somato-psychic gap. Analysis of the consciousness embryogenesis shows impossibility to indicate a moment of the emergence of the initial psychic phenomenology and its connection with (relation to) any somatic event throughout the whole development cycle.

The unprecedental advantage of the Gurwitsch's field theory is its wide descriptive capacity: by the SAME postulates, the theory is efficient when applied to somatic processes, on one hand, and manifestations of consciousness, on the other. Its descriptive capacity was tested on different levels of biological organization: molecular (metabolism), cellular (mitosis, differentiation and histogenesis) and organismic (morphogenesis, neuro-muscular system, brain structure and functioning), on one hand, and different above-described manifestations of the consciousness (feeling, stream of incoherent chaotic thoughts, memory and recollection), on the other hand. This makes it possible to understand and comprehend the consciousness as a BIOLOGICAL reality described by the common language with the other biological phenomena. Gurwitsch's theory provides notions and concepts, like **integral actual field, field configuration, unbalanced protein molecular constellations (molecular continuum), "structured process"** and some others more specific notions which are efficient instruments for the adequate description of the biological AS WELL AS psychological phenomena in the common language.

5. Supervenience and the Absence of Isomorphic Identity

The vitalistic conclusion can be derived by using the recently developed supervenience concept (Horgan, 1982; Kim, 1984, 1987, Chalmers, 1996). In the limits of this concept, the species-specific morphology of the WHOLE can be considered as the high level property (B-properties), which supervene on the properties of the lower level - the cells constituting this morphology (A-properties) (Chalmers, 1996). According to Chalmers, "the notion of supervenience formalizes the intuitive idea that one set of facts can fully determine another set of facts. The physical facts about the world seem to determine the biological facts, for instance, in that once all the physical facts about the world are fixed, there is no room for the biological facts to vary. (Fixing all the physical facts will simultaneously fix which objects are alive). This provides a rough characterization of the sense in which biological properties supervene on physical properties. In general, supervenience is a relation between two sets of properties: B-properties - intuitively, the **high-level** properties - and A-properties, which are the more basic **low-level** properties" (Chalmers, 1996, p. 32-33). Similarly, the properties of the morphological level (high-level properties) must be considered as supervenient on the cellular level (low level properties). However, the Driesch's experiments (Driesch, 1891) clearly demonstrated that the morphogenesis is not realized by the unequivocal connections between the cells participating in the morphogenesis - the latter is realized neither by the Lego toy (puzzle) mosaic principle, nor by the stimulus-response reactivity chain. Consequently, such morphological supervenience is disproved by the Driesch's experiments. The logical inference is that the morphogenesis is not determined by the internal properties of the cells but is determined by the laws associated with the developing WHOLE which is an expression of the vitalistic principle (Driesch, 1891, 1908, 1915).

Using a modern language, the vitalistic principle can be inferred from the notion of isomorphism. Namely, it is possible to claim that a robust fact inherent to any living system is the ABSENCE OF ISOMORPHIC IDENTITY between the living state in its macro-expression and the corresponding processes occurring on the molecular level including current chemical (enzymatic) reactions and any other, no matter how much and how far micro- and submicro-scopic, phenomena. Correspondingly, the absence of the isomorphic identity can be seen in any interplay between the levels, namely: (a) between intracellular "macrophenomena" and the processes occurring on the molecular level; and (b) between the morphological (organismic) characteristics and the "macrophenomena" observed on the cellular level (and, in turn, the "microphenomena" occurring on the molecular level).

Thus, there is a crucial conceptual difference between a however super-complex and hyper-complicated but definitely dead system and however "primitive" living species (amoeba, bacterium, chlamidium etc). The former can be exhaustively and unequivocally identified with (reduced to, explained by, shown to be supervenient on) the corresponding constituting it physical entities - molecules, atoms, "quantum mechanical structure of the world" (Hardcastle, 1996, p.9), while the latter cannot. Namely, in the latter case, a **number (rather a high number) of possible states on the molecular level will correspond to the SAME living macro-phenomenon**. The important point is that such diversity is NOT a result of **statistical distribution** which fluctuates around a principal rigid pivot-like regularity. Phenomena observable on the cellular level may be realized via a high variety of the corresponding processes observable on the molecular level. This means that the macrophenomena observed on the cellular level are NOT determined (only are influenced) by molecular processes occurring on the microlevel. Besides, it should be taken into account that any living system (at macrolevel) is naturally irreversibly changing with time, e.g. it is **aging**, while the notion of "age" does not belong to any physico-chemical components (molecules, compounds). Instead, in the glossary of physico-chemical parameters, there are strict notions like stability or half-life time, and so on. Therefore, the living system can be considered as a certain WHOLE which organizes the movement and distribution of constituting it molecules. While the word "organizes" as having an antropomorphic aura immediately induces "anti-vitalistic" reaction, the word "determines" seems to be innocent. However, the term "determination" means a rigid unequivocal (isomorphic) connection between the "influencing" and "influenced" and, hence, does not reflect the above-described very loose influence of the living macro-system (the whole) upon "non-living" molecules. Therefore, Gurwitsch introduced a special term which is called "Normierung" in German (Gurwitsch, 1930) and "normirovka" in Russian (Gurwitsch and Gurwitsch, 1948), having no linguistic equivalent in English. This term, contrary to the term "determination", means that the influence of an upper level of biological organization on a lower one is realized not by a rigid unequivocal (isomorphic) regularity but within a certain "norm", i.e. within certain limits permitting a rather high freedom for molecules' "behavior" within these limits without distorting specificity of the process observed on the macro-level.

In the frame of the suggested theory, the mode of the up-down causation (the effect of the influence of the living whole on the constituting it molecules) is defined as an action against chaotic movements of the involved molecules, i.e. specific orientation is put over the chaotic movement. The orientation is determined by the resulting vector of the field configuration at any particular locus within the whole. The field issues from the whole, i.e. it is not generated by the physical properties of the molecules themselves which are within the field influence, as opposed to the electrical field which is formed by the elements themselves (their charges).

The above conclusions can be illustrated by Laplace's demon allegory vividly described by Chalmers (1996). "We can imagine that a hypothetical superbeing - Laplace's demon, say, who knows the location of every particle in the universe - would be able to straightforwardly "read off" all the biological facts, once given all the microphysical facts. The microphysical facts are enough for such a being to construct a model of the microscopic structure and dynamics of the world throughout space and time, from which it can straightforwardly deduce the macroscopic structure and dynamics. Given all that information, it has all the information it needs to determine which systems are alive, which systems belong to the same species, and so on. As long as it possesses the **biological concepts** and has a full specification of the microphysical facts, no other information is relevant" (Chalmers, 1996, p. 35). The logical inconsistency within this citation can be easily found. Namely, these "biological concepts" destroy the logical elegance of the whole Laplace's demon idea. On one side, "the microphysical facts *are enough* (my italics) for such a being to construct a model of the microscopic structure and dynamics of the world throughout space and time from which it can *straightforwardly* (my italics) deduce the macroscopic structure and dynamics. Given all the information, it has *all* (my italics) the information it needs to determine which systems are alive, which systems belong to the same species, and so on". However, just the next sentence claims: "As long as it possesses the *biological concepts* (my italics) and has a full specification of the microphysical facts, no other information is relevant". Hence, the possession of the "biological concepts" is needed! This means that the knowledge about the "location of every particle in the universe" is NOT enough, and this immediately deprives the Demon of all his demonic capacity. Now, the result of the Demon's activity will depend on the sort of the biological concept(s) which would be in his possession. But suppose he will possess such biological concepts like harmonic regulations, equipotentiality and equifinality - he would easily become "vitalistically oriented" with all the consequences.

Thus, the facts demonstrating the absence of isomorphic identity between the processes observable on the upper (cellular and morphological) and the lower (molecular) levels make the above allegory invalid in respect to the living systems. The Laplace's demon unless he is a vitalist will be puzzled by a queer asymmetry in relationships between the "location of every particle" and the ability "to straightforwardly read off all the biological facts". In the best case, the comprehensive knowledge about all the "microphysical facts" about, say, developing embryo would permit the demon straightforwardly to deduce the embryo macroform, but the SAME macroform will be deduced from a drastically **different** pattern of the microphysical facts. A high number of such drastically different micro-patterns for the same species-specific living macroform would confuse the demon if not to drive him crazy.

6. Definitions of Life and Consciousness

This ambitious task is connected with the recently proclaimed 'hard problem' of consciousness (Chalmers, 1995). In a general abstract form, the problem is formulated by the question: 'How can a physical system of ANY degree of complexity be aware of itself?' From neurobiological grounds, the problem is reduced to the question: 'How and why can performance of any form of neural activity give rise to subjective experience?' The conceivable alternative may easily be that all the information-processing which determines behavior and survival might go on 'in dark'. By critical consideration of a number of works on consciousness of theoretical significance (among them: Jackendoff, 1987; Baars, 1988; Edelman, 1989; Crick, 1994; Dennett, 1991; Penrose, 1994; Hameroff, 1994), Chalmers demonstrates their insolvency concerning the hard problem and comes to conclusion that "the emergence of experience goes beyond what can be derived from the physical theory" (Chalmers, 1995, p. 208). This dualistic conclusion does not concern Life which is considered by Chalmers as being a part of the physical world: the vitalistic principle is rejected as 'obsolete' based on naive persuasion of a non-biologist that All-Mighty DNA has solved, in principle, all the Life problems and riddles.

As to the consiousness definition, according to the tacit consent of experts, it is not worth hurrying with this by the reason that "search for the most adequate definition would be paralysing or lead to the evocative, but hardly satisfactory suggestion, that consciousness can only be 'defined in terms of itself' (Angell, 1904)" [Bisiach, 1988, p. 102].

As to the definition of Life, according to the majority of the familiar to me authors dealing with Consciousness problem, including "New Mysterians" (Chalmers, 1995, 1996), "Liberal Naturalists" (Rosenberg, 1996), devoted materialists (Dennett, 1991; Hardcastle, 1996), the Life is **not defined but described** by a SET of life manifestations. There is a mere enumeration of arbitrarily chosen life manifestations, the choice depending on the authors' educational background, taste and emotional inclination, for example: "DNA, adaptation, reproduction, and so on" (Chalmers, 1995), in one case, and "reproduction, development, growth, self-repair, immunological self-defence and the like" (Dennett, 1996), in another. Such kind of definition is intolerable in Physics whose basic definitions are grounded on consideration of mathematically expressed "necessary" and "sufficient" conditions. Therefore, besides the semantic meaning, the syntax - the hierarchy between and amongst the connected notions and their attributes - determines the necessity and sufficiency of the definition conditions.

6.1. DEFINITION OF LIFE

As opposed to the above invalid "set-like" kind of the life definition, I suggest the one which is as follows. LIFE (ANY LIVING SYSTEM) IS EVOLVING AND AGING SPECIES-SPECIFIC NON-EQUILIBRIAL (i.e. NEEDING INCESSANT ENERGY INFLUX) SELF-REPRODUCING AND SELF-PRESERVING **GEOMETRICAL FORM** WHICH IS CONTINUOUSLY REFILLED BY SPECIFIC SUBSTANCES. From the formal point of view, the advantage of this definition, as compared to the above-mentioned "set-like" ones, is its syntax. Namely, instead of a dull enumeration of manifestations of the Life as the defined subject, the definition has ONE predicate ("FORM") around which there are necessary attributes (including a subordinate clause) hierarchically subordinated to the predicate. Together with this, the suggested definition is neither tautological, nor synonymic. In this definition, however, there is a certain inevitable logical "disbalance" between the attributes to the predicate "Form", namely, some of the attributes are related to the notion of Life as a living individual while the others are related to the general notion of Life as a cosmic phenomenon. This is manifested, especially, by the attributes "evolving and aging" clearly concerning any living INDIVIDUAL which is necessarily (at any moment of observation) found to be at a certain point of its ascending-descending life-time curve. On the other side, the attribute "non-equilibrial" is related to the general notion of Life as a realm. However, the "individualization" (individual parcellation) is a fundamental natural attribute of Earth Life although at least a theoretical alternative to the individualization is conceivable (Lem, 1976).

It could seem, nevertheless, that some of the "basic" manifestations of the Life enumerated in the above "set-like" definitions are not covered by the suggested definition, for example, heredity ("DNA", "gene"), from the Chalmers' and Dennett's list. However, these properties, like all the others from the "living set" (adaptation, development, growth, self-repair, immunological self-defence "and the like" "and so on" (Chalmers, 1995; Dennett, 1996), are particular functional mechanisms of realization of the life attributes given in my definition. For example, the attribute "self-reproducing" is realized through the genetic mechanisms, e.g. DNA triplet code and transcription, RNA translation, protein post-translational modifications, gene expression during cell differentiation etc. Therefore, I believe that the suggested definition of Life has a claim to be necessary and sufficient.

Apart from the syntactic estimation, the crucial point of the above definition - the MORPHIC principle - is not trivial. Usually, other Life manifestations, like metabolism, heredity, reproduction, internal mileau constancy, evolution, and, accordingly, different principles, like biochemical, biophysical, genetic, energetic, thermodynamic, synergetic (self-organization) are regarded as of paramount importance (Loeb, 1906, 1912; Lotka, 1925; Szent-Gyorgyi, 1957; Monod, 1971; Dawkins, 1976; Cairns-Smith, 1982; Caplan and Essig, 1983; Dennett, 1995; Prigogine and Stengers, 1984; Haken, 1977; Babloyantz, 1986; Waechterhauser, 1988; Eigen, 1992, 1993; Elitsur, 1994; Lifson, 1987, Kauffman, 1996). These principles are clearly associated with reductionistic trend of explanation contrary to the morphic principle: the FORM in my definition is that biological entity which cannot be derived from the physical fundamentals while it is exhaustively described and explained by the Gurwitsch's field theory. If the essence of Life is based on the morphic principle, the question is HOW THE EVOLVING SPECIES-SPECIFIC GEOMETRICAL FORM OF A LIVING SYSTEM IS MADE UP IF *A PRIORI* IT IS **NOT** DETERMINED BY (NOT SUPERVENIENT ON) THE MATERIAL STUFF CONSTITUTING (FILLING) IT. This "HOW"-question becomes entangled with the "WHY"-question, namely, why the molecules entering the living species-specific system behave differently as compared with their "*in vitro* behaviour" determined by their canonic (physico-chemical) properties. In such case, the EXPLANATORY GAP is between the molecules' physico-chemical properties determining their **chaotic movement in solution,** on one hand, and the molecules' "*in vivo* behaviour" characterized by **VECTORIZED mode** of their **movement within the liquid medium** of **a living system**, on the other hand. Accordingly, the molecules entering the living system display an additional quality of moving along certain preferrable directions which are non-linear and dynamically changing. The trajectories of coordinated movements of the whole totality of the involved molecules will provide somatic (morphological) species-specific appearance of the living system as well as its coherent functioning.

6.2. DEFINITION OF CONSCIOUSNESS

By analogy with the Life definition, the Consciousness also can be described by a list of its manifestations, like cognition, qualia, intention etc, etc, but imperfection of such way is as evident as that in the above-discussed case of the Life definition. Therefore, syntactically valid non-tautological definition is to be formulated, being intimately connected with the accepted conceptual framework. The latter is based on the Gurwitsch's theory of biological field as a working vitalistic principle.

Contrary to the classic dualism which establishes a principal gap between CONSCIOUS AND NON-CONSCIOUS realms, the vitalism establishes a principal gap between LIVING AND NON-LIVING realms where the **Consciousness** is an immanent attribute of the **Life**, the development of somatic and mental spheres going in inalienable association. The only initial point of the individual psychic development is the individual's zygote under primary cleavage. The embryo's rudimentary "**experience**" is postulated as the embryo's current **knowledge** about its own continuously developing morphological contours ("**geometrical feeling**"). This rudimentary ("**pre-cerebral**") consciousness is an integral part of the primordial **psychic** act which is expressed by that the embryo "KNOWS" ("EXPERIENCES") **geometry** of its form (which is the expression of the Gurwitschian integral field) and "ACTS" **morphogenetically** smoothing any disturbances of (experimental intervention into) the "normal plan" of the development determined by the species-specific field anisotropy.

Given this as a basis, the analysis (in reductionistic sense) of the CONSCIOUSNESS *PER SE* must be done by formulating a notion of elementary PROTO-CONSCIOUSNESS which is inherent to any, no matter how primitive, living system. In accordance with this, such elementary Consciousness is defined as follows.

ELEMENTARY CONSCIOUSNESS IS THE EMBODIED INHERENT CAPACITY OF ANY LIVING SYSTEM **TO FEEL** ITS OWN EVOLVING DYNAMICALLY FLUCTUATING SPECIES-SPECIFIC morphology (synonyms for this capacity: MORPHIC SENSATION, 'GEOMETRICAL FEELING'). This 'geometrical feeling' of the living system is intrinsically associated with the urge to preserve the species-specificity of the living form by reacting morphogenically to any effects caused by any disturbing factors changing or damaging it. Consequently, the combimation: **'geometrical feeling'** - **'morphogenic reaction'** is a rudimentary psychic act (**'morphological mind'**). Consequently, the ontogenetic development of a human individual's consciousness includes a chain of stages: from the 'pre-cerebral' slowly reacting **'morphological mentality'** - through the quickly reacting neurophysiological (instinctive or reflective) behaviour - to the 'cerebral' consciousness *per se* in which case the individual 'feels' ('experiences') the disharmonies within the Gurwitschian field of the brain continuum which are caused by the incessant stream of the neural network-mediated stimulation. The analogous stages could be retraced during animal phylogenetic development of consciousness which are expressed by the same triad: 'morphological consciousness' - instinctive and reflective physiological behaviour - 'cerebral' consciousness *per se.*

Thus, in accordance with the idea of a new basic fundamental ('extra ingredient', by Chalmers, 1995) associated with the above-suggested definition of Life, the elementary consciousness as an unalienable attribute of life is expressed in that any species-specific living system, whenever it is alive, **feels its own geometry**. The latter concerns not only external morphological contours: the geometrical feeling transpierces through the whole system, *i.e.* each geometrical (**stereometrical**) dot of the three-dimensional whole is felt, sensed, awared of by the system, **no matter what material entity is localized at this geometrical dot**.

7. Epilogue

In the frame of the above Life and Consciousness definitions, the reduction of the Consciousness to the Life is achieved. Postulation of the 'geometrical feeling' as an inalienable attribute of any living system and (although hypothetical) derivation of the consciousness *per se* from the primordial 'morphological mind' is in agreement with the ontogenetic and phylogenetic reality. The main advantage of this reduction is the transfer through the Brain-Mind Explanatory Gap, or, correctly, moving this Gap up to the common basis of the initial unquestionable fundamentals. In this light, the hard problem of consciousness loses its 'hardness', e.g. desappears. However, if as a result of this reduction the Hard Problem of Consciousness is exchanged for the Hard Problem of Life, the advantage of such reduction may be doubtful. In order to analyze the situation, both the Hard Problems should be confronted by the same Chalmers' stylistic mode: the "WHY of Consciousness" (Hardcastle, 1996) should be compared with the "WHY of Life". The "Why of Consciousness" means why all this information-processing does not go on 'in the dark', free of any inner feel (Chalmers, 1995). In this respect, the question "Why of Life" loses any sense since it has no analogical connection with the concept of unconscious Zombie perfectly functioning "in the dark". The Life is always alive - there is no "Living Gadget", nor "Mechanical Life". The general difference between both the Hard Problems is that the Consciousness arises from the living brain while any form of Life comes only from the same species of Life. The everyday reality is that any (human) consciousness ontogenetically originates within a developing, proceeding from zygote individual, i.e. the consciousness originates from the life, while any new (progeny) life originates only from the pre-existing (ancestral) life WITHOUT INTERRUPTION, i.e. without any explanatory or time gap. The inference from the suggested theory is that there is NO Hard Problem of Life and, as far as the Consciousness is reducible to the Life, there is NO Hard Problem of Consciousness.

8. References

Angell, J.R. (1904) *Psychology: an Introductory Study of the Structure and Functions of Human Consciousness*, Holt, New York.

Babloyantz, A. (1986) *Molecules, Dynamics and Life*, John Wiley & Sons, New York.

Baars, B.J. (1988) *A Cognitive Theory of Consciousness*, Cambridge University Press, Cambridge.

Bisiach, E. (1988) The (haunted) brain and consciousness, in A. Marcel & E. Bisiach (eds), *Consciousness in Contemporary Science*, Oxford University Press, Oxford, pp. 101-120.

Cairns-Smith, A.G. (1982) *Genetic Takeover and the Mineral Origin of Life*, Cambridge University Press, Cambridge.

Caplan, S.R. and Essig, A. (1983) *Bioenergetics and Linear Nonequilibrium Thermodynamics*, Harvard University Press, Cambridge, Massachusetts.

Chalmers, D.J. (1995) Facing up to the problem of consciousness, *Journal of Consciousness Studies* **2**, 200-219.

Chalmers, D.J. (1996*) The Conscious Mind: In Search of a Fundamental Theory*, Oxford University Press, New York - Oxford.

Crick, F.H.C. (1994) *The Astonishing Hypothesis - The Scientific Search for the Soul*, Simon and Schuster, London.

Damasio, A.R. (1989) The brain binds entities end events by multiregional activation from convergence zones, *Neural Computation* **1**, 123-132.

Dawkins, R. (1976) *The Selfish Gene*, Oxford University Press, Oxford; New edition (1989) Oxford University Press, Oxford.

Dennett, D.C. (1991) *Consciousness Explained*, Little-Brown & Co.; 2nd edition (1993), Penguin Books, Harmondsworth.

Dennett, D.C. (1995) *Darwin's Dangerous Idea*, Simon and Schuster, New York.

Dennett, D.C. (1996) Facing backwards on the problem of consciosness, *Journal of Consciousness Studies* **3**, 4-6.

Driesch, H. (1891) Entwicklungsmechanische Studien. 1. Der Werth der beiden ersten Furchungszellen in der Echinodermenentwicklung Experimentelle Erzeugung von Theil- und Doppelbildungen, *Zeitschrift fuer Zoology* **53**, 160-178.

Driesch, H. (1908) *Science and Philosophy of the Organism*, Adam & Charles Black, London.

Driesch, H. (1915) *Vitalism: Its History and System*, Russian Edition, A.G.Gurwitsch (ed) the authorized translation from German into Russian by A.G. Gurwitsch (In Russion).

Edelman, G. (1989) *The Remembered Present: A Biological Theory of Consciousness*, Basic Books, New York.

Eigen, M. (1992) *Steps Towards Life - A Perspective on Evolution*, Oxford University Press, Oxford.

Eigen, M. (1993) The origin of genetic information: viruses as models, *Gene* **135**, 37-47.

Elitsur, A.C. (1994) Let there be life: Thermodynamic reflections on biogenesis and evolution, *Journal of Theoretical Biology* **164**, 429-459.

Gurwitsch, A.G. (1930) *Die histologischen Grundlagen der Biologie*, G. Fischer Verlag, Jena.

Gurwitsch, A.G. (1937) Degradational radiation from central nerval system, *Arkhiv Biologicheskikh Nauk (Archives of Biological Sciences)* **45,** 53-57 (In Russian).

Gurwitsch, A.G. (1944) *The Theory of the Biological Field.* "Sovetskaya Nauka" Publishing House, Moscow (In Russian).

Gurwitsch, A.G. and Gurwitsch, L.D. (1948) *Introduction into the doctrine of mitogenetic radiation*, USSR Academy of Medical Sciences Publishing House, Moscow (In Russian).

Gurwitsch, A.G. (1954) *Principles of Analytical Biology and the Theory of Cellular Fields*, Manuscript, Moscow (In Russian).

Gurwitsch, A.G. and Gurwitsch, L.D. (1959) *Die mitogenetische Strahlung*, G. Fischer Verlag, Jena.

Gurwitsch, A.G. (1991) *Principles of Analytical Biology and of the Theory of Cellular Fields.* Nauka Publishers, Moscow (In Russian).

Haken, N. (1977) *Synergetics: An Introduction*, Springer, Heidelberg.

Hameroff, S. (1994) Quantum coherence in microtubules: A neural basis for an emergent consciousness? *Journal of Consciousness Studies* **1**, 91-118.

Hardcastle, V.G. (1994) Psychology's binding problem and possible neurobiological solutions, *Journal of Consciousness Studies* **1**, 66-90.

Hardcastle, V.G. (1996) The why of consciousness: a non-issue for materialists, *Journal of Consciousness Studies* **3**, 7-13.

Horgan, T. (1982) Supervenience and microphysics, *Pacific Philosophical Quaterly* **63**, 29-43.

Hubel, D.H. and Livingstone, M.S. (1987) Segregation of form, color, and stereopsis in primate area 18, *The Journal of Neuroscience* **7**, 3378-3415.

Hume, D. (1735/1958) *A Treatise on Human Nature*, Oxford University Press, Oxford - New York.

Hume, D. (1777/1975) *Enquiries Concerning Human Understanding and Concerning the Principles of Morals*, Oxford University Press, Oxford - New York.

Jackendoff, R. (1987) *Consciousness and the Computational Mind.* MIT Press, Cambridge Massacusetts.

Kauffman, S. (1996) *At Home in the Universe: the Search for Laws of Self-Organization and Complexity*, Viking, New York.

Kim, J. (1984) Concepts of supervenience, *Philosophy and Phenomenological Research* **65**, 153-176.

Kim, J. (1987) 'Strong' and 'global' supervenience revisited, *Philosophy and Phenomenological Research* **68**, 315-326.

Lem, S. (1976) *Solaris*, Progress, Moscow (In Russian).

Levine, J. (1983) Materialism and qualia: the explanatory gap, *Pacific Philosophical Quaterly* **64**, 354-361.

Li, K. H. (1992a) Coherence in physics and biology, in F.-A. Popp, K. H. Li & Q. Gu (eds), *Recent Advances in Biophoton Research and its Applications*, World Scientific, Singapore - New Jersey - London - Hong Kong, pp. 113-156.

Li, K. H. (1992b) Coherent radiation from DNA molecules, in F.-A. Popp, K. H. Li & Q. Gu (eds), *Recent Advances in Biophoton Research and its Applications*, World Scientific, Singapore - New Jersey - London - Hong Kong, pp. 157-196.

Li, K. H. (1995) Coherence - a bridge between micro- and macro-systems, in L.V. Beloussov & F.-A. Popp (eds), *Biophotonics: Non-Equilibrium and Coherent Systems in Biology, Biophysics and Biotechnology*, Bioinform Services Co., Moscow, pp. 99-114.

Li, K. H. (1996a) Coherence in physics and biology, in C. Zhang, F.-A. Popp & M. Bischof, (eds), *Current Development of Biophysics*, Hangzhou University Press, Hangzhou, pp. 41-91.

Li, K. H. (1996b) Coherence - a bridge between micro- and macro-systems, in C. Zhang, F.-A. Popp & M. Bischof (eds), *Current Development of Biophysics*, Hangzhou University Press, Hangzhou, pp. 287-305..

Li, K. H. and Popp, F.-A. (1992) Coherence and some quantum paradoxes, in F.-A. Popp, K. H. Li & Q. Gu (eds), *Recent Advances in Biophoton Research and its Applications*, World Scientific, Singapore - New Jersey - London - Hong Kong, pp. 421-438.

Li, K. H. and Popp, F.-A. (1996) Coherence, De Broglie paradox and EPR paradox, in C. Zhang, F.-A. Popp & M. Bischof (eds), *Current Development of Biophysics*, Hangzhou University Press, Hangzhou, pp. 287-305.

Lifson, S. (1987) Chemical selection, diversity, teleonomy and the second law of thermodynamics: Reflections of Eigen's theory of self-organization of matter, *Biophysical Chemistry* **26**, 303-311.

Lipkind, M. (1987a) A.G. Gurwitschs Theorie vom biologischen Feld. I. Major prerequisites, origin, logical structure and conceptual development, *Fusion* (German edition) **8**, 30-49 (In German).

Lipkind, M. (1987b) A.G. Gurwitschs Theorie vom biologischen Feld. II. Analysis of different phenomena in living systems by means of the theory of the vectorial biological field, *Fusion* (German edition) **8**, 53-65 (In German).

Lipkind, M. (1992) Can the vitalistic entelechia principle be a working instrument? (The theory of the biological field of Alexander G. Gurwitsch). in F.-A. Popp, K. H. Li & Q. Gu (eds), *Recent Advances in Biophoton Research and its Applications,* World Scientific, Singapore - New Jersey - London - Hong Kong, pp. 469-494.

Lipkind, M. and Beloussov, L. (1995) Alexander Gurwitsch: personality and scientific heritage. in L.V. Beloussov & F.-A. Popp (eds), *Biophotonics: Non-Equilibrium and Coherent Systems in Biology, Biophysics and Biotechnology*, Bioinform Services Co, Moscow, pp. 15-32.

Livingstone, M.S. and Hubel, D.H. (1987) Psychophysical evidence for separate channels for the perception of form, color, movement, and depth, *The Journal of Neuroscience* 7, 3416-3468.

Loeb, J. (1906) *The Dynamics of Living Matter*, Columbia University Press, New York.

Loeb, J. (1912) *The Mechanistic Conception of Life*, University Chicago Press, Chicago.

Lotka, A. (1925) *Elements of Physical Biology*, Williams and Wilkins, Baltimore; Republished in 1956 as *Elements of Mathematical Biology*, Dover, New York.

Monod, J. (1971) *Chance and Necessity: An Essay on the Natural Philosophy of Modern Biology*, Knopf, New York.

Penrose, R. (1994) *Shadows of the Mind*, Oxford University Press, Oxford.

Popp, F.-A. (1989) Coherent photon storage of biological systems, in F.-A. Popp et al. (eds), *Electromagnetic Bio-Information*, Urban & Schwarzenberg , Muenchen - Vienna - Baltimore, pp. 144-167.

Popp, F.-A. (1992a) Some essential questions of biophoton research and probable answers, in F.-A. Popp, K. H. Li & Q. Gu (eds) *Recent Advances in Biophoton Research and its Applications*, World Scientific, Singapore - New Jersey - London - Hong Kong, pp. 1-46.

Popp, F.-A. (1992b) Some remarks on biological consequences of a coherent biophoton field. in F.-A. Popp, K. H. Li & Q. Gu (eds), *Recent Advances in Biophoton Research and its Applications*, World Scientific, Singapore - New Jersey - London - Hong Kong, pp. 357-374.

Popp, F.-A. (1992c) Evolution as the expansion of coherent states. in F.-A. Popp, K. H. Li & Q. Gu (eds), *Recent Advances in Biophoton Research and its Application*, World Scientific, Singapore - New Jersey - London - Hong Kong, pp. 445-456.

Popp, F.-A. (1995) Modern physical aspects of mitogenetic radiation (biophotons), in L.V. Beloussov & F.-A. Popp (eds), *Biophotonics: Non-Equilibrium and Coherent Systems in Biology, Biophysics and Biotechnology*, Bioinform Services Co, Moscow, pp. 85-98.

Popp, F.-A. (1996a) Some remarks on biological consequences of a coherent biophoton field, in C. Zhang, F.-A. Popp & M. Bischof (eds), *Current Development of Biophysics*, Hangzhou University Press, Hangzhou, pp. 22-40.

Popp, F.-A. (1996b) Evolution as the expansion of coherent states, in C. Zhang, F.-A. Popp & M. Bischof (eds), *Current Development of Biophysics*, Hangzhou University Press, Hangzhou, pp. 252-264.

Popp, F.-A. and Li, K.H (1993) Hyperbolic relaxation as a sufficient condition of a fully coherent ergodic field, *International Journal of Theoretical Physics* **32**, 1573-1583.

Popp, F.-A., Chang, J.J., Gu, Q. and Ho, M-W. (1994) Nonsubstantial biocommunication in terms of Dicke's theory, in M.-W. Ho, F.-A. Popp & U. Warnke (Eds), *Bioelectrodynamics and Biocommunication*, World Scintific, Singapore - New Jersey - London - Hong Kong, pp. 293-318.

Prigogine, I. and Stengers, I. (1984) *Order out of Chaos*, Bantam, New York.

Ramachandran, V.S. (1990) Visual perception in people and machines, in A. Balke & T. Trosciankopp (eds). *AI and the Eye*, John Wiley and Sons Ltd, New York, pp. 21-77.

Ramachandran, V.S. and Anstis, S.M. (1986) The perception of apparent motion, *Scientific American* **254**, 101-109.

Rosenberg, G.H. (1996) Rethinking nature: hard problem within the hard problem, *Journal of Consciousness Studies* **3**, 76-88.

Treisman, A. (1986) Features and objects in visual processing, *Scientific American* **254**, 114-125.

Von der Malsburg, C. (1987) Synaptic plasticity as basis of brain organization, in J.-P. Changeux & M Konishi (eds), *The Neural and Molecular Bases of Learning*, John Wiley and Sons, New York, pp. 411-432.

Waechterhauser, G. (1988) Before enzymes and templates: theory of surface metabolism, *Microbiological Reviews* **52**, 452-484.

Zeki, S. (1992) The visual image in mind and brain, *Scientific American* **254**, 68-77.

Zhang, C.L. (1994) Log-normal distribution of physiological parameters and coherence of biological systems, in M.-W. Ho, F.-A. Popp & U. Warnke (Eds), *Bioelectrodynamics and Biocommunication*, World Scientific, Singapore - New Jersey - London - Hong Kong, pp. 355-364.

Zhang, C.L. and Popp, F.-A. (1996) Log-normal distribution of physiological parameters and coherence of biological systems, in C. Zhang, F.-A. Popp & M. Bischof (eds), *Current Development of Biophysics*, Hangzhou University Press, Hangzhou, pp. 22-40.

HOLISM AND FIELD THEORIES IN BIOLOGY

Non-molecular Approaches and their Relevance to Biophysics

M.BISCHOF

International Institute of Biophysics
Station Hombroich, Vockrather Straße, D-41472 Neuss, Germany

1. Introduction

Today's biology, dominated by the molecular approach developed since about 1940, is suffocated by an immense number of experimental data on molecular aspects of biological functions which present an extremely fragmented view of the living state. The very success of this approach is now becoming biology's greatest enemy; the low prestige awarded for such efforts has made unifying concepts a rarity and has not favored the emergence of any comprehensive picture of the living state. There is neither a conceptual framework to unify the findings into a theory of biology, nor even one to guide further research.

In this situation the necessity of balancing the molecular bias by non-molecular approaches is seen by an increasing number of authors (see, f.i., Albrecht-Buehler, 1990; Bischof, 1995; Laszlo, 1995, 1996; Popp et al., 1992; Primas, 1992, 1993; Welch and Smith, 1990; Welch, 1992; Wolkowski, 1985; Zhang et al., 1996). Thus, it may be useful to look at the history of such approaches, especially those developed earlier in this century. For those already working on the development of contemporary approaches to a holistic biophysics, such as biophoton research, a review of the historical antecedents may be of not less value.

2. The Development of Early 20th Century Holism

At the end of the last century, a counter-movement to the mechanist-reductionist program of the Berlin school of physically oriented physiologists that had dominated the second half of the 19th century (Bischof, 1996), began with the revival of vitalism by Hans Driesch, Henri Bergson, and others. This was followed in the first decades of the 20th century by a series of proposals of a non-vitalistic alternative to mechanistic biology under the name of holism or *organicism* (Haraway, 1976; Harrington, 1996). Both groups (in different ways) referred to the first antireductionist and teleological vision of organic wholeness in the early 19th century, by Goethe and other Romantic biologists such as Alexander von Humboldt, Carl Gustav Carus, Johann Friedrich Blumenbach, Karl Ernst von Baer and Johannes Mueller (Meyer-Abich, 1949),

J. J. Chang et al. (eds.), Biophotons, 375-394.

reactivating the old discussion on the laws of development of organic form and structure, the relationship between part and whole in the organism and the demarcation between life and inorganic matter.

The starting point of this development was the controversy between **Wilhelm Roux** and **Hans Driesch** about the interpretation of their embryological experiments around 1890 (Mocek, 1974; Allen, 1975). In his experiments with frog eggs in 1888 that mark the transition from descriptive to experimental morphology, Roux had killed one of the two first cleaving cells with a hot needle, but without separating it from the unharmed half. Roux interpreted the subsequent development of the surviving cell into a half frog as evidence for the local (internal) predetermination of the fate of the parts, and developed his mechanistic *„mosaic theory of development"* on this basis. He postulated that the early embryo is like a mosaic whose parts develop independently. Intent upon saving biology from its inferiority to physics and confident in the machine paradigm, Roux tried in his *„Entwicklungsmechanik"* (developmental mechanics) to find fully causal, physical and mathematical explanations for embryology.

Although Roux can be credited with the merit of having formulated the central questions which would lead to the breaking of a too rigid mechanistic thinking, the internal contradictions of his basic assumptions soon were revealed. In 1891, Hans Driesch repeated Roux's experiment with sea-urchin eggs, but separated the two first blastomeres from each other by shaking the eggs in sea-water (Driesch, 1891). The isolated cleavage cells first developed as in Roux' experiment till the formation of the half-blastula, but then, the half balls grew into whole ones and formed intact, if much smaller urchin larvae. Driesch noted that this faculty of regulation showed that - at least at this early stage of development - a whole could develop, in principle, from each part; the first cleavage cells contained many more potencies than they realized in the course of their actual development. He defined these potencies as the *„prospective potency"* of the cell, while his term of *„prospective significance"* denotes that which actually is developed. While in Roux' mosaic theory all parts of the organism developed from the beginning along their predetermined pathways, without influencing each other, in Driesch's holistic view *„the whole is more than the sum of its parts"* (Aristotle). The state of each part of the organized whole of the organism was dynamically co-determined by the neighbouring parts, such that there is a relational structure between the parts that is specific for the organism. Function was dependent on the position within the whole, not on the mechanical preformation of parts. Disturbances of the organized whole, such as the removal, adddition or shifting of parts, resulted in a restitution of the whole. As the agency guiding this and all other processes of development and regulation according to a plan, Driesch postulated the Aristotelian *„entelechy"* which he defined as an entity beyond space and time and thus not accessible to scientific investigation. The entelechy

> *„is affected by and acts upon spatial causality as if it came out of an ultra-spatial dimension; it does not act* in *space, it acts* into *space; it is not in space, it only has points of manifestation in space"* (Driesch, 1908).

Entelechy (Greek for „that which contains its own goal") also accounted for the teleological characteristic of *„equifinality"* - i.e., in organismic processes the same final state may be reached from different initial conditions and by different pathways.

The debate between Roux and Driesch was a continuation of the old controversy between preformationists and epigeneticists. Roux represented a contemporary version of the preformationist doctrine that the source of an organism's form must already be present within the fertilized egg in material form. Weisman had resurrected it in the form of his doctrine of the germ plasm. Driesch - and many holists with him - adhered to the epigeneticist view that development involved the emergence of material structures that were not there before - a fact already observed by earlier biologists such as C.F.Wolff and K.E. von Baer. A related dichotomy was that of dependent versus independent differentiation - was differentiation caused by factors internal, or external to the organism ?

Driesch's challenge caused a fundamental crisis in embryology which led to a reformulation of basic concepts in experimental embryology and cell biology. At first, a number of outstanding biologists followed Driesch by assuming one of several kinds of neo-vitalistic stances, operating with some more or less vague, untestable metaphor such as *„nostalgia"* (J.S.Haldane), *„impulses"* (J. von Uexküll), *„hormé"* (C. von Monakow) or *„mneme"* (R. Semon). But by 1930, a number of approaches had emerged that formed a clear alternative to either vitalism or mechanism, by formulating a definitive program of experiment and conceptual framework for organismic, or holistic, biology. This group of approaches, far from being marginal, played a considerable role in European and American biology from the 1920's to the 1950's.

3. The early evolution of holism

The stage for the scientific study of biological wholeness was set by Roux and Driesch, and their school of developmental mechanics. In the early 1900's, **Hans Spemann**, starting as a strong proponent of Roux' approach, tried to find answers by investigating later stages of development of the embryo, while Roux and Driesch had studied the initial stages (Allen, 1975). He showed on amphibia how in the formation of organs in a series of stepwise *„inductions"* certain regions in neighbouring tissue layers caused the orderly sequence of events in differentiation. In each sequence the induced tissue from one step became the inducer for the next one.

In the early 1920's he performed a series of experiments to test the hypothesis that there must be a *primary inducer* that initiates all subsequent inductions. He found that the transplantation of tissue from the animal half of an early gastrula stage embryo to a different region of another young blastula embryo leads to a differentiation according to the region of implantation, while transplantation from a later gastrula stage leads to differentiation according to provenience. For the gastrula stage therefore Driesch's *prospective potency* is greater than the *prospective significance* of the germinal regions.

However, when Spemann transplanted tissue from the dorsal lip of an early gastrula to the ventral region (or any other region) of another young embryo, it would not only not change its fate and develop according to its new environment, but instead modify the fates of the surrounding parts, inducing them to form a second axial system leading to the formation of a second nervous system. Spemann called such a transplant from the dorsal lip „*organizer*" and the area from which it was taken, *organization center*.

Spemann's work gradually led him to a more holistic view of development. The following quote from his „*Experimentelle Beiträge...*" (1936) shows him groping for a holistic description.

> „*It is mainly the unity of the process which is so striking and amazing. One has the impression as if the movements of the single parts composing the overall movement were joined in an inextricable connection, as in a current which is generated in a point within a drop of water*".

In the late 1920's and early 1930's, the work of Spemann and his school inspired two branches of inquiry (Allen, 1975). On the one hand, the subsequent search to identify and isolate an „*organizer substance*" led to the establishment of the new field of biochemical embryology. Important work in this field was done by Needham, Child, Huxley and De Beer. However, at this time all attempts to identify any specific organizer substance failed.

A second direction of investigation, spawned by the more holistic side of Spemann's work, was the attempt, by men such as Harrison, Weiss, Needham, and Waddington, of identifying the organizing principles in developing systems by more carefully analyzing them on the level of tissues. This more holistic side of Spemann's work inspired by Spemann ultimately led to the foundation of embryological field theory.

4. The High Time of Holism and the Role of Field Theories

Beginning in the 1920's, the new paradigm of organismic biology crystallized out in the work of a number of mainly British and American workers inspired by Spemann's work, such as Alexander Gurwitsch, Ross Harrison, Paul Weiss, Joseph Needham, and Conrad Waddington. In the work of these scientists, the concept of a *morphogenetic field* emerged as a central metaphor for uniting both the organizer and the organized tissues and for understanding the integrating and organizing principles in organisms. Implicitly, the field properties of organisms were already contained in Driesch's 1891 suggestion that the whole embryo is a „*harmonious equipotential system*".

These workers were intent on the concretization of the organismic paradigm by experimental work on the level of tissues and metabolism in the tradition of developmental mechanics. In order to understand the concrete content of a holistic model, the complex and dynamic web of development had to be resolved into component processes; the use of words such as *organization*, or *field*, should not replace analysis, but rather provide guiding metaphors for analytic work within the holistic framework (Weiss and Moscona, 1958). Process should be revealed through the clear perception of structure. In contrast to vitalists such as Driesch, the organicists did not devalue or ignore the material substrate of the organism and its biochemistry, but rather tried to „*bridge the ancient gulf between physico-chemical processes and their morphological manifestations*" (Needham, 1931) by the use of various concepts of supramolecular ordering. By doing this, they paved the way for molecular biology, although the holistic premises from which they operated would be largely lost on later generations of molecular biologists. Thus, Allen's two branches of Spemann's influence are that intimately intertwined in the work of the same people that we cannot describe them speparately.

Organicists such as Child and Needham made important contributions to the new field of biochemical embryology. One of the most influential concepts in this field has been that of „*gradients*" (Haraway, 1976). The notion of physiological gradients and polarities was introduced in 1901 by Boveri (Boveri, 1901). Many animals, at least in the embryonic stage, possess one or several axes along which there exists a gradient of some sort. The course of development depends on these gradients to a high degree.

Gradient theory came into full prominence with the work of **Charles Mannings Child** (1929) based on experiments with hydroids and planaria, and was generalized and popularized into „*gradient fields*" by Huxley and DeBeer (1934). Child's concept of gradients was restricted to gradients of metabolic intensity. In Child's view, the organizer region was the high end of a metabolic gradient; morphological structures arose as the result of physiological processes (Child, 1924). The gradient pattern and the relations of dominance and subordination between different regions of the organism - a concept central to Child's thinking - arising from it constituted the wholeness of the organism. However, in his attempt to bridge morphology and biochemistry, Child neglected to appreciate the intermediate level of tissue and cell structures. But still, his concept came to be an important element in the development of organicism, even if in considerably modified form.

The biochemist **Joseph Needham**, who in 1932-33 worked in the German lab of the Spemann group with Waddington, wanted to build a bridge between biochemistry and morphology (Haraway, 1976). Needham began studying the biochemistry of development in 1923. In the early 1930's he focused on the chemical nature of the amphibian organizer, at the same time when he became one of the founders of the Oxford „Theoretical Biology Club", a group who was central in the formation of the organismic, or holistic, paradigm, and also included C.H.Waddington, J.H.Woodger, J.D.Bernal, and L.L.Whyte. Like other organicists, he tried to define an intermediate level of organization in which structure and function came together; some kind of supramolecular ordering was involved. Like Harrison, he started from a mechanistic perspective, but moved on to an organismic view.

With the concept that the „*evocator*" - the hypothetical organizer substance that he and many others searched at the time - and the whole *hierarchy* of inductors would stimulate irreversible distortions and re-orientations of molecular systems, Needham approached a field-like solution to developmental organization. He was convinced that field thinking was a „*powerful aid to the codification of 'Gestaltungsgesetze', the rules of morphogenetic order*" (Needham, 1936). The dynamic description of the establishment of order and form during development in terms of field laws was a definite advance over the static anatomical description in embryology. Fields provided a rich source of images for the experimental task of concretization. He hoped that his work on the evocator, together with an advanced understanding of proteins and paracrystalline structures, would give a solid biochemical content to morphogenetic fields. He saw the work of D'Arcy Thompson (1917) as an important, but chemically unspecified root of legitimate field thinking. He rejected the geometrical fields of Gurwitsch and relied on the theoretical work of Weiss and Waddington.

Together with Spemann and his school, Yale embryologist **Ross Harrison** belonged to the first generation laying the experimental groundwork for organicism and for the field concept, with his work on the emergence of order and coordination in developmental processes (Haraway, 1976). Rooted in Roux' developmental mechanics, he rejected the vitalistic resolutions of the problem of biological hierarchy; Lloyd Morgan's theory of *emergence* (1928) as too vague and general, and the holism of General Smuts (1926) as obscuring the issue of the whole instead of making unity a subject of experimental investigation. He maintained it was impossible to develop science wholly from the top down or from the bottom up. The investigator entered where he could gain a foorthold by whatever means were available. In his work, among other things the analysis of the establishment of the axes of symmetry in the developing limb bud and inner ear of the newt and the study of the regional differentiation of the egg, he started from the mechanistic crystal analogy, but gradually transformed it into a concept of hierarchies of dynamical molecular structures, and finally fields. Among organicists, Ross Harrison was one of the first to explicitly introduce the field concept into embryology (Haraway, 1976). His study of the newt limb (1910-1930) is a classic instance of analysis of the structure of a field.

> *„The egg and the early embryo consist of fields - gradients or differentiation centers in which the specific properties drop off in intensity as the distance from the field center increases, but in which any part within limits may represent any other"* (Harrison, 1969).

Paul Weiss, trained as an engineer and well grounded in the physical sciences, was the first to introduce the concept of *„system"* into biology in 1924, and originated, together with Gurwitsch (1922), the concept of the biological field (Haraway, 1976). Inspired by Koehler's *Gestalt* approach, he first used the field concept in his *„Morphodynamik"* (1926) to explain the results of his and other's work on limb regeneration in amphibians, but later generalized it to ontogeny as a whole. In 1925-1930 he investigated bone regeneration in amputated urodele limbs and demonstrated that the regeneration blastema (germinal tissue) of the limb stump had equipotential character, in the same way as the embryonic organ rudiments in normal development. They are an organized assembly of undifferentiated equipotential cells with the potential to differentiate into different directions. This differentiation, Weiss came to think, is directed by a *„limb field"* in the stump. As in ontogeny, the limb field was a property of the field district as a whole, and not of any particular discrete group of elements. In the 1940's, Weiss concluded from studies of nerve orientation in repair and growth processes that tissue behaves as a coherent unit (Haraway, 1976). He found a selective contact guidance of nerve fiber tips by the matrix of surrounding structures, according to the type of nerve, and a specific affinity of peripheral nerve fibers to the corresponding type of end organ, which could not be explained as crude mechanical phenomena. In studies of chondrogenesis (cartilage formation) in the 1950's, he specified the role of the tissue environment in general, and of mechanical stresses in particular, in the differentiation of mesenchyme cells into cartilage. Isolated pre-cartilaginous blastema cells of different types developed in cell culture into the respective specific varieties of cartilaginous structure according to the *in vivo* pattern. Therefore they possessed highly specific fields with

> „*distinctive morphogenetic properties determining the particular patterns of cell grouping, proliferation, and deposition of ground substance which, in due course, lead to the development of a cartilage of a distinctive and typical shape*" (Weiss and Moscona, 1958).

Weiss concluded from his experimental work that the emergence of organ and tissue structuring during development implied that the component unities assumed patterned space relations which show themselves in geometric features of position, proportion, orientation and the like. The ordering processes responsible for this were best referred to as *field actions*. However, this should not be taken to be an explanation. Weiss defined the field as

> „*a condition to which a living system owes its typical organization and its specific activities. These activities are specific in that they determine the character of the formations to which they give rise. (...) Inasmuch as the action of fields does produce spatial order, it becomes a postulate that the field factors themselves possess definite order. The three-dimensional heterogeneity of developing systems, that is, the fact that these systems have different properties in the three dimensions of space, must be referred to a three-dimensional organization and heteropolarity of the originating fields*" (Weiss, 1939).

The fields were specific, i.e., each species of organism had its own morphogenetic field. Within the organism there were subsidiary fields within the overall field of the organism, thus producing a nested hierarchy of fields within fields.

Although the first to introduce the field concept into biology (Gurwitsch, 1922), **Alexander Gurwitsch** is not usually counted among organicists. Like his older friends Roux and Driesch, the Russian histologist and embryologist was not satisfied with the merely descriptive embryology of his time, but did also not like Roux' approach of developmental mechanics (Lipkind and Beloussov, 1995; Bischof, 1995). He was more inspired by Driesch's holistic ideas, but - and here at least he was in line with the organicists - could not accept Driesch's 'unconstructive' postulate that the morphogenetic factor was not experimentally verifiable. He was determined to put his hypothesis of a morphogenetic factor to the experimental test. Considering himself a „*practical vitalist*" , he defended the right of the biologist to introduce what he called „*the chief postulate of my own 'vitalism' - spatial but immaterial factors of morphogenesis*" (Gurwitsch, 1915). He first called this hypothetic morphogenetic factor the *„dynamically preformed morphe"*, but in 1922 replaced it by *„embryonal, or morphogenetic field*" (Gurwitsch, 1922), a concept that he constantly redefined and modified throughout his working life, the last time in a posthumously published work (Gurwitsch, 1991). In 1922 Gurwitsch wrote

> *„the place of the embryonal formative process is a field (in the usage of the physicists) the boundaries of which, in general, do not coincide with those of the embryo but surpass them. Embryogenesis, in other words, comes to pass inside the fields (...) Thus what is given to us as a living system would consist of the visible embryo (or egg, respectively) and a field"*. (Gurwitsch, 1922).

and later he called his field *„an overall state of tissues which not necessarily need to be a function of cellular activities"*, and *„regions of space in which there exist certain anisotropic states that are essentially factors of the coordinates of the field"*, and again *„states which force the particles contained within them into certain trajectories"* (Gurwitsch, 1923). While Gurwitsch originally stressed the importance of the overall field of the whole organism, he later also investigated local fields and tried to find an intermediate level bridging morphogenesis and biochemistry. He suggested the field produces non-equilibricity in the molecular ensembles of cells and tissues, and, on the other hand, the potential energy released by the decomposition of such excited non-equilibrium molecular complexes can be transformed into kinetic energy which leads to directed movement of substance. With this hypothesis, Gurwitsch became one of the early proponents, together with his colleague Ervin Bauer, of what is today known as dissipative structures.

5. Theoretical Aspects of Holism

Apart from these protagonists of the holistic approach, in the first decades of the century many others from diverse fields of inquiry such as neurology and neuropsychiatry, behavioural zoology, and physiology contributed experimental evidence and conceptual elements to a holistic theory of the organism.

Of seminal influence was the experimental and theoretical work in Gestalt psychology (Ash, 1995) by such men as Christian von Ehrenfels (Weinhandl, 1978), Wolfgang Köhler (1920, 1922), Max Wertheimer, and Kurt Goldstein (1939). Their experiments demonstrated that the organism did not just react to single stimuli, but rather to relationships among stimuli, to overall patterns and to stimuli perceived as part of a given context. They postulated the organism initially perceives overall patterns, and only subsequently becomes sensitive to their component parts, and assumed that the nervous system is organized in terms of neural fields, operating across wide regions of the cortex. Similarly, holistic features of brain functioning, postural reflex patterns, sensory-motor integration and behaviour, and the self-regulating capacities of the organism in the integrated actions of the circulatory, hormonal and nervous systems were demonstrated.

Among the many different approaches in holistic thinking we can distinguish between holism in the narrower sense of the holistic (or organismic) biology with which we are dealing here, and the wider movement of holistic thinking in all sciences. In the same period of the first decades of this century when organicism developed, holistic thinkers were as well at work in other fields such as philosophy, psychology (Gestalt psychology has been mentioned), sociology, law, economics, the history of art, medicine, and physics.

In philosophy, important contributions came from Smuts, Ehrenfried, Alfred North Whitehead, Othmar Spann, Emile Boutroux, Nicolai Hartmann, Hermann Friedmann, Adolf Meyer-Abich, and Jean Gebser. In psychology, besides Ehrenfried, Koehler, Wertheimer, Krueger and Schilder, we must mention Kurt Lewin with his field theory for social science (1951). In sociology, besides Lewin, Spann, who also contributed to economic holism. In physics, holistic approaches originate from quantum theory with Planck, Bohr, David Bohm, and the more recent work on non-local EPR correlations.

Austrian economist and philosopher, **Othmar Spann**, in his *„Kategorienlehre"* (1939) gave a full analysis of wholeness in the Aristotelian tradition. He analysed the logical relation of the whole and the part based on the category of rank, which he defined as the degree of essentiality, or nearness to the whole, that characterizes the parts.

Historian of science and natural philosopher, **Adolf Meyer-Abich**, the main proponent of biological holism in Germany, maintained that *„All the sciences of the future will be concerned with wholes and therefore be holistic, or they will not be at all"* (Meyer-Abich, 1949). He postulated „holism" as an

> *„active principle which preserves and supports the wholeness of the system and its normal functioning, reproduces it ontogenetically from its hereditary factors, causes phylogenetical development of the organism and causes transformation of existent and formation of not yet existing hereditary factors"* (Meyer-Abich, 1964-65).

By stating that in the process of evolution, complex holisms issue from more simple holisms by *„holobiosis"* which was the most important causal holism of phylogenetic development, Meyer-Abich (1948, 1964-65) became one of the early pioneers of what today is known as „symbiotic evolution".

Whitehead's philosophy of *„organic mechanism"* was of paramount influence upon organismic biologists. In *„Science and the Modern World"* (1925) he postulated that the foundation for the unity of science must be sought elsewhere than in reductionism, because chemistry and physics themselves were outgrowing mechanistic models. The concept of organism was fundamental for all nature, because all nature, including atoms etc., was alive, and had its organizing principle within itself. The basic unit of nature therefore was the organism, to whom he gave primacy over the atom. The unity of science was based on organic *„events"* (processes) rather than simple atoms. Whitehead subscribed to a field theory of reality; he considered the world as a continous process. All things were modifications of a field-like space-time, with a focal region, but extending non-locally. Their influence diminished gradually through successive *„envelopes"* in every direction. Organisms were *„structures of activity"* of various complexity, *„enduring patterns"* which were evolving. Like all events, organisms and their parts were at the same time „related as a whole to parts and as joint parts within some whole", i.e. at each level of organization, there were organized wholes, made up of parts which were themselves organized wholes. And at each level, the whole was more than its parts; it had its irreproducible integrity.

It is especially significant that holistic thinking was intrinsic to quantum theory from the beginning. **Max Planck** (1929) wrote, for instance, that

> *„according to this principle [the quantum principle] one can only arrive at an understanding of physical laws if one does not dissect the system under consideration into its individual spatial parts, but observes it as a whole.*

More recently, **quantum theoretical holism** has experienced a strong impetus from the discussion and experimental tests of *quantum nonlocality* as developed in the wake of the *Einstein-Podolsky-Rosen (EPR) paradox* (Kafatos and Nadeau, 1990; Primas, 1992, 1993; Atmanspacher, 1996). Einstein, with Podolsky and Rosen (Einstein et al., 1935), had postulated the existence of holistic correlations in quantum mechanics between systems (f.i., particles) that do not interact physically.

David Bohm's (1980) concept of the *„implicate order"*, clearly inspired by Whitehead and developed from his analysis of the EPR problem, is probably one of the most influential formulations of the holistic world picture arising from the philosophical implications of quantum and relativity theories. The latter necessitates, according to Bohm, the *„giving up of the notion that the world is constituted of basic objects or 'building blocks'; rather, one has to view the world in terms of universal flux of events and processes"*. Bohm proposed that the objective world of everyday perception and scientific measurement, the „explicate order", *„in which things are unfolded in the sense that each thing lies only in its own particular region of space (and time) and outside the regions belonging to other things"*, must be seen as arising from the universal background of a multidimensional, holographic, non-local „implicate order" of quantum potentials, or *„holomovement"*, in which *„everything is enfolded into everything else"*. The explicate order is a *„relatively independent, recurrent, stable, sub-totality"* of the implicate order, which Bohm regards as the primary, fundamental level of reality from which as well matter as also consciousness are arising as secondary manifestations.

The existence of EPR correlations was confirmed in experiments by Alain Aspect and others in the last few decades, on the basis of further concretization of the question by John Bell (Bell, 1987). According to Primas (1992), quantum theory is the first logically consistent and fully formalized holistic theory. In the quantum theoretical picture, the world is, according to Primas, an undivided whole which has no parts. Parts, i.e., classical systems, are arising as a consequence of the loss of the EPR correlations by objectivation through the *'Heisenberg cut'* which divides the system into an observer system and an object system (Atmanspacher, 1996) - a process similar to the emergence of *Gestalten* in perception (Ammann, 1996). To describe the original unbroken world, many different complementary ways dissecting reality are possible and ontologically equivalent, but exclude each other epistemologically (Primas, 1992). Accordingly, Primas defines the whole as that which can only be described by a manifold of complementary descriptions. Primas predicts that quantum holism will play a key role in a future theory of nature; Atmanspacher (1996) writes that holism, in order to be complete, must also include consciousness. Attempts to develop such an extension of quantum theoretical holism have been made by Pauli (Atmanspacher et al., 1995), Primas (1992, 1993) and Atmanspacher (1995) on the bases of the Pauli-Jung collaboration and of Jean Gebser's work (Gebser, 1986). Primas concludes that also the overestimation of rationality will have to be revised, and holism must include „the non-rational as an equal to the rational", as well as a reconsideration of the elimination of finality and its replacement by chance.

Theoretical concepts in biological holism of the period were characterized by the common intention of substituting metaphysics by some *'metabiology'* which was supposed to remedy the biases of both vitalism and mechanism, by establishing a comprehensive theory of life (Gloy, 1996). It was agreed that the vitalism-mechanism controversy must be overcome, and believed that holistic approach will bring a solution. Biology must cease to be a subdivision of physics, and become again an autonomous science based on specific biological principles, with its own epistemology (Meyer-Abich, 1934).

Quite a few holistic thinkers (among them J.S.Haldane, A.Meyer-Abich, H.Friedmann) even postulated that, all reality being of organismic nature (Whitehead), biological laws must be considered more fundamental than physicochemical ones, and therefore biology should even be considered supreme to physics whose laws rightly must be considered special, restricted cases of the biological ones. According to the *Gestalt* philosopher and mathematician, Hermann Friedmann (1949), the reason for the supremacy of biology was the fact that measurable and countable quantities must be considered as limiting cases of form or *Gestalt*, the basic concepts of biology. Science up to now had been active in the field of *'haptics'*, investigating sensory data gained by the tactile senses; the new era of science would be based on *'optics'*, operating with the holistic faculty of *„perceiving* Gestalten *as higher forms of reality"* (Friedmann, 1930).

A further central element of holistic theory concerns the irreducibility of higher to lower levels. The whole is not only indivisible and has no parts, but only *'members'* (Spann, Meyer-Abich) - it also necessarily must be self-differentiated and have an internal structure. Nature is stratified into multiple levels which are categorically heterogenous (Hartmann, 1950); each level possesses its own specific laws which cannot be reduced to or deducted from those of levels below or above it (Boutroux, 1874; Hartmann, 1950). According to Boutroux (1874) there are logical breaks, or *contingencies*, between the natural laws pertaining to the different domains or levels of reality, but also between the various domains of biological study, where the fundamental phenomena of one domain (the *'modals'* of Helmholtz) cannot continually be extended into the next domain by the simple addition of new characteristics. Therefore, the various internally consistent domains cannot be subordinated into a comprehensive overall domain with a unitary law, but appear to be ordered quantum-like. Each attempt to master a new domain with the means of the preceding one will thus lead to Heisenberg-like uncertainties (Meyer-Abich, 1950). Each domain has to be investigated on its own level. Vitalism and mechanism are guilty of *'categorial transgression'* (Hartmann, 1950) by either applying the physicochemical principles determining anorganic existence to organic existence (the bottom-up understanding of mechanism), or by using categories of the psychical and spiritual realms determining psychological life for the explanation of organic existence (top-down understanding of vitalism).

6. Contemporary Developments in Biological Field Theory

In the period of 1920 to 1950, biological field theories, together with the diverse organicist and holistic approaches, were much more commonly accepted among biologists as working models than they are today. As an undercurrent of the dominant trend in the life sciences since the 1950's which did not favour such holistic approaches, they have survived up to the present.

Conrad Waddington attempted to clarify the field concept with the idea of *"individuation fields"* active in the formation of organs (Waddington, 1934). In the 1950s he extended the idea in his concept of the *"chreode"*, or developmental pathway, whose goal he described as an attractor (Waddington, 1957). **René Thom** later developed Waddington's idea in mathematical models in which the end points towards which systems develop are represented by attractors, or basins of attraction within morphogenetic fields (Thom, 1975, 1983).

More recently, morphogenetic fields, although conceptualized in a way much in contrast with the organismic approach, have acquired considerable public attention in the form of **Rupert Sheldrake**'s *'morphic fields'* (Sheldrake, 1988). According to his *'hypothesis of formative causation'*, the nature of all things depends on fields, called morphic fields. Each kind of natural system has its own field. Sheldrake defines morphic fields as *'fields of information'*, *„non-material regions of influence extending in space and continuing in time. They are localized within and around the systems they organize"*. They disappear with the system they organize, but *„can appear again physically in other times and places, whenever and wherever conditions are appropriate. When they do so they contain within themselves a memory of their physical existences"* (Sheldrake, 1988). In contrast to Platonic forms which are transcendent blueprints, morphic fields are described, like Aristotelian entelechies, as immanent blueprints evolving within the realm of nature and influenced by the past of the system; there is a two way flow of influence from fields to organisms and from organisms to fields.

In contemporary biology, it is mainly *'structuralist biology'* (Webster and Goodwin, 1982) that has embraced the concept of the biological field. **Brian Goodwin** et al. (1987) claim that the evolutionary paradigm does not enable a satisfactory theoretical biology to be constructed. The Rational Morphologists with their field approach came, in their opinion, somewhat closer to presenting a satisfactory basis for a unified theoretical biology. Field properties of organisms underlie both reproduction and regeneration, which share the essential feature that from a part a whole is generated. According to Goodwin et al. (1987) reproduction is not to be understood in terms of Weismann's germ plasm or DNA, but rather as a process arising from field properties of the living state. Goodwin (1987a) postulates a feedback loop between morphogenetic fields and gene activity. The fields generate ordered spatial heterogeneities that can influence gene activities. Gene products, in turn, can influence the fields, destabilizing certain patterns and stabilizing others.

Welch and Smith (1990) and Welch (1992) propose a relativistic field concept for understanding the organization of cell metabolism based on their findings on organized multienzyme systems in the cell. They assume the field-like collective behaviour of enzymes and other particles in cell metabolism to be structured by a metabolic gauge-field which organizes the structure of biological space-time itself. According to Welch (1992), the living state may even be considered to be continous with, and a special domain within, the more comprehensive background field of Bohm's implicate order of nature itself. He suggests that

> *„matter is „biotic" (or „alive") when it exists in a self-propagating, internally-negentropic region of space-time structured (or „curved") by global (boundary) free-energy dissipation".*

7. The Nature of Biological Fields

As Sheldrake (1988) writes, there are three main perspectives on biological fields. In the *Platonic perspective*, biological fields are eternal, changeless transcendent forms or ideas of essentially mathematical, or geometrical, nature, but nevertheless constituting an objective, mathematical reality. Gurwitsch's fields, but also Thom's and Goodwin's concepts belong here; genetic theory as seen by Darwinist fundamentalism (Dawkins etc.) is of a similar nature. In the *Aristotelian view*, the entelechies, or whatever the fields are called, are organizing principles immanent to the organism that evolve with the organisms and play a causal role in organizing the material systems under their influence. Not only Driesch's entelechy falls into this class of field concepts, but also Sheldrake's morphogenetic field. There is a two-way flow of influence between these fields and the organism, which gives them a memory of past states of the organism. From the *nominalist, or empiricist, position* finally, the fields are just convenient ways of describing the phenomena of morphogenesis; they have no reality outside of our minds. Many organicists basically fall into this last category. Many, however, among them Weiss, Gurwitsch, and Waddington have been ambiguous or oscillating between two of those viewpoints.

Although the field concepts of organicists such as Harrison, Needham and Weiss were rooted in the field concept of physics, even those organicists tending to the Aristotelian position usually preferred not to go beyond the analogy and to leave open the exact nature of the biological fields. The notion of real electrical, electromagnetic or otherwise physical, fields of long-range force carried too clearly vitalist implications for them. On the other hand, to their taste, the ideal, strongly geometrical fields of Gurwitsch were too dematerialized and too far from biochemical reality (Haraway, 1976). For Weiss (1935), the field concept was also a means to remain open as to the nature of the organizing factor(s) as long as the tools for determining it were not adequate.

Although the concept of the biological field lends itself easily to to an electromagnetic interpreation to which all of its features fit excellently, in the early decades of the 20th century, when Gurwitsch and the organicists developed their field concepts, the knowledge on electromagnetic fields did not allow a consideration of *electromagnetic fields* as candidates for the biological field they postulated. Meanwhile the situation has changed. In recent decades, a number of scientific advances have made such an interpretation increasingly feasible.

(1) Electrodynamics and optics have considerably evolved; especially the new fields of nonlinear electrodynamics and optics (maser, raser and laser) are opening new vistas for the understanding of bioelectromagnetic phenomena. Dicke's (1954) alternative formulation of the laser principle shows the holistic behaviour of light-matter interaction in microscopic domains.

(2) it is now established that living beings are reacting sensitively to the impact of electromagnetic fields (Presman, 1970; Polk and Postow, 1995); the likely influences of various types of endogenous physical fields on cellular organisation and morphogenesis has been reviewed by O'Shea (1988).

(3) We know that several kinds of electromagnetic fields, including optical frequencies (biophotons), are emitted from living beings (Bigu, 1976; Popp et al., 1992).

(4) There is evidence that weak endogenous electrical currents in living tissue are involved in regeneration and growth of new tissue (Jaffe, 1982; Borgens et al, 1989); recently, the role of ionic currents (Nuccitelli, 1986) in morphogenesis has also been demonstrated.

(5) Communication by electromagnetic fields is established for fishes (Moller, 1976; Kramer, 1990), birds (Warnke, 1989), and insects (Warnke, 1989; Callahan, 1990), which suggests this may even be a more general phenomenon.

(6) the insight that all interactions relevant in biology, including the chemical ones, basically are electromagnetic ones.

The growing appreciation of the holistic implications of quantum theory, together with other influences, such as the increasing ecological awareness, and the growing knowledge and practice of methods from traditional Asiatic medicine, based on a kind of field-theoretic model of the organism, have also contributed to to a change of perspective in the life sciences and to the increased plausibility of field approaches.

In the course of time, a number of electromagnetic field theories for biology have been advanced. Early proposals were made by Keller (1918), Crile (1936), Lund (1947), (1947), and Lakhovsky (1951). The hypothesis of **Burr and Northrop** (Burr and Northrop, 1935), based on Burr's work on bioelectric potentials (Burr, 1973), is summarized in the following quote:

> *The pattern of organization of any biological system is established by a complex electro-dynamic field, which is in part determined by its atomic physico-chemical components and which in part determines the behaviour and orientation of those components. This field is electrical in the physical sense and by its properties it relates the entities of the biological system in a characteristic pattern and is itself in part a result of the existence of those entities. It determines and is determined by the components. More than establishing pattern, it must maintain pattern in*

> *the midst of a physico-chemical flux. Therefore, it must regulate and control living things, it must be the mechanism the outcome of whose activity is "wholeness", organization and continuity. The electro-dynamic field then is comparable to the entelechy of Driesch, the embryonic field of Spemann, the biological field of Weiss".* (Burr and Northrop, 1935).

The groundbreaking work of **Alexander Presman** (1970) marks the beginning of modern EM field theories for biology. Presman argued that environmental EM fields have played some, if not a central, role in the evolution of life and also are involved in the regulation of the vital activity of organisms. Living beings behave as specialized and highly sensitive antenna systems for diverse parameters of weak fields of the order of strength of the ambient natural ones. According to Presman, EM fields serve as mediators for the interconnection of the organism with the environment as well as between organisms, and EM fields produced by the organisms themselves play an important role in the coordination and communication of physiological systems within living organisms. In all these biological functions, in addition to energetic interactions, informational interactions play a significant (if not the main) role. These bioinformational capacities of electromagnetic fields occur, thus Presman's conclusion from experimental evidence, in their fullest state of development only in the organism as a whole, and are either not present, or not present in comparable form, at the molecular level. As these capacities seem to be a function of the complexity of organisms, it may be useless to attempt to investigate EM interactions with biological systems on the molecular level.

However, the concept of electromagnetic biofields only achieved maturity with the development of the concept of *coherence*, which may be called the biophysical base of modern holism. **Prigogine** (Nicolis and Prigogine, 1977) demonstrated it at the level of the coordinated collective behaviour of particles, while **Fröhlich** (1968, 1975) investigated the coherence of the electromagnetic fields coupled to these particles and connecting them. **Glauber** (1963) finally established the bases for our knowledge of quantum coherence. These achievements became important elements in the *'biophoton theory'* developed by Popp (Popp, 1979, 1992; Nagl and Popp, 1983; Popp and Nagel, 1983; Bischof, 1995) on the basis of measurements of ultraweak photon emission from organisms. Here living organisms are conceptualized as biological lasers of optically coupled emitters and absorbers operating at the laser threshold. The highly coherent, holographical biophoton field constituted in this way is hypothesized to be the basis of communication on all levels of organization; the components of the organism are seen to be connected by phase relations of the field in such a way that they are at all times instantly informed about each other. The biophoton field is also postulated to be the basis of memory and the regulation of biochemical and morphogenetic processes. By oscillating around the laser threshold, the field can take advantage of both the chaotic and the coherent regimes and thus establishes a feedback cycle of growth and differentiation. The process of Bose condensation which leads to an expansion of coherent states, is proposed to be the basis of the evolution of matter and consciousness (Popp, 1992a). Evolution is seen to be the result not of a struggle, but of a striving for improved cooperation and communication. Increasing complexity is generated by a feedback cycle between matter and radiation, whereby longer and longer wavelengths and more and more units are incorporated.

8. The Question of Informational Fields

Like Presman, Wolkowski (1995) points out that the biological effects of electromagnetic fields probably depend less on the quantity of energy than on the quality of information they transmit. Their role may be more that of a signal triggering a redistribution of the energy in the system. Likewise, in the biological field, even if we assume it is a real physical field like an electromagnetic one, the informational content is essential. A highly coherent field such as the biophoton field proposed by Popp (Popp et al., 1992) has a great capacity for information storage and transmission. If biological fields are conceptualized as electromagnetic fields, we may also ask if they should be classical fields, or quantum fields. While classical fields are causal, i.e., producing predictable interactions, and local, i.e., changes in them propagating the speed of light or below it, quantum fields are neither causal nor local (Laszlo, 1996). They are probabilistic fields; the behaviour of objects in them cannot be described in classical terms, as obeying causal laws and having single, determined locations in space. They merely state the potentials for the manifestations of physical effects. Bohm postulates quantum potentials as *'guiding fields'* (Einstein's *'Führungsfelder'*) which form a non-material matrix that guides and channels quantum events - an effect depending entirely on the structure, not the energy of the holographic implicate order. Biophoton fields - by virtue of their very low intensity and high coherence - are non-classical fields. Popp's concept of *'potential information'* in the form of a *'non-measurable, coherent carrier field'*, from which the *'actual information'* of the measurable biophoton field emerges, can be interpreted as a quantum potential.

There is no agreeement yet as to whether quantum potentials are just theoretical devices or physically real fields. However, an increasing number of scientists are proposing the existence of a *'fifth field'* besides the recognized four universal interactions of gravitational, electromagnetic, strong nuclear and weak nuclear fields, to account for the EPR correlations in terms of a physically real field (Laszlo, 1995). This fifth field, which, according to Laszlo (1995) may be the medium that could provide the universal interconnections necessary in a holistic worldview, would also have non-classical properties and is either assumed to be the vacuum of space-time itself or arising from it. Bohm's *'implicate order'* basically is nothing else than the vacuum. Laszlo's *'psi field'* (1993, 1995, 1996), Bearden's *'scalar fields'* (Bearden, 1991, 1991a, 1993), and Akimov and Shipov's *'torsion fields'* (Shipov, 1992; Akimov, 1995) are such concepts of a fifth field based on properties of the vacuum, while Heim (Auerbach and Ludwiger, 1992) in his six-dimensional world-model proposes two virtual *'transdimensions'* where potential structural patterns are selected from the *'entelechial field'* and actualized into space-time. These transdimensions are roughly equivalent to Bohms 'holomovement'. Laszlo maintains that in order to treat the interactions that govern bodily functions, structure, and chemistry, the introduction of an electromagnetic biofield is necessary - but beyond that the understanding of life also necessitates the recognition that all living organisms are embedded in and constantly interacting with the holographic field of the quantum vacuum. Laszlo's fifth field is a fundamental information or memory field created by constant feedback between the quantum vacuum and the objects and events of the observable world. It both generates and carries form, but also interconnects and 'informs' all elements in the organism, including consciousness, of each other and the environment; it is the basis of evolution and evolves with matter and life. To sum up, in principle all biological fields, being described as topological characteristics of space-time, could be interpreted as vacuum states.

9. Conclusion

The holistic approach to biological studies is a necessary complement to the contemporary practice, and even a more detailed historical study of the work of the organicist and holistical biologists of the earlier decades of the 20th century may therefore be useful. Field theories, as a central element of holistic models, possibly will be dominant models in the next future; especially those based on quantum theory and allowing the inclusion of consciousness. However, the approach of the organicists teaches an important lesson: concretization. Taking the holistic path requires keeping a delicate balance between the extremes of vague generalizations and untestable hypotheses on the one hand, and getting lost in the jungle of molecular detail without any unifying vision, on the other hand. Concretization necessitates the use of combined field-particle models, requires the integration of field perspectives with the wealth of knowledge accumulated in last decades in biochemistry and molecular biology, by means of the information we have about regulations and integrative processes on the levels of the whole organism, organ systems and tissues. Concretization will also make the acceptance of biophysical field theories easier for biologists, biochemists and the like, by establishing more links to those concrete biological processes and mechanisms that he knows from his own work, which will enable him to see the possible relevance of the concept for his work and put it to the test. In biophoton research, much more work is now needed to concretize the relationships of the biophoton field to specific biological structures and processes at various levels. The use of holistic and field concepts also requires pragmatism combined with openness. Processes of each level have to be investigated and conceptualized according to their own laws, and not prematurely related to those of other levels. We also conclude that epistemologic considerations, and philosophical ones in general, should be paid more attention in biology. Even top-down causality and non-causal models should be considered for their usefulness. The use of the field model by the organicists shows how one can operate successfully with concepts that earlier-on would have been condemned as „metaphysical“ entities, by using them as metaphors which facilitate concretization by experiment and experience, and at the same time allow to keep open all options concerning the ultimate nature of the 'substrate' underlying the concept, until appropriate tools are available to decide the issue.

10. References

Akimov, A.E., ed. (1995) *Consciousness and the Physical World*, CISE-VENT/Yakhtsmen, Moscow.

Albrecht-Buehler, G. (1990) In defense of 'nonmolecular' cell biology, *Int.Rev.Cytol.* **120**, 191-241.

Allen, G.E. (1975) *Life Science in the Twentieth Century*, Wiley, New York.

Ammann, A. (1996) Gestalten in der Quantenmechanik, *Zeitschrift für Parapsychologie und Grenzgebiete der Psychologie* **38** (1/2), 46-60.

Ash, M. (1995) *Gestalt Psychology in German Culture 1890-1967. Holism and the Quest for Objectivity.* Cambridge University Press, Cambridge.

Atmanspacher, H.; Primas, H.; Wertenschlag-Birkhäuser, E., eds. (1995) *Der Pauli-Jung-Dialog und seine Bedeutung für die moderne Wissenschaft*, Springer, Berlin.

Atmanspacher, H. (1996) Erkenntnistheoretische Aspekte physikalischer Vorstellungen von Ganzheit, *Zeitschrift für Parapsychologie und Grenzgebiete der Psychologie* **38** (1/2), 20-45.

Auerbach, T; Ludwiger, I. von (1992) Heim's theory of elementary particle structures, *J. Scientific Exploration* **6**, 217-231.

Bearden, T.E. (1991) On a testable unification of electromagnetics, General Relativity, and quantum mechanics, *Proc.26th Intersoc. Energy Conversion Eng.Conf., Boston, August 4-9, 1991.*

Bearden, T.E. (1993) Information content of the field: dynamic EM structuring of the scalar potential and its use by biological systems, Paper presented at *2nd Estrian Workshop on „The Organization of Information Transfer in Living Systems*", August 26-28, University of Sherbrooke, Canada.

Bell, J.S. (1987) *Speakable and Unspeakable in Quantum Mechanics*, Cambridge University Press, Cambridge.

Bigu, J. (1976) On the biophysical basis of the human 'aura', J. Res. Psi Phenomena **1** (2), 8-43.

Bischof, M. (1995) *Biophotonen*, Zweitausendeins, Frankfurt.

Bischof, M. (1995a) Vitalistic and mechanistic concepts in the history of bioelectromagnetics, in L.V.Beloussov and F.A.Popp (eds.), *Bio-Photonics - Non-Equilibrium and Coherent Systems in Biology, Biophysics and Biotechnology*, Bioinform Services, Moscow, pp.3-14.

Bischof, M. (1996) Some remarks on the history of biophysics and its future, in C.L.Zhang, F.A.Popp and M.Bischof (eds.), Current Development in Biophysics, Hangzhou University Press, Hangzhou, pp.10-21.

Bohm, D. (1980) *Wholeness and the Implicate Order*, Routledge and Kegan Paul, London.

Borgens, R.B.; Robinson, K.R.; Vanable, J.W.; McGinnis, M.E., eds. (1989) *Electric Fields in Vertebrate Repair: Natural and Applied Voltages in Vertebrate Regeneration and Healing*, Alan R.Liss, New York.

Boutroux, E. (1874) *De la contingence des lois de la nature*, Paris.

Boveri, Th. (1901) Über die Polarität des Seeigeleis, *Verhandlg. Phys.-Med. Ges. Würzburg* **NF 34**, 145-176.

Burr, H.S.; Northrop, F.S.C. (1935) The electro-dynamic theory of life, *Quarterly Review of Biology* **10**, 322-333.

Burr, H.S. (1973) *The Fields of Life*, Ballantine Books, New York.

Callahan, P.S. (1990) Nonlinear infrared coherent radiation as an energy coupling mechanism in living systems, in R.K.Mishra (ed.), *Molecular and Biological Physics of Living Systems*, Kluwer, Dordrecht, pp.239-273.

Child, C.M. (1929) The physiological gradients, *Protoplasma* **5**, 447-476.

Crile, G.W. (1936) *The Phenomena of Life - A Radio-Electric Interpretation*, W.W.Norton, New York.

Dicke, R.H. (1954) Coherence in spontaneous radiation processes, *Physical Review* **93**, 99-110.

Driesch, H. (1891) Entwicklungsmechanische Studien I. Der Wert der beiden ersten Furchungszellen in der Echinodermen-Entwicklung. *Zeitschrift für wissenschaftliche Zoologie* **53**, 160-178.

Driesch, H. (1908) *The Science and Philosophy of the Organism*, Black, London.

Einstein, A.; Podolsky, B.; Rosen, N. (1935) Can quantum-mechanical description of physical reality be considered complete ? *Physical Review* **47**, 777-780.

Friedmann, H. (1930) *Die Welt der Formen*, C.H.Beck, Munich.

Friedmann, H. (1949) *Wissenschaft und Symbol*, C.H.Beck, Munich.

Fröhlich, H. (1968) Long range coherence and energy storage in biological systems, *International Journal of Quantum Chemistry* **2**, 641-649.

Fröhlich, H. (1975) Evidence for Bose condensation-like excitation of coherent modes in biological systems, *Physics Letters* **51A** , 21.

Gebser, J. (1986) *The Ever-Present Origin*, Ohio University Press, Athens.

Glauber, R. J. (1963) The quantum theory of optical coherence. *Physical Review* **130**, 2529-2539.

Gloy, K. (1996) *Die Geschichte des ganzheitlichen Denkens.* (Das Verständnis der Natur, Vol.II), C.H.Beck, Munich.

Goldstein, K. (1939) *The Organism*, American Book Company, New York.

Goodwin, B.C. (1987) Developing organisms as self-organizing fields, in F.E.Yates (ed.), *Self-Organizing Systems*, Plenum Press, New York, pp.167-180.

Goodwin, B.C. (1987a) The stability of nature, *Speculations in Science and Technology* **10** (4), 325-338.

Goodwin, B.C.; Webster, G.; Smith, J.W. (1987) The 'evolutionary paradigm' and 'constructional' biology, *Explorations in Knowledge* **4** (1), 29-40.

Gurwitsch, A.G. (1915) On practical vitalism, *American Naturalist* **49**, 763-771.

Gurwitsch, A.G. (1922) Über den Begriff des embryonalen Feldes, *Archiv für Entwicklungsmechanik der Organismen* **51**, 383-415.

Gurwitsch, A.G. (1923) Versuch einer synthetischen Biologie, *Schaxel's Abhandlungen zur theoretischen Biologie* **17**, 1-87.

Gurwitsch, A.G.; Gurwitsch, L. (1936) Der Feldbegriff in seiner Anwendung auf das Problem der Zellteilung, *Acta Biotheoretica*, Series A, **2**, part 2, 77-92.

Gurwitsch, A.G. (1991) *Principles of Analytical Biology and the Theory of Cell Fields*, Nauka, Moscow (in Russian).

Haraway, D.J. (1976) *Crystals, Fabrics, and Fields: Metaphors of Organicism in Twentieth-Century Developmental Biology*, Yale University Press, New Haven.

Harrington, A. (1996) *Reenchanted Science: Holism in German Culture from Wilhelm II to Hitler*, Princeton University Press, Princeton.

Harrison, R.G. (1969) *Organization and Development of the Embryo*, S.Wilens, ed. Yale University Press, New Haven.

Hartmann, N. (1950) *Philosophie der Natur*, Berlin.

Huxley, J.S; De Beer, G.R (1934) *The Elements of Experimental Embryology*, Cambridge University Press, London.

Jaffe, L.F. (1982) Developmental currents, voltages, and gradients, in S.Subtelny (ed.), *Developmental Order: its Origin and Regulation*, Alan R.Liss, New York.

Jammer, M. (1980/81) Feld (field, champ), *Arch. Gesch. Naturwiss.* **2**, 113-121.

Kafatos, M.; Nadeau, R. (1991) *The Conscious Universe - Part and Whole in Modern Physical Theory*, Springer, New York.

Keller, R. (1918) *Die Elektrizität in der Zelle*, Braumüller, Vienna.

Köhler, W. (1920) Die physischen Gestalten in Ruhe und im stationären Zustand, in W.D.Ellis, ed., *Sourcebook of Gestalt Psychology*, Routledge & Kegan Paul, London. 1938.

Köhler, W. (1922) Gestaltprobleme und Anfänge einer Gestalttheorie, *Jahresberichte über die gesamte Physiologie und experimentelle Pharmakologie für das Jahr 1922* **3**, 512-539.

Lakhovsky, G. (1951) *The Secret of Life*, 2nd ed. True Health Publishers, Stockwell (orig.French ed.Paris 1929).

Laszlo, E. (1993) *The Creative Cosmos*, Floris Books, Edinburgh.

Laszlo, E. (1995) *The Interconnected Universe - Conceptual Foundations of Transdisciplinary Unified Theory*,World Scientific, Singapore.

Laszlo, E. (1996) *The Whispering Pond - A Personal Guide to the Emerging Vision of Science*, Element Books, Rockport, Mass.

Lewin, K. (1951) *Field Theory in Social Science*, Harper and Brothers, New York.

Lipkind, M.A., Beloussov, L.V. (1995) Alexander Gurwitsch: Personality and scientific heritage, in L.V.Beloussov and F.A.Popp (eds.), *Bio-Photonics - Non-Equilibrium and Coherent Systems in Biology, Biophysics and Biotechnology*, Bioinform Services, Moscow, pp.15-32.

Lund, E.J. (1947) *Bioelectric Fields and Growth*, University of Texas Press, Austin.

Meyer-Abich, A. (1934) *Ideen und Ideale der biologischen Erkenntnis*, J.A.Barth, Leipzig.

Meyer-Abich, A. (1948) *Naturphilosophie auf neuen Wegen*, Hippokrates, Stuttgart.

Meyer-Abich, A. (1949) *Biologie der Goethezeit*, Hippokrates, Stuttgart.

Meyer-Abich, A. (1950) Zur Logik der Unbestimmtheitsbeziehungen, in W.Heinrich (ed.), *Die Ganzheit in Philosophie und Wissenschaft*, Braumüller, Vienna, pp.47-76.

Meyer-Abich, A. (1964-65) Gedanken zur Theorie und Philosophie des Organismus, *Leopoldina* (Series 3) **10**, 151-204; **11**, 179-224.

Mocek, R. (1974) *Wilhelm Roux - Hans Driesch. Zur Geschichte der Entwicklungsphysiologie der Tiere*. Gustav Fischer, Jena.

Morgan, C.L. (1928) *Emergent Evolution*, Henry Holt, New York.

Moller, P. (1976) Electric signals and schooling behaviour in a weakly electric fish, *Marcusenius cyprinoides* L. (Mormyriformes), *Science* **193**, 697-699.

Nagl, W.; Popp, F.A. (1983) A physical (electromagnetic) model of differentiation. I. Basic considerations. *Cytobios* 37 , 45-62.

Needham, J. (1931) *Chemical Embryology*, Cambridge University Press, London.

Needham, J. (1936) *Order and Life*. Cambridge University Press, Cambridge/Yale University Press, New Haven.

Needham, J. (1942) *Biochemistry and Morphogenesis*, Cambridge University Press, London.

Nuccitelli, R. (1986) *Ionic Currents in Development*, Alan R. Liss, New York.

O'Shea, P.S. (1988) Physical fields and cellular organisation: field-dependent mechanisms of morphogenesis, *Experientia* **44**, 684-694.

Planck, M. (1929) *Das Weltbild der modernen Physik*, Leipzig.

Polk, C.; Postow, E., eds. (1995) *CRC Handbook of Biological Effects of Electromagnetic Fields*, 2nd ed. CRC Press., Boca Raton.

Popp, F.A..; Becker, G.; König, H.L.; Peschka, W., eds. (1979) *Electromagnetic Bio-Information*, Urban & Schwarzenberg, München-Wien-Baltimore.

Popp, F.A.; Nagl, W. (1983) A physical (electromagnetic) model of differentiation. 2) Application and examples. *Cytobios* **37**, 71-83.

Popp, F.A.; K.H.Li; Q.Gu, eds. (1992) *Recent Advances in Biophoton Research and its Applications*, World Scientific, Singapore.

Popp, F.A. (1992a) Evolution as the expansion of coherent states, in Popp, F.A.; K.H.Li; Q.Gu, eds., *Recent Advances in Biophoton Research and its Applications*, World Scientific, Singapore.

Presman, A.S. (1970) *Electromagnetic Fields and Life*, Plenum Press, New York.

Nicolis, G. / Prigogine, I. (1977) *Self-Organization in Non-Equilibrium Systems*, Wiley, New York.

Primas, H. (1992) Umdenken in der Naturwissenschaft, *GAIA* **1** (1), 5-15.

Primas, H. (1993) Ein Ganzes, das nicht aus Teilen besteht. Komplementarität in den exakten Naturwissenschaften, *Mannheimer Forum*, **92/93**, 81-111.

Sheldrake, R. (1988) *The Presence of the Past*, Collins, London.

Shipov, G.I. (1992) *The Theory of the Physical Vacuum*, CISE-VENT, Moscow (in Russian).

Smuts, J.C. (1926) *Holism and Evolution*, Macmillan, New York.

Spann, O. (1939) *Kategorienlehre*, Gustav Fischer, Jena (2nd rev. ed.).

Spemann, H.C. (1936) *Experimentelle Beiträge zu einer Theorie der Entwicklung*, Berlin.

Spemann, H.C.(1938) *Embryonic Development and Induction*, Yale, New Håven.

Thom, R. (1975) *Structural Stability and Morphogenesis*. Benjamin, Reading, Mass.

Thompson, D'Arcy W. (1917) *On Growth and Form*, Cambridge.

Waddington, C.H. (1957) *The Strategy of the Genes*. Allen & Unwin, London.

Waddington, C.H. (1966) Fields and gradients, in M.Locke (ed.), *Major Problems in Developmental Biology*, Academic Press, New York, pp.105-124.

Warnke, U. (1989) Information transmission by means of electrical biofields, in F.A.Popp, U.Warnke, H.L.Koenig, and W.Peschka (eds.), *Electromagnetic Bio-Information*, Urban and Schwarzenberg, Munich-Vienna-Baltimore, 2nd ed., pp.74-101

Weiss, P. (1926) *Morphodynamik*, Borntraeger, Berlin.

Weiss, P. (1939) *Principles of Development*, Holt, New York.

Weiss, P.; Moscona, A. (1958) Type-specific morphogenesis of cartilages developed from dissociated limb and scleral mesenchyme in vitro, *J.Embryol.Exp.Morph.* **6**, 238-246.

Welch, G.R.; Smith, H.A. (1990) On the field structure of metabolic space-time, in R.K.Mishra (ed.) *Molecular and Biological Physics of Living Systems*, Kluwer, Dordrecht, pp.53-85.

Welch, G.R. (1992) An analogical 'field' construct in cellular biophysics: history and present status, *Progress in Biophysics and Molecular Biology* **57**, pp.71-128.

Whitehead, A.N. (1926) *Science and the Modern World*, Cambridge University Press, Cambridge.

Wolkowski, Z.W. (1995) Recent advances in the phoron concept: an attempt to decrease the incompleteness of scientific exploration of living systems, in L.V.Beloussov and F.A.Popp (eds.), *Bio-Photonics - Non-Equilibrium and Coherent Systems in Biology, Biophysics and Biotechnology*, Bioinform Services, Moscow, pp.33-50.

COHERENCE IN ART AND IN THE PHYSICAL BASIS OF CONSCIOUSNESS

F. FRÖHLICH

Honourary Member of the International Institute of Biophysics, Station Hombroich, Neuss, Germany

13 Greenheys Road, Liverpool. L8 0SX, U.K.

If one thinks of collective phenomena in which discrete constitutive individuals are modified in their behaviour - the whole becoming more than and different from their parts - living organisms would seem to be an ideal example: "The Universe is an animal", as Plato said in "*The Timaeus*". At this degree of apparent complexity, relevance becomes an essential cognitive property, and some sort of overall insight into the whole situation becomes necessary in order to reveal the underlying simplicity amidst the more superficial mind-stunning complexity.

In my contribution entitled "Collective Phenomena", to a book dedicated to Herbert Fröhlich called "*Cooperative Phenomena*", edited by Hermann Haken and Max Wagner, I wrote over 30 years ago:

"When one attempts to introduce into biology different forms of explanation, or even to reformulate problems in terms of collective phenomena, the approach is often dismissed as superfluous, or even felt to be mysterious and vitalistic. There may here even be a certain distant echo of the opposition to Newton's 'mystical' forces. Thus, paradoxically, there has arisen, from the programme based on applications of the scientific method to the explanation of biological phenomena by means of the laws of physics, the contrary result that modern physics, in its role of possible coordinating explanations of biological systems, is rendered suspect of being vitalistic. This prejudice inhibits such potentially illuminating methods of enquiry of a dynamic nature, as measurements of the rate of various processes; furthermore, it blurs the perhaps essential differences between multicellular, differentiated organisms - in which movements over long distances must take place relatively rapidly and in a highly coordinated fashion - and very much smaller monocellular organisms where diffusion could plausibly serve as an adequate mechanism of movement."

Somewhat later, I extended the concept of 'collective phenomena' to apply a group of painters working together on the same surface, one gesture being relevant to, and often motivated by, the gestures already made.

J. J. Chang et al. (eds.), Biophotons, 395-404.

About myself

After studying philosophy in Chicago with Rudolph Carnap, and then in Oxford with Sir Peter Strawson, but finding certain aspects unsatisfactory, I became a painter; only more recently I have returned to philosophy in the quest of attempting to discover the physical basis of consciousness.

Whilst studying etching with the etcher and painter S.W. Hayter - who numbered among his students and colleagues in Paris: Picasso, Viera de Silver, Max Ernst, and Giocometti, and in New York: Jackson Pollack, Mark Rothko, and indeed most of the Abstract Expressionism movement - I became fascinated by his work in "experimental drawing"; here, although the process is known, what arises out of it is *not* - indeed radically different results must in general be anticipated for different participants. Subsequently, (and later in collaboration with Sylvie le'Sea'ch) I began writing a book about this, entitled

"*Experimental Drawing,*
Systems of Oscillating Perceptual Fields,
Conscious and Unconscious Dialogue"

In the present paper, I mention only two relevant aspects of "Experimental Drawing" - insofar as colour is relevant to bio-luminescence - and suggest not only experiments which can actually be performed, but also more radical ones concerning the origins of colour in the eye itself; it is the interaction of these two which determines the colour we actually sée. By intensely viewing a surface saturated with a single colour, Hayter could see clearly the after-image created around it: a blue field, for instance, would create yellow after-images [depending on the *exact* colour of the blue] shifting around it, sometime coinciding to make green. He then took this further, suggesting that by pressing quite hard against the eyes, one would observe a variety of colours originating from this pressure - *i.e.* one would enjoy both a visual *and* a visionary experience.

If we could see colours in the same way that we hear sounds, then a *counterpoint* of colour would arise, in which develops the analogue of individual melodies and their supposition as chords in transparent colour (as opposed to opaque colour), in which differently coloured areas can be superimposed yielding ambiguous, strange new colours without known names.

Hayter had many profound conversations with Herbert Fröhlich about the dimensions of space and about science's relation to art, but, strangely, he never discussed colour with him; if he had, ideas about a laser-like process involving enzymes in the brain and in the human eye (which is sensitive to a single quantum of light) could well have arisen which would have been acutely relevant here!

We now proceed to consider the other aspect of Hayter's ideas which is more connected to 'collective phenomena' as a cooperation of several artists, all working together on the same surface, with gestures developing out of each others gestures, according to visual counterpoint. After the first free automatic gestures [this painting is abstract], all subsequent ones are to some degree relevant to the entire configuration; eventually, it develops so much complexity that we must discuss in what way a coherent, final composition could be realized, as if it had been done by one super-person.

Whilst we were working on the "*Conscious and Unconscious Dialogue*" section of our book, Sylvie le Se"ach and I experimented with the games of the Surrealists [among whom Hayter himself had figured] had played. One such was the visual equivalent of the game of making a "poem", when one word or phase is hidden from the other participants, only leaving the last letter and requiring that the next word begin with that letter -- 'Cadavre Exquis'. In visual terms, this consisted of concealing all but the last spot and requiring the next person to go on from there. One of us made a free gesture, leaving only the last point protruding and the other proceeded from there; it was not, of course, coherent. Seeking coherence, we tried interposing our gestures on the same surface, finding that, through visual counterpoint, one of us would make a gesture and the other would counter it with a relevant response. From this we found that a project involving several people working together on the same surface can lead to a new kind of coherence.

Borrowing the term '*Collective Phenomena*' from the paper mentioned above, we proceeded to involve a number of other painters - from America, Liverpool, London, Paris, Rome and China.

The following is the manifesto that we wrote, in the more dramatic French manner, for our exhibition and collaborative event - five people working on two surfaces, accompanied by Lawrence Ball, the musician, who improvised according to our movements, whilst we, in turn, reacted to his music; this event was held in Paris under the auspices of the publisher, John Calder.

"Let us end our fear of the unconscious creativity of others and replace it with a dialogue between different unconsciousness. Let us bring to an end the arbitrary in art in favour of the rigorous and mysterious structure of the unconscious. There is a whole unconscious structure that must be explored without fearing the dialogue with infinity of an unconsciousness that is different from one's own. That fear underlies the differences that separate artists from each other.

The question here is how to explore those structures of the unconscious that can enable work to take place simultaneously or sequentially on the same space, developing in visual terms a form of counterpoint. Put another way, a linkage is established between what one person has just done and what another is about to do.

Communal work very quickly eliminates the personal unconsciousness and allows it to become the ***collective unconsciousness****. In practise, in order for there to be a dialogue between the structures of different unconsciousness, each one has to be highly aware of what the other is doing, as in a chess game, where the desire to be competitive is excluded. On the one hand, each one must always feel a need to observe all the possible paths that can lead to progressive establishment of a certain kind of coherence. It no longer consists of simply asking oneself who has done what, but of concentrating instead on what has already been committed by the other on the paper.*

At first, different types of coherence appear and can be metamorphosed during the course of the work. This, in practise, takes the form of a process coming into being, that, as it approaches its completion, provokes analysis, a reconsideration and a collective decision. In this way a certain kind of unconscious meaning can consciously lead to a unity. It is sometimes necessary to re-immerse oneself into the unconscious in order to find a new solution - one which will bring about a total coherence. A collective decision must determine, in the final analysis, the path that is chosen. It is also possible that the unity that is sought will appear by itself in an unexpected way, which brings about a sensation of harmony.
To sum up, collective creation has to do with a spontaneity that comes from the unconscious, which is succeeded by a common attitude. The dialectic between the conscious and the unconscious has nothing to do with chance.

This means that the dialogue between the different unconsciousness dissolves the fear of entering into the unconsciousness of others. The force and the mystery of the unconscious avoids what is arbitrary. Therefore, the dialectic between the collective unconscious and the collective consciousness overcomes the barrier between them and allows dialogue to take place, which perpetually renews itself, as is suggested by the idea of an apparition contained in the word Phenomenon."

Here, "*Collective Phenomena*" was chosen as the name for this group because it emphasises its generality - referring both to the way in which we artists collaborate, whilst also having resonances with collective phenomena of a wider nature - from social to scientific: for instance a single molecule would follow the random laws of motion (action equals reaction), but once the concepts of temperature and pressure have been introduced - concepts which are inapplicable to a single particle, but only to a collection of particles - the individual molecules find themselves subjected to the Gas Laws. At a higher level of complexity, where atomic forces of attraction and repulsion come into play, coherence is even more involved, extending through the inorganic - in superconductivity for instance - and beyond, to biological beings in long-range interaction involving long-range coherence; such coherence can involve enzymes and substrates in the brain itself. It might even extend, through a metaphysical hypothesis, or plausible picture, to the relation between brain and mind. This takes us directly to the second aspect of coherence to be investigated - namely, the coherence underlying the basis of the operation of brain cells, and, through a metaphysical hypothesis [or plausible picture], to the representation of mind itself.

Hyland has already told you how Fröhlich applied the concept of *long-range phase correlations* - which he had learnt about and used to great effect in the low-temperature, *equilibrium* phenomena such as superconductivity - to dissipative living systems at ambient temperatures, which through their metabolic activity are *far* from thermal equilibrium, but are nevertheless structurally stable. This novel application of quantum mechanics to biological systems took place at l' Institutede la Vie, during a meeting of leading physicists -- Bardeen, Cooper, Wigner, Haken etc., chemists: Prigogine and Onsager, and biologists: Monod, Crick, Edelman, which took place every two years in Versailles over a period of twenty years until 1987. On this occasion, Fröhlich [2] started the discussion by putting together two seemingly irrelevant pieces of his former research - thereby creating the brilliant new concept which appeared to offer a quite unique possibility for understanding the orderly functioning of active biosystems. He was of the opinion that he would present this exciting new concept to biologists, and that they could pursue it with enthusiasm, leaving him free to return to his pure (non-biological) physics; but it wasn't to be, as you well know. He remained very involved with his bio-theory and with supporting experiments until two years before his death when he decided he would (and did) return to pure physics.

Now it is surprising and encouraging that many of these scientists from very different domains at the beginning of "l' Institutede la Vie" meetings - for instance: Edelmann, an immunologist, Cooper, the theoretical physicist, Crick, the biologist, and now even Haken, another theoretical physicist - have concentrated their efforts more and more on the brain, as did Fröhlich himself many years before, writing in 1977 in the issue of the Neuroscience Research Bulletin [3]; this was, however, strangely ignored until Hyland (who had been continuing Fröhlich's pure physics theory of Dirac particles) and I resurrected it [4]. Fröhlich's model for the generation of brain waves is based on self-sustaining collective chemical oscillations between the excited and ground states of a system of enzymes distributed throughout the greater membrane of the brain, possibly *via* their attachment to the microtubles. The energy required to activate these enzymes comes from interaction with surrounding substrate molecules. In their excited state, the enzymes possess an electric dipole which is absent in their ground state. The well-known correlation between the *EEG* and varying degrees on consciousness effectively places Fröhlich's Brain-Wave model at the interface between mind and brain -- permitting clarification of the parameters which control the observed spectra and hence of the associated different states of consciousness.

In the meantime, Michael Lockwood, an Oxford philosopher, cited [5] at great length in his book "*Mind, Brain and the Quantum*", Fröhlich's theory of bio-coherence at room temperature. Penrose, on the other hand, said that there should be vibrational effects within the active cells which would resonate with certain microwave electromagnetic radiation as a result of quantum coherence, asserting this [6] in his book "*Shadows of the Mind*". Thereupon, Penrose linked this quantum mechanical coherence to consciousness itself through the work of Hammeroff, an anaesthetist, who had asserted that it was in the *microtubules* that the distinction between consciousness and unconsciousness could be controlled by the administration of anaesthetics [7]. However, the conscious experience is surely much more complex than a simply not being *unconscious* - despite the fact that it can admittedly be turned off using anaesthetics. Consciousness, for instance, can move about, as in the process of breathing *consciously*, rather than doing it automatically; or playing the piano unconsciously, by rote or in learning a new piece of music, or in the automatic function of the liver contrasted with an acute pain in the liver. Consciousness skips around the body, now here, now elsewhere. One would need to expand this definition of the conscious - unconscious dichotomy so as not to identify the unconscious with simply the anaesthetized, but to broaden the range of unconscious activities to the automatic - such as playing the piano by rote, or even to the automatic drawing by which we start out in *our* realization of *collective phenomena.* But it is good that the conscious -- unconscious distinction has, in this way, been brought together with the physical basis of consciousness, even though one might object to its limitations. Thus the possibility of linking the physical basis of consciousness through a scientific experiment [the application an anaesthetic] with conscious/unconsciousness is an almost metaphysical step.

Indeed, consciousness has a different discipline relating to it, namely intentional logic, which asks radically different questions - questions about the contents of the mind and its intentions. These are coinciding, but very different, enquiries made by the scientist and by the philosopher. Both forms of enquiry are intimately involved in the same process but appear to be irrelevant, belonging to completely different disciplines, such as cause and effect, and reasons for a particular action. They might, however, be linked by a meta-physical or meta-intentional step.

In the past, these two - the physical and the mental - were integrated by philosophers; Spinoza for instance, asserted that they were two aspects - namely Extension and Thought - of the same substance, whilst Leibnitz individualised this substance into the monads, each containing the world of thought and matter from its own point of view; they both shared a dualistic approach.

In more recent times, Alfred North Whitehead - who, with Bertrand Russell, wrote "*Principia Mathematicia*", and who himself developed a theory of relativity having the same experimental consequences as Einstein's, but which is more logically complex [for which reason it has been largely ignored], wrote (in 1929) his metaphysical work, "*Process and Reality*" [8], in which he developed the idea of the "actual entity in concrescence" - his fundamental unit consisting of both a 'physical' and a 'mental' pole, describing how they were integrated at each phase of the process into an organic unity. To show how this view is appealing even now, we quote from David Chalmers' "*The Conscious Mind: in Search of a Fundamental Theory*" [9].

"If this view is correct, consciousness does not come in sudden jagged spikes with isolated complex systems producing rich complex experience. Rather it is a more uniform property of the universe, with very simple systems having very simple phenomenology and complex systems having complex. This make consciousness 'less special' in the same way and so more reasonable".

What I propose to do here is to use Whitehead's own somewhat inscrutable terminology, but "translate" it into terms which are more familiar to scientists and to non-Whiteheadian philosophers; whilst some of the more radical insights might in the process be lost, the results are more apt here. "*The actual entity in the process of concrescening*" is here identified with elementary bio-physical events and elementary experience becoming coherent within an organic unity. This combines the physical pole with the mental pole in a way which is perpetually illuminating. I talked about these ideas with David Bohm, who was at the time highly influenced by Whitehead, long before he had written "*Wholeness and the Implicit Order*".

The following is a hypothesis, or as Plato would say, 'a likely story', which might provide a philosophical means of combining the physical with the mental in some sort of organic, symbiotic way, in which they would both constantly influence each other.

Although Whitehead did not himself illustrate his concepts with diagrams, I propose the following very rudimentary diagram (Fig.1) which, while misleading in some instances, might be illuminating in others - such as in illustrating the relationship between physical and mental poles. Thus the physical pole - the physical basis - is represented by a straight (continuous) line, (*A*), which converges into a triangle as it meets and interacts with the mental pole, which is here represented by a dotted line (*B*) - its concavity indicating choice. The labile mental pole converges (downwards) towards the physical pole, maintaining interaction with it at all times. This diagram represents something akin to Leibnitz's 'monads'. The mental pole and the physical pole are integrated in various ways. *B* indicates the choice available to the mental pole either to conform - which leads to a solid object - or to initiate change - which leads to novelty. Such interactions between the mental and physical pole are not yet sufficient (except in certain cases, such as when a single electron can initiate the change). Generally they must consist of a "society" - which can be either temporal or spatial - to create a resonance whereby they make themselves perceived.

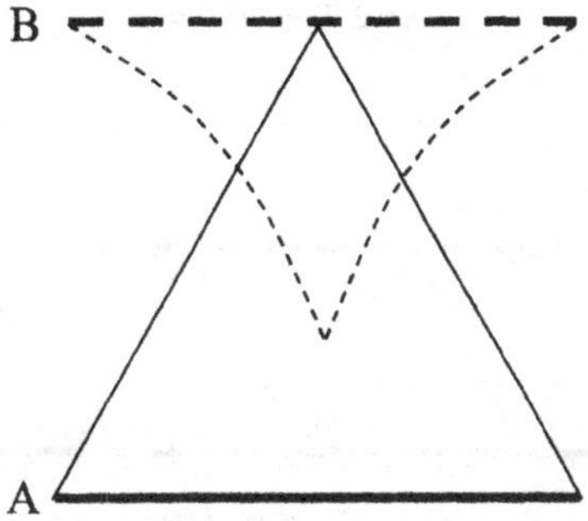

Figure 1. Relationship between physical and mental poles.

For instance, in the case of a human being, one must remember the beginning of word in order to complete it meaningfully. In a temporal society one must be aware of what ones has remembered in the mental pole. A vastly simplified illustration of *temporal* order is shown in Fig.2, in which for clarity full and dashed lines are used, respectively, to depict the physical and mental poles.

In order to illustrate *spatial* order (Fig.3), the system is repeated *horizontally*, thereby creating a space for resonance. Of course, in any real organism these two - the spatial and the mental - must combine.

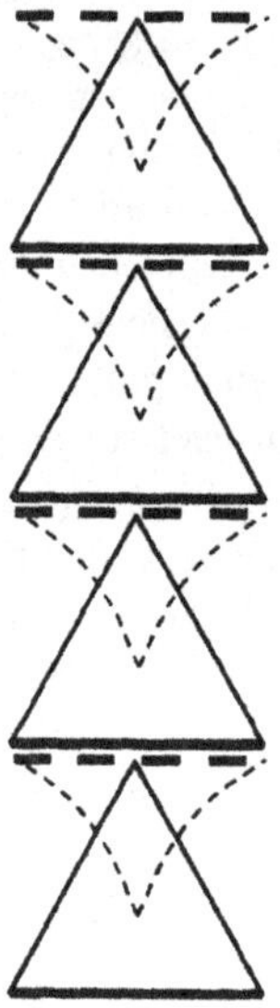

Figure 2. Temporal order.

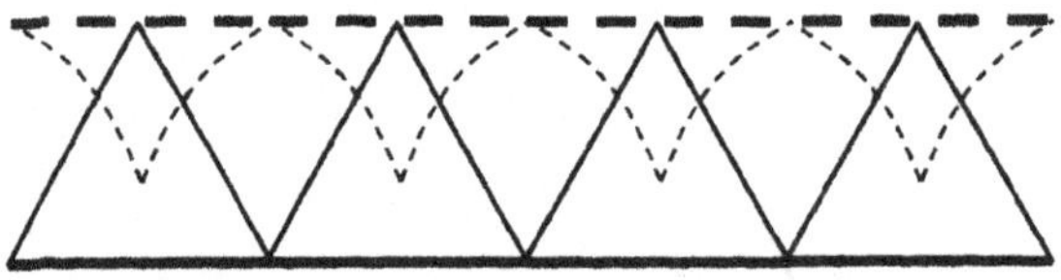

Figure 3. Spatial order.

As an absurd example, consider a machine or a table, for which there is a minimum mental pole representing conformity - each successor of the mental pole resembling its predecessors - except when the machine refuses to go on working or the table disintegrates - the "society" of components then breaking down or going beyond the verge of collapse. Thus the distinction applies to objects - the mental pole conforming until the machine does not function or the table is in the process of disintegration.

Intention arises in the mental pole within a "personal order" embedded in a "society" - *i.e.* within an organic whole - for instance, such as a human individual. However, the more primitive embodiment is in the actual entity which involves cause and effect interacting intimately with intentions at every stage.

These two, the social order and the personal order, can this be superimposed to give a diagrammatic idea of how the mental pole enters (at least minimally) at each stage into a conversation with the physical pole, thus making the mental pole [consciousness] more general and less mysterious.

Conclusion

A reference was made to a new kind of '*collective phenomena*' as exemplified not in groups of non-living beings - such as in the case of superconductivity, for instance - but specifically in *living* entities.

Two examples of collective phenomena were given:

1. The unusual acting together of several artists on the same surface, in which one gesture having been made, all others must have some relevance to it until the time comes when the complexity is too great, and the painters have to discuss what path might lead to coherence; for this, '*Collective Phenomena*', with all its ramifications, was borrowed as a title.

2. Coherence in brain processes, including the step to consciousness itself. Two examples were presented - Hyland"s analysis of Fröhlich's model for the generation of brain-waves, which takes into account certain empirical aspects (chaos) of neurological *EEG*, data, and R. Penrose's use of biological quantum coherence plus Hameroff's 'switch-off' of consciousness through anaesthetics in his book "*Shadows of the Mind*". There would seem, however, to be a meta-physical and meta-logical gap between these two - which are even represented by different logics: the cause and effect logic [modified by quantum mechanics] for scientists, and the intentional logic - the contents of the mind - for a philosophers.

A suggestion was made following A.N. Whitehead's treatise, "*Process and Reality*", as to how this gap might be overcome by the mental and physical poles of an actual entity working together intimately at every stage and integrating cause - and - effect logic with intentional logic.

References

Fröhlich, F.: Collective Phenomena, in H. Haken and M. Wagner (eds.), *Cooperative Phenomena*, Springer Verlag, Berlin, 1971.

Fröhlich, H.:, in M. Marois (ed.),*Theoretical Physics and Biology*, Proceedings of the First International Conference on Theoretical Physics and Biology, Versailles, 1967, l'Institutede la Vie, North Holland, Amsterdam.

Fröhlich, H.:, *Neuroscience Res. Prog. Bull.* **15**(1977), 67-72.

Fröhlich, F., and Hyland, G.J.: in , J. King and K.H. Pribram (eds.), *Scale in Consciousness Experience*, Lawrence Erlbaum, New Jersey, 1995, pp.407 - 438.

Lockwood, M.: *Mind, Brain and Quantum, The Compound "I"*, Blackwell, 1989.

Penrose, R.: *Shadows of the Mind*, Oxford University Press, 1994.
Hameroff, S.R.:, *J. Consciousness Studies* **1** (1994), 9-18.
Whitehead, A.N.: *Process and Reality*, Macmillan, 1929.
Chalmers, D.: *The Conscious Mind*, Oxford University Press, 1996.
Hyland G.J., Lewis C., and Ucar A.: - to be published, 1998.

AUTHOR INDEX

SUBJECT INDEX

Made in United States
North Haven, CT
04 November 2026

21743223R00239